Lecture Notes in Physics

Volume 868

For further volumes:
www.springer.com/series/5304

The Lecture Notes in Physics

The series Lecture Notes in Physics (LNP), founded in 1969, reports new developments in physics research and teaching—quickly and informally, but with a high quality and the explicit aim to summarize and communicate current knowledge in an accessible way. Books published in this series are conceived as bridging material between advanced graduate textbooks and the forefront of research and to serve three purposes:

- to be a compact and modern up-to-date source of reference on a well-defined topic
- to serve as an accessible introduction to the field to postgraduate students and nonspecialist researchers from related areas
- to be a source of advanced teaching material for specialized seminars, courses and schools

Both monographs and multi-author volumes will be considered for publication. Edited volumes should, however, consist of a very limited number of contributions only. Proceedings will not be considered for LNP.

Volumes published in LNP are disseminated both in print and in electronic formats, the electronic archive being available at springerlink.com. The series content is indexed, abstracted and referenced by many abstracting and information services, bibliographic networks, subscription agencies, library networks, and consortia.

Proposals should be sent to a member of the Editorial Board, or directly to the managing editor at Springer:

Christian Caron
Springer Heidelberg
Physics Editorial Department I
Tiergartenstrasse 17
69121 Heidelberg/Germany
christian.caron@springer.com

Cecilia Flori

A First Course in Topos Quantum Theory

Springer

Cecilia Flori
Perimeter Institute for Theoretical Studies
Waterloo, Ontario,
Canada

ISSN 0075-8450 ISSN 1616-6361 (electronic)
Lecture Notes in Physics
ISBN 978-3-642-35712-1 ISBN 978-3-642-35713-8 (eBook)
DOI 10.1007/978-3-642-35713-8
Springer Heidelberg New York Dordrecht London

Library of Congress Control Number: 2013933014

Printed on acid-free paper

Springer is part of Springer Science+Business Media (www.springer.com)

To my parents Elena Romani and Luciano Flori, for believing in me

Acknowledgement

The Author would like to thank Perimeter Institute. Research at Perimeter Institute is supported by the Government of Canada through Industry Canada and by the Province of Ontario through the Ministry of Research and Innovation.

Acknowledgment

Contents

Chapter 1
Introduction

"We can't solve problems by using the same kind of thinking we used when we created them."
(Einstein)

The great revolution of the 20th century started with the theory of special and general relativity and culminated in quantum theory. However, up to date, there are still some fundamental issues with quantum theory that are, yet, to be solved. Nonetheless, a great deal of effort in fundamental physics is spent on an elusive theory of quantum gravity, which is an attempt to combine the two above mentioned theories which seem, as they have been formulated, to be incompatible. In the last five decades, various attempts to formulate such a theory of quantum gravity have been made, but none has fully succeeded in becoming *the* quantum theory of gravity. One possibility of the failure for reaching an agreement on a theory of quantum gravity might be the presence of unresolved fundamental issues already present in quantum theory, first pointed out by John Bell in [1, 13] and subsequently by others. Since then the scientific community has been split in two depending on whether or not the interpretation of quantum theory is regarded as problematic. In fact, many new interpretations of quantum theory started gaining ground, each motivated from the belief that the conceptual framework of quantum theory was wrong or incomplete. The major examples of alternative formulations of quantum theory are: (i) *many-worlds*, (ii) *many minds*, (iii) *consistent histories* and (iv) *hidden-variables*.

The *many-worlds* interpretation of quantum theory, which assumes that all evolutions are governed by Schrödinger's equation was first put forward by Everett in [2], where he postulated that the subjective states of each observer were correlated with various parts of the external world. However followers of Everett proposed different interpretations of this theory, the most famous ones are (i) the interpretation put forward by DeWill and Graham [3] in which they assume that the universe branches into distinct universes (hence the *many-worlds* interpretation) (ii) the *many-minds* interpretation put forward by Lockwood [8]. Related to the *many-worlds* interpretation is the *consistent histories approach* in which histories rather than events at a single time are the main elements of the theory. Therefore one does not consider the branching of the universe but simply considers all possible histories of the universe

C. Flori, *A First Course in Topos Quantum Theory*, Lecture Notes in Physics 868,
DOI 10.1007/978-3-642-35713-8_1, © Springer-Verlag Berlin Heidelberg 2013

at once. In this framework probabilities are assigned at a given time. Such an interpretation was developed by Griffiths [57, 62], Omnès [58–61, 63] and Gell-Mann and Hartle [54–56] (See Chap. 17).

A less bizarre interpretation of quantum theory is given by the *hidden-variables* interpretation which assumes that the formalism of quantum theory is incomplete. In particular they postulate the existence of a set of hidden variables which belong to some space. Such a set of hidden variable together with the state ψ of a system allow us to uniquely determine the value of all quantities of that system [4–7].

However, notwithstanding the many "alternative" interpretations of quantum theory, the majority of physicists adhere to the "standard" Copenhagen interpretation and do not regarded it as neither incomplete nor problematic. For example Nico van Kampen [9], Rudolf Peierls [10], and Kurt Gottfried [11] assert that quantum theory "provides a complete and adequate description of the observed physical phenomena on the atomic scale" and the presupposed interpretational problems are mainly due to the fact that one is trying to force classical concepts onto quantum systems, i.e. if one tries to reason classically about quantum phenomena then one is bound to find contradictions. In their view the measurement problem is solved by the process of *decoherence* with the environment, i.e. the interactions between system and surrounding environment lead to an apparent reduction of the state vector of the system.[1] When considering a pair of systems, the process of decoherence is expressed mathematically by tracing over the degrees of freedom of one of the systems (apparatus for example), thus obtaining a mixed stated for the remaining system (electron for example). Although both the original entangled state and the reduced state give the same predictions for physical quantities pertaining to the electron, however they disagree when considering quantities regarding both the apparatus and the electron. This led some physicists to believe that such an interpretation could not provide an epistemic interpretation of probabilities such that the discrepancies between results obtained for the entangled state and the reduced state can be ascribed to our ignorance of the macroscopic object [12].

Notwithstanding this ongoing debate regarding the interpretation of quantum theory, most approaches to quantum gravity adopt standard quantum theory as their starting point, with the hope that the unresolved issues of the theory (if there are any) will get solved along the way. For the supporters of the claim that indeed there are conceptual problems present in quantum theory, it seems reasonable to try solving fundamental issues *before* attempting to define a quantum theory of gravity.

If one adopts this point of view the questions that come next are:

(i) which are the main conceptual issues in quantum theory?
(ii) How can these issues be solved within a new theoretical framework of quantum theory?

[1] This view is also supported by the fact that, effectively, isolated systems are idealisations and are not very frequent in practice, unless of course one is studying cosmology where the universe itself is an isolated system.

We will now follow the ideas put forward by C. Isham, A. Döring, J. Butterfield and others who have proposed that the main issues in the standard quantum formalism are: (A) the use of critical mathematical ingredients, which seem to assume certain properties of space and/or time which are not entirely justified. In particular, it could be the case that such a priori assumptions of space and time are not compatible with a theory of quantum gravity. (B) The instrumental interpretation of quantum theory that denies the possibility of talking about systems without reference to an external observer. A consequence of this issue is the problematic notion of a closed system in quantum cosmology.

A possible way to overcome the above mentioned issues is through a reformulation of quantum theory in terms of a different mathematical framework called topos theory [44]. The reason for choosing topos theory is that it 'looks like' sets and is equipped with an internal logic. As we will explain in detail in the following chapters, both these features are desirable, because they will allow for a reformulation of quantum theory, which is more realist[2] (thus solving issue (B)) and which does not rest on a priori assumptions about the nature of space and time.

The hope is that such a new formulation of quantum theory will shed some light on how a quantum theory of gravity should look like.

It should be pointed out that the adoption for a more realist interpretation should be considered only as a possible strategy to overcome issue (B) above and to make contact with the interpretation of classical physics but *not* as a universal criterion for a more "acceptable" theory. It could well be that ultimately once should not strive for a realist interpretation of quantum theory but a more "general" one which can then be reduced to a realist interpretation in the limit applicable to classical physics. Indeed, as we will see in the course of the book, this is what happens in the topos approach to quantum theory in which certain conditions for a realist interpretation have to be relaxed resulting in the so called *neo-realist* interpretation. This interpretation, however, reduces to the standard realist interpretation when considering classical systems.[3]

Therefore if one accepts the viewpoint that quantum theory has certain conceptual problems in its standard formulation, then redefining it in a more realist way reveals itself a fruitful strategy to overcome these problems [21, 22].

The main idea in the topos formulation of normal quantum theory [14–16, 26, 27, 29–33, 35] (summary can be found in [42, 50]) is that using topos theory to redefine the mathematical structure of quantum theory leads to a reformulation of quantum theory in such a way that it is made to 'look like' classical physics. Furthermore, this reformulation of quantum theory has the key advantages that (i) no fundamental role is played by the continuum; and (ii) propositions can be given truth values

[2] By a 'realist' theory we mean one in which the following conditions are satisfied: (i) propositions form a Boolean algebra; and (ii) propositions can always be assessed to be either true or false. As it will be delineated in the following, in the topos approach to quantum theory both of these conditions are relaxed, leading to what Isham and Döring called a *neo-realist* theory.

[3] The reason is, roughly, because the internal logic of the topos used to describe quantum theory is *multi-valued* while the internal logic of the topos used to describe classical theory is *bi-valued*.

without needing to invoke the concepts of 'measurement' or 'observer'. Let us analyse the reasons why such a reformulation is needed in the first place. These concern quantum theory in general and quantum cosmology in particular.

- As it stands, quantum theory is non-realist. From a mathematical perspective, this is reflected in the Kocken-Specher theorem.[4] This theorem implies that any statement regarding state of affairs, formulated within the theory, acquires meaning counterfactually, i.e., after measurement. This implies that it is hard to avoid the Copenhagen interpretation of quantum theory, which is intrinsically non-realist.
- Notions of 'measurement' and 'external observer' pose problems when dealing with cosmology. In fact, in this case there can be no external observer since we are dealing with a closed system. But this then implies that the concept of 'measurement' plays no fundamental role which, in turn, implies that the standard interpretation of probabilities in terms of relative frequency of measurements breaks down.
- The existence of the Planck scale suggests that there is no *a priori* justification for the adoption of the notion of a continuum in the quantum theory, used in formulating quantum gravity.

These considerations led Isham and Döring to search for a reformulation of quantum theory that is more realist than the existing one. It turns out that this can be achieved through the adoption of topos theory as the mathematical framework by which to reformulate quantum theory.

One approach to reformulating quantum theory in a more realist way is to re-express it in such a way that it 'looks like' classical physics, which is the paradigmatic example of a realist theory. This is precisely the strategy adopted by the authors in [27, 29–31] and [32]. Thus the first question is: what is the underlying structure which makes classical physics a realist theory?

The authors identified this structure with the following elements:

1. The existence of a state space S.
2. Physical quantities are represented by functions from the state space to the reals. Thus each physical quantity, A, is represented by a function

$$f_A : S \to \mathbb{R}. \tag{1.1}$$

3. Propositions are of the form "$A \in \Delta$" ("The value of the quantity A lies in the subset $\Delta \subseteq \mathbb{R}$"). There are represented by subsets of the state space S: namely, those subspaces for which the proposition is true. For example for the proposition "$A \in \Delta$", this is just

$$f_A^{-1}(\Delta) = \{s \in S \mid f_A(s) \in \Delta\}. \tag{1.2}$$

The collection of all such subsets forms a Boolean algebra, denoted $Sub(S)$.

[4] Roughly the *Kochen-Specker Theorem* asserts that, if the dimension of $\mathcal{H}$ is greater than 2, then it is not possible to assign values to all physical quantities, at the same time, in a consistent way. See Chap. 3.

4. States ψ are identified with Boolean-algebra homomorphisms

$$\psi : Sub(S) \to \{0, 1\} \tag{1.3}$$

from the Boolean algebra $Sub(S)$ to the two-elements $\{0, 1\}$. Here, 0 and 1 can be identified as 'false' and 'true', respectively.

The identification of states with such maps follows from identifying propositions with subsets of S. Indeed, to each subset $f_A^{-1}(\{\Delta\})$, there is associated a characteristic function $\chi_{A\in\Delta} : S \to \{0, 1\} \subset \mathbb{R}$ defined by

$$\chi_{A\in\Delta}(s) = \begin{cases} 1 & \text{if } f_A(s) \in \Delta; \\ 0 & \text{otherwise.} \end{cases} \tag{1.4}$$

Thus each state s either lies in $f_A^{-1}(\{\Delta\})$ or it does not. Equivalently, given a state s every proposition about the values of physical quantities in that state is either true or false, thus (1.3) follows.

The first issue in finding quantum analogues of 1, 2, 3, and 4 is to consider the appropriate mathematical framework in which to reformulate the theory. As previously mentioned, the choice fell on topos theory. There were many reasons for this, but a paramount one is that in any topos (which is a special type of category) distributive logic arises in a natural way: i.e., a topos has an internal logical structure that is similar, in many ways, to the way in which Boolean algebra arises in set theory. This feature is highly desirable, since requirement 3 implies that the sub-objects of our state space (yet to be defined) should form some sort of logical algebra.

The second issue is to identify which topos is the right one to use. Isham et al. achieved this by noticing that the possibility of obtaining a 'neo-realist' reformulation of quantum theory lied in the idea of a *context*. Specifically, because of the Kocken-Specher theorem, the only way of obtaining quantum analogues of requirements 1, 2, 3 and 4 is by defining them with respect to commutative sub-algebras (the 'contexts') of the non-commuting algebra, $\mathcal{B}(\mathcal{H})$, of all bounded operators on the quantum theory's Hilbert space. Thus 'locally', with respect to these contexts, quantum theory effectively behaves classically. So the idea is to try and define each quantum object locally in terms of these abelian contexts. The key feature, however, is that the collection of all these *contexts* or *classical snapshots* form a category ordered by inclusion. This implies that although one defines each quantum object locally, the global information is never lost, since it is put back into the picture by the categorical structure of the collection of all these classical snapshots.

Hence the task is to find a topos which allows you to define a quantum object as (roughly speaking) a collection of classical approximations. As we will see this can be done through the topos of presheaves over the category of abelian sub-algebras.

In terms of this topos of presheaves, quantum theory can be re-defined so that it retains some realism and its interpretation is not riddled with the above mentioned conceptual problems.

Chapter 2
Philosophical Motivations

This introductory chapter will deal with what the main interpretational problems of canonical quantum theory are.

In particular, we will analyse how the mathematical formalism of quantum theory leads to a non-realist interpretation of the theory. The focus will lie on understanding and analysing the Kochen-Specker theorem (K-S theorem), which can be thought of as the main mathematical underlying reason why quantum theory is non realist. The interpretation which comes out is the well-known Copenhagen interpretation of quantum theory, which is an instrumentalist interpretation. However, such an interpretation leads to many conceptual problems.

Topos quantum theory is a way of overcoming such problems by re-defining quantum theory in the novel language of topos theory. The advantage of this language is that it renders the theory more realist, thus solving the above mentioned problems. In the process, however, one ends up with a multivalued/intuitionistic logic rather than a Boolean logic.

2.1 What Is a Theory of Physics and What Is It Trying to Achieve?

A theory of physics can be seen as a mathematical model which tries to answer three of the fundamental questions humanity has been and still is struggling to answer:

I What is a thing? (Heidegger)
II How are "things" related to one another?
III How do we get to know (1) and (2)?

The first two questions are related to ontological[1] issues while the third is of an epistemological[2] nature.

---

[1]Ontology comes from the Greek word meaning "being, that which is" and indicates the study of what things are in themselves and what can be said to exist.

[2]Epistemology comes from the Greek word meaning "knowledge" and it is the study concerning what is knowledge and how do we gain knowledge.

C. Flori, *A First Course in Topos Quantum Theory*, Lecture Notes in Physics 868,
DOI 10.1007/978-3-642-35713-8_2, © Springer-Verlag Berlin Heidelberg 2013

The two main physics theories which presuppose to answer the above questions are

- Classical physics.
- Quantum theory.

The way in which these two theories have answered the above questions is by defining a mathematical model which is supposed to describe nature. The interpretation of this mathematical model then, in turn, gives rise to a philosophical view of the world which provides an answer to the question. In classical physics, the mathematical model developed is in accordance with our common believes about the world. In fact it is arguable whether our common believes modelled classical theory, but we will not delve into this issue. On the other hand, things are not so straightforward in quantum theory in which, as we will see is Sect. 2.3, the mathematical formalism of the theory seems, at time, to defy our common sense.

In any case, in order to fully understand how a philosophical picture of the world can be derived through the mathematical formulation of a theory of physics, we need to refine the questions 1–3 defined at the start of this section (see [21] for an in depth discussion). In particular, any theory of physics, worthy of that name, should address the following issues:

1. What is the system under investigation?
2. What is the ontological status of physical terms?
3. What is the epistemological status of physical terms?
4. How physical statements can be verified or falsified?
5. What is the relation between the mathematical model and the physical world?

As we will see, the different answers given to the above issues by classical theory and quantum theory, respectively, will highlight the radical differences between the two theories and the different interpretations which each of them gives to the "outside world".

Thus, summarising, the main idea exposed in this section is that the mathematical tools used to describe a physical system encode a philosophical position regarding the world.

In the following subsection, we will briefly analyse how questions 1–5 are dealt with in classical physics.

2.2 Philosophical Position of Classical Theory

The philosophical position of classical physics is that of a *realist theory*. Here, by *realist theory* we mean a theory in which

(i) properties can be ascribed to a system at any given time and do not depend on the act of measuring;

(ii) the underlying logic is Boolean (classical) logic[3] which is the same logic we employ in our language.

The realism of classical theory implies that a *thing* is defined[4] in terms of a bundle of properties, which are said to belong to the *thing* (system). The type of properties that we are dealing with are of two kinds:

1. *Internal properties* which belong exclusively to the system, for example the mass, charge etc.
2. *External properties* which define relations to other systems, for example position, velocity etc.

When defining what a thing is one usually considers internal properties. Thus, in a realist theory, the outside world (physical world) exists independently of us and physical terms represent[5] actual things existing 'out there '. This answers questions (1) and (2) of Sect. 2.1.

Regarding the epistemological question (3) and (4) of Sect. 2.1 one can say that in classical physics, knowledge is acquired through the process of measurement. In fact, *measurement* enables us to know the values of a given system. However, in this context, measurement is just another form of interaction, i.e. no special role is ascribed to measurement since both object and subject, as viewed from a classical perspective, exist out there independently of one another.

Finally it is worth mentioning that, generally, classical theory is thought of as being a deterministic theory,[6] i.e. given initial state of a system at a given time, it is possible to predict with certainty the state of the system at a subsequent time.

From the above discussion it emerges that classical theory provides a mathematical model which accurately describes the 'outside' world. Thus one says that there is a bijective correspondence between the 'outside' world (physical world) and the mathematical model put forward by classical theory.

We have thus answered all questions (1–5) of Sect. 2.1 for classical theory. We would now like to identify which mathematical constructs (and in particular the way in which they are defined) render classical theory a realist theory.

The answer to this question will be given in detail in subsequent Chapters, but for now we will restrict ourselves tn answering it in a very conceptual way, so as to give a general idea of the relation between mathematical constructs and induced philosophical ideas.

---

[3]Boolean logic will be described later on in the book. For now we will simply say that a Boolean logic is the logic we use in our every-day thinking and in our language. Such a logic is characterised by the fact that (i) it is distributive, (ii) it only has two truth values $\{true, false\}$ and (iii) the logical connectives are our linguistic logical connectives: "and", "exclusive or", "not", "if then".

[4]It is worth noting that our own language reflects a realist view of the world: "The tree *is* three meters tall".

[5]Care should be taken at this point since, quite often, physical terms only represent idealisations of real physical systems.

[6]In a stochastic approach the realist conditions (i) and (ii) at the beginning of the section still hold.

The elements/concepts whose mathematical description render classical theory a realist theory are the following:

1. *State space*. In classical physics, the state space S is defined to be the collection of all states $s_i \in S$ of the system such that each s_i at a given time t_i, embodies all the properties of the system at that time.
2. *Definition of physical quantities*. A physical quantity in classical physics is identified with the collection of values it can have for a given system. This is a consequence of the fact that in classical physics quantities do have values and are characterised by these values.
3. *Definition of propositions*. The type of propositions dealt with in classical physics are of the form "$A \in \Delta$", meaning "the physical quantity A takes its values in the interval $\Delta \subseteq \mathbb{R}$". In classical physics such a proposition is identified with the collection of states $s_i \in S$ for which the quantity A does have values which lie in the interval Δ. Hence "$A \in \Delta$" is identified with a subset of the state space, precisely that subset for which the proposition is true.
4. *Boolean logic*. The logic governing classical propositions is Boolean logic which is a distributive logic and admits only two truth values: $\{true, false\}$. Verification of such truth values is done through the measurement interaction.
5. *Probabilities*. Epistemic interpretation of probabilities which, in a sense, represent what we know of the system. In this view the discrepancy between our knowledge and the actual state of the system can be made arbitrarily small by refining either our measurements or the description of the system.

Later in the book we will describe, in details, how the above classical concepts are mathematically represented but, for now, it suffices to say that it is precisely the way in which the above elements of the theory are mathematically expressed which renders classical theory a realist theory. In fact, when considering quantum theory we will see how the same elements are mathematically described in a very different way. This will induce a different conceptual understanding of such elements which, in turn, will imply a different philosophical interpretation of the theory.

2.3 Philosophy Behind Quantum Theory

If we analyse the mathematical formalism of quantum theory, we immediately realise that the theory is non-realist (with the definition of realist given above). In fact the above conditions [(i), (ii)] at the start of Sect. 2.2 do not strictly hold[7] in quantum theory. In fact the formalism of quantum theory implies a clear distinction between measuring apparatus and measuring system, such that the act of measuring gets ascribed a special status. In this setting, measurement becomes a means for assigning a probabilistic spread of outcomes rather than a means to determine properties of the system. Thus the very concept of properties ceases to have its common sense

[7] We will clarify this later on in the book.

meaning, since its definition is now intertwined with the act of measurement. It is as if properties acquire the status of latent attributes which are brought into existence by the act of measurement, but which can not be said to exist independently of such measurement. Therefore it becomes meaningless to talk about a physical system as possessing properties. The interpretation that results is the so called *instrumentalist interpretation* of quantum theory which is a non-realist interpretation [21]. This interpretation of quantum theory makes it very difficult to give answers to questions (1–5) of Sect. 2.1. The reason being that in quantum theory (as it has been expressed so far) there is no distinction between ontological and epistemological status of physical terms. In fact, very roughly, one can say that quantum theory is a mathematical model which, in a way, gives rise to "things" through the measuring process. Thus a "thing" becomes simply a result of a measurement which, as such, only describes what we assume exists 'out there'. In turn, physical statements seem only to represent our knowledge of events rather than events themselves. Thus the philosophical position of quantum theory can be summarised by the following statements:

1. Properties can not be said to be possessed by a system a priori. All that can be said is that after a measurement is performed, the system "acquires" the "latent" properties (state-vector reduction).
2. Any statement regarding 'states of affairs' about a system can only be made a posteriori after measurement. However such statements can not be regarded as describing properties of the system, on the contrary, they describe probabilities of possible measurement outcomes.
3. Measurement becomes a very special type of interaction.
4. Clear distinction between observer and observed system. By this we mean a qualitative distinction, not merely a quantitate one. In fact it is true that in any branch of physics which requires experiments, there is *ipso fatto* a distinction between the physics and the system they are studying. However in a classical regime this distinction is temporary and quantitative, since classical physics is governing both observer and observed system. However in the quantum case the distinction is qualitative since the two systems undergo different *types* of physics.
5. States are not seen as bearers of physical properties but are simply the most efficient tools to enable one to determine/compute predictions for possible measurements, i.e. predictions of probabilities of outcomes, not outcomes themselves.
6. Quantum theory is deterministic *but* what evolves are now predicted probabilities of measurement results, *not* actual measurements.
7. Relative frequency interpretation of probabilities. In this interpretation certain uncertainties can not be eliminated by refinements of our measurement.

The above features of quantum theory, which directly derive from the mathematical representation of the theory, imply a non realist interpretation of quantum theory.

Such an interpretation, although it works for some situations, causes various conceptual problems in the context of quantum gravity and quantum cosmology.

2.4 Conceptual Problems of Quantum Theory

The canonical mathematical formulation of quantum theory leads to an interpretation which has many conceptual obstacles for a fully coherent theory. These problems can be summarised as follows:

- Notions of 'measurement' and 'external observer' pose problems when dealing with cosmology. In fact, in this case there can be no external observer since we are dealing with a closed system. But this then implies that the concept of 'measurement' plays no fundamental role which, in turn, implies that the standard definition of probabilities, in terms of relative frequency of measurements, breaks down.
- The existence of the Planck scale suggests that there is no *a priori* justification for the adoption of the notion of a continuum in the quantum theory used in formulating quantum gravity.
- Standard quantum theory employs, in its formulation, the use of a fixed spatio-temporal structure (fixed background). This is needed to make sense of its instrumentalist interpretation, i.e. it needs a space-time in which to make a measurement. This fixed background seems to cause problems in quantum gravity where one is trying to make measurements of space-time properties. In fact, if the action of making a measurement requires a space time background, what does it mean to measure space time properties?
- Given the concept of superposition present in quantum theory, by applying such a concept to quantum gravity we would have to account for the occurrence of quantum superpositions of eigenstate properties of space, time and space-time.

Chapter 3
Kochen-Specker Theorem

In this chapter, we will analyse one of the main theorems (another one would be Bell's inequality) which states the impossibility of quantum theory, as it is canonically expressed, to be a realist theory. This theorem is know as the Kochen-Specker theorem and will be explained in details in Sect. 3.3. However, in order to fully understand this theorem, we first of all have to introduce the concept of a *valuation function*. We will do so first for classical theory and then quantum theory.

3.1 Valuation Functions in Classical Theory

Roughly speaking, a *valuation function* is a map which assigns, to each physical quantity its value, such that certain 'consistency' conditions are satisfied. Thus, as a first step in defining a valuation function is to understand how a physical quantity is described in classical physics.

In classical physics, physical quantities are represented by functions from the state space to the reals. Thus, each physical quantity, A, is represented by a function $f_A : S \to \mathbb{R}$, such that for each state of the system $s_i \in S$, $f_A(s_i) \in \mathbb{R}$ represents the value of A given the state s_i. This association of physical quantities with real-valued functions on the state space is 1:1 (one-to-one: for each quantity A there corresponds one and only one function f_A). This association is injective.

Given the definition of physical quantity in terms of maps on the state space, the definition of valuation function in classical physics is straightforward. In particular, a valuation function is defined, for each state s_i in the state space S, as a map

$$V_{s_i} : \mathcal{O} \to \mathbb{R} \tag{3.1}$$

from the set of observables (physical quantities) $\mathcal{O}$ to the reals, such that for each $A \in \mathcal{O}$ we obtain:

$$A \mapsto V_{s_i(A)} := f_A(s_i) \tag{3.2}$$

where $V_{s_i(A)}$ represents the value of the physical quantity A given the state s_i. This equation defined V_{s_i} uniquely.

C. Flori, *A First Course in Topos Quantum Theory*, Lecture Notes in Physics 868,
DOI 10.1007/978-3-642-35713-8_3, © Springer-Verlag Berlin Heidelberg 2013

A condition such a valuation function has to satisfy is the so called *functional composition condition* (FUNC) which is defined as follows:

$$\text{for any } h:\mathbb{R}\to\mathbb{R}, \quad V_{s_i}\big(h(A)\big)=h\big(V_{s_i}(A)\big). \tag{3.3}$$

In this equation $h(A)\in\mathcal{O}$ and is defined in terms of composition of functions:

$$h(A):=h\circ f_A : S \xrightarrow{f_A} \mathbb{R} \xrightarrow{h} \mathbb{R}. \tag{3.4}$$

If A represents the physical quantity *energy*, and h is a function which defines the square, i.e. $h(A)=A^2$, what (3.3) would mean is: "the value of the energy squared is equal to the square of the value of the energy".

3.2 Valuation Functions in Quantum Theory

If we were to mimic classical theory then we would define a valuation function as follows:
for each state $|\psi\rangle\in\mathcal{H}$ (where $\mathcal{H}$ is a finite dimensional Hilbert space) the valuation function is a function from the set of self-adjoint operators (quantum analogues of physical quantities) to the reals

$$V_{|\psi\rangle}:\mathcal{O}\to\mathbb{R} \tag{3.5}$$

such that, for each state $|\psi\rangle$, $V_{|\psi\rangle}$ assigns to each self-adjoint operator $\hat{A}\in\mathcal{O}$ a real number $V_{|\psi\rangle}(\hat{A})\in\mathbb{R}$, that represents the value of $\hat{A}$ for the state $|\psi\rangle$ of the system.

However this definition of a valuation function only makes sense if $|\psi\rangle$ is an eigenvector of $\hat{A}$. Other than that special case, the above definition of a valuation function does not really make sense. So the question is how to generalise it for an arbitrary state $|\psi\rangle$? A possible generalisation is the following:

Definition 3.1 A valuation function for quantum theory is a map $V:\mathcal{O}\to\mathbb{R}$ which satisfies the following two conditions:

(i) For each $\hat{A}\in\mathcal{O}$, $V(\hat{A})\in\mathbb{R}$ represents the value of the operator $\hat{A}$ and it belongs to the spectrum of $\hat{A}$.
(ii) FUNC: for all $h:\mathbb{R}\to\mathbb{R}$ the following holds:

$$V\big(h(\hat{A})\big)=h\big(V(\hat{A})\big). \tag{3.6}$$

Any function satisfying the above conditions is a valuation function.

At this point it is worth understanding, explicitly, what $h(\hat{A})$ is.

What Is $h(\hat{A})$? Given a self-adjoint operator $\hat{A}$ we have two situations:

(i) Let $|\psi\rangle\in\mathcal{H}$ be an eigenvector of $\hat{A}$, i.e. $\hat{A}|\psi\rangle=a|\psi\rangle$. It is then straightforward to define the following:

$$\hat{A}^2|\psi\rangle=a^2|\psi\rangle; \qquad \hat{A}^3|\psi\rangle=a^3|\psi\rangle. \tag{3.7}$$

Thus, generalising for any polynomial function Q, we obtain

$$Q(\hat{A})|\psi\rangle = Q(a)|\psi\rangle \tag{3.8}$$

provided $Q(a)$ is well defined. Given the above we are justified in defining, for any function $h : \mathbb{R} \rightarrow \mathbb{R}$ the following:

$$h(\hat{A})|\psi\rangle = h(a)|\psi\rangle \tag{3.9}$$

provided $h(a)$ is well defined (for example not infinite).

(ii) We now would like to generalise it to arbitrary states, not just eigenvectors. To this end, we recall that the set of eigenvectors of a self-adjoint operator forms an orthonormal basis for $\mathcal{H}$. This means that any state $|\psi\rangle \in \mathcal{H}$ can be written in terms of such an orthonormal basis. Thus, considering the case in which $\hat{A}$ has a discrete spectrum (all that follows can be easily generalised for the continuum case)[1] the *spectral decomposition* of $\hat{A}$ is

$$\hat{A} := \sum_{m=1}^{M} a_m \hat{P}_m \tag{3.10}$$

where $\{a_1 \dots a_m\}$ is the set of eigenvalues of $\hat{A}$, while each $\hat{P}_m$ is the projection operator onto the subspace of eigenvectors with eigenvalue a_m. In particular,

$$\hat{P}_m := \sum_{j=1}^{d(m)} |a_m, j\rangle\langle a_m, j|. \tag{3.11}$$

Here $j = 1 \dots d(m)$ labels the degenerate eigenvectors with common eigenvalue a_m. In this setting any state $|\psi\rangle$ can be written as follows

$$|\psi\rangle = \sum_{m=1}^{M} \sum_{j=1}^{d(m)} \langle a_m, j|\psi\rangle |a_m, j\rangle. \tag{3.12}$$

Keeping this in mind, and inspired by case (i) above we define

$$h(\hat{A})|\psi\rangle := \sum_{m=1}^{M} \sum_{j=1}^{d(m)} h(a_m)\langle a_m, j|\psi\rangle |a_m, j\rangle$$

[1] In Chap. 13 we will explain, in detail, what a spectral decomposition is, but for now we will simply state that each self-adjoint operator $\hat{A}$ can be written as $\hat{A} = \int_{\sigma(A)} \lambda d\hat{E}^{\hat{A}}_{\lambda}$. Such an expression is called the spectral decomposition of $\hat{A}$. Here $\sigma(A) \subseteq \mathbb{R}$ represents the spectrum of the operator $\hat{A}$ and $\{\hat{E}^{\hat{A}}_{\lambda} | \lambda \in \sigma(\hat{A})\}$ is the spectral family of $\hat{A}$. In the discrete case we would have $\hat{A} = \sum_{\sigma(A)} \lambda \hat{P}^{\hat{A}}_{\lambda}$, where the projection operators $\hat{P}^{\hat{A}}_{\lambda}$ project on subspaces of the Hilbert space for which the states ψ have value λ for A.

$$= \sum_{m=1}^{M} h(a_m)\hat{P}_m|\psi\rangle. \tag{3.13}$$

Obviously this makes sense only if $h(a_m)$ is well defined. Since definition (3.13) is valid for all $|\psi\rangle \in \mathcal{H}$ it follows that

$$h(\hat{A}) := \sum_{m=1}^{M}\sum_{j=1}^{d(m)} |a_m, j\rangle\langle a_m, j| = \sum_{m=1}^{M} h(a_m)\hat{P}_m. \tag{3.14}$$

Now that we have defined what $h(\hat{A})$ means we will analyse the FUNC condition on the valuation function.

3.2.1 Deriving the FUNC Condition

The FUNC condition is a direct consequence of three assumptions and a principle present in quantum theory:

(1) **Statistical functional compositional principle**: given a self-adjoint operator $\hat{A}$ that represents an observable A and a function $f : \mathbb{R} \to \mathbb{R}$, then for an arbitrary real number a we have the following equality:[2]

$$prob\big[V\big(f(\hat{A})\big) = a\big] = prob\big[f\big(V(\hat{A})\big) = a\big].$$

In order to prove the above principle we have to define the relation between projection operators and their respective characteristic functions.

Let us consider the following characteristic function χ_r defined by

$$\chi_r(t) = \begin{cases} 1 & \text{if } t = r \\ 0 & \text{otherwise.} \end{cases}$$

It then follows that, given a self-adjoint operator $\hat{A}$, whose spectral decomposition (assume discrete) contains the spectral projector $\hat{P}_m$, one can write:

$$\chi_r(\hat{A}) := \sum_{m=1}^{M} \chi_r(a_m)\hat{P}_m = \begin{cases} \hat{P}_m & \text{if } a_m = r \\ 0 & \text{otherwise.} \end{cases} \tag{3.15}$$

What (3.15) uncovers is that $\chi_r(\hat{A}) = \hat{P}_m$ iff r is the eigenvalue a_m of $\hat{A}$. Moreover, given a function $f : \sigma(\hat{A}) \to \mathbb{R}$ (where $\sigma(\hat{A})$ represents the spectrum of $\hat{A}$) we have:

$$\chi_r\big(f(\hat{A})\big) = \chi_{f^{-1}(r)}(\hat{A}). \tag{3.16}$$

---

[2] Note that in the following we will use probabilistic valuation functions of the form $V : \mathcal{O} \to P([\mathbb{R}])$ where $P([\mathbb{R}])$ represents the space of probability distributions.

We know that Born rule [21] for projection operators is

$$prob\big(V(\hat{A}=a_m)\big)=\mathrm{Tr}(\hat{P}_m\hat{P}_{|\psi\rangle}) \tag{3.17}$$

where $\hat{P}_{|\psi\rangle}:=|\psi\rangle\langle\psi|$. This means that if a measurement of an observable A is made on a system in state $|\psi\rangle$, then the probability of obtaining as a result the eigenvalue a_m is given by (3.17).

Therefore from (3.15) and (3.17) we get

$$prob\big(V(\hat{A})=a_m\big)=\mathrm{Tr}\big(\chi_{a_m}(\hat{A})\hat{P}_{|\psi\rangle}\big).$$

We can now prove the statistical functional compositional principle.

Proof Using (3.15), (3.17) and (3.16) we can write the statistical algorithm for projector operators as follows:

$$\begin{aligned}prob\big(V\big(f(\hat{A})\big)=b\big)&=\mathrm{Tr}\big(\chi_{f^{-1}(b)}(\hat{A})\hat{P}_{|\psi\rangle}\big)\\&=\mathrm{Tr}(\hat{P}_{f^{-1}(b)}\hat{P}_{|\psi\rangle})\\&=prob\big(V(\hat{A})=f^{-1}(b)\big)\end{aligned}$$

but

$$V(\hat{A})=f^{-1}(b)\quad\Leftrightarrow\quad f\big(V(\hat{A})\big)=b$$

therefore

$$prob\big(V\big(f(\hat{A})\big)=b\big)=prob\big(f\big(V(\hat{A})\big)=b\big).$$ □

(2) **Non-contextuality**: the value of observables is independent of the measurement context, i.e. the value of each observable is independent of any other observables evaluated at the same time.
(3) **Value definiteness**: observables possess definite values at all times.
(4) **Value realism**: to each real number α, such that $\alpha=prob(V(\hat{A})=\beta)$, for an operator $\hat{A}$ there corresponds an observable A with value β.

From the above conditions (1), (2), (3) and (4) the FUNC condition follows:

Proof Consider an observable B represented by the self-adjoint operator $\hat{B}$. From (3) we deduce that $\hat{B}$ possesses a value: $V(\hat{B})=b$. Given a function $f:\mathbb{R}\to\mathbb{R}$ we obtain the quantity $f(V(\hat{B}))=f(b)=a$. Applying (1) we get $prob[f(V(\hat{B}))=a]=prob[V(f(\hat{B}))=a]$ which means that there exists a self-adjoint operator of the form $f(\hat{B})$. From (4) it then follows that the corresponding observable for $f(\hat{B})$ has value a, therefore $f(V(\hat{B}))=V(f(\hat{B}))$. From (2) this result is unique, therefore FUNC follows. □

From the above discussion it follows that the property of values being "mutually exclusive and collectively exhaustive"[3] is a consequence of the FUNC condition.

[3]Values are said to be mutually exclusive and collectively exhaustive if one and only one gets assigned the value 1 (true), while the rest gets assigned the value 0 (false).

3.2.2 Implications of the FUNC Condition

The conditions on the valuation function implied by FUNC are:

(a) The *sum rule*

$$V(\hat{A}+\hat{B}) = V(\hat{A}) + V(\hat{B}) \tag{3.18}$$

where $\hat{A}$ and $\hat{B}$ are such that $[\hat{A}, \hat{B}] = 0$.

Proof To prove the above result we need the following theorem:

Theorem 3.1 *Given any pair of self-adjoint operators* $\hat{A}$ *and* $\hat{B}$*, such that* $[\hat{A}, \hat{B}] = 0$ *and two functions* $f, g : \mathbb{R} \rightarrow \mathbb{R}$*, then there exists a third operator* $\hat{C}$ *such that* $\hat{A} = f(\hat{C})$ *and* $\hat{B} = g(\hat{C})$*.*

Given two commuting operators $\hat{A}$ and $\hat{B}$ from the above theorem it follows that $\hat{A} = f(\hat{C})$ and $\hat{B} = g(\hat{C})$, therefore there exists a function $h = f + g$ where $((f+g)(x) := f(x) + g(x))$, such that $\hat{A} + \hat{B} = h(\hat{C})$, therefore

$$\begin{aligned} V(\hat{A}+\hat{B}) &= V\big(h(\hat{C})\big) \\ &= h\big(V(\hat{C})\big) \\ &= f\big(V(\hat{C})\big) + g\big(V(\hat{C})\big) \\ &= V\big(f(\hat{C})\big) + V\big(g(\hat{C})\big) \\ &= V(\hat{A}) + V(\hat{B}). \end{aligned}$$

□

(b) The *product rule*

$$V(\hat{A}\hat{B}) = V(\hat{A})V(\hat{B}) \tag{3.19}$$

where $\hat{A}$ and $\hat{B}$ are such that $[\hat{A}, \hat{B}] = 0$.

Proof Given $\hat{A} = f(\hat{C})$ and $\hat{B} = g(\hat{C})$ there exists a function $k = fg$ where $(fg(x) := f(x)g(x))$ such that $\hat{A}\hat{B} = k(\hat{C})$, therefore

$$\begin{aligned} V(\hat{A}\hat{B}) &= V\big(k(\hat{C})\big) \\ &= k\big(V(\hat{C})\big) \\ &= f\big(V(\hat{C})\big)g\big(V(\hat{C})\big) \\ &= V\big(f(\hat{C})\big) \cdot V\big(g(\hat{C})\big) \\ &= V(\hat{A})V(\hat{B}). \end{aligned}$$

□

As a consequence of the product and sum rules we obtain the following equalities:

$$\begin{aligned} V_{|\psi\rangle}(\hat{1}) &= 1 \\ V_{|\psi\rangle}(\hat{0}) &= 0 \\ V_{|\psi\rangle}(\hat{P}) &= 0 \text{ or } 1. \end{aligned} \tag{3.20}$$

Proof 1. Given any physical quantity B (with associated self-adjoint operator $\hat{B}$), from the product rule we have that, for $\hat{A} := \hat{1}$, the following relation holds:

$$V(\hat{1}\hat{B}) = V(\hat{B}) = V(\hat{1})V(\hat{B}) = V(\hat{B}). \tag{3.21}$$

This implies that $V(\hat{1}) = 1$, for $V(\hat{B}) \neq 0$.

2. Given any physical quantity B (with associated self-adjoint operator $\hat{B}$), from the sum rule we have that, for $\hat{A} := \hat{0}$, the following relation holds:

$$V(\hat{0} + \hat{B}) = V(\hat{B}) = V(\hat{0}) + V(\hat{B}) = V(\hat{B}). \tag{3.22}$$

This implies that $V(\hat{0}) = 0$.

3. Given a projection operator $\hat{P}$ we know that $\hat{P}^2 = \hat{P}$, therefore

$$V(\hat{P})^2 = V\big(\hat{P}^2\big) = V(\hat{P}). \tag{3.23}$$

It follows that

$$V(\hat{P}) = 1 \text{ or } 0. \tag{3.24}$$

□

Since quantum propositions can be expressed as projection operators (the reason will be explained later on in the book), what the last result implies is that, for any given state $|\psi\rangle$, the valuation function can only assign value *true* or *false* to propositions.

The set of all eigenvectors of a self-adjoint operator $\hat{A}$ forms an orthonormal basis for $\mathcal{H}$, thus we can define the resolution of unity in terms of the projection operators corresponding to the eigenvectors:

$$\hat{1} = \sum_{m=1}^{M} \hat{P}_m. \tag{3.25}$$

From (3.18), (3.19), (3.20) and (3.25) we conclude (for discrete case but it can easily be extended to the continuous case)

$$V(\hat{1}) = V\left(\sum_{m=1}^{M} \hat{P}_m\right) = \sum_{m=1}^{M} V(\hat{P}_m) = 1. \tag{3.26}$$

What this equation means is that one and only one of the projectors that form the resolution of unity gets assigned the value 1 (true), while the rest gets assigned the value 0 (false), i.e. the value assignment is said to be " mutually exclusive and collectively exhaustive" [22]. However, the Kochen-Specker theorem shows that it is impossible to give simultaneous values to all observables associated with a set of self-adjoint operators, in such a way that the values are "mutually exclusive and collectively exhaustive".

3.3 Kochen Specker Theorem

The Kochen-Specker theorem derives from the incompatibility of two assumptions regarding observables in quantum theory, namely [21, 22]:

1. The need of assigning simultaneous values to all observables in $\mathcal{O}$ (collection of all self-adjoint operators on $\mathcal{H}$).
2. The need for the values of observables to be "mutually exclusive and collectively exhaustive"[4] [22].

The precise definition of the Kochen-Specker theorem is the following:

Theorem 3.2 (Kochen-Specker Theorem) *If the dimension of $\mathcal{H}$ is greater than 2 then, there does not exist any valuation function $V : \mathcal{O} \rightarrow \mathbb{R}$ from the set $\mathcal{O}$ of all bounded self-adjoint operators $\hat{A}$ of $\mathcal{H}$ to the reals $\mathbb{R}$, such that the functional composition principle is satisfied for all $\hat{A} \in \mathcal{O}$.*

Another way of stating the theorem which seems more useful for developing a proof later on is the following:

Theorem 3.3 (Kochen-Specker Theorem) *Given a Hilbert space $\mathcal{H}$, such that $dim(\mathcal{H}) > 2$ and a set $\mathcal{O}$ of self-adjoint operators $\hat{A}$ which represent observables, then the following two statements are contradictory*:

1. *All observables associated with projectors in $\mathcal{O}$ have values simultaneously, i.e. they are mapped uniquely into the reals.*
2. *The values of observables follow the functional composition principle* (*FUNC*).

It follows that, in quantum theory, the Kochen-Specker theorem is related to the existence of a value function $V : \mathcal{O} \rightarrow \mathbb{R}$ from the set $\mathcal{O}$ of self-adjoint operators (which are the quantum analogues of physical quantities) to the Reals.

---

[4]Mutually *exclusive* means that only one value of an observable can be realised at a given time, while collectively *exhaustive* means that at least one of the values has to be realised at a given time.

3.4 Proof of the Kochen-Specker Theorem

There are various proofs of the Kochen-Specker Theorem. We will report a simplified version of the proof due to Kernaghan (1994) [23].

In the previous section we saw that the properties of the valuation function implied that V can only assign the value *true* or *false* to any projection operator $\hat{P}$ in such a way that this assignment is mutually exclusive and collectively exhaustible.

A special case would be when $\hat{P}_i := |e_i\rangle\langle e_i|$ where $\{|e_1\rangle, |e_2\rangle, \ldots, |e_n\rangle\}$ is an orthonormal basis (ONB) of the Hilbert space $\mathcal{H}^n$. In this setting, the valuation function must assign the value 1 to only one of the projection operators and zero to all the rest. Moreover, if the same projection operator belongs to two different ONB, the value assigned to this projection operator by V has to be the same, independently of which set it is considered to belong to. This is what is meant by *non-contextuality*. Kernaghan, in his proof of the K-S theorem, considers a real 4-dimensional Hilbert space $\mathcal{H}^4$ (there is no loss in generality, in considering the Hilbert space to be real). He then chooses 11 sets of 4 orthogonal vectors. Each vector is contained in either 2 of these sets or 4, so that there are some correlations among the ONB. The Kochen-Specker theorem is then reduced to a colouring problem, i.e. "within every set of orthogonal vectors in $\mathcal{H}^4$ exactly one must be coloured white (1, true), while the remaining ones black (0, false)". Writing down this collection of vectors, we would end up with the following table where each column denotes a set of 4 orthogonal vectors forming a particular ONB:

$\lvert e_1\rangle$	1, 0, 0, 0	1, 0, 0, 0	1, 0, 0, 0	1, 0, 0, 0	−1, 1, 1, 1	−1, 1, 1, 1
$\lvert e_2\rangle$	0, 1, 0, 0	0, 1, 0, 0	0, 0, 1, 0	0, 0, 0, 1	1, −1, 1, 1	1, 1, −1, 1
$\lvert e_3\rangle$	0, 0, 1, 0	0, 0, 1, 1	0, 1, 0, 1	0, 1, 1, 0	1, 1, −1, 1	1, 0, 1, 0
$\lvert e_4\rangle$	0, 0, 0, 1	0, 0, 1, −1	0, 1, 0, −1	0, 1, −1, 0	1, 1, 1, −1	0, 1, 0, −1

$\lvert e_1\rangle$	1, −1, 1, 1	1, 1, −1, 1	0, 1, −1, 0	0, 0, 1, −1	1, 0, 1, 0
$\lvert e_2\rangle$	1, 1, −1, 1	1, 1, 1, −1	1, 0, 0, −1	1, −1, 0, 0	0, 1, 0, 1
$\lvert e_3\rangle$	0, 1, 1, 0	0, 0, 1, 1	1, 1, 1, 1	1, 1, 1, 1	1, 1, −1, −1
$\lvert e_4\rangle$	1, 0, 0, −1	1, −1, 0, 0	1, −1, −1, 1	1, 1, −1, −1	1, −1, −1, 1

We now want to assign value true (colour white) to one and only one projection operator associated to each vector in each column. This requirement represents condition (3.26) above. However, it is easy to see from the table that such condition (3.26) cannot be satisfied. In fact, if it were satisfied we would end up with 11 entries being coloured white, since each column would have exactly one entry coloured white and there are 11 columns. But, since each vector appears two or four times, we end up with an even number of white entries which cannot be 11. Therefore, we conclude that it is impossible to obtain a colouring of a set of orthogonal vectors that is consistent with condition (3.26). Remember that we have assumed a non-contextual assignment of the entries, i.e. we are assuming that same vectors get assigned the same colour, independently of the column they belong to.

Although this is a very simplified version of the proof of the Kochen-Specker theorem, the main idea is the same as the main idea in the original proof, namely: given a set of orthogonal vectors in $\mathcal{H}^n$ it is impossible to assign to each of them a number $\{1, 0, 0, 0 \ldots 0\}$ where only one entry is equal to 1, i.e. it is impossible to give simultaneous values to all observables while respecting the FUNC condition which, as shown further on, implies non-contextuality.

3.5 Consequences of the Kochen-Specker Theorem

The implications of the Kochen-Specker theorem is that one or both of the following two assumptions must be dropped:

(i) The set of truth values is represented by $\{0, 1\}$.
(ii) The functional composition principle.

In the topos approach, we abandon the idea that the set of truth values is only $\{0, 1\}$. In fact, in this approach we utilise a multivalued logic which, in turn, will imply the adoption of an intuitionistic logic.

On the other hand, abandoning FUNC would entail abandoning some or all of the three following assumptions:

1. *Non-contextuality*
2. *Value Definiteness*
3. *Value Realism*

from which it derives.

In particular, if the FUNC principle gets dropped, then quantum theory turns out to be contextual and non-realist. To understand this situation we first have to introduce the notion of simultaneously measurable observables:

Definition 3.2 Given two observables $\hat{A}$ and $\hat{B}$ we say that they are **simultaneously measurable** iff $[\hat{A}, \hat{B}] = 0$.

Let us now consider two observables: $\hat{A}$, $\hat{B}$, such that $[\hat{A}, \hat{B}] \neq 0$, and $\hat{A}$ and $\hat{B}$ have a common projection in their spectral decomposition

$$\begin{aligned} \hat{A} &= a_1 \hat{P} + a_2 \hat{P}_{a_2} + a_3 \hat{P}_{a_3} \\ \hat{B} &= b_1 \hat{P} + b_2 \hat{P}_{b_2} + b_3 \hat{P}_{b_3}. \end{aligned} \tag{3.27}$$

From (3.15) it follows that $\hat{P}$ can be expressed in terms of $\hat{A}$ or of $\hat{B}$, i.e.:

$$\begin{aligned} \hat{P} &= \chi_{a_1}(\hat{A}) := \sum_{m=1}^{M} \chi_{a_1}(a_m) \hat{P}_{a_m} \\ \hat{P} &= \chi_{b_1}(\hat{B}) := \sum_{m=1}^{M} \chi_{b_1}(b_m) \hat{P}_{b_m}. \end{aligned}$$

Since commuting operators correspond to orthogonal projections, then, if we choose to express $\hat{P}$ in terms of $\hat{A}$ i.e. $\hat{P} = \chi_{a_1}(\hat{A})$, the relevant operators commuting with $\hat{P}$ are:[5] $\hat{P}_{a_2}$, $\hat{P}_{a_3}$, $\hat{P} \vee \hat{P}_{a_2}$, $\hat{P} \vee \hat{P}_{a_3}$ and $\hat{P}^{\perp} := \hat{1} - \hat{P}$. If, instead, we choose to express $\hat{P}$ in terms of $\hat{B}$, i.e. $\hat{P} = \chi_{b_1}(\hat{B})$, then the commuting operators of $\hat{P}$ would be: $\hat{P}_{b_2}$, $\hat{P}_{b_3}$, $\hat{P} \vee \hat{P}_{b_2}$, $\hat{P} \vee \hat{P}_{b_3}$ and $\hat{P}^{\perp}$.

Now, as a consequence of the FUNC condition we obtain that

$$\begin{aligned} V(\hat{P}) &= V\big(\chi_{a_1}(\hat{A})\big) = \chi_{a_1}\big(V(\hat{A})\big) \\ V(\hat{P}) &= V\big(\chi_{b_1}(\hat{B})\big) = \chi_{b_1}\big(V(\hat{B})\big) \end{aligned} \tag{3.28}$$

which implies that $\chi_{a_1}(V(\hat{A})) = \chi_{b_1}(V(\hat{B}))$.

However, if the FUNC condition does not hold, then we might have that

$$\chi_{a_1}\big(V(\hat{A})\big) \neq \chi_{b_1}\big(V(\hat{B})\big). \tag{3.29}$$

What this implies is that the value of $\hat{P}$ will depend on whether $\hat{P}$ is considered as belonging to the spectral decomposition of $\hat{A}$ or that of $\hat{B}$. In fact, let us assume that $\chi_{a_1}(V(\hat{A})) = 1$, which means that $V(\hat{A}) = a_1$. a_1 is the eigenvalue with corresponding projection operator $\hat{P}$, then it follows that $V(\hat{P}) = 1$, since $\hat{P}$ projects on the subspace of eigenvectors which have eigenvalue precisely a_1. This means that the proposition associated[6] with $\hat{P}$ is true. However, if $\chi_{a_1}(V(\hat{A})) \neq \chi_{b_1}(V(\hat{B}))$, then $\chi_{b_1}(V(\hat{B})) \neq 1$, which must imply (since it is a characteristic function) that $\chi_{b_1}(V(\hat{B})) = 0$. If this is the case then $\hat{P}$ is false, i.e. $V(\hat{P}) = 0$.

As a consequence of the above, a physical quantity A is not represented by an operator with unique *meaning* in quantum theory. On the contrary, each operator has different *meaning* depending on what other operators are considered at the same time. This implies that the quantisation map $A \mapsto \hat{A}$ is one to many.

The contextuality derived from dropping FUNC has great impact on the 'realism' of quantum theory. In fact, when one says that a given quantity has a certain value, we mean that, that quantity "possesses" that value, and the concept of "possession" is independent of the context chosen. However, if our theory is contextual, what does it exactly mean that a quantity has a given value? It would seem that in a contextual theory, there is no room for a realist interpretation. In fact, if we measured two pairs of quantities (A, B) and (A, C) and obtained the values (a, b) and (a', c), respectively, such that $a \neq a'$, then what value does the quantity A actually posses?[7]

[5]Here $\hat{P}_1 \vee \hat{P}_2 = \hat{P}_1 + \hat{P}_2 - \hat{P}_1\hat{P}_2$.

[6]As we will explain in detail later on, in standard quantum theory, propositions are represented by projection operators. In particular the proposition "$A = a_1$", indicating that the value of the physical quantity is a_1, will be represented by the projection operator $\hat{P}$ which projects on the subspace of eigenvectors which have eigenvalue precisely a_1.

[7]It should be noted that the probabilistic predictions of quantum theory, are not affected by the notion of contextuality. In fact, the result of measuring a property A of a system does not depend on what else is measured at the same time, since the probability of obtaining a_m as the value of A will always be $\langle\psi|\hat{P}_{a_m}|\psi\rangle$.

The above reasoning implies that is not possible to answer the question of what it means for a quantity to have a given value.

Therefore, if one drops the FUNC principle we will end up with a non-realist contextual interpretation of quantum theory, which clashes with the realism of classical physics and our common sense. Moreover, as previously mentioned, such an interpretation is riddled with conceptual problems.

A question then arises: what if, instead, we dropped the first assumption, namely if we allowed for the truth values to be in some larger set other than the set $\{0, 1\}$?

This is precisely what is done in topos quantum theory. In fact, in this setting the FUNC principle is conserved, but the set of truth values is replaced by some larger set than simply $\{0, 1\}$ leading to a multivalued logic. The interpretation we end up with is not strictly realist, due to the multivalued nature of the resulting logic. However we reach a more realist interpretation of the theory since now it makes sense to say that values are possessed by quantities in a context independent way.

At this stage it should be pointed out that in the topos formulation of quantum theory there will be the notion of *contextuality*, albeit its interpretation will be very different, thus it will not impinge on the notion of realism (of the interpretation).

Chapter 4
Introducing Category Theory

Roughly speaking a category is a collection of objects and relations between these objects. These relations are required to satisfy certain properties which make the set of all such relations 'coherent'. Given a category, it is not the case that every two objects have a relation between them, some do and others don't. For the ones that do, the number of relations can vary depending on which category we are considering.

Working with category theory means adopting an external description of mathematical objects. Normally in set theory we are used to an internal description of objects, i.e. objects are defined according to a set of communal property all of their elements share. This description however is not adequate for category theory. In this context objects are defined in terms of relations they have with other objects (in the same category). Thus, in order to work with category theory, one has to perform a change in perspective, from internal to external.

In this chapter we will explain, in details, what this change in perspective really is. Moreover, to illustrate the difference we will describe how the axioms of group theory can be given in both an internal and an external way.

We will then give an axiomatic definition of what a category is, introducing also the concept of a subcategory. This will be augmented with various examples of categories and a list of the most common categories which appear in physics.

We then conclude with the concept of duality, which is a very important concept in category theory and central for what will be explained in the rest of the book.

4.1 Change of Perspective

> "Category Theory allows you to work on structures without the need first to pulverize them into set theoretic dust" (Corfield).

The above quotation explains, in a rather pictorial way, what *category theory* and, in particular, *topos theory* are really about. In fact, *category theory* and, in particular, *topos theory* allow to abstract from the specification of points (elements of a set) and functions between these points to a universe of discourse in which the basic elements are arrows, and any property is given in terms of compositions of arrows.

C. Flori, *A First Course in Topos Quantum Theory*, Lecture Notes in Physics 868,
DOI 10.1007/978-3-642-35713-8_4, © Springer-Verlag Berlin Heidelberg 2013

The reason for the above characterisation is that the underlining philosophy behind category theory (and topos theory) is that of describing mathematical objects from an external point of view, i.e. in terms of relations.

This is in radical contrast to *set theory*, whose approach is essentially internal in nature. By this we mean that the basic/primitive notions of what sets are and the belonging relations between sets, are defined in terms of the elements which belong to the sets in question, i.e. an internal perspective.

In order to be able to implement the notion of external definition we first need to define two important notions: (i) the notion of a map or arrow, which is simply an abstract characterisation[1] of the notion of a function between sets; (ii) the notion of an "equation" in categorical language. We will first start with the notion of a map.

Given two general objects A and B (not necessarily sets) an arrow f is said to have domain A and codomain B if it goes from A to B, i.e. $f : A \rightarrow B$. It is a convention to denote $A = dom(f)$ and $B = cod(f)$.

We will often draw such an arrow as follows:

$$A \xrightarrow{f} B. \tag{4.1}$$

Given two arrows $f : A \rightarrow B$ and $g : B \rightarrow C$, such that $cod(f) = dom(g)$, we can *compose* the two arrows obtaining $g \circ f : A \rightarrow C$. The property of *composition* is drawn as follows:

$$A \xrightarrow{f} B \xrightarrow{g} C. \tag{4.2}$$

For each object there always exists[2] an *identity arrow* $id_A : A \rightarrow A$. When A is a set we have:

$$id_A : A \rightarrow A, \quad \text{such that} \quad id_A(a) = a \quad \forall a \in A. \tag{4.3}$$

The collection of arrows between various objects satisfies two laws:

(i) *Associativity law*: given three arrows $f : A \rightarrow B$, $g : B \rightarrow C$ and $h : C \rightarrow D$ with appropriate domain and codomain relations, we then have

$$h \circ (g \circ f) = (h \circ g) \circ f. \tag{4.4}$$

(ii) *Unit law*: given $f : A \rightarrow B$, $id_A : A \rightarrow A$ and $id_B : B \rightarrow B$, the following holds:

$$f \circ id_A = f = id_B \circ f. \tag{4.5}$$

In the next step we try to define the analogue of an equation in an abstract categorical language. This is done through the notion of commutative diagrams. So, what is a diagram? A diagram is defined as follows:

---

[1]By abstract characterisation here we mean a notion that does not depend on the sets or objects between which the arrow is defined.

[2]The symbol ∀ means " for all".

Definition 4.1 A graph is a collection of vertices $\bullet_{x_i}$ and directed edges $e : \bullet_{x_1} \rightarrow \bullet_{x_2}$ where $\bullet_{x_1}$ is the source vertex while $\bullet_{x_2}$ is the target. If the vertices are labelled by objects X_i and the edges are labelled by arrows, such that each $e : \bullet_{x_1} \rightarrow \bullet_{x_2}$ is now labelled as $f : X_1 \rightarrow X_2$, then we say that the graph is actually a diagram.

A typical diagram will be either a triangle or a square

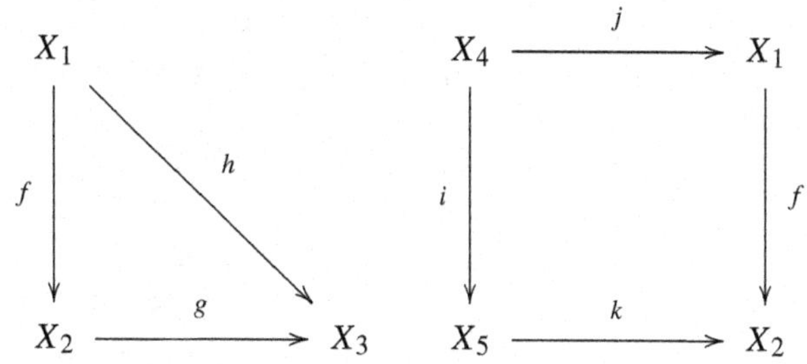

An "equation" is then given by the concept of commutativity, i.e. we say that the above diagrams commute iff

$$g \circ f = h; \qquad f \circ j = k \circ i. \tag{4.6}$$

Care should be taken since commutativity is not as strict a condition as one might think. In particular, if we have the following commuting diagram

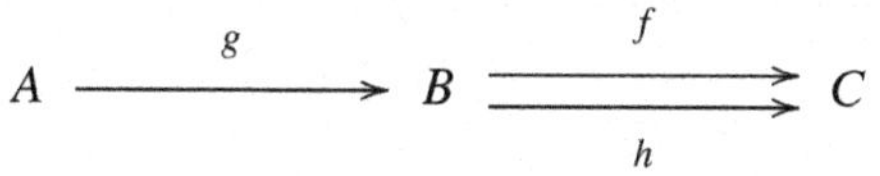

we can only infere that $f \circ g = h \circ g$, but not that $f = h$.

Obviously diagrams can be combined together to form a bigger diagram, as long as the rules pertaining composition of arrows hold. Thus, considering the commuting diagrams in the above example we can combine them to obtain

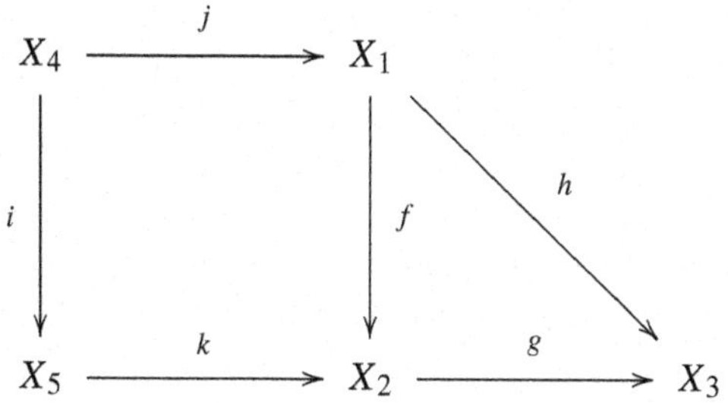

Now that we know how arrows and "equations" are defined, we are ready to give some examples of how the same concepts can be described both internally, using set theory and, externally, using a categorical language. We will first list them in the table below and then proceed in explaining them in details.

	Internal		External
Element	$a \in S$	$\overset{a=f(*)}{\leftrightarrow}$	$\{*\} \xrightarrow{f} S$
Subset	$A \subseteq S$	$\leftrightarrow$	$S \to \{0, 1\}$
Associative binary operation	$a \cdot (b \cdot c) = (a \cdot b) \cdot c$ $\forall a, b, c \in S$	$\leftrightarrow$	$\mu : S \times S \to S$, such that

$$\begin{array}{ccc} S \times S \times S & \xrightarrow{\mu \times id_S} & S \times S \\ {\scriptstyle id_S \times \mu}\downarrow & & \downarrow{\scriptstyle \mu} \\ S \times S & \xrightarrow[\text{commutes}]{\mu} & S \end{array}$$

Element The internal description of an element of a set is straightforward, it is simply given in terms of belonging. On the other hand, externally, an element of a set S is identified with a map from the singleton[3] $\{*\}$ to the set itself. In fact, any map from the singleton to a set can only pick 1 element of the set since its domain is only a singleton. Hence for each element a there is associated a unique map $f_a : \{*\} \to S$ such that $f_a(*) = a$.

Thus if we consider a simple set with three elements $S = \{a, b, c\}$ we would obtain the following identification of elements:

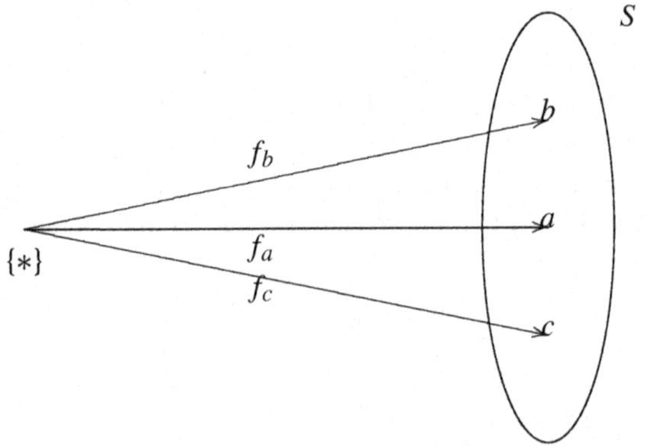

Subset The external way of describing a subset A of a set S is by identifying a collection of elements with a communal property. Thus we say that the subset A is

[3]A singleton is a set with only 1 element.

defined by $A := \{x \in S | x \text{ has property } r\}$. It is clear from the definition of A that $A \subseteq S$.

On the other hand, the external description of a subset is through a map $S \to \{0, 1\}$. The object $\{0, 1\}$ is called the sub-object classifier of sets. This is a very important object since, essentially, it represents the collection of truth values. The detailed description of the sub-object classifier, which is denoted as Ω, will be dealt with in subsequent chapters, for now it suffices to say that it is the object representing truth values.

In order to understand why a subset S is identified via a map $S \to \{0, 1\}$, we will make use of the following (not very mathematically precise) diagram:

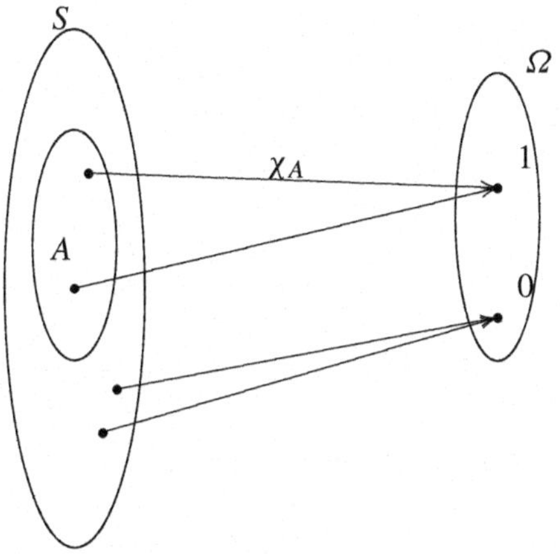

Here the map χ_A is the characteristic function of A defined as

$$\chi_A(x) = \begin{cases} 0 & \text{if } x \notin A, \\ 1 & \text{if } x \in A. \end{cases} \tag{4.7}$$

So, roughly, what a sub-object classifier does is to "classify" sub-objects according to which elements belong to the sub-object in question. It is in this sense that $A \subseteq S$ is identified with the map $\chi_A : S \to \{0, 1\}$.

Associativity will be dealt with when explaining the group operations.

It is interesting how the definition and the axioms of a group can be described in an external way. In particular, we have

1. *Associativity*: the internal (set theoretic) way of describing associativity is through the following condition:

$$\forall g_1, g_2, g_3 \in G, \quad g_1(g_2g_3) = (g_1g_2)g_3. \tag{4.8}$$

On the other hand, the external description of this concept is through the commutativity of the following diagram:

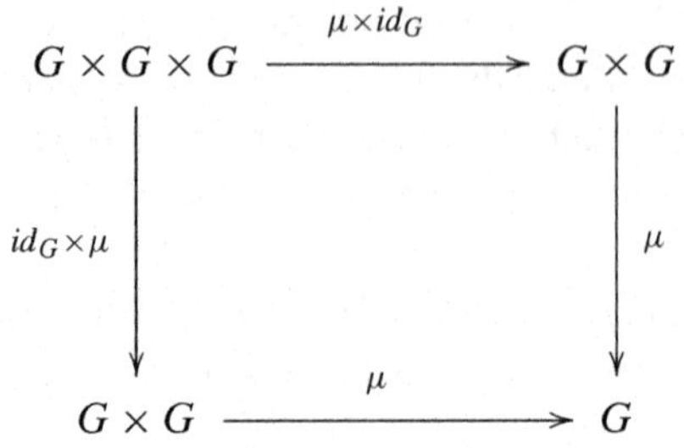

where

$$\begin{aligned} \mu : G \times G &\to G \\ \langle g_1, g_2 \rangle &\mapsto g_1 g_2 \end{aligned} \tag{4.9}$$

and

$$\begin{aligned} id_G : G &\to G \\ g &\mapsto g. \end{aligned} \tag{4.10}$$

Commutativity of the above diagram means that

$$\mu \circ (id_G \times \mu) = \mu \circ (\mu \times id_G). \tag{4.11}$$

If we then chose any element $(g_1, g_2, g_3) \in G \times G \times G$ and applied first the right hand side of (4.11) we would obtain

$$\mu \circ (\mu \times id_G)(g_1, g_2, g_3) = \mu\big((g_1 g_2), g_3\big) = (g_1 g_2) g_3. \tag{4.12}$$

On the other hand, the left hand side of (4.11) gives

$$\mu \circ (id_G \times \mu)(g_1, g_2, g_3) = \mu\big(g_1, (g_2 g_3)\big) = g_1 (g_2 g_3). \tag{4.13}$$

Commutativity means that

$$(g_1 g_2) g_3 = g_1 (g_2 g_3). \tag{4.14}$$

2. *Identity Element*
In set theory, the condition for the existence of the identity element is given by the equation

$$\forall g \in G, \quad ge = eg = g. \tag{4.15}$$

The external description of such a condition is given by the following commuting diagram

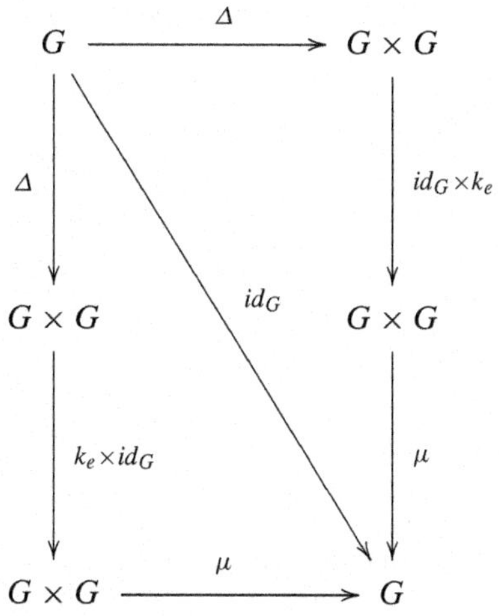

where

$$\begin{aligned} k_e : G &\to G \\ g &\mapsto k_e(g) := e \end{aligned} \tag{4.16}$$

is the constant map which maps each element to the identity element and Δ is the diagonal map

$$\begin{aligned} \Delta : G &\to G \times G \\ g &\mapsto (g, g). \end{aligned} \tag{4.17}$$

Commutativity means

$$\mu \circ (id_G \times k_e) \circ \Delta = \mu \circ (k_e \times id_G) \circ \Delta = id_G. \tag{4.18}$$

Thus given any element $g \in G$, applying the middle part of the above equation we obtain

$$\mu \circ (k_e \times id_G) \circ \Delta(g) = \mu \circ (k_e \times id_G)(g, g) = \mu(e, g) = eg. \tag{4.19}$$

Applying the left hand side we get

$$\mu \circ (id_G \times k_e) \circ \Delta(g) = \mu \circ (id_G \times k_e)(g, g) = \mu(g, e) = ge. \tag{4.20}$$

Commutativity then means that $ge = eg = g$.

3. *Inverse*
   The internal set theoretic definition of the axiom of inverses is

$$\forall g \in G,\ \exists g^{-1} \quad \text{such that} \quad gg^{-1} = g^{-1}g = e. \tag{4.21}$$

The external description of this condition is given by the commuting diagram

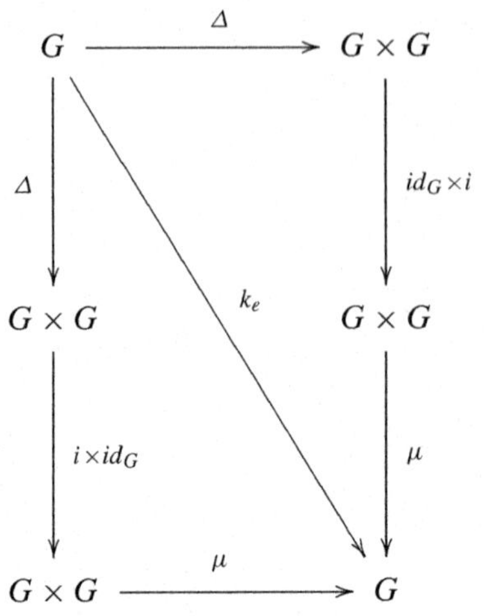

where i is the inverse map defined as follows:

$$\begin{aligned} i : G &\to G \\ g &\mapsto g^{-1}. \end{aligned} \tag{4.22}$$

Commutativity of the above diagram means that

$$\mu \circ (id_G \times i) \circ \Delta = \mu \circ (i \times id_G) \circ \Delta = k_e. \tag{4.23}$$

Thus applying this equation to any element $g \in G$ we obtain

$$\mu \circ (id_G \times i) \circ \Delta(g) = \mu \circ (id_G \times i)(g, g) = \mu(g, g^{-1}) = gg^{-1} \tag{4.24}$$

and

$$\mu \circ (i \times id_G) \circ \Delta(g) = \mu \circ (i \times id_G)(g, g) = \mu(g^{-1}, g) = g^{-1}g \tag{4.25}$$

such that

$$gg^{-1} = g^{-1}g = e. \tag{4.26}$$

4.2 Axiomatic Definition of a Category

Definition 4.2 [20] A (small[4]) category $\mathcal{C}$ consists of the following data:

[4] A category $\mathcal{C}$ is called small if $Ob(\mathcal{C})$ is a Set.

1. A collection $Ob(\mathcal{C})$ of $\mathcal{C}$-objects.
2. For any two objects $a, b \in Ob(\mathcal{C})$, a set $Mor_{\mathcal{C}}(a, b)$ of $\mathcal{C}$-arrows (or $\mathcal{C}$-morphisms) from a to b.
3. Given any three objects $a, b, c \in \mathcal{C}$, a map which represents composition operation:

$$\begin{aligned} \circ : Mor_{\mathcal{C}}(b, c) \times Mor_{\mathcal{C}}(a, b) &\rightarrow Mor_{\mathcal{C}}(a, c) \\ (g, f) &\mapsto g \circ f. \end{aligned} \tag{4.27}$$

   Composition is *associative*, i.e. for $f \in Mor_{\mathcal{C}}(a, b)$, $g \in Mor_{\mathcal{C}}(b, c)$ and $h \in Mor_{\mathcal{C}}(c, d)$ we have

$$h \circ (g \circ f) = (h \circ g) \circ f \tag{4.28}$$

   which in diagrammatic form is the statement that the following diagram commutes

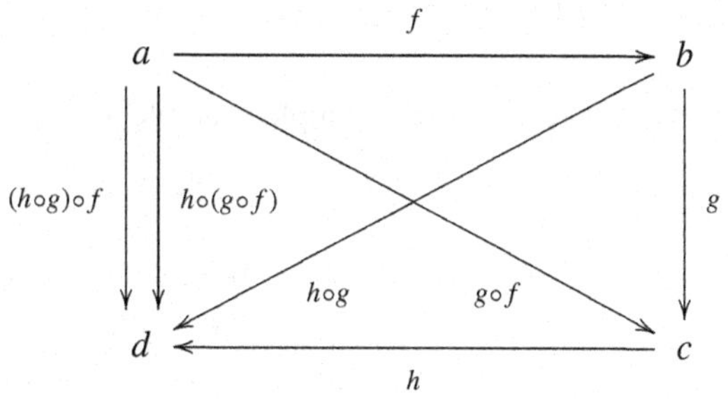

4. For each object $b \in \mathcal{C}$ an identity morphism $id_b \in Mor_{\mathcal{C}}(b, b)$, such that the following *Identity law* holds: for all $f \in Mor_{\mathcal{C}}(a, b)$ and $g \in Mor_{\mathcal{C}}(b, c)$ then $g = g \circ id_b$ and $f = id_b \circ f$. In diagrammatic form this is represented by the fact that the diagram

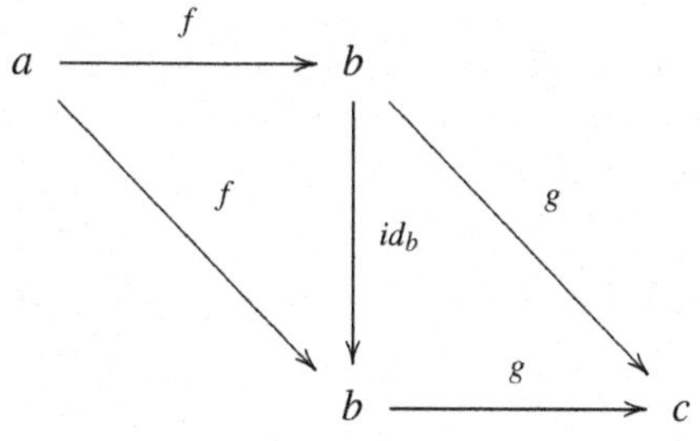

   commutes.

So, a category is essentially a collection of diagrams[5] for which certain "equations" (commutative relations) hold.

Definition 4.3 $\mathcal{D}$ is a *subcategory* of $\mathcal{C}$, denoted $\mathcal{D} \subseteq \mathcal{C}$, if:

(i) $Ob(\mathcal{D}) \subseteq Ob(\mathcal{C})$.
(ii) For any two objects $c, d \in Ob(\mathcal{D})$, then $Mor_{\mathcal{D}}(c,d) \subseteq Mor_{\mathcal{C}}(c,d)$.

Thus a subcategory is a sub-collection of objects with a sub-collection of graphs containing these objects.

Definition 4.4 $\mathcal{D}$ is a *full subcategory* of $\mathcal{C}$ if an extra requirement is satisfied:

(iii) for any $\mathcal{D}$-objects a and b, then $Mor_{\mathcal{D}}(a,b) = Mor_{\mathcal{C}}(a,b)$.

Keeping with our graph description, a full subcategory is a sub-collection of that has the same collection of graphs containing these objects.

4.2.1 Examples of Categories

Example 4.1 (Two Objects Category) A simple example of a two objects category is the following:

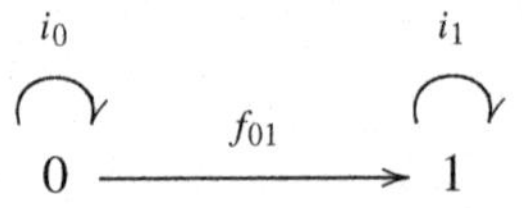

This category has 3 arrows:

- $i_0 : 0 \to 0$ identity on 0.
- $i_1 : 1 \to 1$ identity on 1.
- $f_{01} : 0 \to 1$.

It is easy to see that the compositions are: $i_0 \circ i_0 = i_0$, $i_1 \circ i_1 = i_1$, $i_1 \circ f_{01} = f_{01}$ and $f_{01} \circ i_1 = f_{01}$.

Example 4.2 (Poset) A partially ordered set (poset) is a set in which the elements are related by a partial order, i.e. not all elements are related to each other. The definition of a poset is as follows:

[5] Note that one can also have the empty diagram or the diagram with only the identity arrow

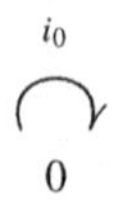

Definition 4.5 Given a set P we call this a poset iff a partial order $\leq$ is defined on it. A partial order is a binary relation $\leq$ on P, which has the following properties:

- Reflexivity: $a \leq a$ for all $a \in P$.
- Antisymmetry: if $a \leq b$ and $b \leq a$, then $a = b$.
- Transitivity: If $a \leq b$ and $b \leq c$, then $a \leq c$.

An example of a poset is any set with an inclusion relation defined on it. Another example is $\mathbb{R}$ with the usual ordering defined on it. This is actually a totally ordered set since any two elements are related via the ordering.

A poset forms a category whose objects are the elements of the poset and, given any two elements p, q, there exists a map $p \to q$ iff $p \leq q$ in the poset ordering. From this definition it follows that the map $p \to q$ is unique. We will be using such a (poset) category quite often when defining a topos description of quantum theory. Thus, it is worth pointing out the following:

Definition 4.6 Given two partially ordered sets P and Q, a map/arrow $f : P \to Q$ is a partial order homomorphism (otherwise called monotone function or order preserving function) if

$$\forall x, y \in P \quad x \leq y \quad \Longrightarrow \quad f(x) \leq f(y). \tag{4.29}$$

Homomorphisms are closed under composition. A trivial example of partial order homomorphisms is given by the identity maps.

Example 4.3 (Comma Category) This category (also called slice category) has as objects arrows with fixed domain or codomain. For example consider a category $\mathcal{C}$ and the reals $\mathbb{R}$, it is possible to form the comma category $\mathcal{C} \downarrow \mathbb{R}$ with:

- Objects: given $A, B \in \mathcal{C}$, the respective objects in $\mathcal{C} \downarrow \mathbb{R}$ are arrows whose codomain is $\mathbb{R}$, i.e. $f : A \to \mathbb{R}$ and $g : B \to \mathbb{R}$, also written as: (A, f) and (B, g).
- Morphisms: given two objects f and g we define an arrow between them as an arrow $k : A \to B$ in $\mathcal{C}$ such that

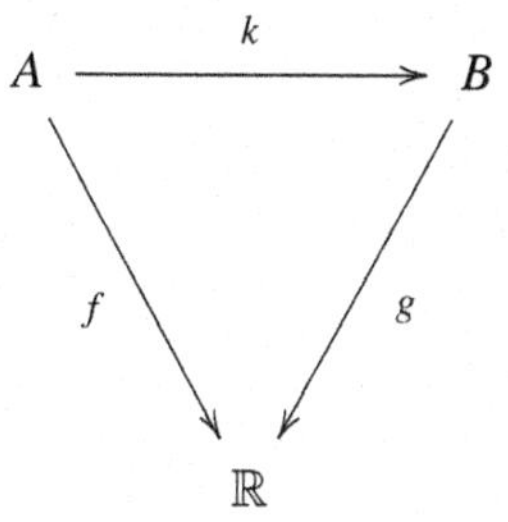

commutes in $\mathcal{C} \downarrow \mathbb{R}$.

The above definition of arrows in $\mathcal{C} \downarrow \mathbb{R}$ implies the following:

- *Composition*: given the two arrows $j : A \to B$ and $i : B \to C$, their composition is defined by the following commutative diagram:

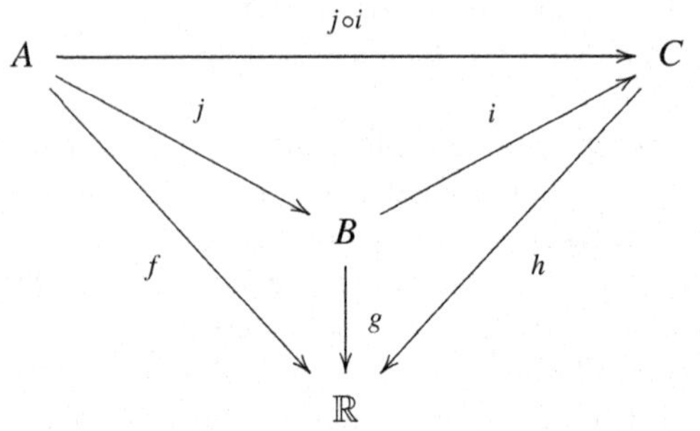

  Basically you just glue triangles together.
- *Identity*: given an element $f : A \to \mathbb{R}$, its identity arrow is: $id_A : (A, f) \to (A, f)$

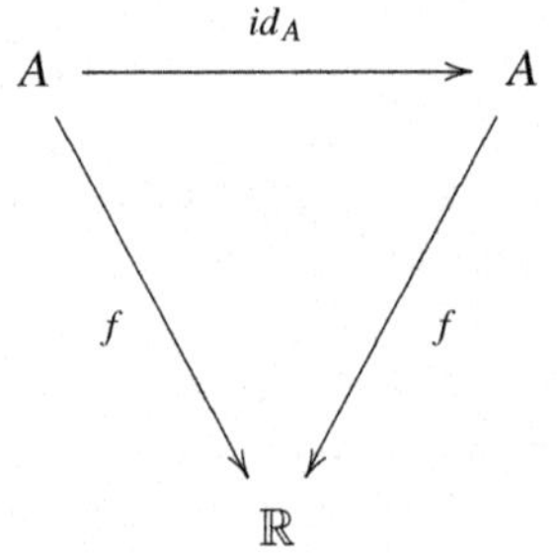

It is interesting to note that, given a category $\mathcal{C}$ for any object $A \in \mathcal{C}$ we can form the comma category $\mathcal{C}/A$ ($\mathcal{C} \downarrow A$) where objects in $\mathcal{C}/A$ are all morphisms in $\mathcal{C}$ with codomain A, while arrows between two objects $f : B \to A$ and $g : D \to A$ are commutative diagrams in $\mathcal{C}$, i.e.

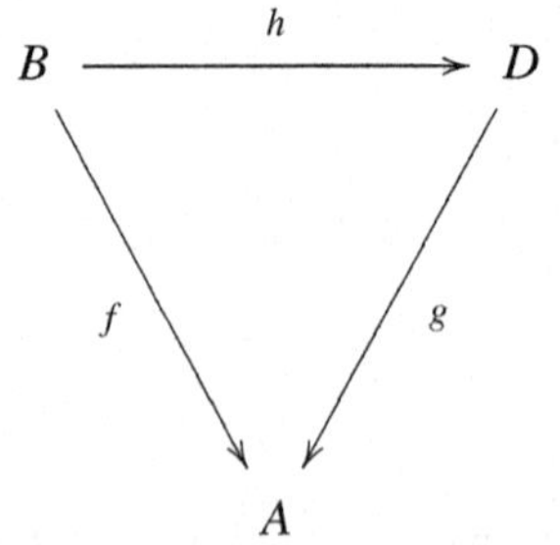

For an object $f : B \to A$ the identity arrow is simply $id_B : B \to B$, i.e.

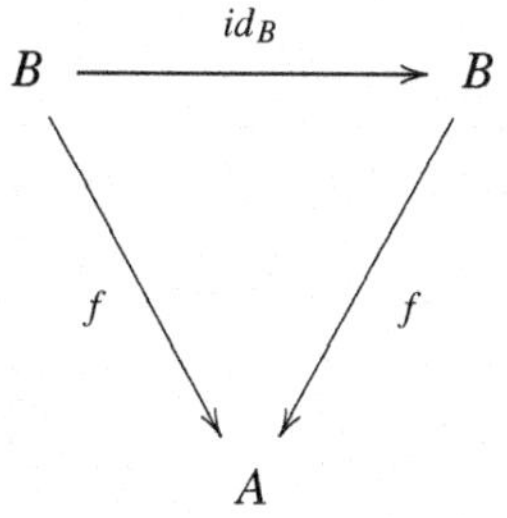

Example 4.4 (Monoid) A *monoid* $\mathcal{M}$ is a one-object category equipped with a binary operation on that object and a unit element. In particular, the definition is as follows:

Definition 4.7 A monoid $\mathcal{M}$ is a triplet $(M, *, i)$ such that,

- M is a Set.
- $*$ is a map $M \times M \to M$ which is associative.
- $i \in M$ is a two-sided identity: $i \star x = x \star i = x \; \forall x \in M$.

The $*$-map can be identified either with the tensor product or with the direct sum or with the direct product, according to which category M one is taking into consideration.

Examples of monoids are $(\mathbb{N}, +, 0)$ and any type of group $(G, \cdot, e)$.

It is also possible to compare two different monoids as follows:

Definition 4.8 Given two monoids M and N a map $h : M \to N$ is said to be a monoid homomorphism iff

$$\forall m_1, m_2 \in M \quad h(m_1 * m_2) = h(m_1) * h(m_2), \qquad h(i) = i. \tag{4.30}$$

We will now give a list of various categories which are frequently used in physics.

Category	Objects	Morphisms
Sets	Sets	Functions
Top	Topological spaces	Continuous maps
Gr	Groups	Homomorphisms of groups
Ab	Abelian groups	Homomorphisms of groups
$\mathbf{Vect}_K$	Vector spaces over a field K	K-linear maps
Man	Manifolds	Smooth maps
Pos	Partially ordered sets	Monotone functions
N	N (as a single object)	Natural numbers
Set as a discrete category	*Set*	Identity arrows
Preorder: P	$x \in P$	At most one arrow between any two objects

4.3 The Duality Principle

A very important notion in category theory is the notion of *duality*. In particular, for any statement (or "equation") Y expressed in categorical language, its dual[6] Y^{op} is obtained by replacing the domain with the codomain and the codomain by the domain for every arrow in Y and reversing the order of arrow composition, i.e. $h = g \circ f$ becomes $h = f \circ g$.

Thus, all arrows and diagrams in Y have the reverse direction in Y^{op}, and the construction /notion described by Y^{op} is said to be dual to the notion described by Y. Moreover, we also have the notion of a dual category.

Definition 4.9 Given a category $\mathcal{C}$ the dual $\mathcal{C}^{op}$ is defined as follows:

$$Ob\big(\mathcal{C}^{op}\big) := Ob(\mathcal{C}), \qquad Mor_{\mathcal{C}^{op}}(a, b) := Mor_{\mathcal{C}}(b, a) \tag{4.31}$$

the composition law is: given $f \in Mor_{\mathcal{C}^{op}}(a, b)$ and $g \in Mor_{\mathcal{C}^{op}}(b, c)$, then

$$g \circ_{\mathcal{C}^{op}} f := f \circ_{\mathcal{C}} g. \tag{4.32}$$

It is easy to see that $(\mathcal{C}^{op})^{op} = \mathcal{C}$ for any category.

Therefore, given the construction Y^{op} referred to a category $\mathcal{C}$, this can be considered as the construction Y applied to the dual category $\mathcal{C}^{op}$.

[6]The index *op* stands for *opposite* and it is synonimus to dual.

The notion of opposite categories leads to the very important notion of *duality principle*, by which a statement Y is true in $\mathcal{C}$ iff its dual Y^{op} is true in $\mathcal{C}^{op}$. This principle allows us to prove various things simultaneously. By this we mean that if we have a statement X, which holds in the category $\mathcal{C}$, then we immediately know that the statement X^{op} holds for $\mathcal{C}^{op}$.

As a very simple example, consider the following diagram

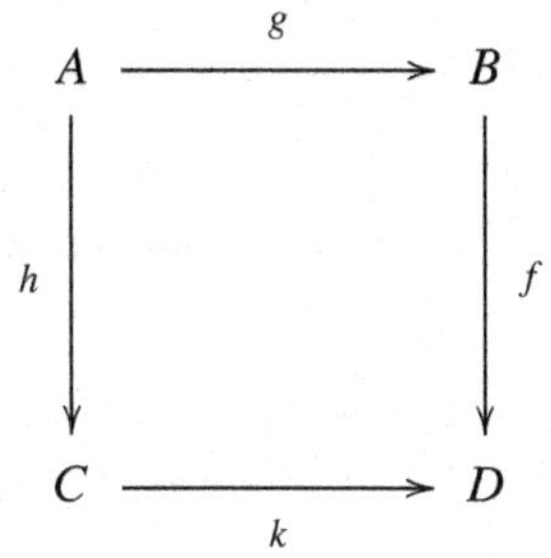

The dual of such a diagram would be

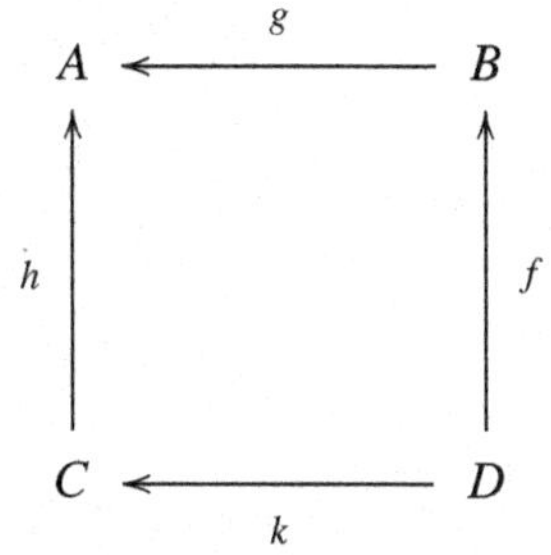

Moreover, if we derive a theorem T from the axioms of category theory, then such a theorem holds for any category $\mathcal{C}$. However, by duality T^{op} holds for every category $\mathcal{C}^{op}$. But each category can be written as the opposite of some other category $((\mathcal{C}^{op})^{op} = \mathcal{C})$, therefore T^{op} holds for all categories. Thus the duality principle allows us to derive a universal theorem from a specific instance of it.

In what follows, we will see many examples of statements, theorems and their duals.

4.4 Arrows in a Category

In this section we will explain the notions of injective, surjective and bijective[7] arrows in a categorical language, i.e. from an external point of view.

[7]A function between two sets $f : A \to B$ is:

4.4.1 Monic Arrows

Monic arrow is the "arrow-analogue" of an *injective function*.

Definition 4.10 Given a category $\mathcal{C}$, a $\mathcal{C}$-arrow $f : a \to b$ is monic if for any pair of arrows $g : c \to a$, $h : c \to a$, the equality $f \circ g = f \circ h$ implies that $h = g$, i.e. f is left cancellable. Monic arrows are denoted as:

$$a \rightarrowtail b$$

We now want to show how it is possible, in **Sets**, to prove that an arrow is monic iff it is injective as a function.

Proof Consider the sets A, B, C, an injective function $f : A \to B$ (i.e. if $f(x) = f(y)$, then $x = y$) and a pair of functions $g : C \to A$ and $h : C \to A$, such that

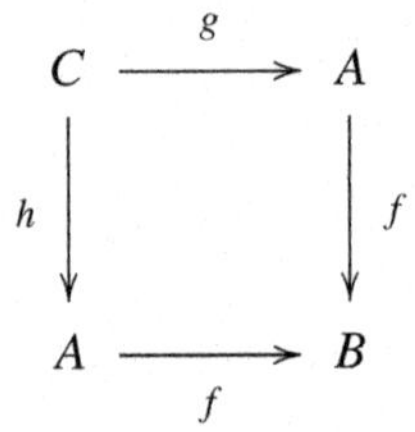

commutes, i.e. $f \circ g = f \circ h$.

Now if

$$x \in C \implies f \circ g(x) = f \circ h(x)$$
$$f\big(g(x)\big) = f\big(h(x)\big).$$

Since f is injective it follows that $g(x) = h(x)$, i.e. f is left cancellable. Vice versa, let f be left cancellable, and consider the following diagram:

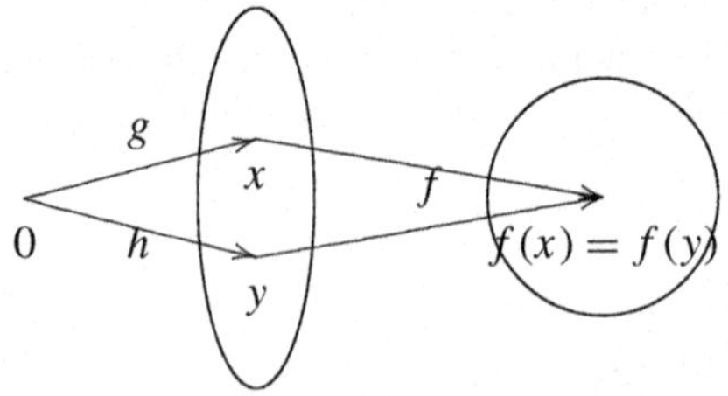

- Injective iff $f(x) = f(y)$ implies that $x = y$ for any two elements $x, y \in A$.
- Surjective iff $\forall y \in B$ there exists an $x \in A$ such that $y = f(x)$.
- Bijective iff f is both injective and surjective.

From the above diagram it is easy to deduce that $f \circ g = f \circ h$, since $f(x) = f(y)$. Given that $x = g(0)$ and $y = h(0)$ by construction and f is left cancellable by assumption, we get: $g = h$, therefore $x = y$ for $f(x) = f(y)$, i.e. f is injective. □

4.4.2 *Epic Arrows*

An epic arrow is the "arrow-analogue" of a *surjective function*.

Definition 4.11 An arrow $f : a \rightarrow b$ in a category $\mathcal{C}$ is epic in $\mathcal{C}$ if, for any pair $g : b \rightarrow c$ and $h : b \rightarrow c$ of arrows, the equality $g \circ f = h \circ f$ implies that $h = g$, i.e. f is right cancellable. Monic arrows are denoted as:

$$a \longrightarrow\!\!\!\!\!\rightarrow b$$

An epic arrow is a dual of a monic arrow. To see this let us dualise the above definition. We obtain

Dual Definition 4.1 An arrow $f : b \rightarrow a$ in a category $\mathcal{C}$ is the dual of an epic arrow if for any pair $g : c \rightarrow b$ and $h : c \rightarrow b$ of arrows, the equality $f \circ g = f \circ h$ implies that $h = g$, i.e. f is left cancellable.

Clearly the above defines a monic arrow.
In **Sets**, the epic arrows are the surjective set functions.

Proof Let us consider three sets A, B, C such that the set function $f : A \rightarrow B$ is surjective but it is not right cancellable, i.e. given two functions $g, h : B \rightarrow C$ although $h \circ f = g \circ f$, $h \neq g$. What this implies is that there exists an element $y \in B$ such that $h(y) \neq g(y)$. However since f is surjective $y = f(x)$ for some $x \in A$, then, $h \circ f(x) \neq g \circ f(x)$ which contradicts the assumption that $h \circ f = g \circ f$.

On the other hand, let us assume that f is epic but not surjective. It is then possible to construct a map $k : B \rightarrow B$ such that k is the identity on the image of f but $k \neq id_B$. We then obtain that $k \circ f = id_B \circ f$. However, since f is epic it follows that $k = id_B$ which contradicts our assumption, hence f is surjective. □

4.4.3 *Iso Arrows*

An iso arrow is the "arrow-analogue" of a *bijective function*.

Definition 4.12 A $\mathcal{C}$-arrow $f : a \rightarrow b$ is iso, or invertible in $\mathcal{C}$ if there is a $\mathcal{C}$-arrow $g : b \rightarrow a$, such that $g \circ f = id_a$ and $f \circ g = id_b$. Therefore, g is the inverse of f, i.e. $g = f^{-1}$.

Theorem 4.1 *g is unique.*

Proof Consider any other g' such that $g' \circ f = id_a$ and $f \circ g' = id_b$, then we have

$$g' = id_a \circ g' = (g \circ f) \circ g' = g \circ (f \circ g') = g \circ id_b = g. \qquad \square$$

An iso arrow has the following properties:

1. *An iso arrow is always monic.*

Proof Consider an iso f, such that $f \circ g = f \circ h$ ($f : a \to b$ and $g, h : c \to a$), then

$$\begin{aligned} g &= id_a \circ g = (f^{-1} \circ f) \circ g = f^{-1} \circ (f \circ g) \\ &= f^{-1} \circ (f \circ h) = (f^{-1} \circ f) \circ h = h \end{aligned}$$

therefore f is left cancellable. □

2. *An iso arrow is always epic.*

Proof Consider an iso f such that $g \circ f = h \circ f$ ($f : a \to b$ and $g, h : b \to c$)

$$\begin{aligned} g &= g \circ 1_b = g \circ (f \circ f^{-1}) = (g \circ f) \circ f^{-1} = (h \circ f) \circ f^{-1} \\ &= h \circ (f \circ f^{-1}) = h \end{aligned}$$

therefore f is right cancellable. □

Note not all arrows which are monic and epic are iso, for example:

1. In *poset*, even though all functions are monic and epic, the only isos are the identity map.
   In fact, consider an arrow $f : p \to q$, this implies that $p \leq q$. If f is an iso, then $f^{-1} : q \to p$ exists, therefore, $q \leq p$. However, from the antisymmetry property $p \leq q$ and $q \leq p$ imply that $p = q$, therefore $f = id_p$ is an identity arrow.

Iso arrows are used to determine isomorphic objects within a given category.

Definition 4.13 Given two objects $a, b \in \mathcal{C}$, we say that they are isomorphic $a \simeq b$ if there exists an iso $\mathcal{C}$-arrow between them.

As we see from the above definitions, we have managed to give an external characterisation for the set-theoretic concepts of injective, surjective and bijective functions.

4.5 Elements and Their Relations in a Category

In this section we will describe certain fundamental constructions or elements present in category theory. While reading this, it is useful to try and understand what the corresponding elements would be in **Sets**.

4.5.1 Initial Objects

Definition 4.14 An *initial object* in a category $\mathcal{C}$ is a $\mathcal{C}$-object 0 such that, for every $\mathcal{C}$-object A, there exists one and only one $\mathcal{C}$-arrow from 0 to A.

An initial object is unique up to isomorphism, i.e. all initial objects in a category are isomorphic. To see this, consider two initial objects $0_1 \in \mathcal{C}$ and $0_2 \in \mathcal{C}$. Being both initial, we have the unique arrows $f_1 : 0_1 \to 0_2$ and $f_2 : 0_2 \to 0_1$. Moreover, the fact that they are initial implies that it is possible to uniquely compose the above arrows obtaining $f_1 \circ f_2 = id_{0_2}$ and $f_2 \circ f_1 = id_{0_1}$. Therefore f_1 and f_2 are isos and $0_1 \simeq 0_2$.

Examples

1. Given a category $\mathcal{C}$ with initial object 0, then the initial object in the comma category $\mathcal{C} \downarrow \mathbb{R}$ of Example 4.3 is $f : 0 \to \mathbb{R}$, such that the following diagram commutes:

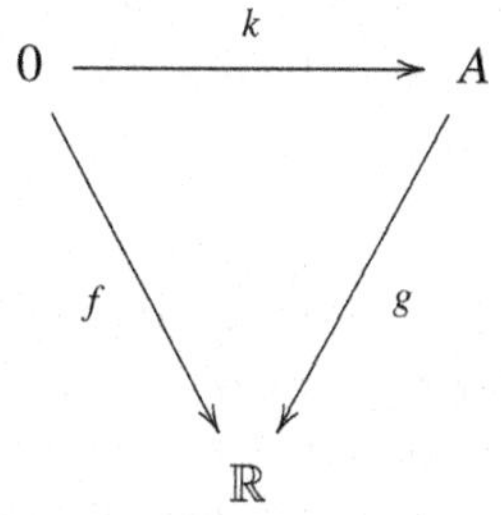

2. In **Sets**, the initial object is the empty set $\emptyset$.
3. In **Pos**, the initial object is the poset $(\emptyset, \emptyset)$.
4. In **Top**, the initial object is the space $(\emptyset, \{\emptyset\})$.
5. In $\mathbf{Vect}_K$, the one-element space $\{0\}$ is the initial object.
6. In a poset, the initial object is the least element with respect to the ordering, if it exists.

4.5.2 Terminal Objects

Definition 4.15 A *terminal object* in a category $\mathcal{C}$ is a $\mathcal{C}$-object 1 such that, given any other $\mathcal{C}$-object A, there exists one and only one $\mathcal{C}$-arrow from A to 1.

Examples

1. In $\mathcal{C} \downarrow \mathbb{R}$ a terminal object is $(\mathbb{R}, id_{\mathbb{R}})$, such that the diagram

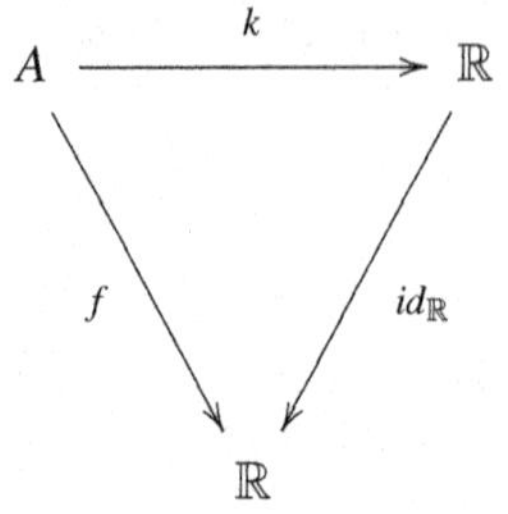

   commutes $(\therefore k = f)$.
2. In **Sets** a terminal object is a singleton $\{*\}$, since given any other element $A \in$ **Sets** there exist one and only one arrow $A \to \{*\}$.
3. In **Pos** the poset $(\{*\}, \{(*, *)\})$ is the terminal object.
4. In **Top** the space$(\{*\}, \{\emptyset, \{*\}\})$ is the terminal object.
5. In $\mathbf{Vect}_k$, the one-element space $\{0\}$ is the terminal object.
6. In a poset the terminal object is the greatest element with respect to the ordering.

A terminal object is dual to an initial object, in fact, if we dualise the definition of terminal object we obtain:

Dual Definition 4.2 The dual of a terminal object is an object x in a category $\mathcal{C}$ such that, given any other $\mathcal{C}$-object A, there exists one and only one $\mathcal{C}$-arrow from x to A.

This states precisely that x is an initial object.

Given the notion of a terminal object we can now define the notion of an element of a $\mathcal{C}$-object. Note that, so far, the definition of every categorical object that was introduced has never rested on specific characteristics of its composing elements. This is because, as stated above, concepts in category theory are defined externally. In fact, it is the case that certain objects in a given category do not have elements. We will return to this later. For now we will give the categorical description of what an element of an object actually is.

Definition 4.16 Given a category $\mathcal{C}$, with terminal object[8] 1, then an element of a $\mathcal{C}$-object B is a $\mathcal{C}$-arrow $x : 1 \to B$.

Example 4.5 In **Sets** the terminal object is the singleton subset $\{*\}$, therefore an element $x \in A$ can be identified with an arrow $\{*\} \to A$ from the terminal object to A.

[8]Note that the choice of terminal object is irrelevant since they are all isomorphic to each other.

4.5.3 Products

We will now give the external/categorical description of the cartesian product. Such a definition will be a general notion of what a product is, which will be valid in any category independent of the details of that category. This is, in fact, one of the powerful aspects of category theory: *an abstract characterisation of objects in terms of universal properties*. In this way, definitions become independent of the peculiarity of individual cases, becoming a more objective, universally valid construction. One can compare the level of abstraction in category theory with the level of abstraction in differential geometry, where one defines objects without the use of a specific coordinate reference frame.

Let us now turn to our task of defining what a product is in categorical language. It should be pointed out that, as with all the other objects defined so far, if a product exists, it is unique up to isomorphism. Given a particular category, we can then verify whether or not products exists in that category. It is only at this point that the particularity of the category in question enters the game, i.e. only into the proof of existence.[9] All the useful properties of the product follow from the general definition. So, what is a product?

Definition 4.17 A *product* of two objects A and B in a category $\mathcal{C}$ is a third $\mathcal{C}$-object $A \times B$ together with a pair of $\mathcal{C}$-morphisms (arrows):

$$pr_A : A \times B \to A, \qquad pr_B : A \times B \to B \tag{4.33}$$

such that, given any other pair of $\mathcal{C}$-arrows $f : C \to A$ and $g : C \to B$, there exists a unique arrow $\langle f, g\rangle : C \to A \times B$, such that the following diagram commutes[10]

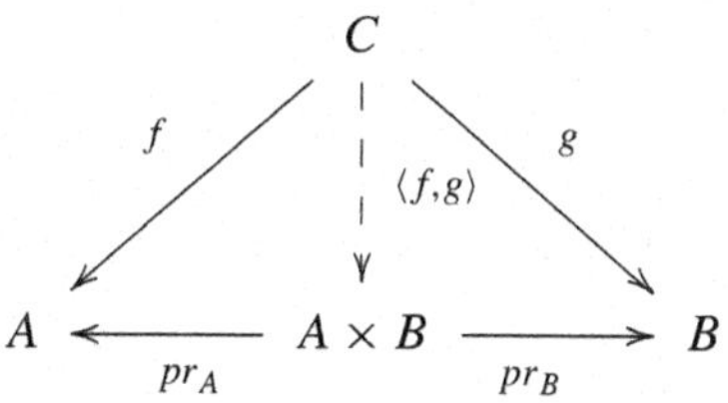

i.e.

$$pr_A \circ \langle f, g\rangle = f \quad \text{and} \quad pr_b \circ \langle f, g\rangle = g.$$

[9]The proof of existence is obtained by constructing an object and verifying that it satisfies the requirements of being (in this case) a product.

[10]Note that an arrow drawn as

$$- - - - - - \to$$

indicates uniqueness of that arrow.

Given two products we would now like to know if and how it is possible to relate them. To this end one needs to introduce the concept of a map between two product objects. Such a map will be called a product map. The definition is straightforward.

Definition 4.18 Consider a category $\mathcal{C}$ in which a product exists for every pair of objects. Then consider two $\mathcal{C}$-arrows $f : A \to B$ and $g : C \to D$. The product map $f \times g : A \times C \to B \times D$ is the $\mathcal{C}$-arrow $\langle f \circ pr_A, g \circ pr_C \rangle$. Such an arrow is the unique arrow which makes the following diagram commute:

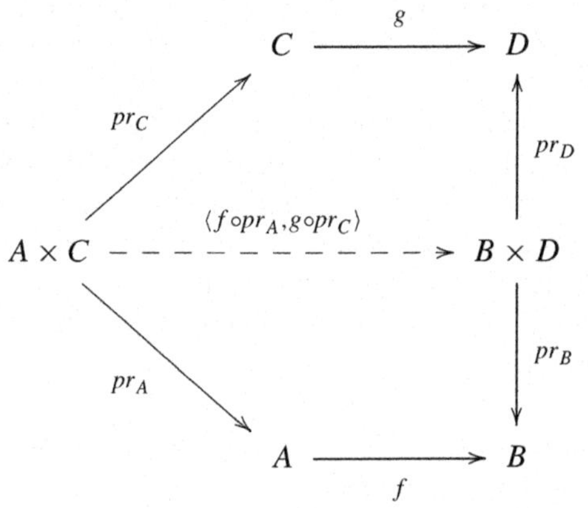

Theorem 4.2 *In* **Sets** *the product of two sets always exists and it is the standard cartesian product with projection maps.*

Proof Given a set R with maps $q_1 : R \to S$, $q_2 : R \to T$, then the map

$$\begin{aligned} \psi : R &\to S \times T \\ r &\mapsto \big(q_1(r), q_2(r)\big) \end{aligned} \tag{4.34}$$

would satisfy the commutativity property of the cartesian product, i.e. $\forall r \in R$

$$\begin{aligned} (p_S \circ \psi)(r) &= q_1(r) \\ (p_T \circ \psi)(r) &= q_2(r). \end{aligned} \tag{4.35}$$

We now need to prove its uniqueness. This is done as follows: if there exists a ϕ which satisfies $p_S \circ \phi = q_1$ and $p_T \circ \phi = q_2$ then, for all $r \in R$, we have

$$\psi(r) = \big(q_1(r), q_2(r)\big) = \big(p_S\big(\phi(r)\big), p_T\big(\phi(r)\big)\big) = \phi(r) \tag{4.36}$$

where the last equality holds, since $(s, t) = (p_S(s, t), p_T(s, t))$ for all $(s, t) \in S \times T$. It follows that ψ is unique. □

We now want to show that the products are commutative, i.e. $A_1 \times A_2 \simeq A_2 \times A_1$. To this end let us consider each product separately. Being products, there exist unique arrows i and j such that the following diagrams commute:

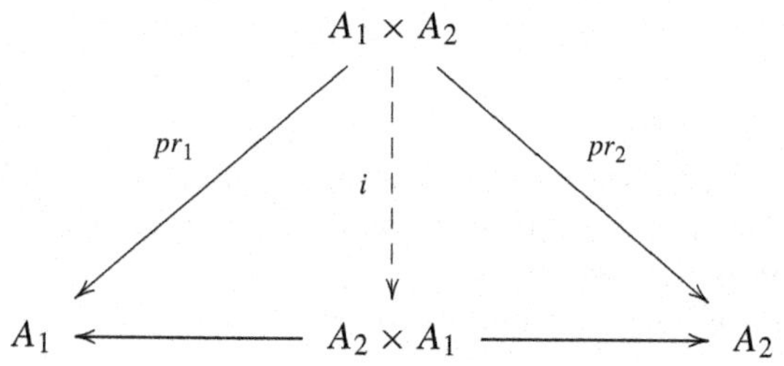

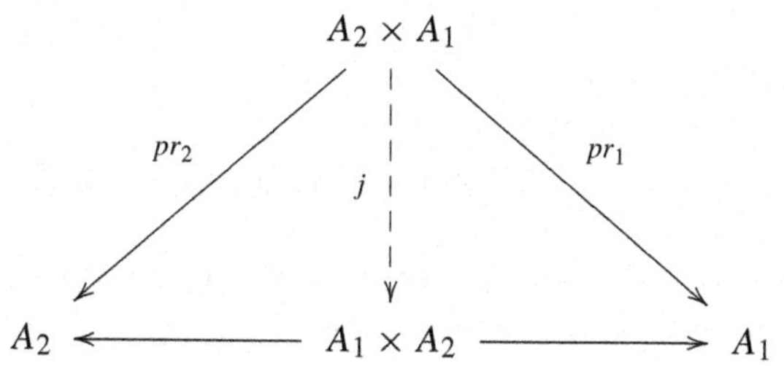

Composition of these diagrams in both orders gives us the following commuting diagrams:

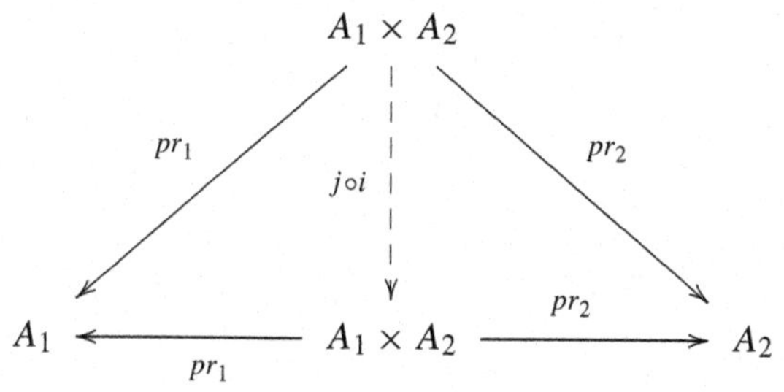

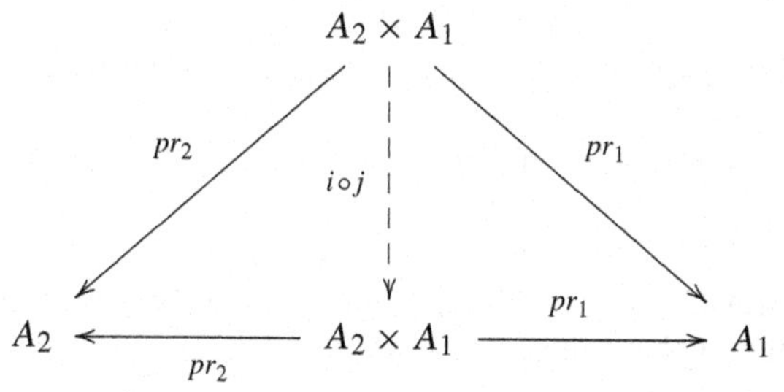

It is clear that $j \circ i = id_{A_1 \times A_2}$ and $i \circ j = id_{A_2 \times A_1}$, thus i and j are isos.

The proof given above can be easily extended to give the associativity of the product operation in general, i.e. for an arbitrary number of factors.

It should be noted that the product of an empty set of objects is just the terminal object, and the product of the family consisting of a single object A is A itself projection $id_A : A \to A$.

Example 4.6

- In **Pos**, products are cartesian products with the pointwise order.[11]
- In **Top**, products are cartesian products with the product topology.
- In **Vect**$_K$, products are direct sums.
- In a poset, products are greatest lower bounds.

4.5.4 Coproducts

We now define the categorical/external definition of disjoint union.

Definition 4.19 A *coproduct* of two objects A and B in a category $\mathcal{C}$ is a third $\mathcal{C}$-object $A+B$ together with a pair of $\mathcal{C}$-arrows:

$$i_A : A \to A + B, \qquad i_B : B \to A + B \tag{4.37}$$

such that, given any other pair of $\mathcal{C}$-arrows $f : A \to C$ and $g : B \to C$, there exists a unique arrow $[f, g] : A + B \to C$ which makes the following diagram commute:

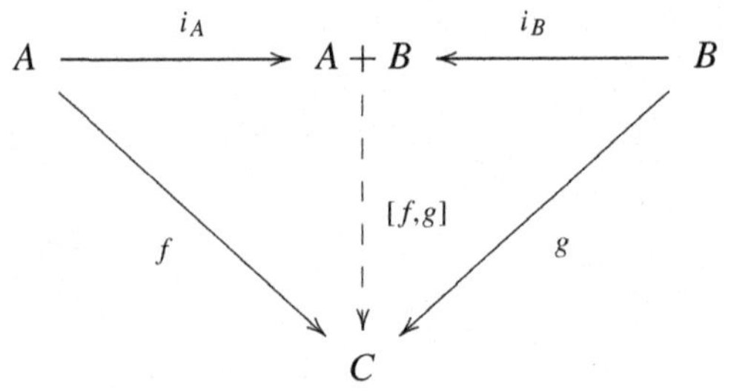

In the above, the arrows i_A and i_B indicate canonical injection maps.

The coproduct is the dual of the product. In fact, dualising the above definition we obtain:

Dual Definition 4.3 Given two objects A and B in a category $\mathcal{C}$ there exists a third $\mathcal{C}$-object D together with a pair of $\mathcal{C}$-arrows:

$$j_A : D \to A, \qquad j_B : D \to B \tag{4.38}$$

[11] In this context pointwise order is defined as follows: $(A, B) \leq (C, D)$ iff $A \leq C$ and $B \leq D$.

such that, given any other pair of $\mathcal{C}$-arrows $f : C \to A$ and $g : C \to B$, there exists a unique arrow $h : C \to D$ which makes the following diagram commute:

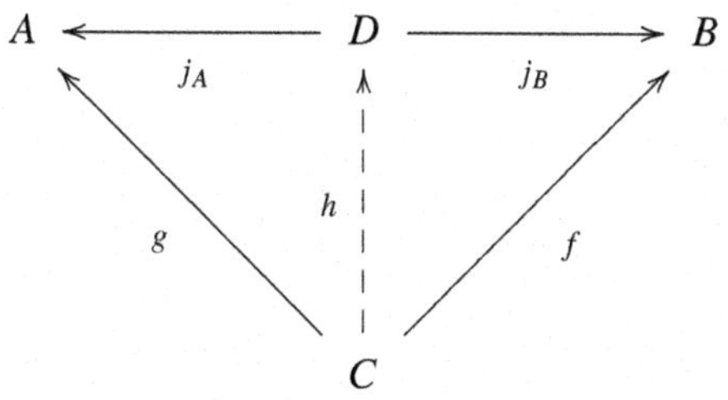

Since the product $A \times B$ satisfies that definition and is unique up to isomorphisms we deduce that $D = A \times B$.

Again, it is possible to define a map between two coproducts. In fact, in defining such a map one can simply dualise the definition of the product map, thus obtaining the following definition:

Definition 4.20 Assuming that coproducts exist in $\mathcal{C}$, we consider two $\mathcal{C}$-arrows $f : A \to B$ and $g : C \to D$. The coproduct map $f + g : A + C \to B + D$ is the unique $\mathcal{C}$-arrow $[i_D \circ f, i_B \circ g]$, such that the following diagram commutes:

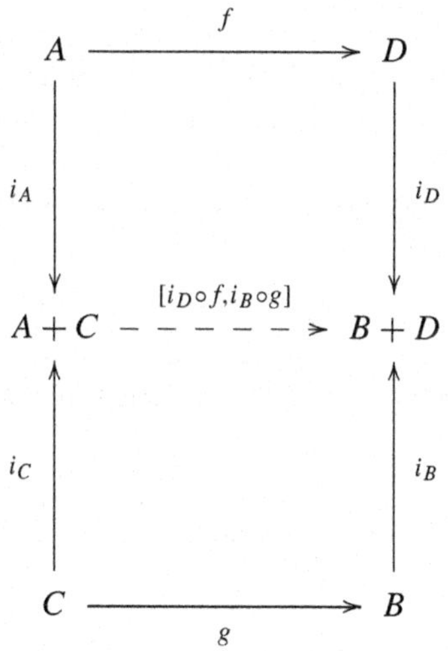

Theorem 4.3 *In* **Sets** *the coproduct of any two elements* $X, Y \in$ **Sets** *always exists and it is the disjoint union*

$$X \coprod Y = \{(x, j) \in (X \cup Y) \times \{0, 1\} | x \in X \textit{ iff } j = 0,\ x \in Y \textit{ iff } j = 1\}. \quad (4.39)$$

We then have

$$i_X : X \to X \coprod Y$$
$$x \mapsto (x, 0) \tag{4.40}$$

and

$$i_Y : Y \to X \coprod Y$$
$$y \mapsto (y, 1). \tag{4.41}$$

It should be noted that the coproduct of an empty family of sets is the initial object.

Example 4.7

- In **Pos**, coproducts are identified with disjoint unions (with the inherited orders).
- In **Top**, coproducts are identified with topological disjoint unions.
- In $\mathbf{Vect}_K$, coproducts are identified with direct sums.
- In a poset, coproducts are identified with least upper bounds.

4.5.5 Equalisers

We now describe the categorical analogue of the concept of the largest set for on which two functions coincide. This is the concept of an equaliser.

Definition 4.21 Given a category $\mathcal{C}$, a $\mathcal{C}$-arrow $i : E \to A$ is an equaliser of a pair of $\mathcal{C}$-arrows $f, g : A \to B$ if

1. $f \circ i = g \circ i$.
2. Given another $\mathcal{C}$-arrow $h : C \to A$ such that $f \circ h = g \circ h$, there is exactly one $\mathcal{C}$-arrow $k : C \to E$ such that the following diagram commutes:

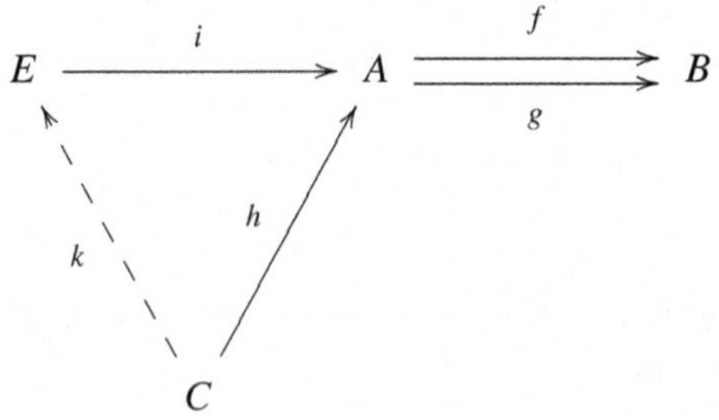

i.e. $i \circ k = h$.

In **Sets**, the equaliser of a pair of maps $f, g : A \to B$ is the largest subset on which the two maps coincide, i.e.

$$E = \{x \in A | f(x) = g(x)\} \subseteq A. \tag{4.42}$$

To see this we show that by defining E as above and i as the inclusion map then conditions (1) and (2) above are satisfied. In particular, the first condition applied to $x \in E$ is

$$f \circ i(x) = g \circ i(x) = f(x) = g(x) \tag{4.43}$$

which is true by definition of E. For the second condition we need to define, for any arrow $h : C \to A$ such that $f \circ h = g \circ g$ then there exists a unique arrow $k : C \to E$ which makes the above diagram commute. Clearly the only k that does that would be h itself.

4.5.6 Coequalisers

Dual to the equaliser, there exists the coequaliser which is defined as follows:

Definition 4.22 Given two $\mathcal{C}$-arrows $f, g : A \to B$, the $\mathcal{C}$-arrow $h : B \to C$ is a coequaliser of f and g if the following conditions hold:

(1) $h \circ f = h \circ g$.
(2) Given any other $\mathcal{C}$-arrow $h' : B \to C'$ such that $h' \circ f = h' \circ g$, there exists a unique $\mathcal{C}$-arrow $u : C \to C'$ such that the following diagram commutes.

Diagram 4.1

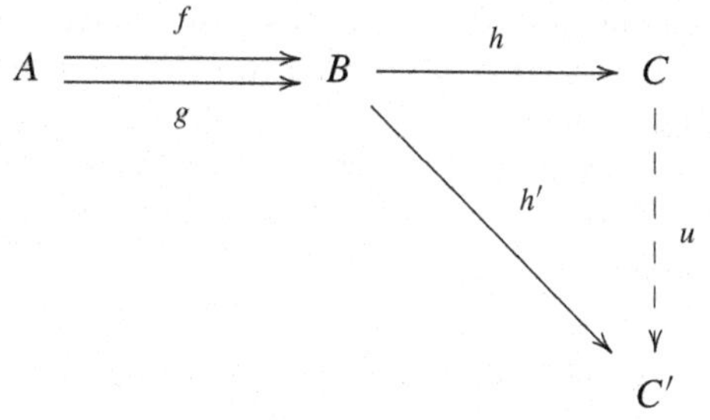

In **Sets**, the coequaliser of a pair of maps $f, g : A \to B$ is the quotient of B by the least equivalence relation[12] for which $f(x) = g(x)$ for all $x \in A$. The condition of

[12]The definition of an equivalence relation is as follows: given a set A a (binary) equivalence relation on A is a subset $R \subseteq A \times A$ defined as $R = \{(a, b) | a \sim_R b\}$. Thus R represents the set of all pairs which are related by the relation $\sim_R$. The relation $\sim_R$ has the following properties:

being the least equivalence relation is required by condition (2) above. The reason we want an equivalence relation is because the coequaliser requirement demands that $h \circ g = h \circ f$. Thus for any $x \in A$, the coequaliser h identifies $g(x)$ with $f(x)$, i.e. $h(f(x)) = h(g(x))$.

Now let us consider the case in which $A := \{(a,b)|a \sim b\}$ is any equivalence relation ($A \subseteq B \times B$) on a fixed set B. We then construct, as an example, the coequaliser of the functions

$$\begin{aligned} f : A &\to B \\ (a,b) &\mapsto a \end{aligned} \tag{4.44}$$

and

$$\begin{aligned} g : A &\to B \\ (a,b) &\mapsto b. \end{aligned} \tag{4.45}$$

For each element $b \in B$ its equivalence class is given by $[b] = \{a : a \sim b\}$. We then define C in the diagram (4.1) to be the quotient of B by this relation, namely $C = B/A$. The coequaliser h would then simply take each element a to its equivalence class $[a]$ and each element in the pair $(a,b) \in A$ to the same equivalence class, namely

$$h(a) = [a] = h(b) = [b]. \tag{4.46}$$

Thus we see that for a given pair $(a,b) \in A$ the first condition of a coequaliser is satisfied since we obtain

$$h \circ f(a,b) = h(a) = [a] = h \circ g(a,b) = h(b) = [b]. \tag{4.47}$$

To satisfy the second condition of being a coequaliser we consider a map h', such that $h' \circ f = h' \circ g$, we then need to define a unique u such that $u \circ h = h'$. Suppose we indeed had such a u but didn't know if it was unique. Given a pair $(a,b) \in A$ we compute $u([a]) = u(h(a)) = h'(a)$ and, similarly, $u([b]) = u(h(b)) = h'(b)$. Since $[a] = [b]$, we can uniquely define u as $u([a_i]) := h'(a_i)$.

From this discussion it transpires, rather clearly, why we would like an equivalence relation on B such that $g(x) = f(x)$. The construction of such an equivalence relation is done as follows:
Consider the set

$$S = \{\langle f(x), g(x)\rangle | x \in A\} \subseteq B \times B. \tag{4.48}$$

This clearly defines a relation on B stating that $f(x) \sim_S g(x)$, but it is not necessarily an equivalence relation. However, one can construct a minimal equivalence rela-

1. Reflexive: for all $a \in A$, $a \sim_R a$.
2. Transitive: if $a \sim_R b$ and $b \sim_R c$ then $a \sim_R c$.
3. Symmetric: if $a \sim_R b$, then $b \sim_R a$.

tion on B which contains S. In particular, such an equivalence relation $R \subseteq B \times B$ would be such that

- $S \subseteq R$.
- Given any other equivalence relation T on B such that $S \subseteq T$, then $R \subseteq T$.

Thus we are looking for the minimal equivalence relation which would contain S above and which would also imply $g(x) \sim f(x)$.

Given a relation $S \subseteq B \times B$ which is not an equivalence relation, there exists a standard procedure to turn it into an equivalence relation. This procedure amount to defining the closure of the set in question in terms of the defining properties of an equivalence relations, namely *transitivity*, *symmetry* and *reflexivity*. In our case the aim is to extend the set S of (4.48) so that it will have the properties of transitivity, symmetry and reflexivity.

To make the set S closed under reflexivity we add the relation $\Delta := \{(b,b)| b \in B\}$ thus obtaining $R_1 = S \cup \Delta$. Such a relation now clearly satisfies the reflexivity property since any element $b \in B$ is now related to itself. In order for the relation to satisfy symmetry we need to add the set $S^{op} := \{(a,b)|(b,a) \in S\}$. We then obtain the relation $R_2 = S \cup \Delta \cup S^{op}$. This, however, still lacks transitivity which is added by augmenting R_2 by the set $S^+ := R_2 \cup (R_2 \cdot R_2) \cup \cdots \cup (R_2 \cdot R_2 \cdot \cdots \cdot R_2)$. We have thus reached the desired minimal equivalence relation R containing S: $R := S \cup \Delta \cup R^{op} \cup S^+ = S^+$.

If we now replace C by B/R in the Diagram 4.1, it can be shown that indeed the map $h : B \to B/R$ sending each element to its equivalence class is the coequaliser of f, g. In fact, given any pair $(a,b) \in R$ then $h(a) = h(b)$. Now let us assume that $(a,b) \in S$ thus $a = f(x)$ and $b = g(x)$. Since $S \subseteq R$, it follows that for all $x \in A$, $h(f(x)) = h(g(x))$, thus $h \circ f = h \circ g$. This proves the first requirement of a coequaliser. To prove the second property we assume we have a map h' such that $h' \circ f = h' \circ g$ and we, then, need to show that there is a unique $u : B/R \to C'$ such that $u \circ h = h'$.

Thus, first of all, we need to show that h' is constant on R. Given any $(a,b) = (f(x), g(x)) \in S$ then clearly $h'(a) = h'(f(x)) = h'(g(x)) = h'(b)$ from the definition of h'. This implies that for $(b,a) \in S^{op}$ then $h'(b) = h'(a)$. Moreover for $(a,a) \in \Delta$ then $h'(a) = h'(a)$. All that remains to show is that h' is constant on S^+. This can be simply done by induction, i.e. if $h'(a) = h'(b)$ and $h'(b) = h'(c)$ then $h'(a) = h'(b)$. This completes the proof that h' is constant on R. It then follows that $h' = u \circ h$ and by the surjectivity of h we conclude that u is unique.

4.5.7 Limits and Colimits

The description of the various objects given so far had all the same form, namely that the object in question satisfied a particular condition and it was the unique one (up to isomorphism) that did so. This type of description is called a *universal construction* since it describes an object as the sole object satisfying a certain property. One then says that the object has a certain *universal property*. In order to clearly understand what a *universal construction* and *universal propery* are we need to introduce the notions of *diagrams*, *limits* and *colimits* of these diagrams.

We now give a more abstract definition of a diagram as the one given in Definition 4.1 in Sect. 4.1:

Definition 4.23 Given a category $\mathcal{C}$, a diagram D in $\mathcal{C}$ is defined to be a collection of $\mathcal{C}$-objects $a_i \in C$ $(i \in I)$ and a collection of $\mathcal{C}$-arrows $a_i \to a_j$ between some of the $\mathcal{C}$-objects above.

Using the notion of graphs given in Definition 4.1 in Sect. 4.1, a diagram can be defined as one graph in the collection of graphs composing a category.

Now a special type of diagram D is the D-cone, i.e. a cone for a diagram D. This consists of a $\mathcal{C}$-object c and a $\mathcal{C}$-arrow $f_i : c \to a_i$ one for each $a_i \in D$, such that

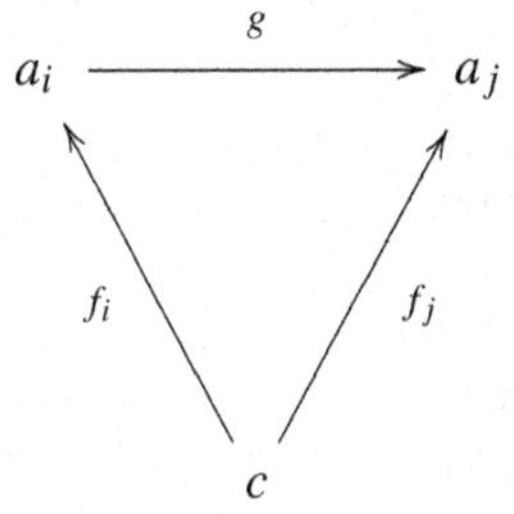

commutes whenever g is an arrow in the diagram D.

A cone is denoted as $\{f_i : c \to a_i\}$ and c is called the vertex of the cone. We now come to the definition of a limit.

Definition 4.24 A limit for a diagram D is a D-cone $\{f_i : c \to a_i\}$ such that, given any other D-cone $\{f_i' : c' \to a_i\}$, there is only one $\mathcal{C}$-arrow $g : c' \to c$ such that, for each $a_i \in D$ the following diagram commutes:

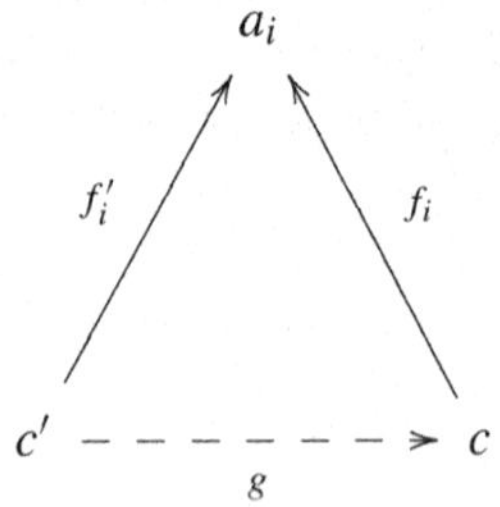

The limiting cone of a diagram D has the *universal property* with respect to all other D-cones, in the sense that any other D-cone factors through the limiting cone.

We will now give some examples of limiting cones.

Example 4.8 The product of two objects A and B in $\mathcal{C}$, defined above, is actually the limiting cone of the diagram containing only two objects A and B and no arrows, i.e. it is the limiting cone of the arrowless diagram

$$A \qquad\qquad B$$

In fact a cone for this diagram is given by any $\mathcal{C}$-object C together with two arrows $f : C \to A$ and $g : C \to B$ giving the cone

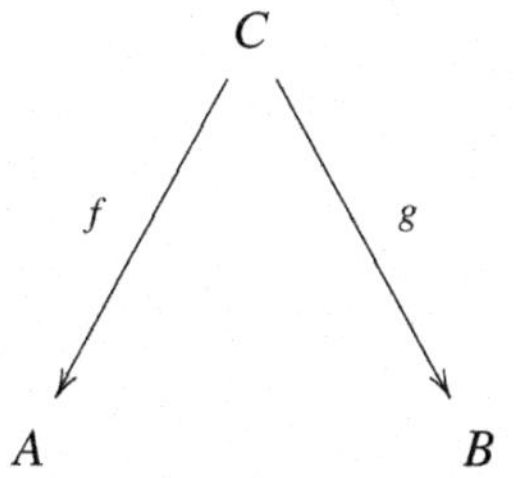

Now, in order for this cone to be a limiting cone, we require that any other cone factors through it. This means that given another cone

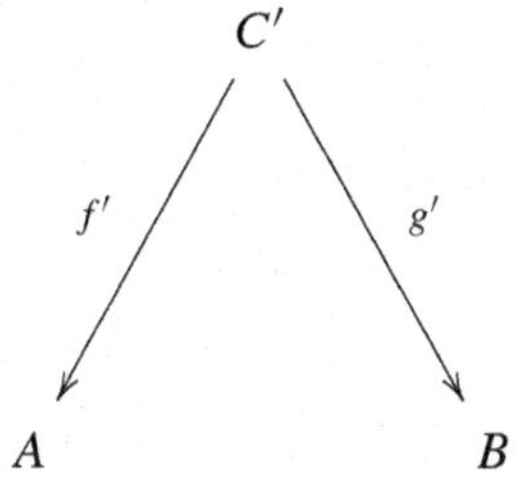

there exists a unique map $h : C' \to C$, such that

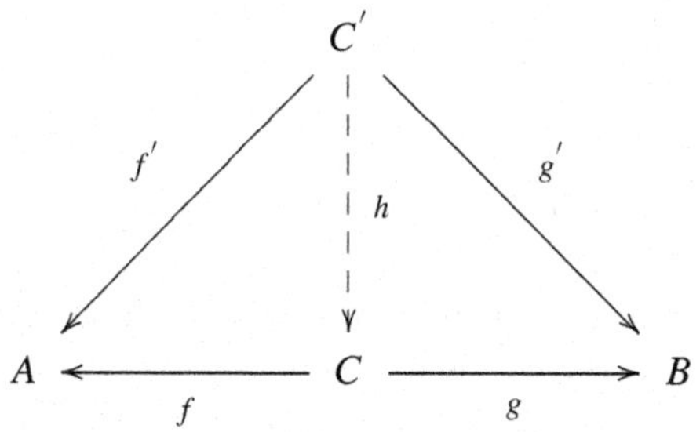

commutes. But this is precisely the definition of the product, i.e. $C = A \times B$.

Example 4.9 The terminal object is the limiting cone of the empty diagram. To see this consider a cone of the empty diagram, this is simply C. Given any other cone C', the condition for C to be a limiting cone implies that D factors though it, i.e. there exists a unique arrow $D \dashrightarrow C'$. This is precisely the definition of a terminal object.

Example 4.10 The equaliser is the limiting cone of the diagram

$$A \underset{g}{\overset{f}{\rightrightarrows}} B$$

In this case a cone for the above diagram is

$$E \xrightarrow{h} A \underset{g}{\overset{f}{\rightrightarrows}} B$$

For it to be a limiting cone, any other cone has to uniquely factor through it. Thus, given a cone

$$C \xrightarrow{k} A \underset{g}{\overset{f}{\rightrightarrows}} B$$

we require that there exists a unique $l : C \dashrightarrow E$, such that the following diagram commutes

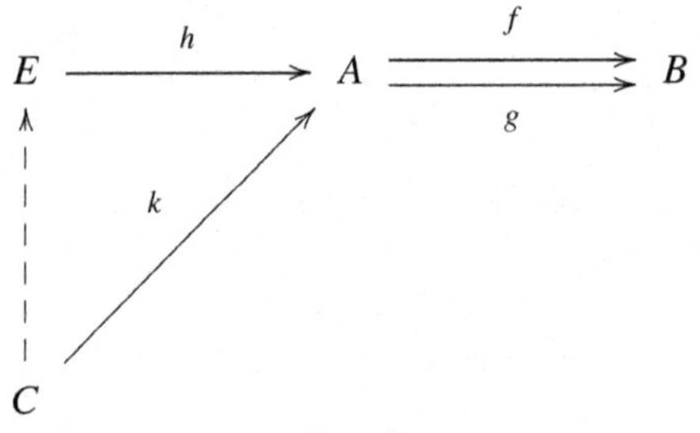

This is precisely the definition of an equaliser.

By duality we also have the notion of a *colimit* whose definition requires (as expected) the notion of a co-cone (dual to a cone).

Given a diagram D, a co-cone consists of an object c and arrows $\{f_i : a_i \to c\}$, one for each object $a_i \in D$. A co-cone is denoted $\{f_i : a_i \to c\}$. With this in mind we now define a colimit as follows:

Definition 4.25 A colimit of D is a co-cone with the (co)-universal property that, given any other D co-cone $\{f_i' : a_i \to c'\}$, there exists one and only one map $f : c \to c'$ such that the following diagram commutes

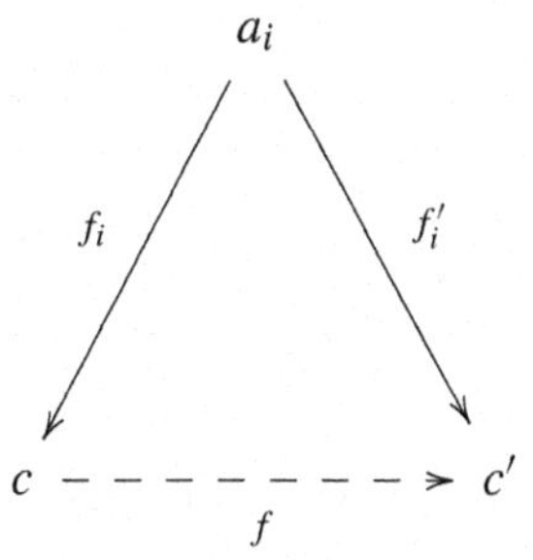

for all $a_i \in D$.

From the duality principle one can deduce what exactly the coproduct, initial object and coequaliser, are. These will be described in the following examples.

Example 4.11 We want to show that the coproduct is the colimit of the diagram

$$A \qquad\qquad B$$

A co-cone for such a diagram is

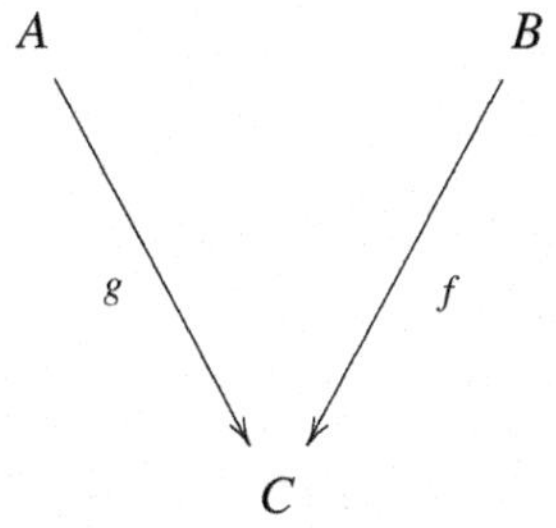

The requirements that this is a colimit implies that given any other co-cone

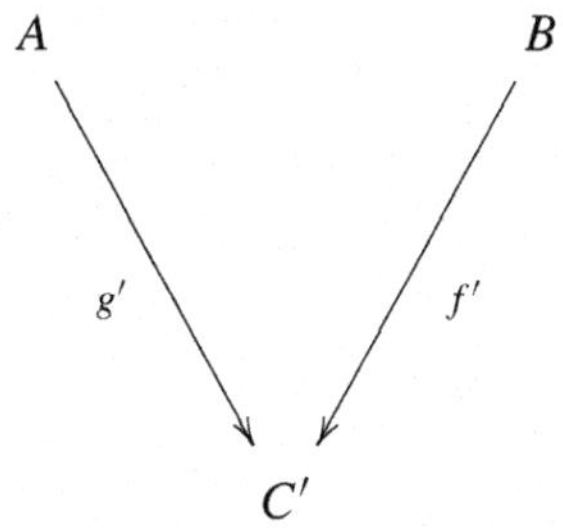

there exists a unique map $h : C \dashrightarrow C'$ such that the following diagram commutes

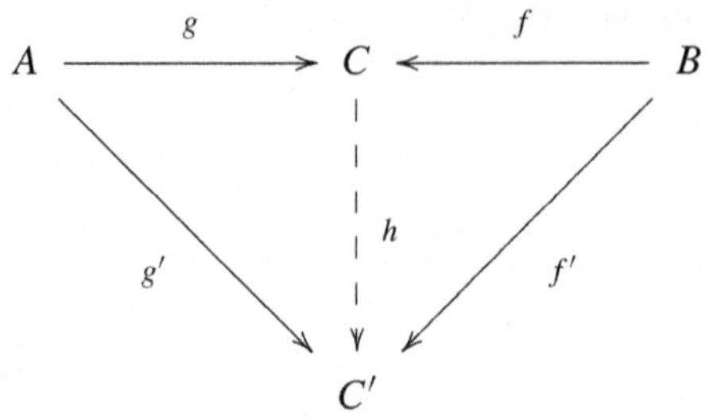

This is precisely the definition of a coproduct $C = A + B$.

Example 4.12 The initial object is the colimit of the empty diagram. For an empty diagram its co-cone is simply C. For this to be a colimit we require that, given any other object C', there exists a unique map $l : C \dashrightarrow C'$. This is precisely the definition of an initial object.

Example 4.13 The coequaliser is the colimit of the diagram

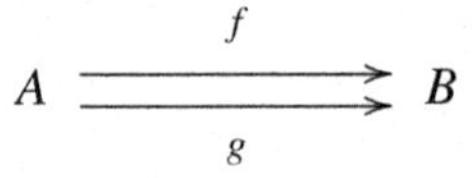

In this case a co-cone would be

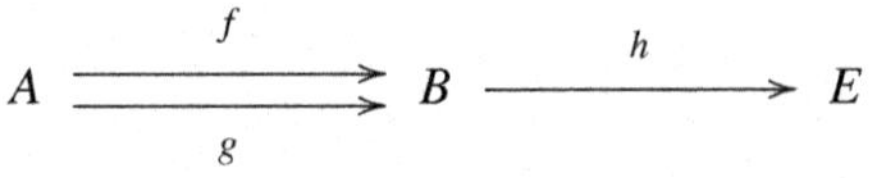

For this to be a colimit we require that any other co-cone of the diagram factors through it. Thus we obtain the following commuting diagram

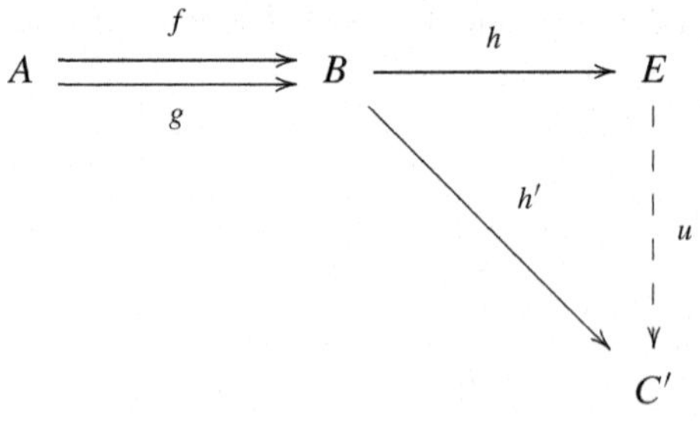

This is a coequaliser diagram.

4.6 Categories in Quantum Mechanics

In this section we will delineate different categories that arise in quantum theory, however we will not go into the details of how each of these categories is used. The aim is simply to show that category theory arises in many more contexts than one can imagine. This list of examples of categories in quantum theory is by no means complete.

4.6.1 The Category of Bounded Self Adjoint Operators

Definition 4.26 [14, 15] The set $\mathcal{O}$ of bounded self-adjoint operators is a **category**, in which

- the objects of $\mathcal{O}$ are the self-adjoint operators;
- given a function $f : \sigma(\hat{A}) \rightarrow \mathbb{R}$ (from the spectrum of $\hat{A}$ to the Reals), such that $\hat{B} = f(\hat{A})$, then there exists a morphism $f_{\mathcal{O}} : \hat{B} \rightarrow \hat{A}$ in $\mathcal{O}$ between operators $\hat{B}$ and $\hat{A}$.

To show that the category $\mathcal{O}$, so defined, is a category, we need to show that it satisfies the identity condition and composition condition. This can be shown in the following way:

- *Identity condition*: given any $\mathcal{O}$-object $\hat{A}$, the identity arrow is defined as the arrow $id_{\mathcal{O}_A} : \hat{A} \rightarrow \hat{A}$ that corresponds to the arrow $id : \mathbb{R} \rightarrow \mathbb{R}$.
- *Composition Condition*: given two $\mathcal{O}$-arrows $f_{\mathcal{O}} : \hat{B} \rightarrow \hat{A}$ and $g_{\mathcal{O}} : \hat{C} \rightarrow \hat{B}$, such that $\hat{B} = f(\hat{A})$ and $\hat{C} = g(\hat{B})$, then, the composite function $f_{\mathcal{O}} \circ g_{\mathcal{O}}$ in $\mathcal{O}$ corresponds to the composite function $f \circ g : \sigma(\hat{B}) \rightarrow \mathbb{R}$.

The category $\mathcal{O}$, as defined above, represents a pre-ordered set.[13] In fact, the function $f : \sigma(\hat{A}) \rightarrow \mathbb{R}$ is unique, therefore it follows that for any two objects in $\mathcal{O}$ there exists, at most, one morphism between them, i.e. $\mathcal{O}$ is a pre-ordered set. However, $\mathcal{O}$ fails to be a poset since it lacks the antisymmetry property. In fact, it can be the case that two operators $\hat{B}$ and $\hat{A}$ in $\mathcal{O}$ are such that $\hat{A} \neq \hat{B}$, but they are related by $\mathcal{O}$-arrows $f_{\mathcal{O}} : \hat{B} \rightarrow \hat{A}$ and $g_{\mathcal{O}} : \hat{A} \rightarrow \hat{B}$ in such a way that:

$$g_{\mathcal{O}} \circ f_{\mathcal{O}} = id_B \quad \text{and} \quad f_{\mathcal{O}} \circ g_{\mathcal{O}} = id_A. \tag{4.49}$$

---

[13] A set S is said to be a pre-ordered set if it is equipped with a binary relation $\leq$ which satisfies the following properties:

- Reflexivity: for all $a \in S$ then $a \leq a$.
- Transitive: if $a \leq b$ and $b \leq c$ then $a \leq c$.

It is possible to transform the set of self-adjoint operators into a poset by defining a new category $[\mathcal{O}]$ in which the objects are taken to be equivalence classes of operators, whereby two operators are considered to be equivalent if the $\mathcal{O}$-morphisms, relating them, satisfy (4.49).

4.6.2 Category of Boolean Sub-algebras

Definition 4.27 [14, 15] The **category** $\mathcal{W}$ of Boolean sub-algebras of the lattice $P(\mathcal{H})$ has:

- as objects, the individual Boolean sub-algebras, i.e. elements $W \in \mathcal{W}$ which represent spectral algebras associated with different operators.
- as morphisms, the arrows between objects of $\mathcal{W}$, such that a morphism $i_{W_1 W_2} : W_1 \rightarrow W_2$ exists iff $W_1 \subseteq W_2$.

From the definition of morphisms it follows that there is, at most, one morphisms between any two objects of $\mathcal{W}$, therefore $\mathcal{W}$ forms a poset under sub-algebra inclusion $W_1 \subseteq W_2$.

To show that $\mathcal{W}$, as defined above, is indeed a category, we need to define the identity arrow and the composite arrow. The identity arrow in $\mathcal{W}$ is defined as $id_W : W \rightarrow W$, which corresponds to $W \subseteq W$, whereas, given two $\mathcal{W}$-arrows $i_{W_1 W_2} : W_1 \rightarrow W_2$ ($W_1 \subseteq W_2$) and $i_{W_2 W_3} : W_2 \rightarrow W_3$ ($W_2 \subseteq W_3$) the composite $i_{W_2 W_3} \circ i_{W_1 W_2}$ corresponds to $W_1 \subseteq W_3$.

Example 4.14 An example of the category $\mathcal{W}$ can be formed in the following way: consider a category consisting of four objects (operators): $\hat{A}$, $\hat{B}$, $\hat{C}$, $\hat{\mathbb{1}}$ such that the spectral decompositions are:

$$\hat{A} = a_1 \hat{P}_1 + a_2 \hat{P}_2 + a_3 \hat{P}_3$$

$$\hat{B} = b_1 (\hat{P}_1 \vee \hat{P}_2) + b_2 \hat{P}_3$$

$$\hat{C} = C_1 (\hat{P}_1 \vee \hat{P}_3) + c_2 \hat{P}_2$$

Then the spectral algebras are the following:

$$W_A = \{\hat{0}, \hat{P}_1, \hat{P}_2, \hat{P}_3, \hat{P}_1 \vee \hat{P}_3, \hat{P}_1 \vee \hat{P}_2, \hat{P}_3 \vee \hat{P}_2, \hat{\mathbb{1}}\}$$

$$W_B = \{\hat{0}, \hat{P}_3, \hat{P}_1 \vee \hat{P}_2, \hat{\mathbb{1}}\}$$

$$W_C = \{\hat{0}, \hat{P}_2, \hat{P}_1 \vee \hat{P}_3 \hat{\mathbb{1}}\}$$

$$W_{\mathbb{1}} = \{\hat{\mathbb{1}}\}.$$

The relation between the spectral algebras is given by the following diagram:

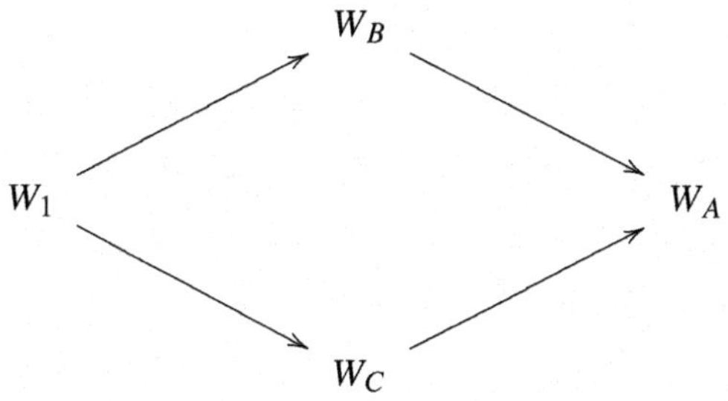

where the arrows are subset inclusions.

Chapter 5
Functors

So far we have talked about individual categories. However, it is possible to relate categories to one another. This is done through maps between categories. There are two types of such maps called *covariant functors* and *contravariant functors*. Both will be described in this chapter. It is also possible to abstract a bit further and define maps between functors themselves. These are called *Natural Transformations*, which are used to define equivalence of categories.

5.1 Functors and Natural Transformations

So far we have introduced the notion of a category, however, if we can not compare categories together we could not do much in category theory. Thus, there must be a way of comparing categories or, at least, define maps between them. This is done through the notion of a functor. Generally speaking, a functor is a transformation from one category $\mathcal{C}$ to another category $\mathcal{D}$, such that the categorical structure of the domain category $\mathcal{C}$ is preserved, i.e. gets mapped onto the structure of the codomain category $\mathcal{D}$.

There are two types of functors:

1. **Covariant Functors**.
2. **Contravariant Functors**.

5.1.1 Covariant Functors

Definition 5.1 : A **covariant functor** from a category $\mathcal{C}$ to a category $\mathcal{D}$ is a map $F : \mathcal{C} \to \mathcal{D}$ that assigns to each $\mathcal{C}$-object A, a $\mathcal{D}$-object $F(A)$ and to each $\mathcal{C}$-arrow $f : A \to B$ a $\mathcal{D}$-arrow $F(f) : F(A) \to F(B)$, such that the following are satisfied:

1. $F(id_A) = id_{F(A)}$.

C. Flori, *A First Course in Topos Quantum Theory*, Lecture Notes in Physics 868,
DOI 10.1007/978-3-642-35713-8_5,

2. $F(f \circ g) = F(f) \circ F(g)$ for any $g : C \to A$.

It is clear from the above that a covariant functor is a transformation that preserves both:

- The domain's and the codomain's identities.
- The composition of arrows, in particular it preserves the direction of the arrows.

A pictorial description of a covariant functor is as follows:

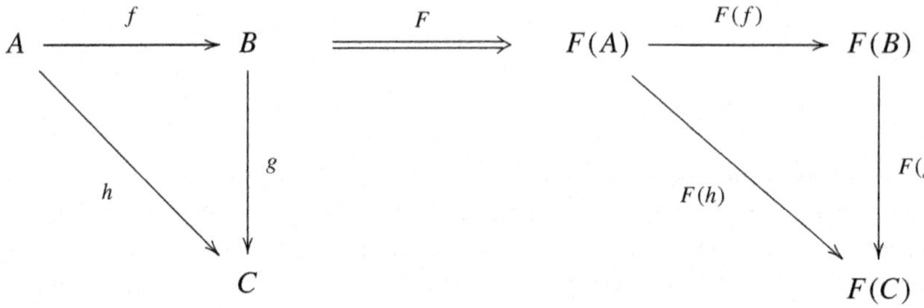

Example 5.1 (Identity Functor) $id_{\mathcal{C}} : \mathcal{C} \to \mathcal{C}$ is such that $id_{\mathcal{C}} A = A$ for all $A \in \mathcal{C}$ and $id_{\mathcal{C}}(f) = f$ for all $\mathcal{C}$-arrows f. Similarly one can define the inclusion functor for any subcategory $\mathcal{D} \subseteq \mathcal{C}$. This is trivially defined as follows

$$\begin{aligned} I &: \mathcal{D} \to \mathcal{C} \\ A &\mapsto A \\ (f : A \to B) &\mapsto (f : A \to B). \end{aligned} \tag{5.1}$$

Given such a definition it follows that

$$\begin{aligned} I(id_A) &= id_{I(A)} \\ I(f \circ g) &= I(f) \circ I(g). \end{aligned} \tag{5.2}$$

Example 5.2 (Power Set Functor) P : **Sets** $\to$ **Sets** assigns to each object $X \in$ **Sets** its power set[1] PX, and to each map $f : X \to Y$ the map $P(f) : PX \to PY$, which sends each subset $S \subseteq X$ to the subset $f(S) \subseteq Y$.

Example 5.3 (Forgetful Functor) Given a category $\mathcal{C}$ with some structure on it, for example the category of groups **Grp**, the *forgetful functor* F : **Grp** $\to$ **Sets** takes each group to its underlining set forgetting about the group structure, and each $\mathcal{C}$-arrow to itself.

[1] Set of all subsets of X.

Example 5.4 (Hom-Functor) Given any $\mathcal{C}$-object A, then the *Hom-functor* $\mathcal{C}(A, -) : \mathcal{C} \to \mathbf{Sets}$ takes each object B to the set of all $\mathcal{C}$-arrows $\mathcal{C}(A, B)$ from A to B, and to each $\mathcal{C}$-arrow $f : B \to C$ it assigns the map

$$\begin{aligned} \mathcal{C}(A, f) : \mathcal{C}(A, B) &\to \mathcal{C}(A, C) \\ g &\mapsto \mathcal{C}(A, f)(g) := f \circ g \end{aligned} \tag{5.3}$$

such that the following diagram commutes

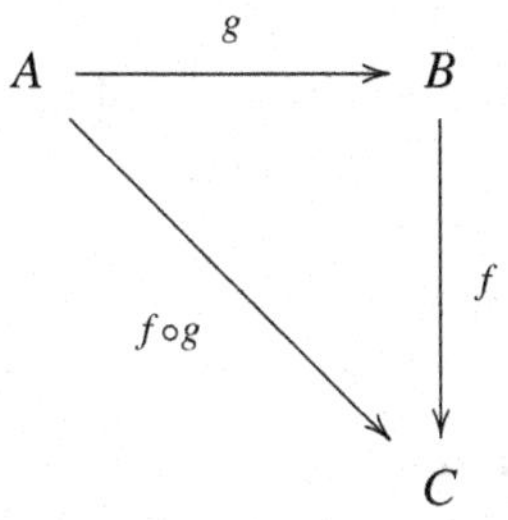

Example 5.5 (Free Group Functor) Given the categories **Sets** and **Grp** the free group functor is a functor $F : \mathbf{Sets} \to \mathbf{Grp}$ which assigns, to each set $A \in \mathbf{Sets}$, the free group[2] generated by A and, to each morphism f, the induced homomorphism between the respective groups which coincides with f on the free generators.

Example 5.6 (Functors Between Pre-ordered Sets) Given two preorders $(P, \leq)$, $(Q, \leq)$. A covariant functor $F : (P, \leq) \to (Q, \leq)$ is defined as a covariant functor $F : P \to Q$ which is order preserving (or monotone), i.e.

$$\forall p_1, p_2 \in P \quad \text{if} \quad p_1 \leq p_2 \quad \text{then} \quad F(p_1) \leq F(p_2). \tag{5.4}$$

It can be easily seen that indeed the above map satisfies the conditions of being a functor. It follows that, in this case, F is simply a monotone map.

Example 5.7 (Functor Between Monoids) Given two monoids $(M, *, 1)$, $(N, *, 1)$, a covariant functor $F : (M, *, 1) \to (N, *, 1)$ is such that it maps M to N. The functoriality condition is istantiates as follows:[3]

$$\forall m_1, m_2 \in M \quad F(m_1 * m_2) = F(m_1) * F(m_2) \quad \text{and} \quad F(1) = 1. \tag{5.5}$$

Hence, a covariant functor between monoids is just a monoid homomorphism.

---

[2] A group G is called free if there exists a subset $S \subseteq G$, such that any element of G can be uniquely (up to trivial equalities such as $xy = xz^{-1}zy$) written as a product of finitely many elements of S and their inverses.

[3] Recall that in a monoid with set M maps have both domain and codomain equal to M (technically these types of morphisms are called endomorphisms), i.e. $m_i : M \to M$ and represent elements of M.

Example 5.8 (Group Action Functor) Given a group $(G, *, 1)$ (which, as we previously saw, can be considered as a monoid), a covariant functor $F : G \to Sets$ represents the action of G on a set $X \in Sets$. In particular $F(G) = X$ and each map $g_i : G \to G$ gets mapped to an endofunction on X, i.e. $F(g_i) := g_1 * - : X \to X$. The functorial condition then amounts to the following:

$$\forall g_1, g_2 \in G \quad F(g_1 * g_2) = F(g_1) \circ F(g_2) \quad \text{and} \quad F(1) = id_X. \tag{5.6}$$

Therefore, given any $x \in X$ the above equations imply

$$(g_1 * g_2)x = g_1 \cdot g_1 \cdot x \quad \text{and} \quad 1 \cdot x = x. \tag{5.7}$$

Thus F defines an action of G on X.

5.1.2 Contravariant Functor

Let us now analyse the other type of functor: *contravariant functor*.

Definition 5.2 A **contravariant functor** from a category $\mathcal{C}$ to a category $\mathcal{D}$ is a map $X : \mathcal{C} \to \mathcal{D}$ that assigns to each $\mathcal{C}$-object A, a $\mathcal{D}$-object $X(A)$ and to each $\mathcal{C}$-arrow $f : A \to B$ a $\mathcal{D}$-arrow $X(f) : X(B) \to X(A)$, such that the following conditions are satisfied:

1. $X(id_A) = id_{X(A)}$.
2. $X(f \circ g) = X(g) \circ X(f)$ for any $g : C \to A$.

A diagrammatic representation of a contravariant functor is the following:

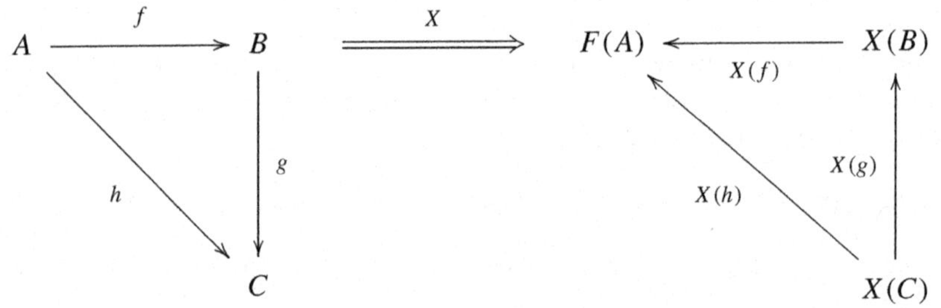

Thus, a contravariant functor in mapping arrows from one category to the next reverses the directions of the arrows, by mapping domains to codomains and vice versa. A contravariant functor is also called a *presheaf*. These types of functors will be the principal objects which we will study when discussing quantum theory in the language of topos theory.

Example 5.9 (Contravariant Power Set Functor) This is a functor $\tilde{P}$: **Sets** → **Sets** which assigns to each set X its power set $\tilde{P}(X)$ and, to each arrow $f : X \to Y$ the inverse image map $\tilde{P}(f) : \tilde{P}(Y) \to \tilde{P}(X)$, which sends each set $S \in P(Y)$ to the inverse image $f^{-1}(S) \in P(X)$.

Example 5.10 (Contravariant Hom-Functor) For any object $A \in \mathcal{C}$, we define the *contravariant hom-functor* to be the functor $\mathcal{C}(-, A) : \mathcal{C} \to$ **Sets**, which assigns to each object $B \in \mathcal{C}$ the set of $\mathcal{C}$-arrows $\mathcal{C}(B, A)$ and, to each $\mathcal{C}$-arrow $f : B \to C$, it assigns the function

$$\begin{aligned} \mathcal{C}(f, A) : \mathcal{C}(C, A) &\to \mathcal{C}(B, A) \\ g &\mapsto \mathcal{C}(B, A)(f) := g \circ f \end{aligned} \tag{5.8}$$

such that the following diagram commutes

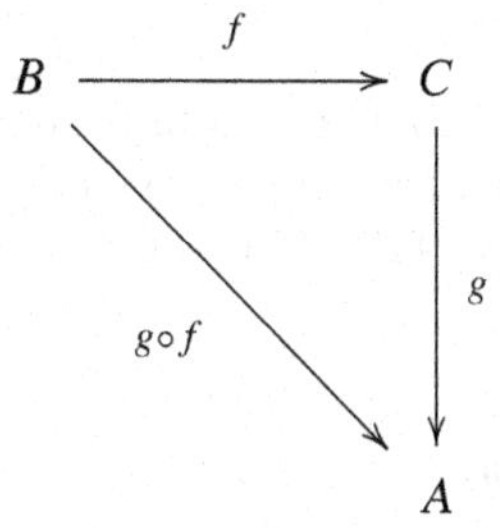

5.2 Characterising Functors

Irrespectively of whether we are talking about covariant or contravariant functors, here are several properties which distinguish different functors. These are the following:

Definition 5.3 A functor $F : \mathcal{C} \to \mathcal{D}$ is called

1. *Faithful* if, $\forall X, Y \in Ob(\mathcal{C})$

$$F : Mor_{\mathcal{C}}(X, Y) \to Mor_{\mathcal{D}}(FX, FY) \tag{5.9}$$

is injective.

2. *Full* if, $\forall X, Y \in Ob(\mathcal{C})$

$$F : Mor_{\mathcal{C}}(X, Y) \to Mor_{\mathcal{D}}(FX, FY) \tag{5.10}$$

is surjective.

3. *Fully faithful* if, $\forall X, Y \in Ob(\mathcal{C})$

$$F : Mor_{\mathcal{C}}(X, Y) \rightarrow Mor_{\mathcal{D}}(FX, FY) \tag{5.11}$$

is bijective.
4. *Forgetful* if, roughly speaking, F takes each $\mathcal{C}$ object to its underlying set, forgetting about any structure which might be present in $\mathcal{C}$, while mapping each $\mathcal{C}$-arrow to itself. Thus, all that is remembered by this forgetful functor is the fact that the $\mathcal{C}$-arrows are set functions.
5. *Essentially surjective*. A functor $F : \mathcal{C} \rightarrow \mathcal{D}$ is *essentially surjective* (or dense) if each object $D \in \mathcal{D}$ is isomorphic to an object of the form $F(C)$ for some object $C \in \mathcal{C}$.
6. *Embedding* if F is full, faithful, and injective on objects.
7. *An equivalence* if F is full, faithful, and essentially surjective.
8. *An isomorphism* if there exists another functor $G : \mathcal{D} \rightarrow \mathcal{C}$ such that

$$G \circ F = id_{\mathcal{C}}, \qquad F \circ G = id_{\mathcal{D}}. \tag{5.12}$$

It should be noted that the definition of a full and faithful functor only requires that there is a bijection between the morphisms of the categories, not between the objects. In fact if $F : \mathcal{C} \rightarrow \mathcal{D}$ is a full and faithful functor, than it could be the case that:

(i) there exists some $Y \in Ob(\mathcal{D})$, such that there is no object $X \in Ob(\mathcal{C})$ for which $Y = F(X)$, i.e. F is not surjetive on objects.
(ii) Given two elements $X_1, X_2 \in Ob(\mathcal{C})$, then if $F(X_1) = F(X_2)$, this does not entail that $X_1 = X_2$, i.e. F is not injective on objects.

So far we have classified functors according to how they act on the collection of objects and morphisms seen as sets. However, one can abstract a little more and try to understand how functors behave on more complex structures, such as properties of arrows. In particular consider a property P of arrows, a functor $F : \mathcal{C} \rightarrow \mathcal{D}$ *preserves* P iff

$$f \text{ satisfies } P \implies F(f) \text{ satisfies } P \tag{5.13}$$

F *reflects* P iff

$$F(f) \text{ satisfies } P \implies f \text{ satisfies } P. \tag{5.14}$$

Theorem 5.1

(i) *Faithful functors reflect monics and epics.*
(ii) *Full and faithful functors reflect isomorphisms.*
(iii) *Equivalences preserve monics and epics.*
(iv) *Every functor preserves isos.*

Proof (i) Given a faithful functor $F : \mathcal{C} \rightarrow \mathcal{D}$ we want to show that if $F(u) : F(A) \rightarrow F(B)$ is monic then so $u : A \rightarrow B$. This means that if $u \circ k = u \circ h$ then

$h = k$. In order to show this we apply the functor F obtaining

$$F(u) \circ F(k) = F(u \circ k) = F(u \circ h) = F(u) \circ F(h). \tag{5.15}$$

From the property of being monic we have that if

$$F(u) \circ F(k) = F(u) \circ F(h) \tag{5.16}$$

then $F(k) = F(h)$. From the property of faithfulness it follows that $k = h$. Thus u is monic.

Similarly you can show that if $F(g) : F(A) \to F(B)$ is epic, then so is g.

(ii) Assume that the functor $F : \mathcal{C} \to \mathcal{D}$ is full and faithful. Given $h : A \to B$, then assume that $F(h) : F(A) \to F(B)$ is an iso arrow in $\mathcal{D}$ then there exists an arrow $F(h)^{-1} : F(B) \to F(A)$ in $\mathcal{D}$. Since F is full then there exists a map $g \in \mathcal{C}$ such that $F(g) = F(h)^{-1}$. From the property of F being a functor we have that $F(h \circ g) = F(h) \circ F(g) = F(h) \circ F(h)^{-1} = id_{F(B)} = F(id_B)$. Since F is faithful, it follows that $h \circ g = id_B$. Similarly one can show that $g \circ h = id_A$.

(iii) We need to show that if $u : A \to B$ is monic then $F(u)$ is monic when F is an equivalence. Thus we assume that we habe two maps $h, g : C \to F(A)$ such that $F(u) \circ h = F(u) \circ g$. Since F is an equivalence, then there exists a $C' \in \mathcal{D}$ such that $C \simeq C'$, i.e $i : F(C') \to C$ is iso. Composing with such an arrow gives $F(u) \circ h \circ i = F(u) \circ g \circ i$. SInce F is full we can find pre-images of $h \circ i$ and $g \circ i$ in $\mathcal{C}$: $F(h') = h \circ i$ and $F(g') = g \circ i$. Substituting we obtain $F(u) \circ F(h') = F(u) \circ F(g')$ which by the functorial property of F becomes $F(u \circ h') = F(u \circ g')$. Since F is faithful $u \circ h' = u \circ g'$, but u is monic hence $h' = g'$ which implies that $F(h') = F(g') = h \circ i = g \circ i$. i is iso therefore $h = g$. It follows that $F(u)$ is indeed monic.

(iv) We need to show that given a map $f : A \to B$ which is iso (thus $f^{-1} : B \to A$) then $F(f)$ is iso. In particular we know that from the functor property $F(f \circ f^{-1}) = F(f) \circ F(f^{-1}) = F(id_B) = id_{F(B)}$. Similarly $F(f^{-1}) \circ F(f) = id_{F(A)}$. Thus indeed $F(f)$ is iso. □

5.3 Natural Transformations

So far we have defined categories and maps between them called *functors*. We will now abstract one step more and define maps between functors. These are called *natural transformations*.

Definition 5.4 Given two contravariant functors X and Y, a **natural transformation** from $Y : \mathcal{C} \to \mathcal{D}$ to $X : \mathcal{C} \to \mathcal{D}$ is an arrow $N : Y \to X$ that associates to each object A in $\mathcal{C}$ an arrow $N_A : Y(A) \to X(A)$ in $\mathcal{D}$ such that, for any $\mathcal{C}$-arrow

$f : A \to B$, the following diagram commutes:

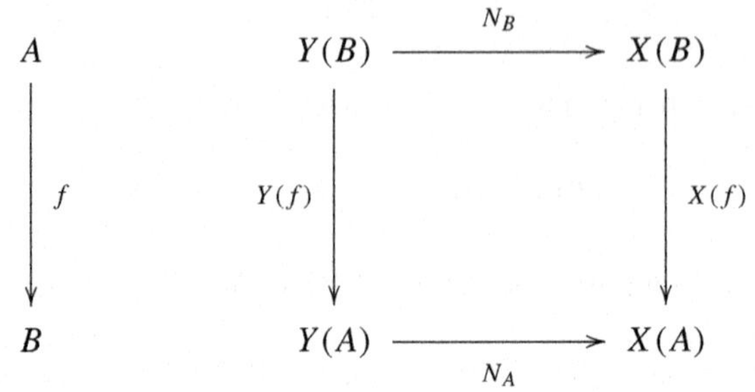

i.e.

$$N_A \circ Y(f) = X(f) \circ N_B.$$

Here $N_A : Y(A) \to X(A)$ are the components of N, while N is the *natural transformation.* An analogous definition holds for natural transformations between covariant functors.

From this diagram it is clear that the two arrows N_A and N_B turn the Y-picture of $f : A \to B$ into the respective X-picture. If each N_A $(A \in \mathcal{C})$ is an isomorphism, then N is a *natural isomorphism*

$$N : Y \overset{\simeq}{\to} X. \tag{5.17}$$

Example 5.11 Consider the operation of taking the dual of a vector space defined over some field K. This operation is actually a functor as follows

$$\begin{aligned} * : Vect_K &\to Vect_K^* \\ V &\mapsto V^* := Hom_k K(V, K). \end{aligned} \tag{5.18}$$

Moreover, given a linear map $f : V \to W$, we obtain a map $f^* : W^* \to V^*$ such that $f^*(\phi) = \phi \circ f$ where $\phi \in W^*$. By repeating this functor we can define a double dual functor as follows:

$$\begin{aligned} ** : Vect_K &\to Vect_K \\ V &\mapsto V^{**} \end{aligned} \tag{5.19}$$

such that $v^{**}(f) = f(v)$ for $f \in V^*$ and $v \in V$.

It is then possible to define a natural transformation between the identity functor $id_{Vect_K} : Vect_K \to Vect_K$ and the double dual functor as follows:

$$N : 1_{Vect_K} \to ** \tag{5.20}$$

whose components are

$$
\begin{array}{ccccc}
V & \quad & id_{Vect_K}(V) & \xrightarrow{N_V} & V^{**} \\
\downarrow f & & \downarrow id_{Vect_K}(f) & & \downarrow f^{**} \\
W & & id_{Vect_K}(W) & \xrightarrow{N_W} & W^{**}
\end{array}
$$

Example 5.12 Given a map $f : A \to B$ in a category $\mathcal{C}$, we obtain a natural transformation between covariant Hom-functors as follows:

$$\mathcal{C}(f, -) : \mathcal{C}(B, -) \to \mathcal{C}(A, -) \tag{5.21}$$

such that for each $C \in \mathcal{C}$ we obtain

$$\begin{aligned} &\mathcal{C}(f, C) : \mathcal{C}(B, C) \to \mathcal{C}(A, C) \\ &\quad (g : B \to C) \mapsto \mathcal{C}(f, C)(g) := (g \circ f : A \to C). \end{aligned} \tag{5.22}$$

To show that $\mathcal{C}(f, -)$ is indeed a natural transformation, we need to show that for all $h : C \to D$ the following diagram commutes

$$
\begin{array}{ccc}
\mathcal{C}(B, C) & \xrightarrow{\mathcal{C}(B,h)} & \mathcal{C}(B, D) \\
\downarrow \mathcal{C}(f,C) & & \downarrow \mathcal{C}(f,D) \\
\mathcal{C}(A, C) & \xrightarrow{\mathcal{C}(A,h)} & \mathcal{C}(A, D)
\end{array}
$$

Chasing the diagram around we have on the one hand

$$\mathcal{C}(A, h)\big(\mathcal{C}(f, C)(g)\big) = \mathcal{C}(A, h)(g \circ f) = h \circ (g \circ f) \tag{5.23}$$

and on the other hand

$$\mathcal{C}(f, D)\big(\mathcal{C}(B, h)(g)\big) = \mathcal{C}(f, D)(h \circ g) = (h \circ g) \circ f \tag{5.24}$$

so that naturality is equivalent to

$$h \circ (g \circ f) = (h \circ g) \circ f. \tag{5.25}$$

This equality is associativity.

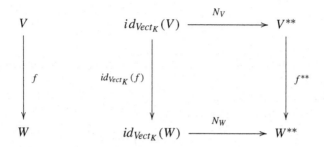

Example 5.13 Given a map $f : A \to B$ in a category $\mathcal{C}$ it is possible to define a natural transformation between contravariant Hom-functors as follows:

$$\mathcal{C}(-, f) : \mathcal{C}(-, B) \to \mathcal{C}(-, A). \tag{5.26}$$

This is indeed a well-defined natural transformation. In fact, given any object $C \in \mathcal{C}$, the action of $\mathcal{C}(-, f)$ is

$$\begin{aligned} \mathcal{C}(C, f) : \mathcal{C}(C, A) &\to \mathcal{C}(C, B) \\ h &\mapsto \mathcal{C}(C, f)(h) = f \circ h. \end{aligned} \tag{5.27}$$

While on an object D we get the component

$$\begin{aligned} \mathcal{C}(D, f) : \mathcal{C}(D, A) &\to \mathcal{C}(D, B) \\ k &\mapsto f \circ k \end{aligned} \tag{5.28}$$

so that the following diagram commutes

$$\begin{array}{ccc} \mathcal{C}(C, A) & \xrightarrow{\mathcal{C}(C,f)} & \mathcal{C}(C, B) \\ {\scriptstyle \mathcal{C}(g,A)} \uparrow & & \uparrow {\scriptstyle \mathcal{C}(g,B)} \\ \mathcal{C}(D, A) & \xrightarrow{\mathcal{C}(D,f)} & \mathcal{C}(D, B) \end{array}$$

Thus $\mathcal{C}(C, f) \circ \mathcal{C}(g, A)(k) = \mathcal{C}(C, f)(k \circ g) = f \circ (k \circ g)$ while $\mathcal{C}(g, B) \circ \mathcal{C}(D, f)(k) = \mathcal{C}(g, B)(f \circ k) = (f \circ k) \circ g$. These two expression are equal by associativity.

We will now state and prove a lemma which is a version of the very important *Yoneda Lemma* which will be analysed in Chap. 8.

Lemma 5.1 *Given a category $\mathcal{C}$ and two objects $A, B \in \mathcal{C}$, then for each natural transformation $t : \mathcal{C}(A, -) \to \mathcal{C}(B, -)$ of covariant functors, there exists a unique $f : B \to A$ in $\mathcal{C}$ such that $t = \mathcal{C}(f, -)$.*

Proof Let us define

$$(f : B \to A) := t_A(id_A). \tag{5.29}$$

Since $t : \mathcal{C}(A, -) \rightarrow \mathcal{C}(B, -)$ is a natural transformation, given any arrow $g : A \rightarrow C$ we have the following commutative diagram

$$\begin{array}{ccc} \mathcal{C}(A, A) & \xrightarrow{\mathcal{C}(A,g)} & \mathcal{C}(A, C) \\ \downarrow{\scriptstyle t_A} & & \downarrow{\scriptstyle t_C} \\ \mathcal{C}(B, A) & \xrightarrow{\mathcal{C}(B,g)} & \mathcal{C}(B, C) \end{array}$$

Chasing the diagram around we have

$$t_C\big(\mathcal{C}(A, g)(id_A)\big) = t_C(g). \tag{5.30}$$

On the other hand

$$\mathcal{C}(B, g)\big(t_A(id_A)\big) = \mathcal{C}(B, g)f = g \circ f \tag{5.31}$$

thus, from the requirement of commutativity

$$t_C(g) = g \circ f. \tag{5.32}$$

However from the definition in (5.22) we know that

$$\mathcal{C}(f, C)(g) := g \circ f \tag{5.33}$$

thus

$$t_C = \mathcal{C}(f, -). \tag{5.34}$$

To prove uniqueness we need to show that if $\mathcal{C}(f, -) = \mathcal{C}(f', -)$ then $f = f'$. To this end consider

$$f = id_A \circ f = \mathcal{C}(f, A)(id_A) = \mathcal{C}\big(f', A\big)(id_A) = id_A \circ f' = f'. \tag{5.35}$$

□

5.3.1 Equivalence of Categories

Now that we have defined maps between categories, i.e. functors and maps between functors, i.e. natural transformations, it is possible to compare two categories and see if they are equivalent or not. To this end we need the notion of isomorphic functors.

Definition 5.5 Two functors $F, G : \mathcal{C} \to \mathcal{D}$ are said to be naturally isomorphic if there exists a natural transformation $\eta : F \to G$, which is invertible.

We can now define equivalent categories in terms of naturally isomorphic functor.

Definition 5.6 Two categories $\mathcal{C}$ and $\mathcal{D}$ are said to be equivalent if there exists functors $F : \mathcal{C} \to \mathcal{D}$ and $G : \mathcal{D} \to \mathcal{C}$, such that the functors $F \circ G \simeq id_{\mathcal{D}}$ and $G \circ F \simeq id_{\mathcal{C}}$ are naturally isomorphic to the identities. In this case F is called an *equivalence of categories*.

It is easy to then show that a functor $F : \mathcal{C} \to \mathcal{D}$ is an equivalence of categories iff F is *full, faithgull* and *essentially surjective* (see Definition 5.3).

Chapter 6
The Category of Functors

From previous paragraphs we have understood what a category is, how to define maps between categories, namely functors, and finally how to define maps between functors, which are natural transformations. It is now possible to combine all these definitions in a coherent way and define the category of functors.

In particular, given two categories $\mathcal{C}$ and $\mathcal{D}$, the collection of all covariant (or contravariant) functors $F : \mathcal{C} \to \mathcal{D}$ is actually a category which will be denoted as $\mathcal{D}^{\mathcal{C}}$. This is called the *category of functors* or *functor category* and has as objects covariant (or contravariant) functors and as map natural transformations between functors. Generally, when we write $\mathcal{D}^{\mathcal{C}}$ we will mean covariant functors, to denote contravariant functors instead we will write[1] $\mathcal{D}^{\mathcal{C}^{op}}$. The category of contravariant functors, for a particular choice of $\mathcal{D}$ and $\mathcal{C}$ will be the main category we will use when defining topos quantum theory.

6.1 The Functor Category

We will now introduce a type of category which is very important for the topos formulation of quantum theory since (i) it is actually a topos,[2] (ii) for an appropriate choice of base category[3] it will be the topos in terms of which quantum theory is defined.

[1] Note that $\mathcal{C}^{op}$ represents the opposite of the category $\mathcal{C}$ as defined in Definition 4.9. Objects in $\mathcal{C}^{op}$ are the same as the objects in $\mathcal{C}$, while the morphisms are the inverse of the morphisms in $\mathcal{C}$, i.e. $\exists$ a $\mathcal{C}^{op}$-morphisms $f : A \to B$ iff $\exists$ a $\mathcal{C}$-morphisms $f : B \to A$. Thus the category $\mathbf{Sets}^{\mathcal{C}^{op}}$ has as objects covariant functors from $\mathcal{C}^{op}$ to **Sets** or, equivalently, contravariant functors from $\mathcal{C}$ to **Sets**.

[2] The precise definition of what a topos is given in Chap. 7. For now it suffices to say that a topos is a category with a lot of 'extra structure' which implies that a topos can be seen as a generalisation of **Sets**, so that many mathematical constructions in **Sets** has a more general topos analogue. In fact **Sets** is a topos, albeit a very special one.

[3] By base category we mean the category representing the codomain of the functor. Thus for example given the functor $F : \mathcal{C} \to \mathcal{D}$ the base category would be $\mathcal{D}$.

C. Flori, *A First Course in Topos Quantum Theory*, Lecture Notes in Physics 868,
DOI 10.1007/978-3-642-35713-8_6, © Springer-Verlag Berlin Heidelberg 2013

The functor category is, in a way, more abstract than any of the categories we have encountered so far, since it has as objects covariant (contravariant) functors and as maps natural transformations. So it is one level higher in abstraction than the previous examples of categories, which had simpler objects with not so much structure and as maps simpler function between these elements. On the other hand the objects in the functor categories are complicated objects being themselves maps between categories, thus they carry a lot of structure. The maps between these objects are now required to preserve such complicated structures. However, on a general level, we still simply have a collection of objects with maps between them, the only difference is that now the hidden level of complexity has to be taken into consideration when defining any categorical construction. So, for example, as we will see in the examples below, when defining the products, the equalisers, etc, the structure of the individual objects of the category has to be taken into consideration.

The definition of the functor category is as follows:

Definition 6.1 Given two categories $\mathcal{C}$ and $\mathcal{D}$, the functor category $\mathcal{D}^{\mathcal{C}}$ has:

- *Objects*: all covariant functors of the form $F : \mathcal{C} \to \mathcal{D}$.
- *Morphisms*: natural transformations between the above mentioned functors.

Given two natural transformations $N_1 : F \to G$ and $N_2 : G \to H$ in $\mathcal{D}^{\mathcal{C}}$ composition is defined as follows:

$$N_2 \circ N_1 : F \to H \tag{6.1}$$

such that, for each $C \in \mathcal{C}$, the individual components are

$$\begin{aligned} (N_2 \circ N_1)_C &: F(C) \to H(C) \\ &: F(C) \xrightarrow{(N_1)_C} G(C) \xrightarrow{(N_2)_C} H(C). \end{aligned} \tag{6.2}$$

Composition is then associative.

Of particular importance for us is the case in which $\mathcal{D}$ is the category **Sets**. In fact $\mathbf{Sets}^{\mathcal{C}}$, for a particular $\mathcal{C}$, will be the category which we will use to describe quantum theory. To be precise, for quantum theory we will be using the category $\mathbf{Sets}^{\mathcal{C}^{op}}$ of contravariant functors or presheaves for an appropriate $\mathcal{C}$. So let us analyse $\mathbf{Sets}^{\mathcal{C}^{op}}$.

6.2 Category of Presheaves

We will now describe in detail the presheaf category $\mathbf{Sets}^{\mathcal{C}^{op}}$ for some category $\mathcal{C}$. The reason we are interested in this category is twofold:

(i) It is a topos.
(ii) For a particular choice of $\mathcal{C}$ it will be the topos in which quantum theory will be defined.

The category $\mathbf{Sets}^{\mathcal{C}^{op}}$ is defined in terms of

- Objects: all contravariant functors $X : \mathcal{C} \rightarrow \mathbf{Sets}$. The pictorial representation of the action of X on a simple category $\mathcal{C}$ with objects $Ob_{\mathcal{C}} = (A, B, C, D)$ is as follows

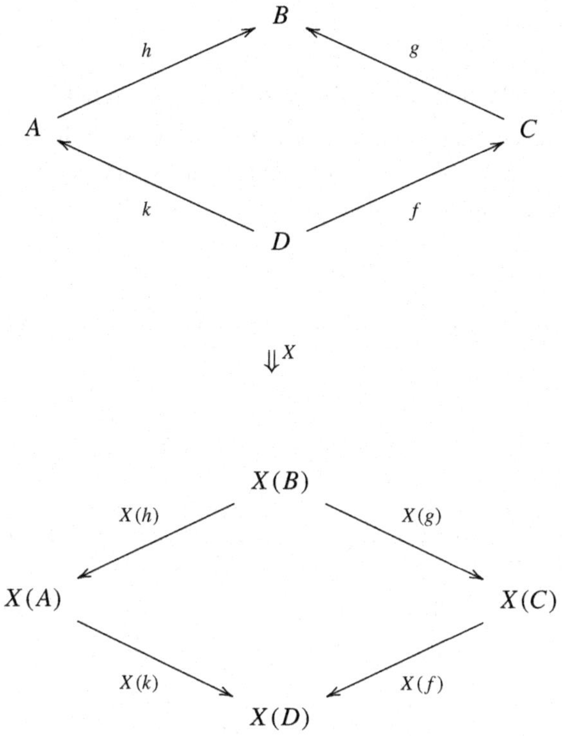

- Morphisms: all natural transformation $N : X \rightarrow Y$ between contravariant functors such that, given a $\mathcal{C}$-arrow $f : D \rightarrow C$, the following diagram commutes

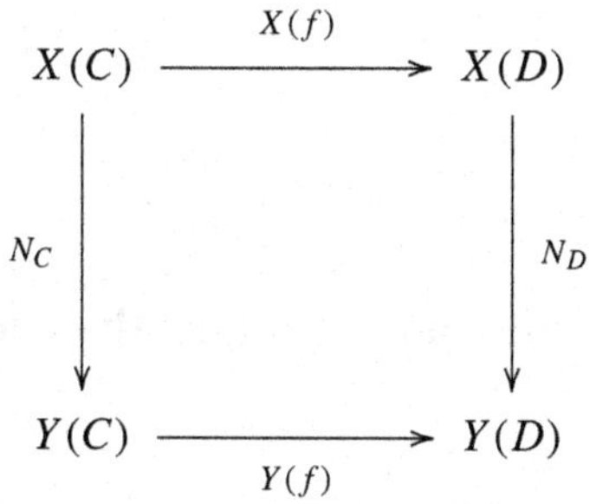

The morphisms satisfy the following conditions:

- *Identity*:

  The identity maps, for each object X in $\mathbf{Sets}^{\mathcal{C}^{op}}$, are identified with natural transformations i_X, whose components i_{X_A} are the identity maps of the set X(A) in **Sets**.

- *Composition*:

  Consider the functors X, Y and Z that belong to $\mathbf{Sets}^{\mathcal{C}^{op}}$, together with natural transformations $X \xrightarrow{N} Y$ and $Y \xrightarrow{M} Z$ between them. We can then form a new map $X \xrightarrow{M \circ N} Y$, whose components would be $(M \circ N)_A = M_A \circ N_A$, i.e. graphically we would have

$$\begin{array}{ccc} X(A) & \xrightarrow{X(f)} & X(B) \\ \downarrow{\scriptstyle N_A} & & \downarrow{\scriptstyle N_B} \\ Y(A) & \xrightarrow{Y(f)} & Y(B) \\ \downarrow{\scriptstyle M_A} & & \downarrow{\scriptstyle M_B} \\ Z(A) & \xrightarrow{Z(f)} & Z(B) \end{array}$$

An intuitive pictorial representation of a presheaf is given in Fig. 6.1, were we have chosen a simple three-object category with only two morphisms and the identity morphisms.

6.3 Basic Categorical Constructs for the Category of Presheaves

In this section we will show how the categorical constructs introduced in Chap. 4 are defined for the category of presheaves.

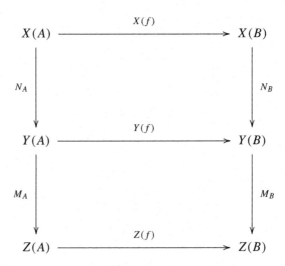

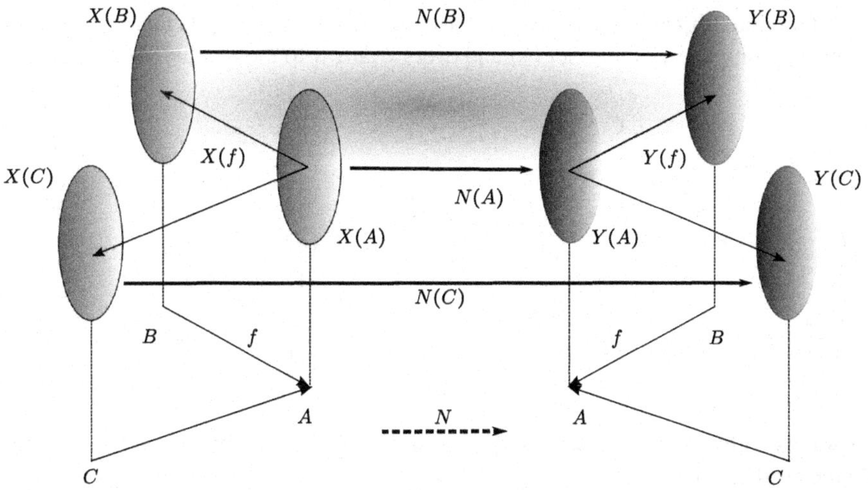

Fig. 6.1 Pictorial representation of the presheaves $X : \mathcal{C} \to \mathbf{Sets}$ and $Y : \mathcal{C} \to \mathbf{Sets}$ where $\mathcal{C}$ has only three objects A, B, C and two arrows $f : B \to A$ and $g : C \to A$. The natural transformation $N : X \to Y$ has components $N(A) : X(A) \to Y(A)$, $N(B) : X(B) \to Y(B)$, $N(C) : X(C) \to Y(C)$. The *shaded area* represents one of the commuting diagrams required for N to be natural

Initial Object An *initial object* in $\mathbf{Sets}^{\mathcal{C}^{op}}$ is the constant functor $0 : \mathcal{C} \to \mathbf{Sets}$ that maps every $\mathcal{C}$-object to the empty set $\emptyset$ and every $\mathcal{C}$-arrow to the identity arrow on $\emptyset$.

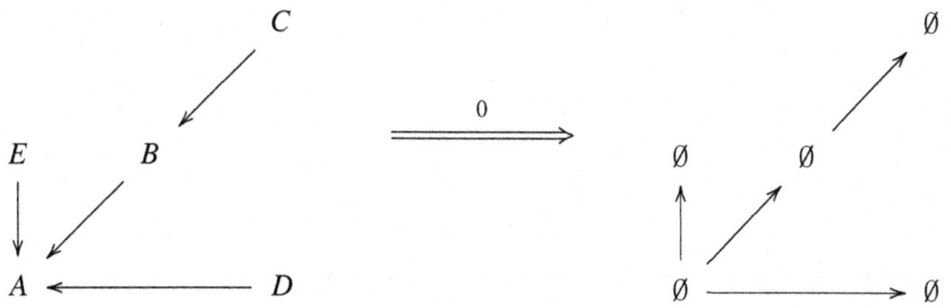

Terminal Object A terminal object in the Topos of presheaves $\mathbf{Sets}^{\mathcal{C}^{op}}$ is given as follows:

A **terminal object** in $\mathbf{Sets}^{\mathcal{C}^{op}}$ is the constant functor $1 : \mathcal{C} \to \mathbf{Sets}$ that maps every $\mathcal{C}$-object to the one-element Set $\{*\}$ and every $\mathcal{C}$-arrow to the identity arrow on $\{*\}$.

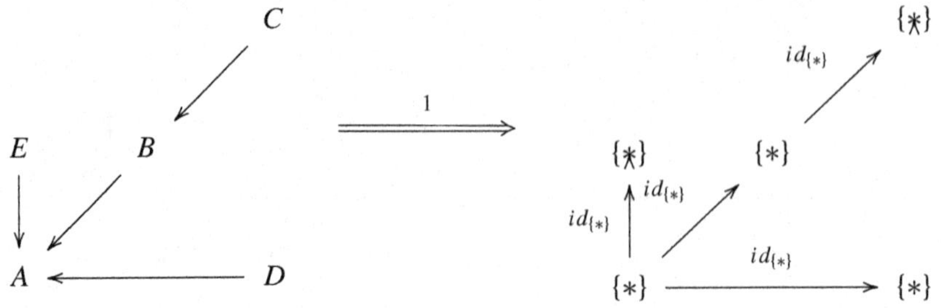

Products Given two presheaves $X, Y \in \mathbf{Sets}^{\mathcal{C}^{op}}$ we define the *product* $X \times Y$ to be the presheaf such that:

(i) for each object $C \in \mathcal{C}$ we obtain the set $(X \times Y)(C) := X(C) \times Y(C)$ which is a product in **Sets**.
(ii) For each $\mathcal{C}$-morphism $f : C \to D$ we obtain the set function $(X \times Y)(f) := X(f) \times Y(f) : X(D) \times Y(D) \to X(C) \times Y(C)$ defined by the diagram

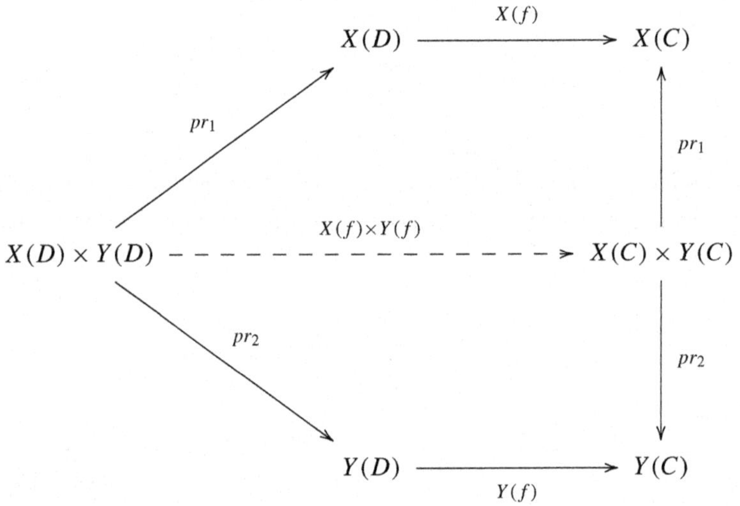

This definition of product has the universal property required of products as defined in Definition 4.17.

It is clear that the product presheaf is a presheaf such that for each object $C \in \mathcal{C}$ we obtain the standard set product.

Coproducts Given two presheaves $X, Y \in \mathbf{Sets}^{\mathcal{C}^{op}}$ we define the *coproduct* $X + Y$ (also denoted as $X \sqcup Y$) to be the presheaf, such that

(i) for each element $C \in \mathcal{C}$ we obtain the set $(X+Y)(C) := X(C)\coprod Y(C)$ which is the disjoint union in **Sets**.
(ii) For each $\mathcal{C}$-morphism $f : C \to D$ we obtain the set function $(X+Y)(f) := X(f)+Y(f) : X(D)\coprod Y(D) \to X(C)\coprod Y(C)$ defined by the diagram

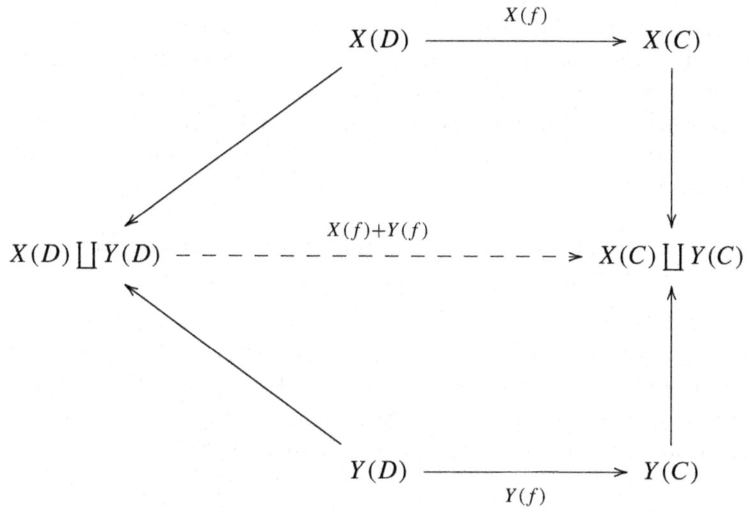

This definition of coproduct satisfies the universality requirements of Definition 4.19.

Similarly, as for the product, the coproduct presheaf is a presheaf such that for each element $C \in \mathcal{C}$ we retrieve the coproduct in **Sets**.

Equaliser Given two natural transformations $N, M : X \to Y$, an equaliser $I : Z \to X$ is a natural transformation such that

(i) $N \circ I = M \circ I$ is satisfied. Thus, for each object $C \in \mathcal{C}$, the components of the natural transformations compose in **Sets**, i.e. $N_C \circ I_C = M_C \circ I_C$.
(ii) Given any other natural transformation $H : W \to X$ which satisfies $N \circ H = M \circ H$, there is a unique natural transformation $K : W \to Z$ such that, for object $C \in \mathcal{C}$ the following diagram commutes in **Sets**

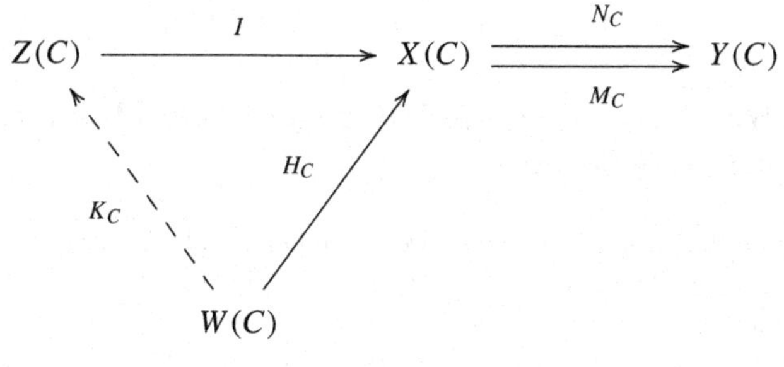

i.e. $I(C) \circ K(C) = H(C)$.

An equaliser in $\mathbf{Sets}^{\mathcal{C}^{op}}$ is a natural transformation such that for each element $C \in \mathcal{C}$ the component of this natural transformation is an equaliser in **Sets**. To see this construct a candidate for an equaliser in $\mathbf{Sets}^{\mathcal{C}^{op}}$ component wise, i.e. for each $C \in \mathcal{C}$, I_C is an equaliser in **Sets**. Therefore, for all $C \in \mathcal{C}$, $N_C \circ I_C = M_C \circ I_C$ which implies that $N \circ I = M \circ I$. On the other hand, for all $C \in \mathcal{C}$, K_C is the unique map which makes the above diagram commute. Since each component K_C for all $C \in c$ is unique it follows that K is unique as a natural transformation. These results put together imply that I is an equaliser in $\mathbf{Sets}^{\mathcal{C}^{op}}$. SInce equalisers are unique up to isomorphism it follows that indeed an equaliser in $\mathbf{Sets}^{\mathcal{C}^{op}}$ is a natural transformation such that for each element $C \in \mathcal{C}$ the component of this natural transformation is an equaliser in **Sets**.

Coequaliser Given two natural transformations $N, M : X \to Y$, a coequaliser $L : Y \to Z$ is a natural transformation such that

(i) $L \circ N = L \circ M$ is satisfied. Thus, for each object $C \in \mathcal{C}$, the components of the natural transformations compose in **Sets**, i.e. $L(C) \circ N(C) = L(C) \circ M(C)$.
(ii) Given any other natural transformation $H : Y \to W$ which satisfies $H \circ N = H \circ N$, there is a unique natural transformation $K : Z \to Y$ such that, for any element $C \in \mathcal{C}$, the following diagram commutes in **Sets**

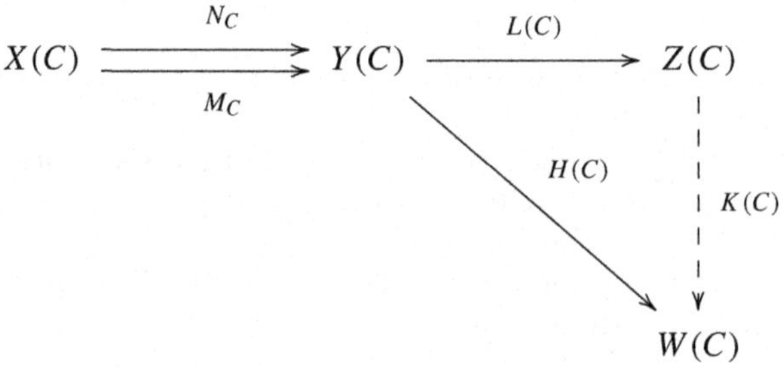

i.e. $I(C) \circ K(C) = H(C)$.

Thus a coequaliser in $\mathbf{Sets}^{\mathcal{C}^{op}}$ is a natural transformation such that, for each element $C \in \mathcal{C}$, the component of this natural transformation is a coequalisers in **Sets**. The proof of this statement is similar to the proof for equalisers in $\mathbf{Sets}^{\mathcal{C}^{op}}$.

6.4 Spectral Presheaf on the Category of Self-adjoint Operators with Discrete Spectra

Depending on the choice of $\mathcal{C}$, there are a number of relevant categories $\mathbf{Sets}^{\mathcal{C}^{op}}$ in quantum theory. Here we will only give one example [14, 15, 26], further examples will be given in Chap. 9.

Consider the category $\mathcal{O}_d$ of self-adjoint operators with discrete spectra. This is a subcategory of the category $\mathcal{O}$ defined in Sect. 4.6.1 of Chap. 4 in which the operators have discrete spectra.[4] If we consider $\mathcal{C}$ to be such a category we obtain the presheaf category $\mathbf{Sets}^{\mathcal{O}_d^{op}}$. This category is important in quantum theory since we can define an object called *spectral presheaf*. The importance of such an object will become clear in Chap. 9 but, for now, it suffices to say that it is a simplified version of the topos analogue of the state space for quantum theory.

Definition 6.2 The **Spectral Presheaf** on $\mathcal{O}_d$, $\Sigma : \mathcal{O}_d \to \mathbf{Sets}$ is defined such that:

1. Objects $\hat{A} \in \mathcal{O}$ get mapped to $\Sigma(\hat{A}) = \sigma(\hat{A})$ where $\sigma(\hat{A})$ is the spectrum of $\hat{A}$.
2. A morphism $f_{\mathcal{O}} : \hat{B} \to \hat{A}$ in $\mathcal{O}_d$, such that $\hat{B} = f(\hat{A})$ ($f : \sigma(\hat{A}) \subseteq \mathbb{R} \to \mathbb{R}$), gets mapped to $\Sigma(f_{\mathcal{O}}) : \Sigma(\hat{A}) \to \Sigma(\hat{B})$, which is equivalent to $\Sigma(f_{\mathcal{O}}) : \sigma(\hat{A}) \to \sigma(\hat{B})$ and is defined by $\Sigma(f_{\mathcal{O}})(\alpha) := f(\alpha)$ for all $\alpha \in \sigma(\hat{A})$.

In order to prove that Σ, as defined above, is indeed a presheaf, we need to prove that, given any function $g_{\mathcal{O}} : \hat{C} \to \hat{B}$ such that $\hat{C} = g(\hat{B})$, then the following equation is satisfied:

$$\Sigma(g_{\mathcal{O}}) \circ \Sigma(f_{\mathcal{O}}) = \Sigma(f_{\mathcal{O}} \circ g_{\mathcal{O}}).$$

Proof If we consider the composite function $f_{\mathcal{O}} \circ g_{\mathcal{O}} = h_{\mathcal{O}} : \hat{C} \to \hat{A}$ from the definition of Σ we have

$$\begin{aligned} \Sigma(f_{\mathcal{O}}) &: \sigma(\hat{A}) \to \sigma(\hat{B}) \\ \Sigma(g_{\mathcal{O}}) &: \sigma(\hat{B}) \to \sigma(\hat{C}) \\ \Sigma(f_{\mathcal{O}} \circ g_{\mathcal{O}}) = \Sigma(h_{\mathcal{O}}) &: \sigma(\hat{A}) \to \sigma(\hat{C}) \end{aligned}$$

therefore

$$\begin{aligned} \Sigma(f_{\mathcal{O}} \circ g_{\mathcal{O}})(\alpha) &= \Sigma(h_{\mathcal{O}})(\alpha) \\ &= h(\alpha) \\ &= (g \circ f)(\alpha) \\ &= g \circ \big(\Sigma(f_{\mathcal{O}})(\alpha)\big) \\ &= \big[\Sigma(g_{\mathcal{O}}) \circ \Sigma(f_{\mathcal{O}})\big](\alpha). \end{aligned}$$

□

Example 6.1 Let us consider a simple category whose objets are defined by

$$\begin{aligned} \hat{A} &= a_1 P_1 + a_2 P_2 + a_3 P_3 \\ \hat{B} &= b_1 (P_1 \vee P_3) + b_2 P_2 \end{aligned}$$

---

[4]The condition of the spectrum being discrete implies that given a Borel function $f : \sigma(\hat{A}) \subseteq \mathbb{R} \to \mathbb{R}$, then $\sigma(f(\hat{A})) = f(\sigma(\hat{A}))$.

$$\hat{C} = c_1 P_3 + c_2(P_1 \vee P_2),$$

where the $\hat{P}_i$ are mutually orthogonal. Clearly such a category is a subcategory of $\mathcal{O}_d$. This can be represented in the following diagram:

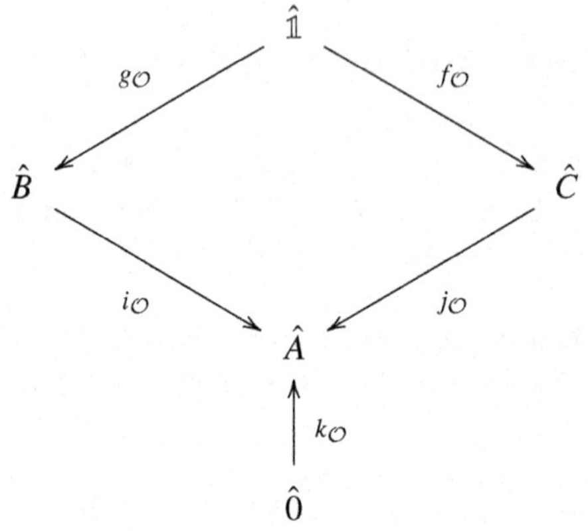

The action of the presheaf Σ is as follows:

$$\sigma(\hat{A}) = \{a_1, a_2, a_3\}$$
$$\sigma(\hat{B}) = \{b_1, b_2\}$$
$$\sigma(\hat{C}) = \{c_1, c_2\}.$$

From Definition 6.2 it follows, for example, that the map $j_{\mathcal{O}} : \hat{C} \to \hat{A}$ gets mapped to $\Sigma(j_{\mathcal{O}}) : \sigma(\hat{A}) \to \sigma(\hat{C})$ such that, component-wise, we get the following mapping:

$$\Sigma(j_{\mathcal{O}})(a_3) = c_1$$
$$\Sigma(j_{\mathcal{O}})(a_1) = c_2$$
$$\Sigma(j_{\mathcal{O}})(a_2) = c_2.$$

Chapter 7
Topos

The very handwavy definition of a topos is that of a category with extra properties. The implications of these extra properties are that they make a topos "look like" **Sets**, in the sense that many mathematical operations which can be done in set theory can be done in a general topos.[1]

In the previous chapter we gave an account (not complete) on how set theoretical structures can be given an external characterisation through category theory. Although it is true that all of the set theoretic constructions can be defined in a categorical language, however, it is not true that all categories have all these set theoretical constructions.

A topos, on the other hand, is a category for which all the categorical versions of set constructs exist and are well-defined. It is precisely in this sense that a topos "looks like" **Sets**.

Before giving the axiomatic definition of what a topos is we, first of all, need to define certain extra constructs of category theory, which are required to be present for a given category to be a topos.

7.1 Exponentials

Given two sets A and B, let us imagine we would like to consider the totality of all arrows between then as an object in itself, i.e., we would then define the set $\{f_i : A \to B\}$. An important property of the set $\{f_i : A \to B\}$ is that

$$\text{if} \quad A, B \in \textbf{Sets} \quad \text{then} \quad \{f_i : A \to B\} \in \textbf{Sets}. \tag{7.1}$$

We will call this object (which in this particular situation is the set $\{f_i : A \to B\}$) an *exponential* and we will denoted it as B^A.

[1] As mentioned in previous chapters, **Sets** is a topos, albeit a very special one.

C. Flori, *A First Course in Topos Quantum Theory*, Lecture Notes in Physics 868,
DOI 10.1007/978-3-642-35713-8_7, © Springer-Verlag Berlin Heidelberg 2013

We now would like to abstract the characterisation of such an object for a general category. Thus we would like to define an object B^A with the properties (i) if $A \in \mathcal{C}$ and $B \in \mathcal{C}$ then $B^A \in \mathcal{C}$; (ii) it represents a 'certain relation' between A and B.

Since in categorical language, objects are defined according to the relations with other objects, we will define B^A in terms of what it 'does' operationally. To this end, let us consider all three objects involved: A, B, B^A. In the case of the category of **Sets**, a possible relation between them can be defined as follows:

$$\begin{aligned} ev : B^A \times A &\to B \\ (f_i, a) &\mapsto ev(f_i, a) := f_i(a). \end{aligned} \tag{7.2}$$

This definition seems very plausible, however, for it to make sense in the categorical world it has to be *universal*, in the sense that any other arrow $g : C \times A \to B$ will factor through ev in a unique way. Thus, in **Sets**, there exists a unique arrow $\hat{g} : C \to B^A$, which makes the following diagram commute:

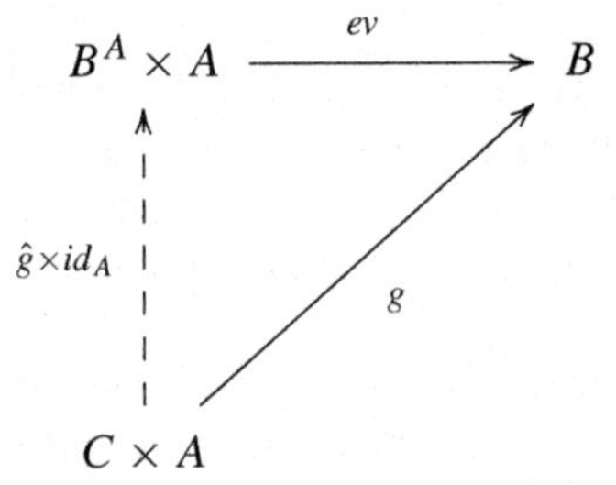

where

$$\begin{aligned} \hat{g} : C &\to B^A \\ c &\mapsto \hat{g}(c) := g(c, -) \end{aligned} \tag{7.3}$$

such that

$$\begin{aligned} g(c, -) : A &\to B \\ a &\mapsto g(c, a). \end{aligned} \tag{7.4}$$

Therefore the map $\hat{g}$ assigns to any $c \in C$ a function $A \to B$ by taking g and keeping the first term fixed at c, while ranging over the elements of A, so for each $c \in C$ we have that $\hat{g}(c) = g(c, -)$.

We are now ready to give the abstract categorical definition of what an exponential is.

Definition 7.1 Given two $\mathcal{C}$-objects A and B, their *exponentiation* is a $\mathcal{C}$-object B^A together with an *evaluation map* $ev : B^A \times A \to B$ with the property that, given any other $\mathcal{C}$-object C and $\mathcal{C}$-arrow $g : C \times A \to B$, there exists a unique arrow

$\hat{g} : C \to B^A$, such that the following diagram commutes:

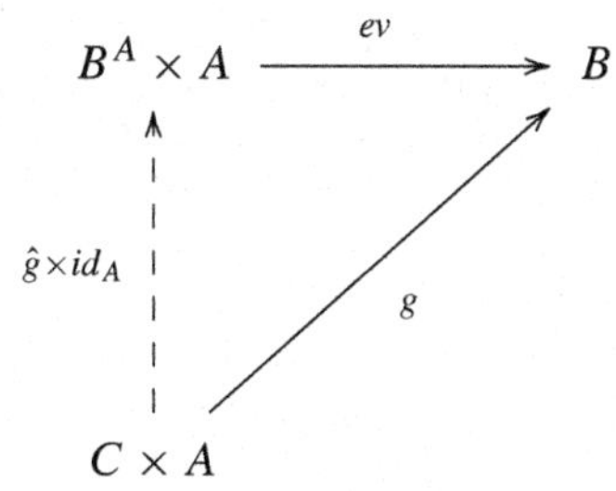

therefore for any $(c, a) \in C \times A$ we get that

$$ev \circ (\hat{g} \times id_A)(c, a) = ev\big(\hat{g}(c), a\big) = g(c, a). \tag{7.5}$$

Now that we have abstractly defined what an exponentiation is, we would like to know what its elements are. Remember that we started with the example in **Sets** and in that case we knew that the objects in B^A where maps $\{f_i : A \to B\}$. We then abstracted from this particular example and defined a general notion of an exponential in terms of the universal property of the exponential map. However, that definition did not rely on what type of elements B^A had, if any. We now would like to go back full circle and see what we can say about the elements of B^A. We know from previous Chapters that an element in any object C is identified with a map $1 \to C$, where 1 is the terminal object defined in Sect. 4.5.2. This correspondence will be used in the following definition.

Definition 7.2 Elements of B^A are in one-to-one correspondence with maps of the form $f : A \to B$.

To see this let us consider the following commuting diagram:

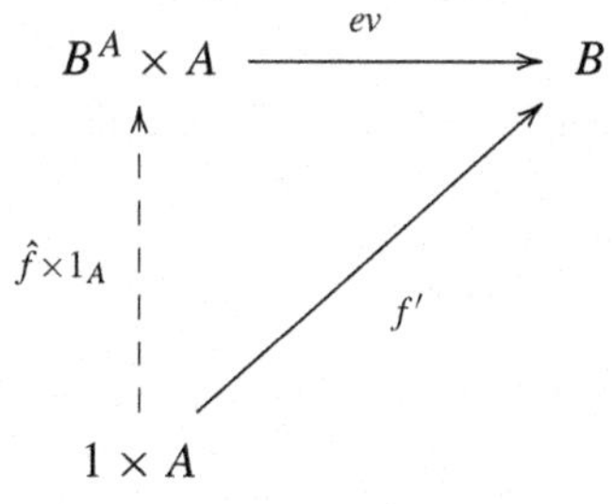

where $f \hat{f} : 1 \to B^A B$ is unique given f. However $1 \times A \equiv A$, therefore to each element of B^A there corresponds a unique function $f : A \to B$. The above definition of *exponentials* allows us to define a subclass of categories called *cartesian closed categories*. The precise definition is as follows:

Definition 7.3 A category C is said to be a cartesian closed category (CCC) if it has a terminal object, products and exponentials.

Examples of CCC are **Sets** and Boolean algebras, seen as categories.

For those who are interested in logic, a Boolean algebra turns out to be a CCC as follows:

- Products are given by conjunctions

$$A \wedge B. \tag{7.6}$$

- Exponentials are implications

$$A \Rightarrow B := \neg A \vee B. \tag{7.7}$$

- Evaluation is Modus Ponens,

$$(A \Rightarrow B) \wedge A \leq B. \tag{7.8}$$

- Universality is the Deduction Theorem,

$$C \wedge A \leq B \quad \Leftrightarrow \quad C \leq A \Rightarrow B. \tag{7.9}$$

We will now give other examples of cartesian closed categories and how the exponentials, in each of them, are formed.

Example 7.1 In **Sets**, given two objects A and B, the exponential B^A is defined as follows:

$$B^A = \{f \mid f \text{ is a function from } A \text{ to } B\}. \tag{7.10}$$

In this case the evaluation map would be the following:

$$ev(\langle f, x\rangle) = f(x) \quad \text{for all } x \in A.$$

Example 7.2 Consider the category **Finord** of all finite ordinals. In this a category, the objects are the numbers $0, 1, \dots, n, \dots$ where

$$0 = \emptyset \tag{7.11}$$

$$1 = \{0\} \tag{7.12}$$

$$2 = \{0, 1\} \tag{7.13}$$

$$3 = \{0, 1, 2\} \tag{7.14}$$

$$\vdots \tag{7.15}$$

$$n = \{1, 2, \dots, n-1\}. \tag{7.16}$$

The maps are then simply the set functions between these ordinals. Given two such objects n and m, then the exponential would be n^m where m is the exponent. Such an element should be considered as a finite ordinal with n^m elements in it.

It is possible to dualise the notion of an exponential obtaining the notion of a *co-exponential*. This is defined as follows:

Definition 7.4 Given two $\mathcal{C}$-objects A and B, their co-exponential is a $\mathcal{C}$-object D together with a map $coev : B \to D + A$, such that for any other $\mathcal{C}$-object C and any map $f : B \to C + A$ there is a unique map $\tilde{f} : D \to C$, such that the following diagram commutes:

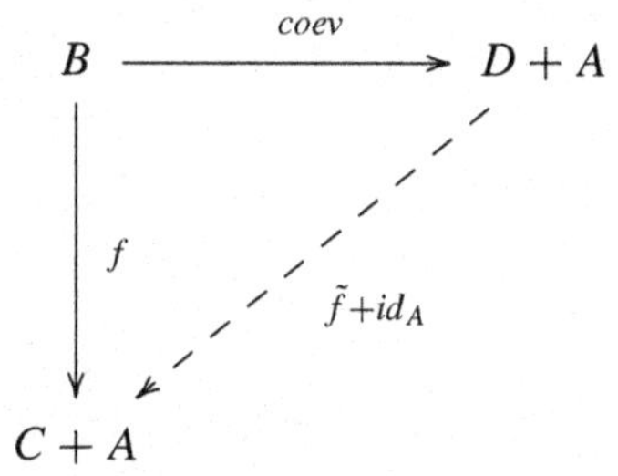

i.e. $f = (\tilde{f} + idA) \circ coev$. Thus the maps $B \to C + A$ are in bijective correspondence with map $D \to C$.

Unfortunately co-exponentials do not exist in **Sets** and they are not required to exist in a general topos, hence we will not analyse them further.

7.2 Pullback

We will now define another construction which is present in any topos. This is the notion of pullback or fiber product.

Definition 7.5 A **pullback** or **fiber product** of a pair of functions $f : A \to C$ and $g : B \to C$ (with common codomain) in a category $\mathcal{C}$ is a pair of $\mathcal{C}$-arrows $h : D \to A$ and $k : D \to B$, such that the following conditions are satisfied:

1. $f \circ h = g \circ k$, i.e. the following diagram commutes:[2]

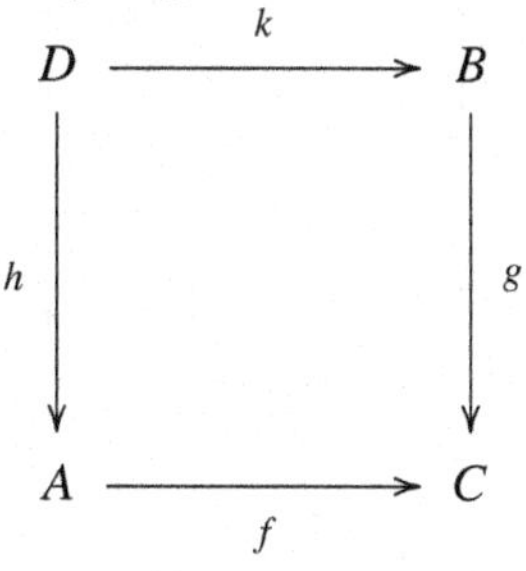

[2]One usually writes $D = A \times_C B$.

2. Given two functions $i : E \to A$ and $j : E \to B$, with $f \circ i = g \circ j$, then there exists a unique $\mathcal{C}$-arrow l from E to D such that the following diagram commutes:

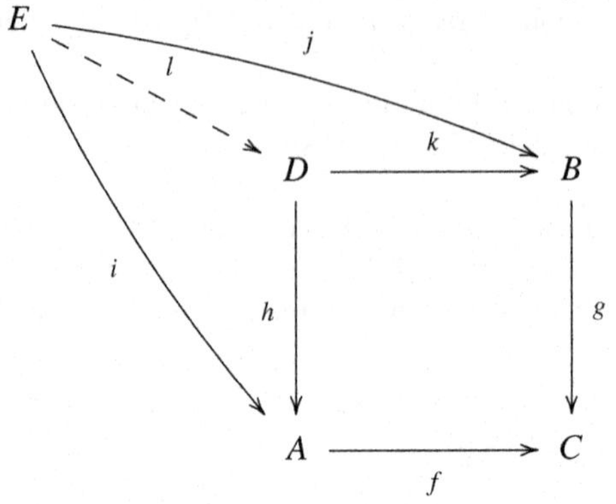

i.e.

$$i = h \circ l, \qquad j = k \circ l.$$

We then say that f (respectively g) has been pulled back along g (respectively f).

A pullback of a pair of $\mathcal{C}$-arrows $A \xrightarrow{f} C \xleftarrow{g} B$ is a limit of the diagram

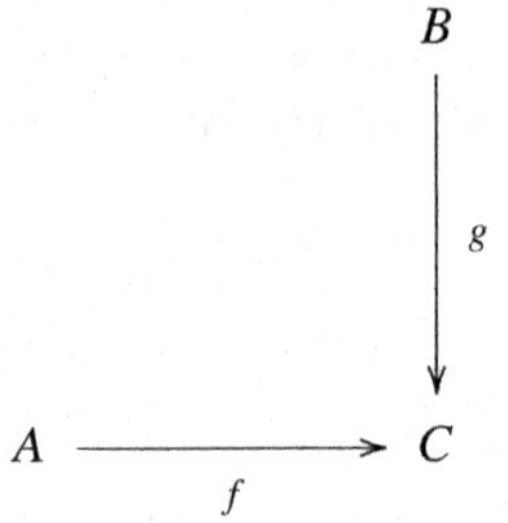

In fact, a cone[3] for this diagram is a pair of $\mathcal{C}$-arrows $A \xleftarrow{i} D \xrightarrow{h} B$ which compose with f (respectively with g) to give the commutative diagram

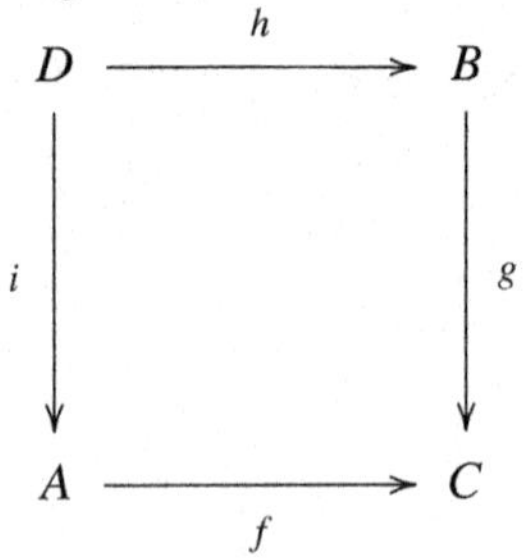

[3]Strictly speaking one has two cones $C \xleftarrow{j} D \xrightarrow{i} A$ and $C \xleftarrow{j} D \xrightarrow{h} B$, but when composing the diagrams it turns out that $j = g \circ h = f \circ i$, thus we can omit the j map.

A limiting cone is such that given any other cone $A \xleftarrow{k} E \xrightarrow{j} B$ it factors uniquely through it, i.e. it gives rise to the following commuting diagram:

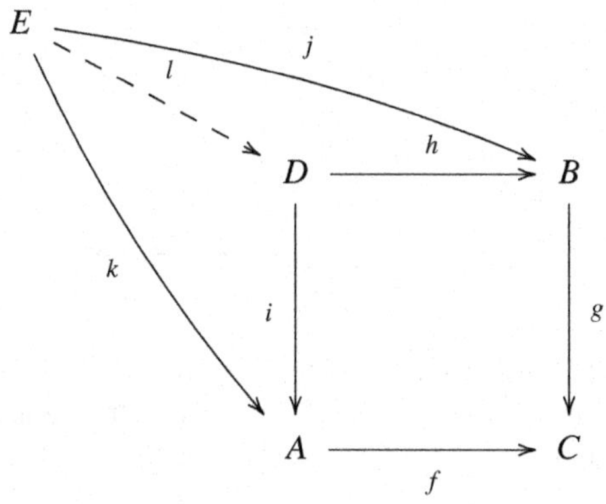

Example 7.3 If A, C, D and B are sets, then

$$D = A \times_C B = \{(a, b) \in A \times B | f(a) = g(b)\} \subseteq A \times B \tag{7.17}$$

with maps

$$\begin{aligned} h : A \times_C B &\to B \\ (a, b) &\mapsto b \end{aligned} \tag{7.18}$$

and

$$\begin{aligned} i : A \times_C B &\to A \\ (a, b) &\mapsto a \end{aligned} \tag{7.19}$$

satisfies the conditions of being a pullback.

Proof Given a set E with maps $j : E \to B$ and $i : E \to A$, then the map $k : E \to A \times_C B; e \longmapsto (k(e), j(e))$ makes the diagram

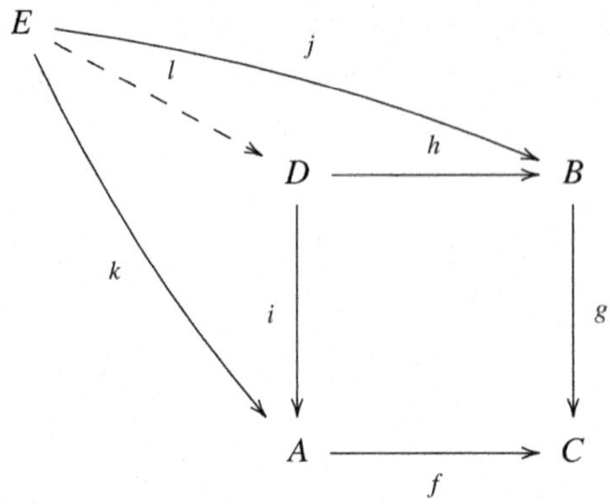

commute. In fact we have the following identities for all $e \in E$, $i \circ l(e) = k(e)$ and $h \circ l(e) = j(e)$. Moreover l is unique since, given any other map $m : E \to A \times_C B$, such that $i \circ m = k$ and $h \circ m = j$, then for all $e \in E$ the following holds:

$$l(e) = \big(k(e), j(e)\big) = \big(i\big(m(e)\big), h\big(m(e)\big)\big) = m(e). \tag{7.20}$$

Therefore l is unique. □

7.3 Pushouts

As usual, any notation in category theory has a dual, thus we will now define the dual of a pullback which is a push out.

Definition 7.6 A **Pushout** or **fiber coproduct** of a pair of arrows $f : A \to B$ and $g : A \to C$ in a category $\mathcal{C}$ is a pair of $\mathcal{C}$-arrows $h : B \to D$ and $k : C \to D$, such that the following conditions are satisfied:

1. $h \circ f = k \circ g$, i.e. the following diagram commutes:[4]

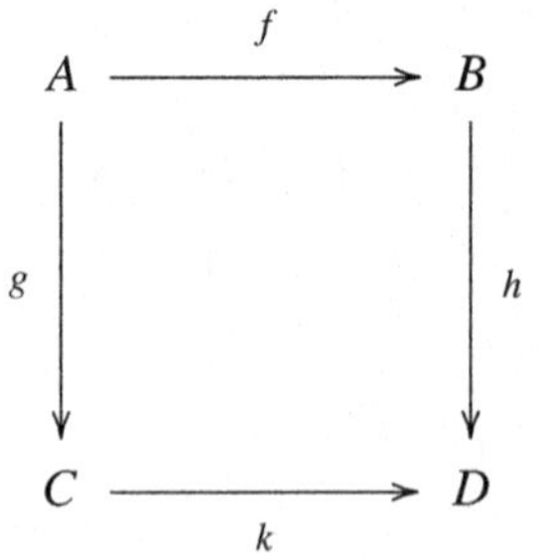

[4]One usually writes $D = C +_A B$.

2. Given two arrows $i : B \to E$ and $j : C \to E$, where $i \circ f = j \circ g$, then there exists a unique $\mathcal{C}$-arrow l from D to E, such that the following diagram commutes

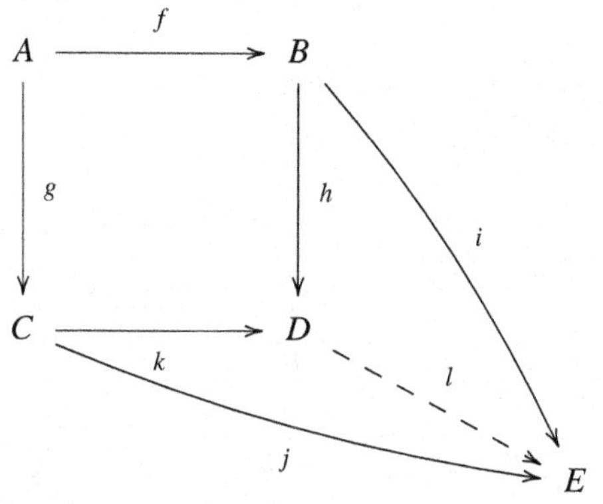

i.e.

$$i = l \circ h, \qquad j = l \circ k.$$

We then say that f (respectively g) has been pushed out along g (respectively f).

Example 7.4 In **Sets**, given three sets A, B, C, the set $D = C +_A B$ always exists. This is constructed by first defining the disjoint union of C and B, i.e.

$$D' = C \amalg B := \left\{ (x, t) \in (C \cup B) \times \{0, 1\} \middle| \begin{cases} x \in B & \text{if } t = 0 \\ x \in C & \text{if } t = 1 \end{cases} \right\}$$

where, in this case, the arrows h' and k' are defined as follows:

$$\begin{aligned} h' : B &\to C \amalg B \\ b &\mapsto (b, 0) \end{aligned} \tag{7.21}$$

and

$$\begin{aligned} k' : C &\to C \amalg B \\ c &\mapsto (c, 1). \end{aligned} \tag{7.22}$$

We then define an equivalence relation on D' as the smallest equivalence relation such that $(f(a), 0) \sim (g(a), 1)$. The pushout is then $D := D'/\sim$.

Proof Given a set E and two maps $j : C \to E$, $i : B \to E$ we define the map $l : D \to E$, such that $[(c, 1)] \longmapsto j(c)$ and $[(b, 0)] \longmapsto i(b)$. We first need to show that such a definition does not depend on the representative chosen in the equivalence class. It suffice to show that $l[(f(a), 0)] = l[(g(a), 1)]$. Applying the definition we get $l[(f(a),0)] = i(f(a))$ while $l[(g(a),1)] = j(g(a))$ because of commutativity it follows that $i(f(a)) = j(g(a))$, thus l is well-defined.

It is then easy to see that the diagram

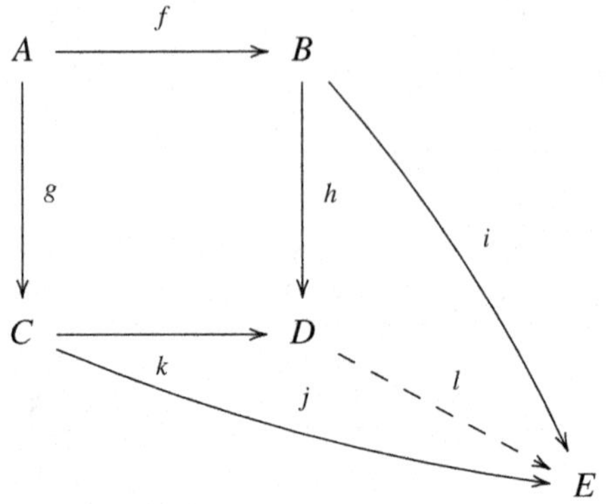

commutes. In fact we have the following:

$$(l \circ h)(b) = l\big[(b,0)\big] = i(b) \tag{7.23}$$

and

$$(l \circ k)(b) = l\big[(c,1)\big] = j(c). \tag{7.24}$$

The second step in the proof is showing that the map l is unique. In fact, given another map $m : D \to E$, such that $m \circ h = i$ and $m \circ k = j$, then we would have the following equality:

$$l\big[(b,0)\big] = i(b) = m\big(h(b)\big) = m\big[(b,0)\big] \tag{7.25}$$

and

$$l\big[(c,1)\big] = j(c) = m\big(k(c)\big) = m\big[(c,1)\big]. \tag{7.26}$$

This shows that l is unique. □

7.4 Sub-objects

The concept of sub-objects is the topos analogue of the notion of sub-sets of a set.

Everyone is familiar with the notion of a subset in the category **Sets**. We now would like to generalise this notion and describe it in categorical language, i.e. in terms of relation.

Let us start with **Sets** and see how much we can abstract from the already known definition of a subset. Consider two sets A and B such that $A \subseteq B$. This means that there is an inclusion map $f : A \hookrightarrow B$. In categorical language, the map f is monic (recall discussion in Sect. 4.4.1). On the other hand, if we consider a monic arrow

$f : A \rightarrowtail B$, this would determine a subset of B, namely $Im\, f := \{f(x)|x \in A\}$. Thus $Im\, f \subseteq B$ and $Im\, f \simeq A$.

What this means is that the domain of a monic arrow is isomorphic to a subset of the codomain of the arrow, i.e. up to isomorphism the domain of a monic arrow is a subset of the codomain.

As you might have noticed, going from a subset of a set to the set itself requires a change of type, namely, going from the domain object to the codomain object. This implies that in the categorical version of subset, there is no feasible way to say that the same element x is in both a set and a subset of the set. All that we can say is that two elements are isomorphic. This might seem striking at first since we are used to thinking in **Sets**-language terms, where objects are determined/defined by the elements which comprise them. Thus, saying that two sets are the same means saying that the elements which compose them are the same. However, in categorical language "the same" now becomes "are isomorphic" since everything is defined in terms of relations, i.e. arrows between objects.

So, tentatively we will define the following:

Preliminary Definition 7.1 A sub-object of a $\mathcal{C}$ object B is an arrow in $\mathcal{C}$ which is monic and which has codomain B.

However, this is not the end of the story, since we are not taking into account that some objects might be isomorphic, which in categorical language means the same, thus we should only really consider them once. Therefore we would like our definition of a sub-object to take into account the existence of equivalent objects. In order to define the notion of equivalent objects we first need the following definition:

Definition 7.7 Given an arrow $f : A \to B$ in some category $\mathcal{C}$, if for some arrow $g : C \to B$ in $\mathcal{C}$ there exists another arrow $h : A \to C$ in $\mathcal{C}$ such that $f = g \circ h$, then we say that f factors through g, since $f = g \circ h$.

Diagrammatically, what factorising means is that the following diagram commutes:

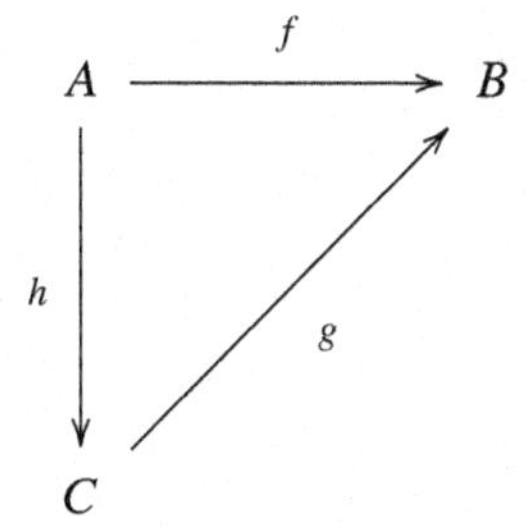

We now consider two sub-objects (according to our preliminary Definition 7.1) $g : C \hookrightarrow B$ and $f : A \hookrightarrow B$, such that they both factor through each other, i.e., the following diagram commutes:

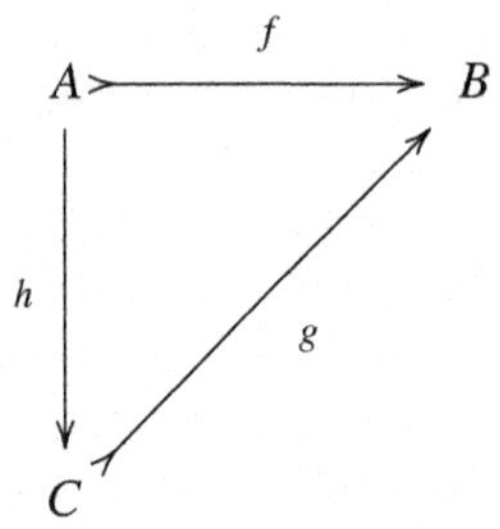

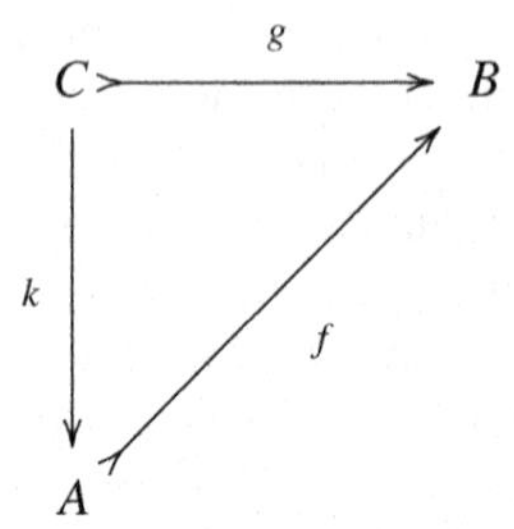

Therefore $g \circ h = f$ and $f \circ k = g$. Substituting for f we obtain $g \circ h \circ k = g \circ id_C$. Since g is monic $h \circ k = id_C$. On the other hand, substituting for g we obtain $f \circ k \circ h = f \circ id_A$. Since f is monic, $k \circ h = id_A$. It follows that k and h are inverse of each other thus iso. These sub-objects (monic arrows) are considered to be equivalent. This leads to the definition of the following equivalence relation.

Definition 7.8 Given two monic arrows f, g with the same codomain we say that they are equivalent $f \sim g$ iff they factor through each other.

Given the above we are now ready to define the categorical version of a sub-object.

Definition 7.9 In a category $\mathcal{C}$, a sub-object of any object in $\mathcal{C}$ is an equivalence class of monic arrows under the equivalence relation $\sim$ defined in Definition 7.8. We will denote such an equivalence class as $[f]$. The sub-object is a proper sub-object if it does not contain id_C.

This definition of sub-objects allows us to define the collection of all sub-objects of a given object as a poset under sub-object inclusion. In particular, we know from set theory that the collection of all subsets of a given set X is a poset, such that there is an arrow between any two such subsets A and B iff $A \subseteq B$. Thus, diagrammatically, we have

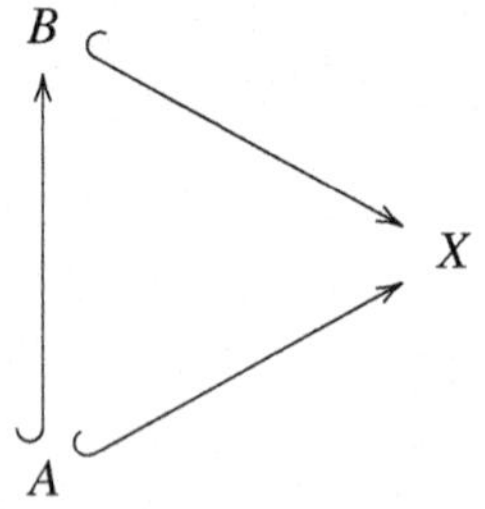

Therefore, abstracting such a definition to categorical language, we say that two monic arrows $f : A \rightarrowtail B$ and $g : C \rightarrowtail B$ are such that $f \subseteq g$ iff there exists an h

such that the following diagram commutes:

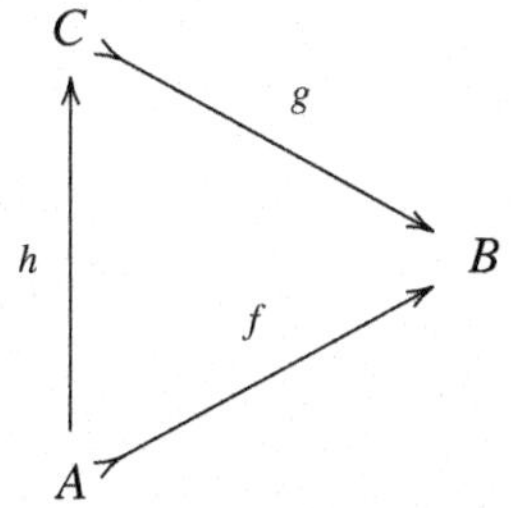

i.e. $f = g \circ h$ (h is automatically monic).[5]

If we now consider all sub-objects of a given object B

$$Sub(B) := \{[f] | f \text{ is monic and } \operatorname{cod} f = B\} \quad \text{where } [f] = \{g | f \sim g\} \tag{7.28}$$

then such a collection of sub-objects forms a poset under subset inclusion defined by

$$[f] \subseteq [g] \quad \text{iff} \quad f \subseteq g. \tag{7.29}$$

Proof • Well-defined: we need to show that the above definition does not depend on the choice of representative of the equivalence class. Consider $f' \in [f]$ and $g' \in [g]$ such that $f' \subseteq g'$ therefore there exists an h' such that $f' = g' \circ h'$. Since $f' \sim f$ and $g' \sim g$, then there exists and iso j such that $f' = f \circ j$ and $f = f' \circ j^{-1}$. Similarly for g' and g' there is an iso i. Substituting for f' and g' we obtain $f' = f \circ j = g' \circ h' = g \circ i \circ h'$. It follows that $f = g \circ i \circ h' \circ j^{-1}$ and therefore $f \subseteq g$.
• Reflexive: $[f] \subseteq [f]$ implies that $f \subseteq f$. The latter is satisfied since

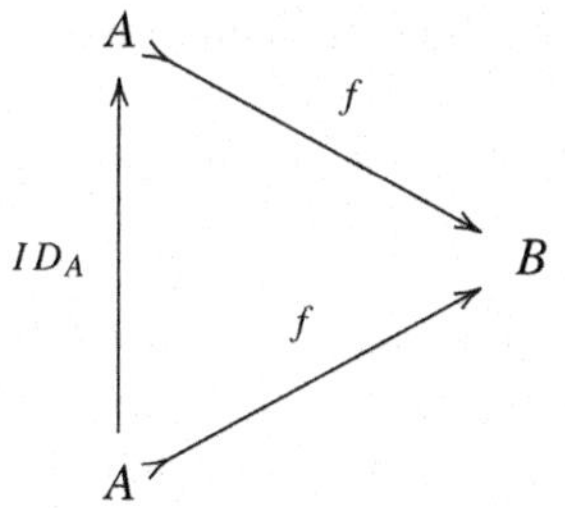

[5]Given $h : A \to C$ and $k', k : D \to A$, with $h \circ k = h \circ k'$. Then consider $g : C \to B$ such that $(g \circ h) = f$ is monic. It follows that:

$$(g \circ h) \circ k = g \circ (h \circ k) = g \circ (h \circ k') = (g \circ h) \circ k' \quad \text{therefore} \quad k = k'. \tag{7.27}$$

• Transitive: if $[f] \subseteq [g]$ and $[g] \subseteq [h]$, then $[f] \subseteq [h]$. The fact that $[f] \subseteq [g]$ and $[g] \subseteq [h]$ implies that there are i and k for which the diagram

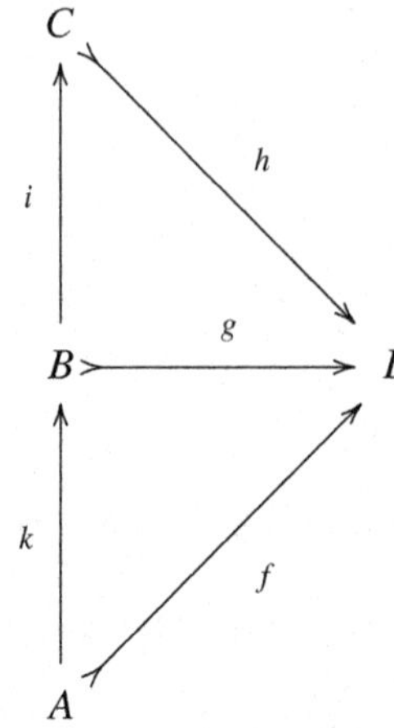

commutes, where $f \in [f]$, $g \in [g]$ and $h \in [h]$. If $g \circ k = f$ and $g = h \circ i$ then it follows that $f = (h\circ)i \circ k$ thus $f \subseteq h$ and $[f] \subseteq [h]$.

• Antisymmetric: if $[f] \subseteq [g]$ and $[g] \subseteq [f]$ then $f \subseteq g$ and $g \subseteq f$. This implies that $f \sim g$ thus $[f] = [g]$.

If we had not considered equivalence classes then $f \subseteq g$ and $g \subseteq f$ would only imply $f \sim g$ but not $f = g$. Which would be satisfied only if the arrow allowing the factorisation of f via g (or the other way round) would be the identity arrow. □

From now on, when we talk about sub-objects it will be implicit that we are referring to equivalence classes of sub-objects, even though it is not explicitly stated.

It is interesting to note that sub-objects, given by a categorical definition, are not the same as subsets, but each subset determines and is determined by a unique sub-object. In fact we have the following definition of a sub-object in **Sets**:

Definition 7.10 In **Sets** a sub-object is an equivalence class of injections (set theoretic equivalence of monic map).

The 1:1 (one to one) correspondence between sub-objects and sets in set theory is given by the statements in Lemmas 7.1 and 7.2.

In particular, Lemma 7.1 shows that given a sub-object O of a set S there corresponds a unique subset I of S. On the other hand Lemma 7.2 shows the reverse, to each subset $I \subseteq S$ there corresponds a unique sub-object O of S.

Lemma 7.1 *Given a set S and a sub-object O (equivalence class of injective maps with codomain S) we have the following*:

(a) *any two injections $f : A \to S$ and $g : B \to S$ which are in O (i.e. they are equivalent) have the same image I in S.*

(b) *The inclusion $i : I \to S$ is equivalent to any injection in O, thus it is an element of O.*
(c) *If $j : J \to S$ is an inclusion of a subset J into S that is in O, then $I = J$ and $i = j$.*
(d) *It follows, from the above statements, that every sub-object O of S contains one and only one injective map which represents the inclusion of a subset of S into S. This subset is the image of any element of O.*

Lemma 7.2 *Given an inclusion map $i : I \to S$ of a subset I into S, we have that:*

1. *i is injective, thus it is an element of some sub-object O of S.*
2. *Any two distinct equivalence classes are disjoint, thus i can not belong to two different sub-objects.*
3. *Hence each subsets of S (and their inclusion maps) defined a unique sub-object of S.*

Proof We now prove both of the above lemmas.

(a) $\mathcal{O} := [f_i]$, $f_i : A \to S$ where $f_1 \sim f_2$ iff there exists g and h making the following diagram commute.

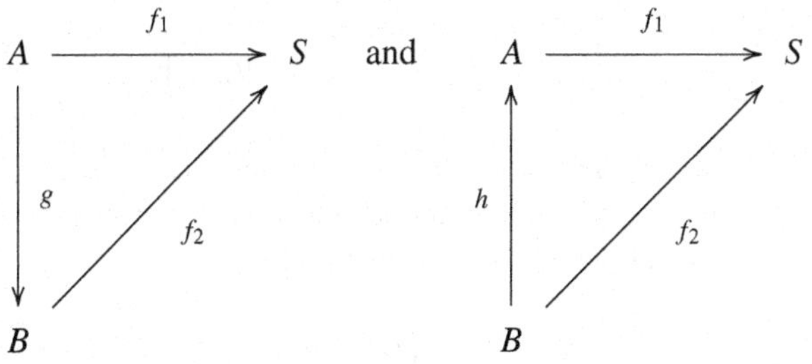

Since $f_1 \circ h = f_2$, then $im(f_1 \circ h) = im(f_2)$ and f_1 and f_2 are monic then h is monic, which in **Sets** means it is injective, thus $im(h) \subseteq A$. It follows that $im(f_2) \subseteq im(f_1)$. On the other hand, $f_2 \circ g = f_1$, thus $im(f_2 \circ g) = im(f_1)$ and g is monic. Therefore $im(g) \subseteq$ B implying that $im(f_1) \subseteq im(f_2)$. Thus $im(f_1) = im(f_2)$.
(c) $j : J \to S$ is an inclusion in $\mathcal{O}$, so an injection, i.e. an element. Each element in $\mathcal{O}$ has the same image, therefore $im(j) = I$. However $im(j) = J$ since $J \subseteq S$ thus $J = I$ and $j = i$.
(b), (d) The proof of statements (b) and (d) is seld evident given (a) and (c). □

The distinction between the categorical definition of a sub-object, as an equivalence class of monic arrows and the standard definition of a subset in set theory is quite important at a conceptual level. However, from an operational point of view, they are equivalent. To understand this consider the integers $\mathbb{Z}$. In standard set theory they are simply a subset of the reals $\mathbb{R}$, such that each integer is actually a real. On the other hand in category theory sub-object relations only require the existence (or definition) of a monic map between $\mathbb{Z}$ and $\mathbb{R}$. Thus, in this case, an integer needs not to be a real number, it could, in principle, be something different. Reiterating, in standard set theory, the image of each integer is represented by the same integer, so that the integers are a subset of the real numbers. From the categorical point of view, instead, this condition can be relaxed, and all that is needed is that there exists a monic map $m : \mathbb{Z} \to \mathbb{R}$. If one wants to recover the standard definition in which such monic arrow picks out the same integer, then an extra condition has to be placed, namely, that m is an equivalent monic, i.e. picks out the same sub-object. However, from an operational point of view the integers, as defined by a monic map $\mathbb{Z} \to \mathbb{R}$ behave in the same way, whether this extra condition is taken into consideration or not.

7.5 Sub-object Classifier (Truth Object)

Now that we have defined what a sub-object is, we would like to understand how to identify sub-objects. To this end, let us consider a specific example in **Sets**. Here we have the following isomorphism which we will prove later:

$$Sub(S) := 2^S \simeq \{S \to 2 = \{0, 1\}\}. \tag{7.30}$$

To understand the above expression consider a subset A of S, i.e. $A \subseteq S$. The notion of being a subset can be expressed mathematically using the so called characteristic function: $\chi_A : S \to \{0, 1\}$, which is defined as follows:

$$\chi_A(x) = \begin{cases} 0 & \text{if } x \notin A \\ 1 & \text{if } x \in A \end{cases} \tag{7.31}$$

(here we interpret $1 = \textit{true}$ and $0 = \textit{false}$). The role of the characteristic function is to determine which elements belong to a certain subset.

Remembering that in any category sub-objects are identified as monic arrows, we define the value true in terms of the following monic map:

$$\begin{aligned} \textit{true} : 1 = \{0\} &\to 2 = \{0, 1\} \\ 0 &\to 1. \end{aligned} \tag{7.32}$$

Given the definition (7.31), it can be easily seen that

$$A = \{x | x \in S \text{ and } \chi_A(x) = 1\} = \chi_A^{-1}(1). \tag{7.33}$$

This equation is equivalent to the statement that the diagram

Diagram 7.1

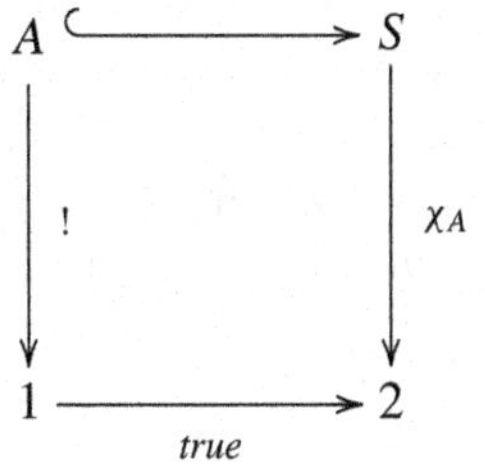

is a pullback, i.e. A is the pullback of $true : 1 \to \{0, 1\}$ along[6] χ_A. In fact, if we consider the following diagram

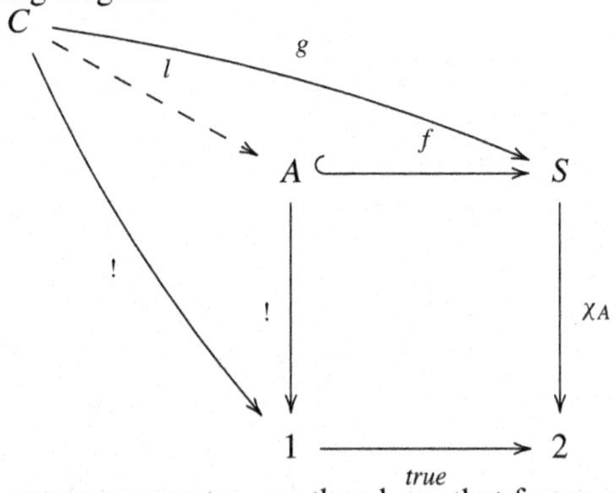

such that the outer square commutes, we then have that for any $c \in C$; $\chi_A(g(c)) = true(!(c)) = 1$. From the definition of A above it follows that $g(c) \in A$. Thus we can define the map l for each $c \in C$ as $l(c) := g(c)$. Obviously such a choice makes the whole diagram commute and it is the only arrow that would do so. It follows that $A \subseteq S$ iff Diagram 7.1 is a pullback. χ_A is the only arrow that makes such a diagram a pullback.

Proof Given the pullback diagram

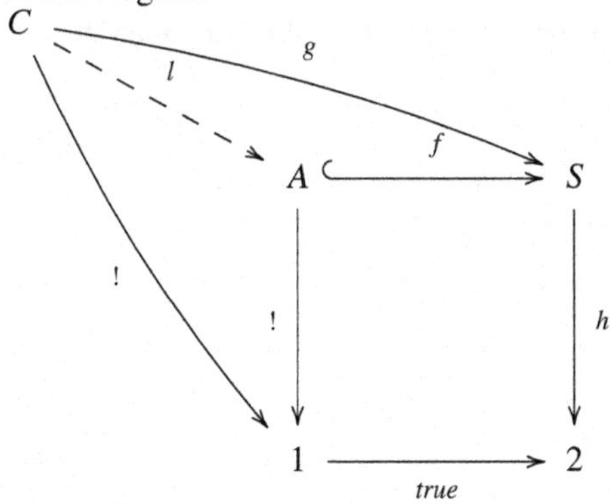

[6]Recall that $! : A \to 1$ is the unique arrow from A to the terminal object 1.

We want to show that $h = \chi_A$. Since the diagram is indeed a pullback and f a monic, we have that for all $x \in A$, $h(x) = true(!(x)) = 1$. Now take $C = h^{-1}(1)$ in the above diagram, it follows that g is monic. Since by the property of the diagram $f \circ l = g$ and both f and g are monic it follows that l is monic, hence $h^{-1}(1) \subseteq A$. Combining these two results we obtain that $A = h^{-1}(1)$. This is precisely the definition of the characteristic map χ_A. □

When pullbacks of the form

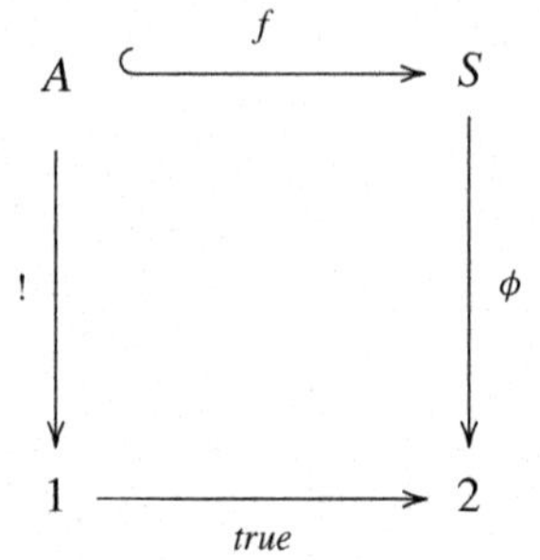

arise we say that ϕ classifies f or the sub-object represented by f.

We can now give an abstract characterisation of what it means to classify sub-objects.

Definition 7.11 Given a category with a terminal object 1, a **sub-object classifier** is an object Ω, together with a monic arrow $\mathcal{T} : 1 \to \Omega$ (topos analogue of the set theoretic arrow *true*) such that, given a monic $\mathcal{C}$-arrow $f : A \to B$, there exists one and only one χ_f arrow, which makes the following diagram

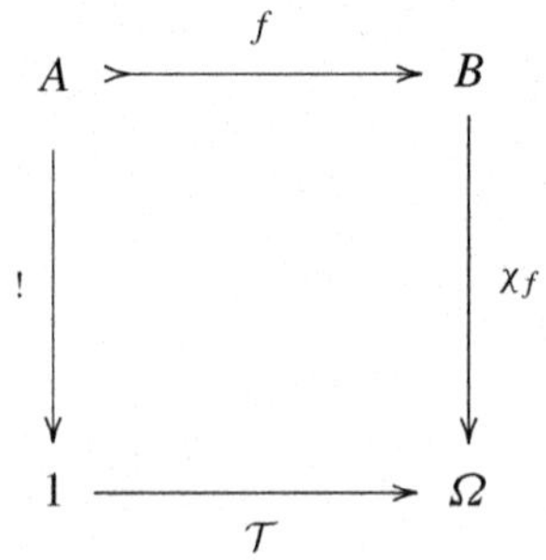

a pullback.

Proposition 7.1 *Given a category $\mathcal{C}$ with sub-object classifier Ω and sub-objects, there exists (in $\mathcal{C}$) an isomorphism*

$$\begin{aligned} y : Sub_{\mathcal{C}}(X) &\simeq \mathcal{C}(X, \Omega) \quad \forall X \in \mathcal{C} \\ f &\mapsto \chi_f. \end{aligned} \tag{7.34}$$

In order to prove the above axiom we need to show that y is iso. Since the proof of the above theorem in a general topos is quite complicated and needs definitions not yet given, we will use an analogous proof in **Sets**, which essentially uses the same strategy as the proof in a general topos, but it is much more intuitive. In **Sets** we can write the above axiom as follows:

Axiom 7.1 *The collection of all subsets of S denoted by $\mathcal{P}(S)$ and the collection of all maps from S to the set $\{0, 1\} = 2$ denoted by 2^S are isomorphic. This means that the function $y : \mathcal{P}(S) \to 2^S$, which in terms of single elements of $\mathcal{P}(S)$ is $A \to \chi_A$, is a bijection.*

Proof Keeping in mind the following diagram

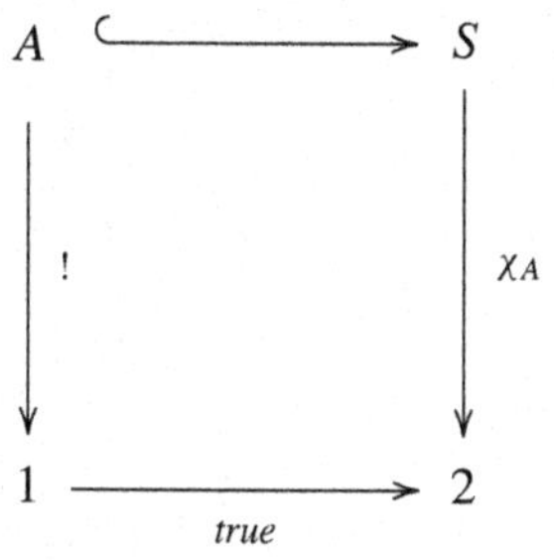

(a) y *is injective* $(1:1)$. Consider the case in which $\chi_A = \chi_B$, where

$$\chi_B(x) = \begin{cases} 1 & \text{iff } x \in B \\ 0 & \text{iff } x \notin B \end{cases}$$

while χ_A was defined as in (7.31).

Since the two functions are the same, they both associate the same domain to the codomain 1, therefore $A = B$ ($A = \chi_A^{-1}(1) = \chi_B^{-1}(1) = B$).

(b) y *is surjective* (*onto*). Given any function $f \in 2^S$, then there must exist a subset A of S, such that $A_f = \{x : x \in S \text{ and } f(x) = 1\}$, i.e. $A_f = f^{-1}(\{1\})$. Therefore $f = \chi_{A_f}$.

(c) The inverse is simply given by the pullback of *true* along χ. □

Corollary 7.1 *The domain of the arrow* true: $1 \to \Omega$ *is always the terminal object.*

Proof Let us assume that instead of the terminal object being the domain of the *true* map we have a general element Q, obtaining $true : Q \to \Omega$. We now want to consider the identity arrow id_A on an object A. Being an identity, such a map is iso, thus monic. We then want to analyse what sub-objects this map defines. We already know the answer since it is simply an identity map but, nonetheless, we apply the sub-object classifier procedure and get the pullback diagram

$$\begin{array}{ccc} A & \xrightarrow{id_A} & A \\ {\scriptstyle h}\downarrow & & \downarrow{\scriptstyle \chi_A} \\ Q & \xrightarrow[true]{} & \Omega \end{array}$$

Thus we obtain that $\chi_A = true \circ h$.

On the other hand, given any arrow $k : A \to Q$, we can define the following diagram:

$$\begin{array}{ccc} & & A \\ & & \downarrow{\scriptstyle k} \\ & & Q \\ & & \downarrow{\scriptstyle true} \\ Q & \xrightarrow[true]{} & \Omega \end{array}$$

and construct its pullback. This we can be done for each diagram separately. For the lower part of the diagram we obtain the classifying diagram for *true*:

$$\begin{array}{ccc} Q & \xrightarrow[id_Q]{} & Q \\ {\scriptstyle id_Q}\downarrow & & \downarrow{\scriptstyle true} \\ Q & \xrightarrow[true]{} & \Omega \end{array}$$

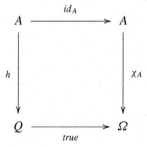

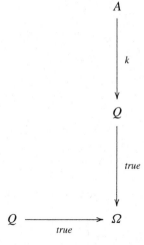

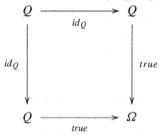

This result follows since the arrow *true* is monic hence the only possible arrow satisfying the pullback condition would be $id_Q : Q \to Q$. We then complete the upper part of the diagram with the trivial pullback

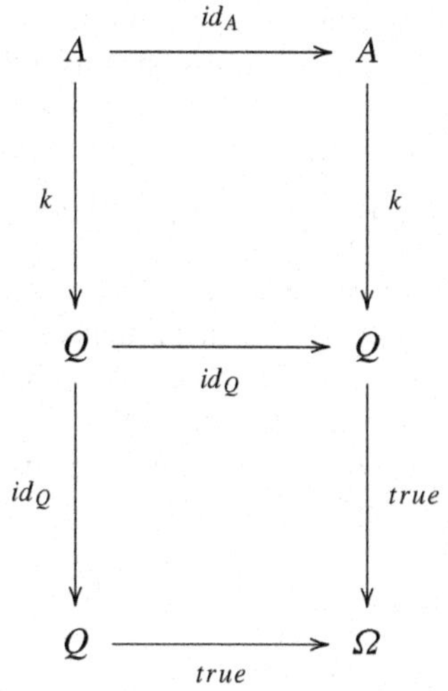

Since both smaller rectangles are pullbacks it follows that the outer rectangle is a pullback, thus it is a classifying diagram for the arrow id_A. However, we know that there exists a unique $\chi_{id_A} : A \to \Omega$ which would make the outer rectangle commute, hence $true \circ k = \chi_{id_A} = true \circ h$. On the other hand *true* is monic, thus $k = h$. What this means is that for a given object (in this case) A there is one and only one arrow from A to Q, this Q must be the terminal object. □

7.6 Elements of the Sub-object Classifier: Sieves

In the case of **Sets**, the sub-object classifier is $\Omega \cong \{0, 1\}$, therefore the elements of Ω are simply 0 and 1, which can be identified with the values false and true in the context of the theory of logic. This is not the case for a general topos in fact, in this case, the elements of the sub-object classifier are *sieves*.

Definition 7.12 A **sieve** on an object $A \in C$ is a collection S of morphisms in C whose codomain is A and such that, if $f : B \to A \in S$ then, given any morphism $g : C \to B$ we have $f \circ g \in S$, i.e. S is closed under left composition:

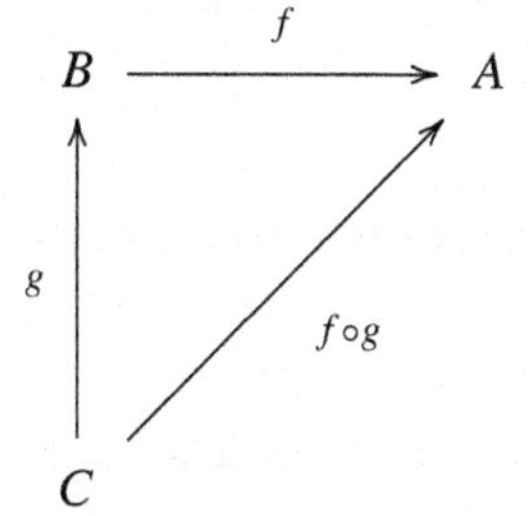

The collection of all sieves on objects of a given category has some structure. In fact it is possible to define maps between different sieves when the objects these sieves are defined on are, in some way, related. For example, if we have a $\mathcal{C}$-arrow $f : A \to B$ then it is possible to define a map from the set of all sieves on B, which we denote $\Omega(B)$ to the set $\Omega(A)$ of all sieves on A as follows:

$$\begin{aligned} \Omega_{BA} : \Omega(B) &\to \Omega(A) \\ S \mapsto \Omega_{BA}(S) &:= \{g|cod(g) = A \text{ and } f \circ g \in S\}. \end{aligned} \tag{7.35}$$

$\Omega_{BA}(S)$ represents the pullback of the sieve S (see below) and can be symbolically writen as $\Omega_{BA}(S) = S \cap \downarrow A$ where $\downarrow A$ is the principal sieve on A, i.e. the sieve that contains the identity morphism of A, therefore it is the biggest sieve on A.[7]

An important property of sieves is the following:

Given a sieve S on A, if $f : B \to A$ belongs to S then the pullback of S by f determines the principal sieve on B, i.e.

$$f^*(S) := \{h : C \to B | f \circ h \in S\} = \{h : C \to B\} = \downarrow B.$$

For example

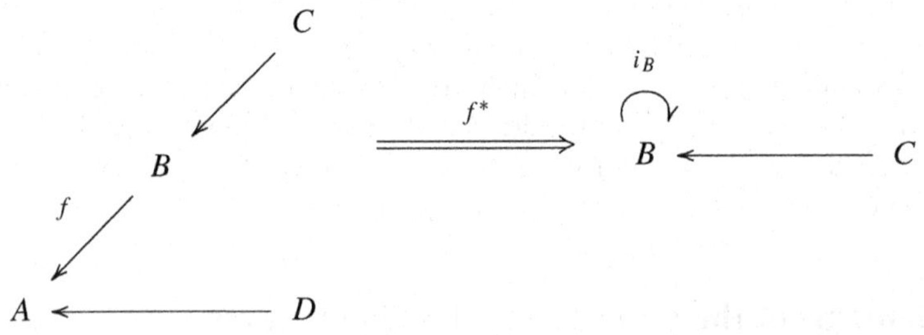

An important property of sieves is that the set of sieves defined on an object forms a Heyting algebra defined in Sect. 7.7 with partial ordering given by subset inclusion. The fact that the set of sieves forms a Heyting algebra is very important since, as we will see later on, such an algebra will represent the logic of truth values. Thus the next question to address is: what is an Heyting algebra? The answer to this question will be the topic of the next section.

7.7 Heyting Algebras

Definition 7.13 A Heyting Algebra $\boldsymbol{H}$ is a relative pseudo-complemented distributive lattice.

---

[7]Essentially $\downarrow A$ is the sieve which contains all possible $\mathcal{C}$-arrows, which have as codomain A, i.e. $\downarrow A := \{f_i \in \mathcal{C} | cod(f_i) = A\}$.

We will explain these attributes one at a time. The definition of a lattice was already given in previous chapter, but for sake of completeness we will nonetheless restate it here.

Definition 7.14 Given a poset $(L, \leq)$, we say that this is a lattice if the following conditions are satisfied.

(1) Given any two elements $a, b \in L$ it is always possible to define a third element $a \vee b \in L$ called the join or least upper bound or supremum.
(2) Given any two elements $a, b \in L$ it is always possible to define a third element $a \wedge b \in L$ called the meet or greatest lower bound or infimum.

Since the elements of the lattice involved in defining join and meet are two, the operations $\vee$ and $\wedge$ are the binary operations of the lattice.

If only the first condition holds we say that L is a join-semilattice, if only the second holds L is a meet-semilattice.

A lattice L is said to be *distributive* if for any $a_i \in L$ the following relations hold:

$$a_1 \wedge (a_2 \vee a_3) = (a_1 \wedge a_2) \vee (a_1 \wedge a_3)$$
$$a_1 \vee (a_2 \wedge a_3) = (a_1 \vee a_2) \wedge (a_1 \vee a_2).$$

In order to understand the property of being a *relative pseudo-complemented lattice* we first of all have to introduce the notion of least upper bound (l.u.b.) (and dually of greatest lower bound (g.l.b.)) of a set. We are already acquainted with the notion of the l.u.b. for two elements a, b in a lattice L with ordering $\leq$. This is simply given by the element $a \vee b$. Similarly, the g.l.b. is $a \wedge b$. But how do we define such notions with respect to a set $A \subseteq L$? The definition is quite intuitive: the l.u.b. of a set A is an element $c \in L$, such that for all $a \in A$, $a \leq c$ and given any other element b with the property $a \leq b$ for all $a \in A$ then $c \leq b$. The condition of c being the l.u.b. of A is denoted by $c \geq A$.

If $c \geq A$ and $c \in A$, we say that c is the *greatest element* of A. Dually the g.l.b. $y \in L$ of A, denoted $A \geq y$ is such that for all $a \in A$, $a \geq y$ and, given any other element $z \in L$, such that $a \geq z$ for all $a \in A$, then $y \geq z$. If in addition $y \in A$ we say that y is the *least element* of A. We can now define the notion of a relative pseudo-complement as follows:

Definition 7.15 L is a relative-pseudo-complemented lattice iff for each two elements $a, b \in L$ there exists a third element c, such that

1. $a \wedge c \leq b$.
2. $\forall x \in L \; x \leq c$ iff $a \wedge x \leq b$

where c is defined as the *pseudo-complement* of a relative to b i.e., the greatest element of the set $\{x : a \wedge x \leq b\}$ and it is denoted as $a \Rightarrow b$, i.e.

$$a \Rightarrow b = \bigvee \{x : x \wedge a \leq b\}. \tag{7.36}$$

If all elements in a lattice L have a relative-pseudo-complement we say that L is a relative pseudo-complemented lattice.

If in the above definition we replace b with the element 0, the least element in the lattice, then we obtain the notion of pseudo-complement.

Definition 7.16 Given a lattice $(L, \leq)$ with a zero element, the pseudo-complement of a is the greatest element of L disjoint from a, i.e. the greatest element of the set $\{x \in L | a \wedge x = 0\}$. The pseudo-complement of a will be denoted as $a \Rightarrow 0$.

The pseudo-complement in a Heyting algebra is identified with the negation operation, i.e. $\neg a := a \Rightarrow 0$. If an element x in a Heyting H algebra has a pseudo-complement, then it is unique and it must be $\neg x$. In fact let us consider an element $x \in H$ with complement a such that $a \wedge x = 0$ and $x \vee a = 1$, where 1 is the upper element of the lattice. It then follows that $a \leq \neg x$ since $\neg x$ is the greatest element of the set $\{s | s \wedge x = 0\}$. On the other hand

$$\neg x = \neg x \wedge (x \vee a) = (\neg x \wedge x) \vee (\neg x \wedge a) = \neg x \wedge a \tag{7.37}$$

which implies that $\neg x \leq a$. Hence $a = \neg x$.

It is interesting to note that, given the pseudo-complement $\neg x$ of x in L, the defining property of pseudo-complement imply the following relation:

$$\forall y \in L \quad y \leq \neg x \quad \text{iff} \quad y \wedge x = 0. \tag{7.38}$$

From the above definition of negation operation $(\neg a : a \Rightarrow 0)$ in a Heyting algebra we obtain the following corollary:

Corollary 7.2 *Given any element s of an Heyting algebra we have*:

$$s \vee \neg s \leq 1. \tag{7.39}$$

Proof Let us consider $s \vee \neg s$. This represents the least upper bound of s and $\neg s$ therefore, given any other element s_1 in the Heyting algebra, such that $s \leq s_1$ and $\neg s \leq s_1$, then, $s \vee \neg s \leq s_1$. But, since for any s we have $s \leq 1$ and $\neg s \leq 1$, it follows that $s \vee \neg s \leq 1$. □

The condition $s \vee \neg s = 1$ is called *tertium non datum* or *Law of excluded middle*. As we will see, this is present in Boolean algebra hence classical logic, but not in a Heyting algebra or intuitionistic logic. What this means in practical terms is that one can not prove things by contradiction. In fact a straightforward consequence of $s \vee \neg s \leq 1$ is the following corollary:

Corollary 7.3 *Given any element s of an Heyting algebra, we have*:

$$s \leq \neg\neg s. \tag{7.40}$$

Proof We know from the definition of negation in a Heyting algebra that $\neg s$ is the greatest element of the set $\{x|x \wedge s \leq 0\}$ thus $s \wedge \neg s \leq 0$. On the other hand $\neg(\neg s)$ is the greatest element of the set $\{x|x \wedge \neg s = 0\}$. However $s \in \{x|x \wedge \neg s = 0\}$ but since $\neg(\neg s)$ is the greatest element it follows that $s \leq \neg\neg s$. □

Example 7.5 A very simple example of a Heyting algebra is given by the collection $\mathcal{O}(T)$ of all open sets in a topological space T. In fact consider such a set $\mathcal{O}(T)$, given an element $S \in \mathcal{O}(T)$, this by definition is open. If we consider the negation $\neg S$ this will be closed, being the complement of an open set, hence $\neg S \notin \mathcal{O}(T)$. To make it belong to the algebra we need to consider the interior $int(\neg(S))$ which is an open set. Clearly $S \cup int(\neg(S)) \leq T$ since it is missing the boundary.

We will now show some important properties of Heyting algebras.

Theorem 7.1 *Given a Heyting algebra H with unit 1 for any two elements $x, y \in H$ the following conditions are satisfied*

1. $x \leq y$ *implies that* $\neg y \leq \neg x$.
2. $\neg\neg\neg x = \neg x$.
3. $\neg 1 = 0$.
4. $\neg 0 = 1$.
5. $\neg(x \vee y) = \neg x \wedge \neg y$.
6. $\neg\neg(x \wedge y) = \neg\neg x \wedge \neg\neg y$.

Proof 1. $x \wedge \neg x = 0$ iff $x \leq \neg\neg x$. However $x \wedge \neg x = \neg x \wedge x = 0$ iff $\neg x \leq \neg x$ iff $\neg x = \neg\neg\neg x \leq \neg x$.

2. There are two ways of proving this. Here we will report both:

(i) $\neg\neg\neg x$ is the greatest element of the set $\{s|s \wedge \neg\neg x = 0\} = S$, however we know that $\neg\neg x$ is the greatest element of the set $\{s|s \wedge \neg x = 0\} = T$, therefore $\neg\neg x \wedge \neg x = 0$ and $\neg x \in S$. We now need to show that $\neg x$ is the l.u.b. of S. Since $x \leq \neg\neg x$ it follows that $x \wedge s \leq \neg\neg x \wedge s$. This implies that if $s \in S$ then $\neg\neg x \wedge s = 0$ and $x \wedge s = 0$ therefore $s \in T' := \{s|s \wedge x = 0\}$. Thus $S \subseteq T'$. What this means is that if $s \in S$ then $s \leq \neg x$. This together with the fact that $\neq x \in S$ implies that $\neg x$ is the greatest element of S hence $\neg x = \neg\neg\neg x$.

(ii) Given 1. ($x \leq y$ implies that $\neg y \leq \neg x$), then $x \leq \neg\neg x$ it follows that $\neg\neg\neg x \leq \neg x$. However replacing x by $\neg x$ in $x \leq \neg\neg x$ we obtain $\neg x \leq \neg\neg\neg x$. Hence $\neg x = \neg\neg\neg x$.

3. $\neg 1$ is the greatest element of the set $\{s|s \wedge 1 = 0\}$. Since 1 is the unit element of H, then, $\forall s \in H$ $s \leq 1$ and the only element with trivial intersection is 0.

4. $\neg 0$ is the greatest element of the set $\{s|s \wedge 0 = 0\} = S$. This is satisfied for all $s \in H$. However we know that $\forall s \in H$ $s \leq 1$, hence 1 is the greatest element of S therefore $\neg 0 = 1$.

5. There are two ways of proving this:

(i) $\neg(x \vee y)$ is the greatest element of the set

$$\begin{aligned}\{s|s \wedge (x \vee y) = 0\} &= \{s|(s \wedge x) \vee (s \wedge y) = 0\} \\ &= \{s|(s \wedge x) = 0\} \cap \{s|(s \wedge y) = 0\}. \end{aligned} \tag{7.41}$$

However $\neg x \geq \{s|(s \wedge x) = 0\}$ and $\neg y \geq \{s|(s \wedge y) = 0\}$ therefore $\neg x \wedge \neg y \geq \{s|(s \wedge x) = 0\} \cap \{s|(s \wedge y) = 0\}$ which implies that $\neg(x \vee y) = \neg x \wedge \neg y$.

(ii) We know that $x \leq x \vee y$. Applying 1. we obtain that $\neg(x \vee y) \leq \neg x$ and similarly for y we obtain $\neg(x \vee y) \leq \neg x$. Since $\neg x \wedge \neg y$ is the g.l.b. of $\neg x$ and $\neg y$ it follows that $\neg(x \vee y) \leq \neg x \wedge \neg y$. On the other hand

$$\begin{aligned}\neg x \wedge \neg y \leq \neg(x \vee y) \quad &\text{iff} \quad (\neg x \wedge \neg y) \wedge (x \vee y) = 0 \\ &\text{iff} \quad (x \vee y) \wedge \neg x \leq \neg\neg y \\ &\text{iff} \quad \neg x \wedge y \leq \neg\neg y \\ &\text{iff} \quad \neg x \wedge y \wedge \neg y = 0 \\ &\text{iff} \quad \neg x \wedge 0 = 0. \end{aligned} \tag{7.42}$$

6. We know that $x \wedge y \leq x$ thus $\neg x \leq \neg(x \wedge y)$ and $\neg\neg(x \wedge y) \leq \neg\neg x$. Applying the same reasoning for y we obtain $\neg\neg(x \wedge y) \leq \neg\neg y$. Hence $\neg\neg(x \wedge y) \leq \neg\neg x \wedge \neg\neg y$. On the other hand

$$\begin{aligned}\neg\neg x \wedge \neg\neg y \leq \neg\neg(x \wedge y) \quad &\text{iff} \quad \neg\neg x \wedge \neg\neg y \wedge \neg(x \wedge y) = 0 \\ &\text{iff} \quad \neg\neg x \wedge \neg(x \wedge y) \leq \neg\neg\neg y = \neg y \\ &\text{iff} \quad \neg\neg x \wedge \neg(x \wedge y) \wedge y = 0 \\ &\text{iff} \quad \neg(x \wedge y) \wedge y \leq \neg\neg\neg x = \neg x \\ &\text{iff} \quad \neg(x \wedge y) \wedge y \wedge x = 0. \end{aligned} \tag{7.43}$$

□

As it can be deduced from the definition of a Heyting algebra, this is nothing but a generalisation of a Boolean algebra.[8] In fact in a Boolean algebra it is required that $s \vee \neg s = 1$ holds as shown in the following definition:

Definition 7.17 A Heyting H algebra is Boolean iff for all $x \in H$, $\neg x \vee x = 1$.

Alternatively we can characterise a Boolean algebra as follows:

Proposition 7.2 *A Heyting H algebra is Boolean iff for all $x \in H$, $\neg\neg x = x$.*

[8]For the sake of clarity we remind the reader that a Boolean algebra B is a distributive lattice with 0 and 1 in which every element $x \in B$ has a compliment $\neg x$ such that $x \wedge \neg x = 0$ and $x \vee \neg x = 1$.

Proof Given a Boolean algebra the complement of an element x is unique and it is $\neg x$ hence $\neg\neg x = x$. On the other hand, given a Heyting algebra H such that for all $x \in H$, $\neg\neg x = x$ we want to show it is Boolean. To this end we note that

$$x \vee \neg x = \neg\neg(x \vee \neg x) \overset{5.}{=} \neg(\neg x \wedge \neg\neg x) = \neg 0 \overset{4.}{=} 1. \qquad (7.44)$$

Since $x \wedge \neg x = 0$ for all x, $\neg x$ is the complement of x and H is Boolean. □

7.8 Understanding the Sub-object Classifier in a General Topos

As previously stated, the role of a sub-object classifier is to identify sub-objects. This is done in terms of associating for each sub-object A, of a given object X, an arrow to the sub-object classifier. For **Sets** this is very straightforward since in this case $\Omega = \{0, 1\}$, which can be interpreted as the values true and false.[9] Thus, in this case, each sub-object A is uniquely identified in terms of the elements x which belong to them, i.e. for which $\chi_A(x) = 1$.

On the other hand, in a general topos, the sub-object classifier is not simply the two-valued set $\{0, 1\}$ as it is for **Sets**. This 'complicates' (depending on the point of view) the situation, however, at an interpretative level, the role of the sub-object classifier is still unchanged and the elements still represent truth values. Since these truth values are not, in general, simply true or false, we will not end up with a Boolean Logic. Instead, we will end up with an intuitionistic logic which, mathematically, is represented by a Heyting algebra as described above.

In particular, in classical logic, either a statement is true or it is false, there is nothing in between. This is not the case for intuitionistic logic. If one thinks about it, examples of intuitionistic logic can be found in our language. In fact, although the way we reason is governed by classical logic, some aspects of our language can, in a way, be considered more intuitionistic in nature. For example, statements regarding more subjective issues. Consider the statement "I am tired" or "I am hungry" the truthfulness of these statements is not simply true or false. In fact, if you asked the question "are you tired?" or "are you hungry?" in most cases you will not get a simple yes or no answer but you could get something more elaborated such as "I am a little tired" ("I am a bit hungry") or "I am not so tired" ("I am not so hungry"), extremely, not at all etc. This is because the expressions tired, hungry, can be 'quantified', i.e. graded. At the extreme points there will be the answers "yes I am tired" (true) and "no, I am not tired" (false) and similarly for "hungry". However, as we have just seen, there will be many different statements in between (different truth values).

---

[9]The fact that in **Sets** the set of truth values is simply $\{0, 1\}$ implies that the logic which is derived is a Boolean logic, i.e. a classical logic. This is basically the logic that each of us adopts when speaking any western language. It is the logic of the classical world and the logic of the western way of reasoning.

Can these truth values be quantified? The answer is yes, as long as we remain in the realm of language since we know that, for example, the expressions "very", "a little", "quite", have the following relations in terms of strength:

$$\text{a little} \leq \text{quite} \leq \text{very}. \tag{7.45}$$

In this case if you are "very hungry" but just "a little tired", then in a way the statement "you are hungry" is more true than the statement "you are tired". However, this analysis relies only on the meaning that language has given to such expressions. The question is if it would be possible to define relations of truthfulness on a more objective ground. To this end let us consider the statement "the 1-litre bottle is full" which we symbolically denote as X. Let us also consider two identical bottles b_1 and b_2, one half full and the other three quarters full, respectively. Obviously, the statement X, when referred to b_1, is less true than when referred to b_2. The objectiveness of such truth values can be defined by measuring how much water there is in each bottle ($1/2$ and $3/4$).

Alternatively, one can also assert how true the proposition X is in a more operational way by measuring how much water needs to be added, such that the proposition X is true in the classical sense, i.e. such that there is 1 litre of water in the bottle. Thus, if proposition X is true when there is exactly one litre of water in a bottle, such that no more water has to be added, it is then possible to define the truth value of X, given a system b_i, as (1 − water to be added). This tells you precisely how far away you are from the truth, i.e. from one litre of water.

Therefore, the truth value of the proposition X, given the 'states' b_1 and b_2 is now given by

$$v(X, b_1) = (1 - 1/2) \leq v(X, b_2) = (1 - 1/4). \tag{7.46}$$

Obviously, the extreme points of such truth values are 0 and 1. For example, when the bottle is empty, 1 litre of water has to be added, thus $v(X, b) = 1 - 1 = 0$. On the other hand when the bottle is full no water has to be added, thus $v(X, b) = 1 - 0 = 1$.

This way of defining truth values is based on the idea of how much one has to *change* the system (fill up the bottles) in order for the *untouched* proposition to be true. However, it is also possible to proceed the other way, i.e. *generalise* the proposition so that the now *unchanged* system satisfies the more general proposition. Therefore, keeping with our water bottle examples, let us assume we have a litre bottle b_1 which has $1/8$ of a litre of water in it. Again, we have the proposition X stating "the bottle is full". We can't say that X is true given b_i, but we can't even say it is false since b_1 is not empty. So the question is: what can we say about the truth value of X given b_1 without changing the system? Well, we simply do the 'inverse' procedure as we did before, i.e. we generalise the proposition X.

To understand this it is convenient to rewrite X as "b_i has 1 litre of water". We then start generalising such a proposition by subtracting amounts of water to the bottle and forming new propositions. For example, we consider the proposition $X_{1/2} =$ "b_i has $(1 - 1/2)$ litres of water". However, given our bottle b_1 which has only $1/8$ of water we still can not say that $X_{1/2}$ is true. So we keep generalising the proposition till we reach the proposition $X_{7/8}$ = "b_1 has $(1-7/8)$ litres of water". Such a proposition is true given b_1.

This example shows how it is possible to identify truth values of propositions, given a state, in terms of how much the original proposition has to be generalised, such that this new coarse grained (generalised) proposition is true (given that state).

Thus, in our case, the truth value of the original proposition X, given the state b_i, is defined in terms of how much we have to generalise X, in this case we coarse grain it to $X_{7/8}$, such that $X_{7/8}$ is now true given b_i. Mathematically this is given by the equation

$$v(X, b_1) := 1 - \inf\{x \mid v(X_x, b_1) = 1\} = 1 - 7/8 = 1/8. \tag{7.47}$$

It is now straightforward to see how truth values can be compared. In particular, if we have a bottle which contains 1/2 litre of water, then its truth value would be

$$v(X, b_2) := 1 - \inf\{x \mid v(X_x, b_2) = 1\} = 1 - 1/2 = 1/2. \tag{7.48}$$

It then follows that $v(X, b_2) \geq v(X, b_1)$.

In this schema, we obviously don't just have 'true' or 'false' as truth values, but we have many more truth values, whose limiting points are precisely 1 (true) and 0 (false).

Although this is a very general description of how different truth values, other than 'true' or 'false', can be obtained, it still helps to shed some light on how intuitionistic logic works. In such a logic, truth values are not restricted to be only true or false but there are many more values in between. Obviously, the two limiting points will then coincide with the classical notions of true and false.

What the above implies is that, differently from a Boolean algebra, in a Heyting algebra the law of excluded middle does not hold: $S \vee \neg S \leq 1$ (in a Boolean logic we would simply have $S \vee \neg S = 1$). Thus, differently from ordinary logic where something is either true or false, in intuitionistic logic this is not the case. As a consequence, the sum of a proposition and its negation doesn't give the 'whole story'.

Above we have given a very intuitive definition of what a sub-object classifier does in a general topos, while the technical definition was given in Definition 7.11. However, the important point to understand is that in a general topos the sub-object classifier, together with the Heyting algebras of sieves, allows us to render mathematically precise the notion of a proposition being nearly true, almost true etc. Moreover, it allows for a well defined mathematical notion of how 'far away' from the truth a proposition is given a state. This 'distance' from the truth is, in the end, what a truth value represents in a general topos.

7.9 Axiomatic Definition of a Topos

Now that we have defined all the above constructions we are ready to give the axiomatic definition of a topos.

Definition 7.18 A **topos** is a category with all finite limits, exponentials and a sub-object classifier.

An alternative but equivalent definition is:

Definition 7.19 A **topos** is a category τ with the following extra properties:

1. τ has an initial object 0.
2. τ has a terminal object 1.
3. τ has pullbacks.
4. τ has pushouts.
5. τ has exponentiation, i.e. τ is such that for every pair of objects X and Y in τ, an exponential Y^X exists.
6. τ has a sub-object classifier.

It is straightforward to see that conditions (2) and (3) above are equivalent to stating that a topos τ has all finite limits, which implies that it has all finite colimits. This requirement is equivalent to conditions (1) and (4). By duality, τ^{op} has all finite limits and colimits. There are various examples of topoi, here we will only give a few.

Example 7.6 The most important in a way 'trivial' example of a topos is **Sets**.

Example 7.7 The category **Sets** × **Sets** is a topos. To see that indeed this is the case consider two diagrams of the same shape D_1 and D_2 then

$$\begin{aligned}\big(\lim(D_1), \lim(D_2)\big) &= \lim(D_1, D_2)\\ \big(\text{colim}(D_1), \lim(D_2)\big) &= \text{colim}(D_1, D_2).\end{aligned} \tag{7.49}$$

It suffices to prove only one of the above assertions, the remaining one holds by duality. This proof relies on the definition of a topos as a category with all finite limits and colimits.

Let us consider $\lim(D_1, D_2)$. By definition this is a cone (C_1, C_2) such that given any other cone (C'_1, C'_2) then the map $(C'_1, C'_2) \to (D_1, D_2)$ uniquely factors through (C_1, C_2), i.e.

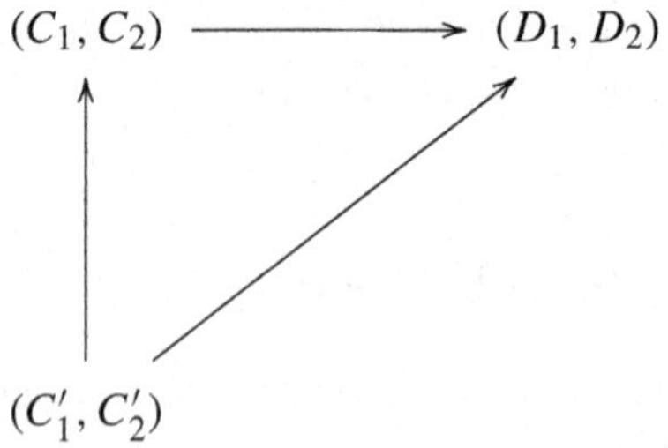

However from the definition of maps in **Sets** × **Sets** the above holds iff there exist unique maps $f : C'_1 \to C_1$ and $g : C'_2 \to C_2$ such that the following factorization occurs:

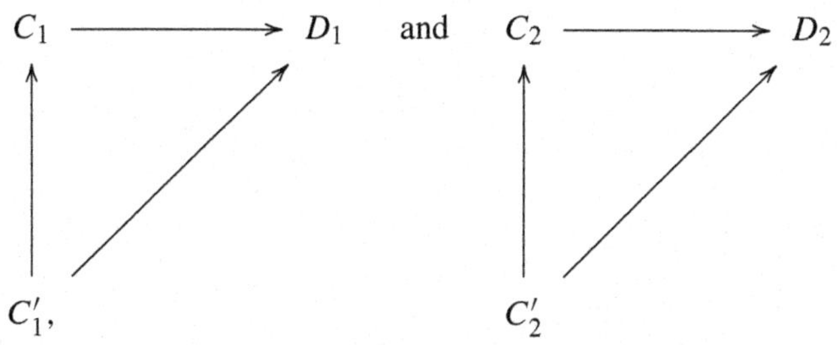

That is $C_1 = \lim(D_1)$ and $C_2 = \lim D_2$. This proves that $(\lim(D_1), \lim(D_2)) = \lim(D_1, D_2)$. By duality $(\mathrm{colim}(D_1), \mathrm{colim}(D_2)) = \mathrm{colim}(D_1, D_2)$. Thus since **Sets** has all limits and colimits so does **Sets** × **Sets**. What remains to show is the existence of exponentials and sub-object classifier. This can be easily verified to be the following:

$$\begin{aligned} \Omega^{\mathbf{Sets}\times\mathbf{Sets}} &:= \left(\Omega^{\mathbf{Sets}}, \Omega^{\mathbf{Sets}}\right) \\ (A, B)^{(C,D)} &:= \left(A^C, B^D\right). \end{aligned} \tag{7.50}$$

Chapter 8
Topos of Presheaves

For an arbitrary category $\mathcal{C}$, the category $\mathbf{Sets}^{\mathcal{C}^{op}}$ of presheaves is actually a topos. This topos is important since, for a particular choice of $\mathcal{C}$, it will be the topos we will utilise to express quantum theory.

We will now show how the categorical constructs, needed to define a topos, are constructed in the category $\mathbf{Sets}^{\mathcal{C}^{op}}$. We recall that the definition of initial and terminal object was given in Sect. 6.3 of Chap. 6, respectively.

8.1 Pullbacks

Pullbacks exist in any category of presheaves $\mathbf{Sets}^{\mathcal{C}^{op}}$. In fact, if $X, Y, B \in \mathbf{Sets}^{\mathcal{C}^{op}}$, then $P \in \mathbf{Sets}^{\mathcal{C}^{op}}$ is a pullback in $\mathbf{Sets}^{\mathcal{C}^{op}}$ iff for any $C \in \mathcal{C}$,

$$\begin{array}{ccc} P(C) & \xrightarrow{k} & Y(C) \\ {\scriptstyle h}\big\downarrow & & \big\downarrow{\scriptstyle g} \\ X(C) & \xrightarrow[l]{} & B(C) \end{array}$$

is a pullback in **Sets**. This implies that $P(C) = (X \times_B Y)(C) \simeq X(C) \times_{B(C)} Y(C)$.[1]

The fact that a pullback in $\mathbf{Sets}^{\mathcal{C}^{op}}$ is defined in terms of a pullback in **Sets** implies that the former always exists, since the latter does. In fact, given any object $C \in \mathcal{C}$ and any three functors X, Y, B, it is always possible to construct (in **Sets**) a diagram

[1] Note that the last product is all in **Sets**

C. Flori, *A First Course in Topos Quantum Theory*, Lecture Notes in Physics 868,
DOI 10.1007/978-3-642-35713-8_8, © Springer-Verlag Berlin Heidelberg 2013

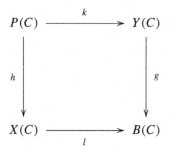

as the one above, where $P(C) = (X \times_B Y)(C) \simeq X(C) \times_{B(C)} Y(C)$. This implies that it is always possible to define a functor $P : \mathcal{C} \to$ **Sets** which assigns, for each $C \in \mathcal{C}$ an object $P(C) = (X \times_B Y)C \simeq X(C) \times_{B(C)} Y(C)$ (a set) and, for each arrow $f : A \to C$ in $\mathcal{C}$ the unique arrow $P(f) : P(C) \to P(A)$ in **Sets**, gives the following diagram in the shape of a cube in which the front and back faces are pullbacks.

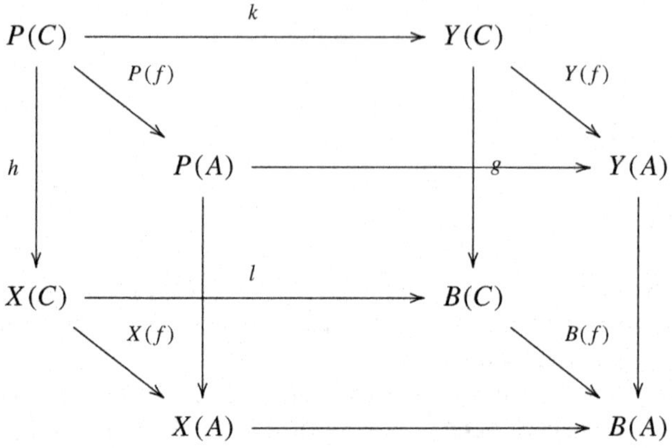

Since the pullbacks are in **Sets** they always exists.

8.2 Pushouts

Similarly, as for pullbacks, pushouts exists in any presheaf category $\mathbf{Sets}^{\mathcal{C}^{op}}$. In fact, given three functors $X, Y, Z \in \mathbf{Sets}^{\mathcal{C}^{op}}$, then a functor W is a pushout in $\mathbf{Sets}^{\mathcal{C}^{op}}$ iff for any $C \in \mathcal{C}$,

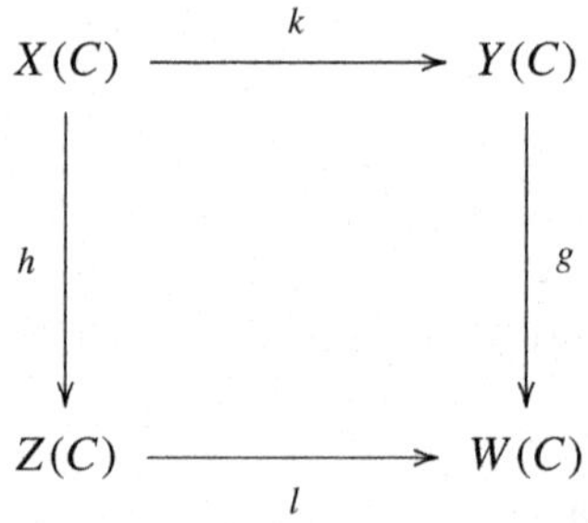

is a pushout in **Sets**. Thus $W(C) = (Z +_X Y)(C) \simeq Z(C) +_{X(C)} Y(C)$. Because of this definition, pushouts always exist in $\mathbf{Sets}^{\mathcal{C}^{op}}$. In fact, similarly as for products, given any three presheaves X, Y, Z the pushout will be the presheaf W such that for each object $C \in \mathcal{C}$, $W(C) := (Z +_X Y)(C) \simeq Z(C) +_{X(C)} Y(C)$ and for each morphisms $f : C \to D$ in $\mathcal{C}$ the corresponding morphism $W(f) : W(D) \to W(C)$

is the unique morphism giving the following diagram in the shape of a cube in which the front and back faces are pushouts:

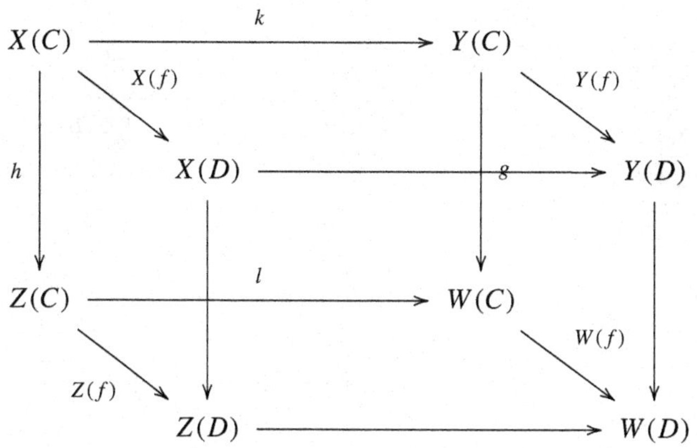

8.3 Sub-objects

A sub-object of a presheaf is defined as follows:

Definition 8.1 Y is a **sub-object of a presheaf** X if there exists a natural transformation $i : Y \to X$ which is defined, component-wise, as $i_A : Y(A) \to X(A)$ and where i_A defines a subset embedding i.e. $Y(A) \subseteq X(A)$.

Since Y is itself a presheaf, the maps between the objects of Y are the restrictions of the corresponding maps between the objects of X as shown in Fig. 8.1.

An alternative way of expressing this condition is through the following commutative diagram:

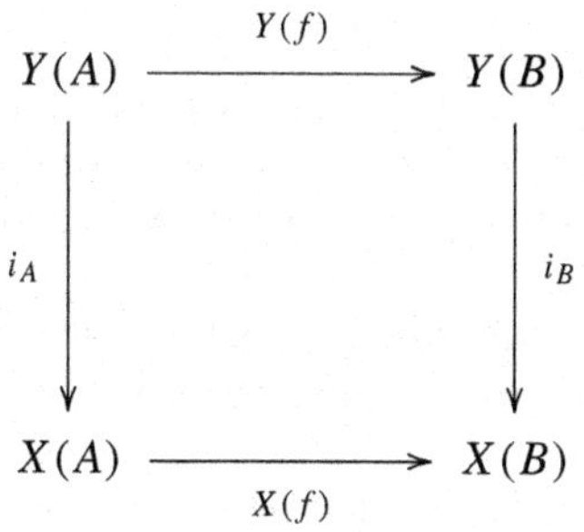

where $f : B \to A$.

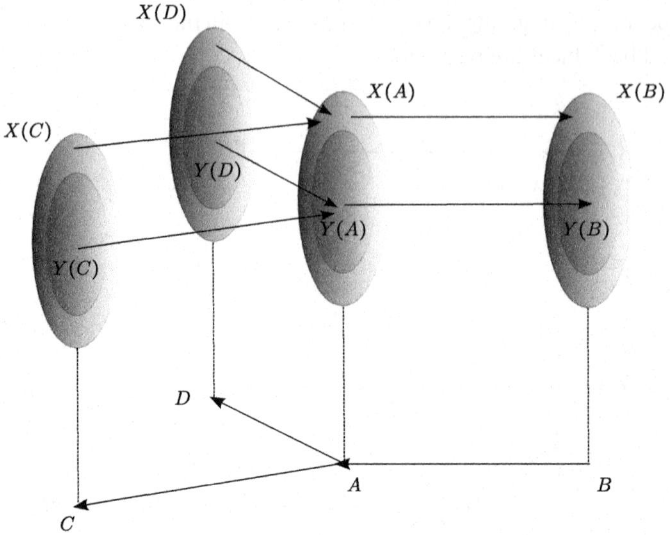

Fig. 8.1 The presheaf $Y : \mathcal{C} \to$ **Sets** is a sub-object of the presheaf $Y : \mathcal{C} \to$ **Sets** when for each object $A \in \mathcal{C}$, then $Y(A) \subseteq X(C)$

8.4 Sub-object Classifier in the Topos of Presheaves

We will now describe the most important object in $\mathbf{Sets}^{\mathcal{C}^{op}}$: the sub-object classifier. As previously mentioned, this will allow us to define truth values in our topos representation of quantum theory.

Definition 8.2 The **Sub-object Classifier** Ω in $\mathbf{Sets}^{\mathcal{C}^{op}}$ is the presheaf $\Omega : \mathcal{C} \to$ **Sets** such that:

- To each object $A \in \mathcal{C}$ there corresponds an object $\Omega(A) \in$ **Sets**, which represents the set of all sieves on A.
- To each $\mathcal{C}$-arrow $f : B \to A$ there corresponds a **Sets**-arrow $\Omega(f) : \Omega(A) \to \Omega(B)$, such that $\Omega(f)(S) := \{h : C \to B | f \circ h \in S\}$ is a sieve on B, where $\Omega(f)(S) := f^*(S)$.

We now want to show that this definition of sub-object classifier is in agreement with Definition 7.11 in Chap. 7. In order to do this, we need to define the analogue of arrow *true* ($\top$) and of the *characteristic function* in a topos.

Definition 8.3 $\top : 1 \to \Omega$ is the natural transformation that has components $\top_A : \{*\} \to \Omega(A)$ given by $\top_A(*) = \downarrow\! A =$ principal sieve on A.

To understand how $\top$ works let us consider a monic arrow $\sigma : F \to X$ in $\mathbf{Sets}^{\mathcal{C}^{op}}$, which is defined component-wise by $\sigma_A : F(A) \to X(A)$ and represents subset inclusion.

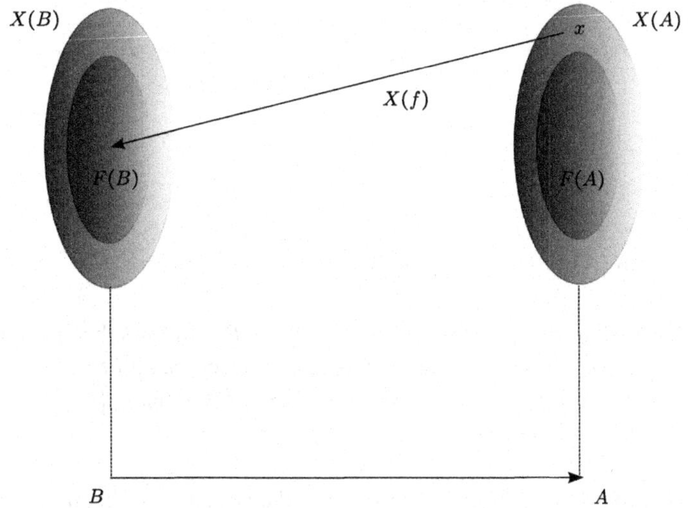

Fig. 8.2 A map $f : B \to A$ belongs to the sieve $\chi^{\sigma}_{A}(x)$ iff $X(f)$ maps x into $F(B)$

Now let us define the character $\chi^{\sigma} : X \to \Omega$ of σ which is a natural transformation between presheaves, such that the components χ^{σ}_{A} represent functions from $X(A)$ to $\Omega(A)$, as depicted in:

$$\begin{array}{ccc} F(A) & \xrightarrow{\ \sigma_A\ } & X(A) \\ \downarrow & & \downarrow \chi^{\sigma}_{A} \\ \{*\} & \xrightarrow{\ \top_A\ } & \Omega(A) \end{array}$$

where $T\{*\} = \downarrow A$.

From the above diagram we can see that χ^{σ}_{A} assigns to each element $x \in X(A)$ a sieve $\chi^{\sigma}_{A}(x) \in \Omega(A)$ on A such that

$$\chi^{\sigma}_{A}(x) := \big\{ f : B \to A | X(f)(x) \in F(B) \big\}. \tag{8.1}$$

This condition is depicted in Fig. 8.2.

Lemma 8.1 $\chi^{F}_{\sigma}(x)$, *as defined by* (8.1), *represents a sieve on* A.

Proof Consider the following commuting diagram which represents sub-object F of the presheaf X:

$$
\begin{array}{ccccc}
F(A) & \xrightarrow{F(f)} & F(B) & \xrightarrow{F(g)} & F(C) \\
\Big\downarrow & & \Big\downarrow & & \Big\downarrow \\
X(A) & \xrightarrow{X(f)} & X(B) & \xrightarrow{X(g)} & X(C)
\end{array}
$$

If $f : B \to A$ belongs to $\chi^\sigma_A(x)$ then, given $g : C \to B$ it follows that $f \circ g$ belongs to $\chi^\sigma_A(x)$, since from the above diagram it can be deduced that $X(f \circ g)(x) \in F(C)$. This is precisely the definition of a sieve, so we have proved that $\chi^\sigma_A(x) := \{f : B \to A | X(f)(x) \in F(B)\}$ is a sieve. □

As a consequence of Axiom 7.1 in Chap. 7 the condition of being a sub-object classifier can be restated in the following way.

Definition 8.4 (Omega Axiom) Ω is a **sub-object classifier** iff there is a "one to one" correspondence between sub-object of X and morphisms from X to Ω.

Given this alternative definition of a sub-object classifier, it is easy to prove that Ω, as defined in Definition 8.2, is a sub-object classifier. In fact, from (8.1), we can see that indeed there is a bijective correspondence between sub-objects of X and characteristic morphisms (characters) χ.

Moreover, for each morphism $\chi : X \to \Omega$ we have

$$F^\chi(A) := \chi_A^{-1}\{1_{\Omega(A)} := \downarrow A\} = \big\{x \in X(A) | \chi_A(x) = \downarrow A\big\} = \text{sub-object of } X(A). \tag{8.2}$$

8.4.1 Elements of the Sub-object Classifier

As mentioned in Sect. 7.6 of Chap. 7, the elements of the sub-object classifier of a presheaf topos are sieves. This fact is derived from the following theorem:

Theorem 8.1 *Given an object $A \in \mathcal{C}$ (where $\mathcal{C}$ is a locally small category),*[2] *every sieve on A can be identified with a sub-object of a representable functor (defined below) $y(A) := Hom_{\mathcal{C}}(-, A) \in$* $\mathbf{Sets}^{\mathcal{C}^{op}}$.

In order to prove the above theorem we need the following lemma:

[2] A category $\mathcal{C}$ is said to be locally small iff every hom-set $\mathcal{C}(A, B)$ forms a proper set.

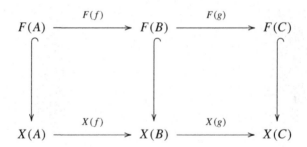

Definition 8.5 (Preliminary) If $\mathcal{C}$ is a locally small category, then each object A of $\mathcal{C}$ induces a natural contravariant functor from $\mathcal{C}$ to **Sets** called a hom-functor $\mathbf{y}(A) := Hom_{\mathcal{C}}(-, A)$.[3] Such a functor is defined on objects $C \in \mathcal{C}$ as

$$\begin{aligned} \mathbf{y}(A) : \mathcal{C} &\to \mathbf{Sets} \\ C &\mapsto Hom_{\mathcal{C}}(C, A) \end{aligned} \tag{8.3}$$

on $\mathcal{C}$-morphisms $f : C \to B$ as

$$\begin{aligned} \mathbf{y}(A)(f) : Hom_{\mathcal{C}}(B, A) &\to Hom_{\mathcal{C}}(C, A) \\ g &\mapsto \mathbf{y}(A)(f)(g) := g \circ f. \end{aligned} \tag{8.4}$$

A very simple graphical example of the above is the following:

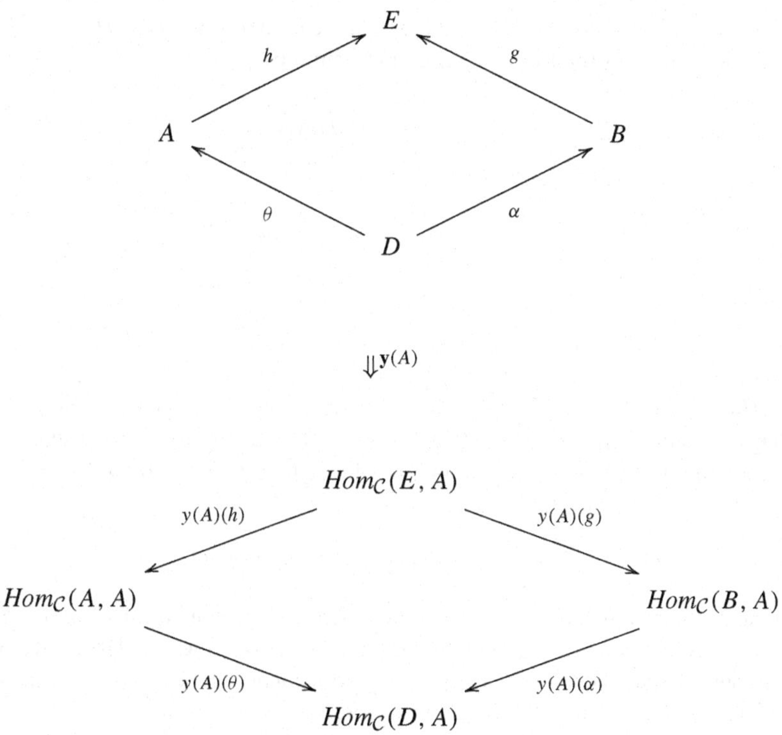

Lemma 8.2 (Yoneda Lemma) *Given an arbitrary presheaf P on* $\mathcal{C}$ *there exists a bijective correspondence between natural transformations* $\mathbf{y}(A) \to P$ *and elements*

[3] We have already encounter this in Example 5.10 of Chap. 5.

of the set $P(A)$ $(A \in \mathcal{C})$ *defined as an arrow*

$$\begin{aligned} \theta : Nat_{\mathcal{C}}\big(\mathbf{y}(A), P\big) &\stackrel{\simeq}{\to} P(A) \\ \big(\alpha : \mathbf{y}(A) \to P\big) &\mapsto \theta(\alpha) = \alpha_A(id_A), \end{aligned} \tag{8.5}$$

where $\alpha_A : \mathbf{y}(A)(A) \to P(A); Hom_{\mathcal{C}}(A, A) \to P(A)$.

A proof of this lemma can be found in [18]. We have now the right tools to prove the above theorem.

Proof Let us consider Ω to be a sub-object classifier of $\hat{C} = \mathbf{Sets}^{\mathcal{C}^{op}}$, i.e. we want Ω to classify sub-objects in $\mathbf{Sets}^{\mathcal{C}^{op}}$. Now if $Q \subset Hom_{\mathcal{C}}(-, C)$ is a sub-functor of $Hom_{\mathcal{C}}(-, C)$, then the set

$$S = \{f \mid \text{for some object } A,\ f : A \to C \text{ and } f \in Q(A)\}$$

is a sieve on C. This can be easily seen from the fact that, since $Q \subseteq Hom_{\mathcal{C}}(-, C)$, then, given $h : B \to A$ the following digram commutes

$$\begin{array}{ccc} Q(A) & \longrightarrow & Hom_{\mathcal{C}}(A, C) \\ & & \Big\downarrow {\scriptstyle Hom_{\mathcal{C}}(h, C)} \\ Q(B) & \longrightarrow & Hom_{\mathcal{C}}(B, C) \end{array}$$

where $Hom_{\mathcal{C}}(h, C)(f) := f \circ h$. Therefore, starting with $f \in Q(A)$, going around the diagram in one direction we obtain $f \circ h \in Hom_{\mathcal{C}}(B, C)$, while going the other way, since the bottom arrow is a monic, it follows that $f \circ h \in Q(B)$. Conversely, given a sieve S on C we define

$$Q(A) = \{f \mid f : A \to C \text{ and } f \in S\} \subseteq Hom_{\mathcal{C}}(A, C)$$

which determines a presheaf $Q : \mathcal{C} \to \mathbf{Sets}$ which is a sub-functor of $Hom_{\mathcal{C}}(-, C)$, i.e. to each object $A \in \mathcal{C}$, Q assigns the set $Q(A) \subseteq Hom_{\mathcal{C}}(A, C)$. The above discussion shows that there exist a bijective correspondence between sub-functors $Q \subseteq Hom_{\mathcal{C}}(-, C)$ and sieves S on C. Therefore

$$\text{Sieve on } C \simeq \text{sub-functor of } Hom_{\mathcal{C}}(-, C).$$ □

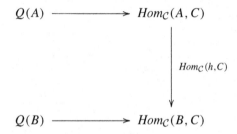

The importance of Yoneda's Lemma is really that it enables us to identify the elements of a sub-object classifier for the topos $\mathbf{Sets}^{\mathcal{C}^{op}}$ as sieves. Since elements of the sub-object classifier are identified with truth values, it follows that in the topos $\mathbf{Sets}^{\mathcal{C}^{op}}$ we will end up with a multivalued logic different from the logic we obtain in **Sets**, where $\underline{\Omega} = \{0, 1\}$, i.e. where the only truth values are true and false.

Considering Definition 8.5 it follows that, for any morphism on $\mathcal{C}$ of the form $f : A \to B$, there exists a natural transformation $\mathbf{y}(A) \to \mathbf{y}(B)$ between the respective presheaves obtained by composition with f, i.e. for all $C \in \mathcal{C}$, $\mathbf{y}(A)(C) \to \mathbf{y}(B)(C)$; $\sigma \mapsto \sigma \circ f$.

We can therefore deduce that $\mathbf{y}$ is actually a functor from the category $\mathcal{C}$ to the set of presheaves defined on $\mathcal{C}$, i.e. $\mathbf{y} : \mathcal{C} \to \mathbf{Sets}^{\mathcal{C}^{op}}$ is such that to each object A in $\mathcal{C}$, $\mathbf{y}$ assigns the Hom-functor $Hom_{\mathcal{C}}(-, A)$, i.e.

$$\begin{aligned} \mathbf{y} : \mathcal{C} &\to \mathbf{Sets}^{\mathcal{C}^{op}} \\ A &\mapsto \mathbf{y}(A) := Hom_{\mathcal{C}}(-, A) \end{aligned} \tag{8.6}$$

where $Hom(-, A)$ corresponds to a Presheaf on $\mathcal{C}$. While, given a $\mathcal{C}$-arrow $f : C \to D$ the induced morphism is

$$\mathbf{y}(f) : Hom_{\mathcal{C}}(-, C) \to Hom_{\mathcal{C}}(-, D) \tag{8.7}$$

whose natural components, for any $A \in \mathcal{C}$ are

$$\mathbf{y}(f)(A) = Hom_{\mathcal{C}}(A, f) : Hom_{\mathcal{C}}(A, C) \to Hom_{\mathcal{C}}(A, D). \tag{8.8}$$

8.5 Global Sections

Other important elements of topos theory are global sections. These represent the categorical analogue of the concept of a point/element as it was explained in Chap. 4.

Definition 8.6 A **global section** or **global element** of a presheaf X in $\mathbf{Sets}^{\mathcal{C}^{op}}$ is an arrow $k : 1 \to X$ from the terminal object 1 to the presheaf X.

What k does is to assign to each object A in $\mathcal{C}$ an element $k_A(\{*\}) \in X(A)$ in the corresponding set of the presheaf X. This assignment is such that, given an arrow $f : B \to A$, the following relation holds

$$X(f)\big(k_A\big(\{*\}\big)\big) = k_B\big(\{*\}\big). \tag{8.9}$$

What (8.9) uncovers is that the elements of $X(A)$, assigned by the global section k, are mapped into each other by the morphisms in X. A pictorial representation of the workings of two local sections is given in Fig. 8.3.

Not all presheaves have global sections thus not all presheaves have elements/points. This fact will be of crucial importance when defining the topos analogue of the Kochen-Specker theorem.

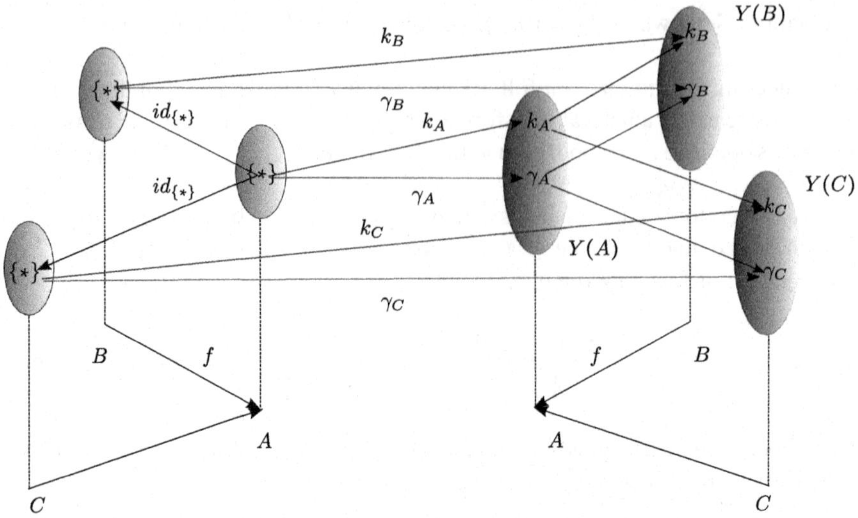

Fig. 8.3 Here we show two global sections k and γ of the presheaf $Y : \mathcal{C} \to$ **Sets** where $\mathcal{C}$ has only three objects A, B, C and two arrows $f : B \to A$ and $g : C \to A$. For notational simplicity we have denoted the elements $k_A(\{*\}) \in Y(A)$ picked up by the section k as k_a and, similarly, for the section γ

A particular important type of global sections are the global elements of Ω. In fact the collection $\Gamma(\Omega)$ of all such global section forms a Heyting algebra and represents the collection of all truth values in a topos logic. We thus obtain, as an internal logic in a topos, a multivalued logic which is of an intuitionistic type.

8.6 Local Sections

Definition 8.7 A **local** or **partial section** of a presheaf X in $\mathbf{Sets}^{\mathcal{C}^{op}}$ is an arrow $\rho : U \to X$ where U is a sub-object of the terminal object 1.

In a presheaf category, a sub-object U of 1 can either be the empty set $\emptyset$ or a singleton $\{*\}$. More precisely, for each object $C \in \mathcal{C}^{op}$ we either obtain the empty set $U(C) = \emptyset$ or a singleton $U(C) = \{*\}$. To each such singleton we then assign an element of $X(C)$. This assignment is said to be "closed downwards", i.e. given $U(A) = \{*\}$ and a $\mathcal{C}$-morphisms $f : B \to A$ then we have $U(B) = \{*\}$, therefore $\rho_A(\{*\}) =: \rho_A \in X(A)$ and $X(f)(\rho_A) = \rho_B$.

To better explain the above, let us consider a category with 4 objects $\{A, B, C, D\}$ such that non-identity morphisms exist between the objects.

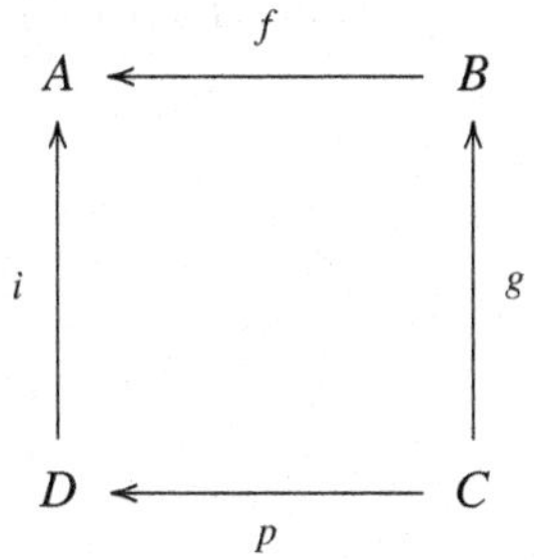

Given a sub-object U of 1 we then have the following relations

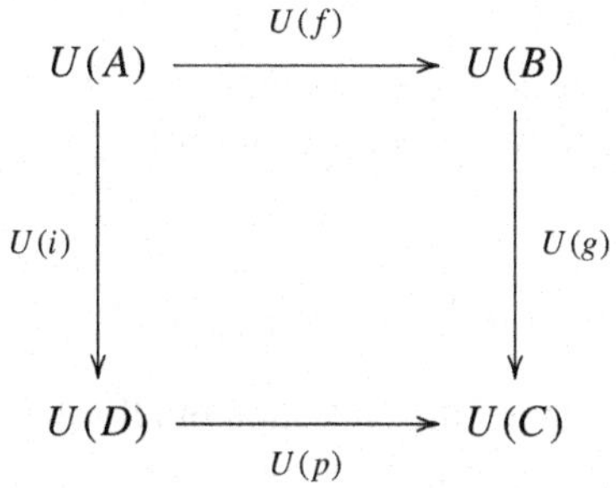

If $U(A) = \emptyset$ then $U(f)$ is either the unique function $\emptyset \to \{*\}$ iff $U(B) = \{*\}$ or $\emptyset \to \emptyset$ iff $U(B) = \emptyset$. If instead $U(A) = \{*\}$ then the only possibility is that $U(B) = \{*\}$ since there does not exist a function $\{*\} \to \emptyset$. Therefore ρ assigns elements ρ_a only to those objects $A \in \mathcal{C}$ for which $U(A) = \{*\}$. These objects A are called the domain of ρ ($dom\,\rho$) and are such that the following conditions are satisfied:

- The domain is closed downwards, i.e. if $A \in dom\,\rho$ and if there exists a map $f : B \to A$ then $B \in dom\,\rho$.
- If $A \in dom\,\rho$ and if there exists a map $f : B \to A$, then the following condition is satisfied:

$$X(f)(\rho_A) = \rho_B.$$

Presheaves with a local or partial section can exist even if they do not have a global section.

8.7 Exponential

In $\mathbf{Sets}^{\mathcal{C}^{op}}$ the exponentiation can be defined as follows:
consider a functor $F \in \mathbf{Sets}^{\mathcal{C}^{op}}$ such that, given an object $a \in \mathcal{C}$, F defines a contravariant functor[4] $F_a : \mathcal{C} \downarrow a \to \mathbf{Sets}$. F_a assigns to each object $f : b \to a \in \mathcal{C} \downarrow a$

[4]Recall from Chap. 4 that the comma category $\mathcal{C} \downarrow a$ has, as objects, all maps in $\mathcal{C}$ with codomain a, i.e. all maps of the form $f : b \to a$. Given two objects $f : b \to a$ and $g : c \to a$ a morphism between them is a $\mathcal{C}$ arrow $h : b \to c$ such that $g \circ h = f$.

an object $F_a(f) := F(b)$ and assigns to each arrow $h : (b, f) \to (c, g)$ for which the diagram

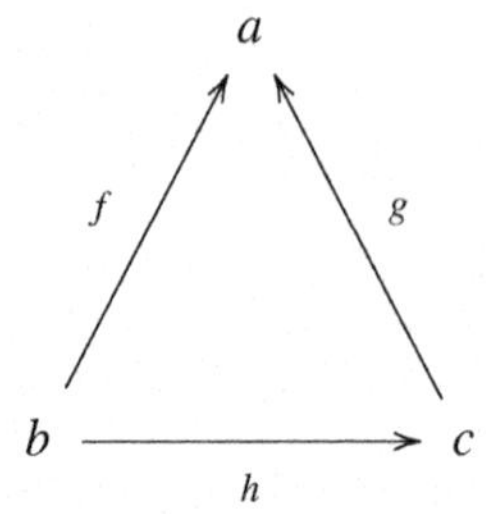

commutes in $\mathcal{C}$, the arrow $F(h) : F(c) \to F(b)$.

Given this context we define the exponential $G^F : \mathcal{C} \to \mathbf{Sets}$ between the contravariant functors $F, G \in \mathbf{Sets}^{\mathcal{C}^{op}}$ as the functor with

- Objects:

$$G^F(a) = Nat[F_a, G_a] \tag{8.10}$$

i.e. the elements of $G^F(a)$ are the collection of all natural transformations from F_a to G_a.[5]
- Morphisms: given an arrow $k : a \to d$ we get

$$G^F(k) : Nat[F_d, G_d] \to Nat[F_a, G_a]. \tag{8.11}$$

To better understand this definition consider the natural transformation $\alpha \in Nat[F_d, G_d]$ and $\theta \in Nat[F_a, G_a]$. The action of $G^F(k)$ can then be illustrated as follows:

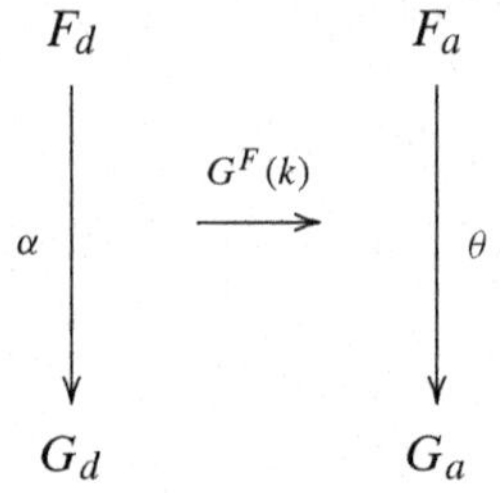

i.e an arrow in $G^F(k)$ assigns, to each natural transformation α from F_d to G_d a natural transformation θ from F_a to G_a. The way in which the natural transformation θ is picked given α can be understood by considering the individual components. In

[5]Here the contravariant functor G_a is defined in an analogous way as F_a.

particular, for any $f : c \to a \in \mathcal{C} \downarrow a$ we need to define

$$\begin{aligned} \theta_f : F_a(f) &\to G_a(f) \\ F(c) &\to G(c). \end{aligned} \tag{8.12}$$

Now, given $k : a \to d \in \mathcal{C}$, we define h in terms of the following commuting diagram

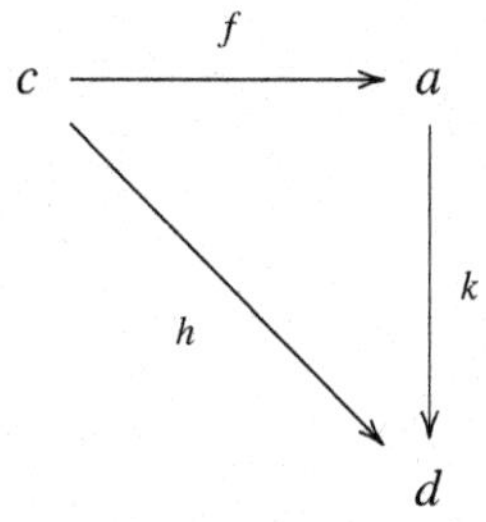

We now define $\theta_f := \alpha_{k \circ f} = \alpha_h : F_d(h) \to G_d(h); F(C) \to G(C)$.

In this formulation the evaluation function is the map: $ev : G^F \times F \to G$ in $\mathbf{Sets}^{\mathcal{C}^{op}}$, such that

$$\begin{aligned} ev_a : G^F(a) \times F(a) &\to G(a) \\ \langle \theta, x \rangle &\mapsto ev_a\big(\langle \theta, x \rangle\big) = \theta_{id_A}(x) \end{aligned} \tag{8.13}$$

where $\theta \in Nat[F_A, G_A]$ and $x \in F(A)$.

Chapter 9
Topos Analogue of the State Space

In this chapter we will describe how topos quantum theory can be seen as a contextual quantum theory, in the sense that each element is defined as a collection of 'context dependent' descriptions. Such context dependent descriptions will turn out to be classical snapshots.

The above mentioned contexts are abelian von Neumann sub-algebras, the collection of which forms a category under sub-algebra inclusion which we denote as $\mathcal{V}(\mathcal{H})$. What this implies is that, although locally quantum theory can be defined in terms of local classical snapshots, the global/quantum information is put back into the picture by the categorical structure of the collection of all such classical snapshots.

The category of von Neumann sub-algebras of[1] $\mathcal{B}(\mathcal{H})$ will be the base category in the topos of presheaves $\mathbf{Sets}^{\mathcal{V}(\mathcal{H})}$ used to define quantum theory. As we will see, in this topos the analogue of the state space of quantum theory is represented by a presheaf called the *spectral presheaf* which, for each context V, assigns its Gel'fand spectrum.

9.1 The Notion of Contextuality in the Topos Approach

In Chap. 3 we have seen how the Knochen-Specker theorem seems to imply that quantum theory is contextual, since values of quantities depend on which other quantities are being measured at the same time. However, that is not the notion of contextuality that we want to address here. In fact, in the topos approach to quantum theory there is another type of contextuality arising, which is fundamental for the formulation of the theory. Surprisingly enough also this notion of contextuality is derived from the Kochen-Specker theorem, but in a very different fashion. In particular, although the K-S theorem prohibits us to define values for all quantities at the same time in a consistent way, it nonetheless allows for the possibility

[1] $\mathcal{B}(\mathcal{H})$ indicates the algebra of bounded operators on the Hilbert space $\mathcal{H}$.

C. Flori, *A First Course in Topos Quantum Theory*, Lecture Notes in Physics 868,
DOI 10.1007/978-3-642-35713-8_9,

of assigning values to commuting subsets of quantities. These commuting subsets can be considered as classical snapshots since all the peculiarities of quantum theory arise from non-commuting operators. Thus, with respect to these classical snapshots (contexts), quantum theory behaves like classical theory.

The idea is then to define quantum theory locally with respect to these classical snapshots but, at the same time, also consider the information coming from the collection of all these classical snapshots. In this way quantum theory could be seen as a collection of local classical approximations.

Although it seems as if one were cheating by doing this, it turns out that this is not the case. The reason being that the collection of all the above mentioned classical snapshots actually forms a category which means that it is always possible to relate (compare) any two contexts.

Summarising, what is happening is the following:

(i) we first consider different contexts which represent classical snapshots.
(ii) We then define our quantum theory locally in terms of such classical snapshots, therefore in a way performing a classical approximation.
(iii) The quantum information, which is lost at the local level is, however, put back into the picture by the categorical structure of the collection of all the classical contexts.

In this way no information is lost and therefore we have not cheat.

The category of classical snapshots we will be utilising is the category of abelian von Neumann sub-algebras of the algebra $\mathcal{B}(\mathcal{H})$ of bounded operators on the Hilbert space.

9.1.1 Category of Abelian von Neumann Sub-algebras

In what follows we will first give the axiomatic definition of what the category $\mathcal{V}(H)$ is and, then, explain through an example what exactly these von Neumann algebras are.

In particular, consider the algebra of bounded operators on a Hilbert space which we denote as $\mathcal{B}(\mathcal{H})$. A quantum system can be represented by a von Neumann algebra $\mathcal{N}$ which is identified with a sub-algebra of $\mathcal{B}(\mathcal{H})$. Thus a von Neumann algebra is a *-algebra of bounded operators.[2] In the following we will make a small digression in the theory of von Neumann algebras and ring theory. However, for a detailed analysis the reader is referred to [28] and references therein.

---

[2]Note that in order to understand the general idea of topos quantum theory it is not necessary to understand in details what a von Neumann algebra is, all that is needed is to roughly understand what they are, how they can be formed and the philosophical implications of their usage in topos quantum theory.

Definition 9.1 A ring is a set X on which two binary operations are defined:

$$\begin{aligned} + : X \times X &\to X \\ (x_1, x_2) &\mapsto x_1 + x_2 \end{aligned} \tag{9.1}$$

and

$$\begin{aligned} \cdot : X \times X &\to X \\ (x_1, x_2) &\mapsto x_1 \cdot x_2 \end{aligned} \tag{9.2}$$

called addition and multiplication. Generally a ring is denoted as $(X, +, \cdot)$ and it has to satisfy the following axioms:

- $(X, +)$ must be an abelian group under addition.
- $(X, \cdot)$ must be a monoid under multiplication.

In the above definition only the addition operation is required to be commutative, while the multiplication is not. However both operations are required to be associative. For this reason rings are often also called associative rings to distinguish them from non-associative rings, which are a subsequent generalisation of the concept of a ring in which $(X, \cdot)$ is not a monoid but, all that is required, is that the multiplication operation be linear in each variable.

Of particular importance to us is the concept of a *-ring which is defined as follows

Definition 9.2 A *-ring is an associative ring with a map $* : A \to A$ s.t.

$$\begin{aligned} (x + y)^* &= x^* + y^* \\ (x \cdot y)^* &= y^* \cdot x^* \\ 1^* &= 1 \\ \left(x^*\right)^* &= x \end{aligned} \tag{9.3}$$

for all x, y in A. We say that $*$ is an anti-automorphism and an involution. Elements such that $x^* = x$ are called self-adjoint or Hermitian.

Definition 9.3 A *-algebra A is a *-ring that is a module[3] over a commutative *-ring R, with the * agreeing on $R \subseteq A$.

A type of $*$-algebras are C^*-algebras:

Definition 9.4 A C^*-algebra $\mathcal{A}$ is a Banach $*$-algebra such that $\|a^* a\| = \|a\|^2$ for all $a \in \mathcal{A}$.

[3]A module over a ring is, roughly, the same thing as a vector space over a filed but now, instead of multiplication by a scalar, one has multiplication by an element of the ring.

Given all the above definitions we are now ready to define what a von Neumann algebra is. Alternatively one can define a von Neumann algebra as:

Definition 9.5 A von Neumann algebra is a *-algebra of bounded operators on a Hilbert space that is closed in the weak operator topology and contains the identity operator.

The above definitions can be trivially extended to abelian von Neumann sub-algebras.

The way in which von Neumann algebras are generated, given a Hilbert space, is through the double commutant theorem. In particular, given an algebra $B \subset \mathcal{B}(\mathcal{H})$ of bounded operators on a Hilbert space $\mathcal{H}$, which contains the identity and is closed with respect to taking the adjoint, then the commutant of such an algebra is

$$B' := \{\hat{A} \in \mathcal{B}(\mathcal{H}) | [\hat{A}, \hat{B}] = 0 \, \forall \hat{B} \in B\} \subset \mathcal{B}(\mathcal{H}). \tag{9.4}$$

The double commutant is then the commutant of B': $(B')' = B''$. This algebra B'' is the von-Neumann algebra generated by B iff $B = B''$. Below we will give a concrete example of how such algebras are generated.

Given a Hilbert space $\mathcal{H}$, the collection of all the abelian von Neumann sub-algebras, denoted as $\mathcal{V}(\mathcal{H})$, forms a category. The importance of this lies in the fact that, although each algebra only gives a partial classical information of the system, the collection of all such algebras retains the full quantum information, since the categorical structure relates information coming from different contexts. In particular, let us consider two contexts V_1 and V_2. If they have a non-trivial intersection $V_1 \cap V_2$ then we have the following relation-arrows:

$$V_1 \leftarrow V_1 \cap V_2 \rightarrow V_2.$$

Now, given any self-adjoint operator $\hat{A}$ in $V_1 \cap V_2$ it can be written as $g(\hat{B})$ for a self-adjoint operator $\hat{B} \in V_1$ and a Borel function $g : \mathbb{R} \rightarrow \mathbb{R}$. On the other hand $\hat{A} = f(\hat{C})$ for $\hat{C} \in V_2$ and f is another Borel function. It follows that $[\hat{A}, \hat{B}] = [\hat{A}, \hat{C}] = 0$, however it is not necessarily the case that $[\hat{B}, \hat{C}] = 0$. Thus, although the elements in $\mathcal{V}(\mathcal{H})$ are abelian, the categorical structure knows about the relation of non commutative operators.

The formal definition of the category $\mathcal{V}(\mathcal{H})$ is as follows:

Definition 9.6 The category $\mathcal{V}(\mathcal{H})$ of abelian von Neumann sub-algebras has

1. *Objects*: $V \in \mathcal{V}(\mathcal{H})$ abelian von Neumann sub-algebras.
2. *Morphisms*: given two sub-algebras V_1 and V_2 there exists an arrow between them $i : V_1 \rightarrow V_2$ iff $V_1 \subseteq V_2$.

From the definition it is easy to understand that $\mathcal{V}(\mathcal{H})$ is a poset, whose ordering is given by subset inclusion.

It is interesting to understand what this poset structure actually means from a physics perspective. In particular, if we consider an algebra V' such that $V' \subseteq V$, then the set of self-adjoint operators present in V', which we denote V'_{sa}, will be smaller than the set of self-adjoint operators in V, i.e. $V'_{sa} \subseteq V_{sa}$. Since self-adjoint operators represent physical quantities, the context V' contains less physical information, so that, by viewing the system from the context V', we know less about it then when viewing it from the context V. This idea represents a type of coarse graining which takes place when going from a context with more information V to a context with less information V'. If we went the reverse direction we would instead have a process of fine graining.

This idea of coarse graining is central in the formulation of the topos quantum theory. We will see, later in the book, how it is actually implemented.

9.1.2 Example

Let us consider a four dimensional Hilbert space $\mathcal{H} = \mathbb{C}^4$. The first step is to identify the poset of abelian von Neumann sub-algebras $\mathcal{V}(\mathbb{C}^4)$. Such algebras are sub-algebras of the algebra $\mathcal{B}(\mathbb{C}^4)$ of all bounded operators on $\mathcal{H}$. Since the Hilbert space is $\mathbb{C}^4$, $\mathcal{B}(\mathbb{C}^4)$ is the algebra of all 4×4 matrices with complex entries which act as linear transformations on $\mathbb{C}^4$. In order to form the abelian von Neumann sub-algebras one considers an orthonormal basis $(\psi_1, \psi_2, \psi_3, \psi_4)$ and projection operators $(\hat{P}_1, \hat{P}_2, \hat{P}_3, \hat{P}_4)$, which project on the one-dimensional subspaces $\mathbb{C}\psi_1$, $\mathbb{C}\psi_2$, $\mathbb{C}\psi_3$, $\mathbb{C}\psi_4$, respectively. One possible von Neumann sub-algebra V is, then, generated by the double commutant of collections of the above projection operators, i.e. $V = lin_{\mathbb{C}}(\hat{P}_1, \hat{P}_2, \hat{P}_3, \hat{P}_4)$. In matrix notation possible representatives for the projection operators are

$$\hat{P}_1 = \begin{pmatrix} 1 & 0 & 0 & 0 \\ 0 & 0 & 0 & 0 \\ 0 & 0 & 0 & 0 \\ 0 & 0 & 0 & 0 \end{pmatrix}, \qquad \hat{P}_2 = \begin{pmatrix} 0 & 0 & 0 & 0 \\ 0 & 1 & 0 & 0 \\ 0 & 0 & 0 & 0 \\ 0 & 0 & 0 & 0 \end{pmatrix},$$

$$\hat{P}_3 = \begin{pmatrix} 0 & 0 & 0 & 0 \\ 0 & 0 & 0 & 0 \\ 0 & 0 & 1 & 0 \\ 0 & 0 & 0 & 0 \end{pmatrix}, \qquad \hat{P}_4 = \begin{pmatrix} 0 & 0 & 0 & 0 \\ 0 & 0 & 0 & 0 \\ 0 & 0 & 0 & 0 \\ 0 & 0 & 0 & 1 \end{pmatrix}.$$

The largest abelian von Neumann sub-algebra generated by the above projectors is $V = lin_{\mathbb{C}}(\hat{P}_1, \hat{P}_2, \hat{P}_3, \hat{P}_4)$, i.e. the algebra consisting of all 4×4 diagonal matrices with complex entries on the diagonal. Since this algebra is the largest, i.e. not contained in any other abelian sub-algebra of $\mathcal{B}(\mathbb{C}^4)$ it is called *maximal*.

Any change of bases $(\psi_1, \psi_2, \psi_3, \psi_4) \rightarrow (\rho_1, \rho_2, \rho_3, \rho_4)$ would give another maximal von Neumann sub-algebra $V' = lin_{\mathbb{C}}(\rho_1, \rho_2, \rho_3, \rho_4)$. In fact there are uncountably many such maximal algebras. If two basis are related by a simple permutation or phase factor, then the abelian von Neumann sub-algebras they generate are the same. Now, considering again our example, the algebra V will have many non maximal sub-algebras which, however, can be divided into two kinds as follows:

$$\begin{aligned} V_{\hat{P}_i \hat{P}_j} &= lin_{\mathbb{C}}(\hat{P}_i, \hat{P}_j, \hat{P}_k + \hat{P}_l) \\ &= \mathbb{C}\hat{P}_i + \mathbb{C}\hat{P}_j + \mathbb{C}(\hat{P}_k + \hat{P}_l) \\ &= \mathbb{C}\hat{P}_i + \mathbb{C}\hat{P}_j + \mathbb{C}(\hat{1} - \hat{P}_i + \hat{P}_j) \end{aligned} \tag{9.5}$$

for distinct $i, j, k, l \in \{1, 2, 3, 4\}$ and

$$V_{\hat{P}_i} = lin_{\mathbb{C}}(\hat{P}_i, \hat{P}_j + \hat{P}_k + \hat{P}_l) = \mathbb{C}\hat{P}_i + \mathbb{C}(\hat{P}_j + \hat{P}_k + \hat{P}_l) = \mathbb{C}\hat{P}_i + \mathbb{C}(\hat{1} - \hat{P}_i) \tag{9.6}$$

for distinct $i, j, k, l \in \{1, 2, 3, 4\}$.

Again there are uncountably many non maximal abelian sub-algebras. It is the case, though, that different maximal abelian sub-algebras have common non-maximal abelian sub-algebras, as could be the case that non-maximal abelian sub-algebras contain the same non-maximal abelian sub-algebra.

Thus, for example, the context V above contains all the sub-algebras $V_{\hat{P}_i \hat{P}_j}$ and $V_{\hat{P}_i}$ for $i, j \in \{1, 2, 3, 4\}$. Now, consider other 4 pairwise orthogonal projection operators $\hat{P}_1$, $\hat{P}_2$, $\hat{Q}_3$, $\hat{Q}_4$, such that the maximal abelian von Neumann algebra $V' = lin_{\mathbb{C}}(\hat{P}_1, \hat{P}_2, \hat{Q}_3, \hat{Q}_4) \neq V$. We then have that

$$V \cap V' = \{V_{\hat{P}_1}, V_{\hat{P}_2}, V_{\hat{P}_1, \hat{P}_2}\}. \tag{9.7}$$

From the above discussion it is easy to deduce that the sub-algebras $V_{\hat{P}_i}$ are contained in all the other sub-algebras which contain the projection operator $\hat{P}_i$ and all the sub-algebras $V_{\hat{P}_i \hat{P}_j}$ are contained in all the sub-algebras which contain both projection operators $\hat{P}_i$ and $\hat{P}_j$.

We should mention that there is also the trivial algebra $V'' = \mathbb{C}\hat{1}$, but we will not consider such algebra when considering the category $\mathcal{V}(\mathcal{H})$ since, otherwise, as it will be clear later on, we will never end up with a proposition being false, but the minimal truth value we would end up would be the trivially true one.

9.1.3 Topology on $\mathcal{V}(\mathcal{H})$

In this section we will explore different types of topologies which can be defined on $\mathcal{V}(\mathcal{H})$. Clearly, since $\mathcal{V}(\mathcal{H})$ is a poset it is naturally equipped with an Alexandroff topology whose basis is given by the collection of all lower sets in $\mathcal{V}(\mathcal{H})$, i.e., by sets of the form $\downarrow V := \{V' \in \mathcal{V}(\mathcal{H}) | V' \leq V\}$, $V \in \mathcal{V}(\mathcal{H})$. However is also possible two define two other topologies:

(i) *The vertical topology*
(ii) *The bucket topology*

The reason we are interested in introducing such new topologies is because, as shown in [45], and explained in details below, given a general group G its action on $\mathcal{V}(\mathcal{H})$ is not continuous with respect to the Alexandroff topology while it is continuous with respect to both the vertical topology and the bucket topology. Let us now describe these different topologies in detail.

Alexandroff Topology We will now analyse the group action on our original category $\mathcal{V}(\mathcal{H})$ equipped with the Alexandroff topology whose basis are all the lower sets $\downarrow V$, for all $V \in \mathcal{V}(\mathcal{H})$.

Let us consider a unitary operator $\hat{U}$ which acts on $\mathcal{V}(\mathcal{H})$. Such an action is defined as

$$\begin{aligned} l_{\hat{U}} : \mathcal{V}(\mathcal{H}) &\rightarrow \mathcal{V}(\mathcal{H}) \\ V &\mapsto \hat{U} V \hat{U}^{-1} \end{aligned} \tag{9.8}$$

where $\hat{U} V \hat{U}^{-1} := \{\hat{U} \hat{A} \hat{U}^{-1} | \hat{A} \in V\}$.

It is easy to see that this action is continuous with respect to the Alexandroff topology since it preserves the partial order on $\mathcal{V}(\mathcal{H})$. Therefore we obtain a representation of the Lie group G of the form

$$\begin{aligned} g \rightsquigarrow l_g : \mathcal{V}(\mathcal{H}) &\rightarrow \mathcal{V}(\mathcal{H}) \\ V &\mapsto l_g V := \hat{U}_g V \hat{U}_{g^{-1}} \end{aligned} \tag{9.9}$$

where each map l_g for $g \in G$ is continuous.

Moreover, the map $G \rightarrow \mathcal{U}(\mathcal{H})$, $g \rightarrow \hat{U}_g$, is strongly continuous, i.e., the map $g \rightarrow \hat{U}_g|\psi\rangle$ is a norm-continuous function for all $|\psi\rangle \in \mathcal{H}$. However, the definition of a proper representation of the topological group G also requires the following map to be continuous:

$$\begin{aligned} \Phi : G \times \mathcal{V}(\mathcal{H}) &\rightarrow \mathcal{V}(\mathcal{H}) \\ (g, V) &\mapsto l_g(V) \end{aligned} \tag{9.10}$$

To prove continuity it suffices to consider only the basis open sets, i.e., the sets $\downarrow V$, $V \in \mathcal{V}(\mathcal{H})$. Thus we consider

$$\begin{aligned} \Phi^{-1}(\downarrow V) &= \{(g, V') | l_g V' \in \downarrow V\} \\ &= \{(g, V') | l_g V' \subseteq V\} \end{aligned} \tag{9.11}$$

A necessary condition for this to be continuous is that, for each $V \in \mathcal{V}(\mathcal{H})$, the induced map

$$\begin{aligned} f_V : G &\rightarrow \mathcal{V}(\mathcal{H}) \\ g &\mapsto l_g(V) \end{aligned} \tag{9.12}$$

is continuous. If we consider the open set $\downarrow V \in \mathcal{V}(\mathcal{H})$ we then have

$$\begin{aligned} f_V^{-1}(\downarrow V) &= \{g \in G | l_g V \in \downarrow V\} \\ &= \{g \in G | l_g V \subseteq V\} \\ &= \{g \in G | l_g V = V\} \\ &=: G_V \end{aligned} \tag{9.13}$$

where G_V is the stabiliser of V, i.e. $G_V := \{g \in G | l_g V = V\}$. The last equality follows since the group action can not transform an algebra into a proper sub-algebra of itself.

Thus in order to show that the action is continuous we need to show that the stability group G_V is open. We know from the result in the Appendix[4] of [45] that if $\mathcal{V}(\mathcal{H})$ is Hausdorff, then G_V is closed, and since a typical Lie group does not have clopen subgroups it follows that the action is not continuous. However $\mathcal{V}(\mathcal{H})$ is not Hausdorff with respect to the Alexandroff topology. In fact, given $V_1, V_2 \in \mathcal{V}(\mathcal{H}) - \mathbb{C}\hat{1}$, with corresponding lower sets $\downarrow V_1, \downarrow V_2$, the smallest neighbourhood containing both is

$$\downarrow V_1 \cap \downarrow V_2 = \begin{cases} \downarrow (V_1 \cap V_2) & \text{if } V_1 \cap V_2 \neq \mathbb{C}\hat{1}; \\ \emptyset & \text{otherwise.} \end{cases} \tag{9.15}$$

Obviously the RHS might not be empty therefore to prove that G_V is closed we will need to use another strategy.

Lemma 9.1 *For each $V \in \mathcal{V}(\mathcal{H})$ the stabiliser G_V is a closed subgroup of the topological group G.*

---

[4]For completeness of exposition we will report both the corollary and the proof.

Corollary 9.1 Let the topological group G act on a Hausdorff topological space X in a continuous way, i.e., the G-action map $G \times X \to X$ is continuous. Then the stabiliser, G_x, of any $x \in X$ is a closed subgroup of G.

Proof For any given $x \in X$, consider the maps $f_x, h_x : G \to X$ defined by $f_x(g) := gx$ and $h_x(g) := x$ for all $g \in G$. The first is continuous since the G-action on X is continuous, and the second is continuous because constant maps are always continuous. Now

$$E(f_x, h_x) = \{g \in G | f_x(g) = h_x(g)\} = \{g \in G | gx = x\} = G_x. \tag{9.14}$$

We then need to show that $E(f_x, h_x)$ is closed. This follows from the fact that, given any two continuous maps $f, h : X \to Y$ between topological spaces X, Y, such that Y is Hausdorff, then $E(f, h) := \{x \in X | f(x) = h(x)\}$ is closed. In fact consider $E^c(f, h) := \{x \in X | f(x) \neq h(x)\}$. Let $x \in E^c(f, h)$. Then since Y is Hausdorff, there exists open neighbourhoods $N_{x,f}$ of $f(x)$ and $N_{x,h}$ of $h(x)$ such that $N_{x,f} \cap N_{x,h} = \emptyset$. Since f, h are continuous $f^{-1}(N_{x,f})$ and $h^{-1}(N_{x,h})$ are open. Thus $f^{-1}(N_{x,f}) \cap h^{-1}(N_{x,h})$ is open and non-empty (since $x \in f^{-1}(N_{x,f}) \cap h^{-1}(N_{x,h})$). In fact, for all $y \in f^{-1}(N_{x,f}) \cap h^{-1}(N_{x,h})$ we have that $f(y) \neq h(y)$. It follows that $E^c(f, h)$ is open, and hence that $E(f, h)$ is closed. This result implies that G_x is closed. □

Proof Given a unitary representation of G on the Hilbert space $\mathcal{H}$, the map $G \to \mathcal{U}(\mathcal{H})$ is strongly continuous, i.e., the map $g \to \hat{U}_g|\psi\rangle$ is a norm continuous function for every $|\psi\rangle \in \mathcal{H}$. Now let g_ν, $\nu \in I$ (a directed index set), be a net of elements of G in G_V, i.e.,

$$\hat{U}_{g_\nu} V \hat{U}_{g_\nu^{-1}} = V \tag{9.16}$$

for all $\nu \in I$. In other words, given any self-adjoint operator $\hat{A} \in V$ we obtain

$$\hat{U}_{g_\nu} \hat{A} \hat{U}_{g_\nu^{-1}} \in V \tag{9.17}$$

for all $\nu \in I$. We assume that the net of group elements converges with $\lim_{\nu \in I} g_\nu = g$. Since the G representation is strongly continuous then $\hat{U}_{g_\nu}$ converges strongly to $\hat{U}_g$. We will denote strong convergence by $\hat{U}_{g_\nu} \mapsto_s \hat{U}_g$. In order to show that G_V is closed we need to show that

$$\hat{U}_g \hat{A} \hat{U}_{g^{-1}} \in V. \tag{9.18}$$

However, operator multiplication is such that if $\hat{U}_{g_\nu} \mapsto_s \hat{U}_g$ then $\hat{U}_{g_\nu}\hat{A} \mapsto_s \hat{U}_g\hat{A}$. Since $\hat{U}^\dagger_{g_\nu} \mapsto_s \hat{U}^\dagger_g$ it follows that

$$\hat{U}_{g_\nu} \hat{A} \hat{U}_{g_\nu^{-1}} \mapsto_w \hat{U}_g \hat{A} \hat{U}_{g_\nu^{-1}} \tag{9.19}$$

where $\mapsto_w$ denotes convergence in the weak operator topology. Von Neumann algebras are weakly closed, thus $\hat{U}_g \hat{A} \hat{U}_{g_\nu^{-1}}$ belongs to V and G_V is closed. □

It follows that the group action in (9.10) is not continuous.

Vertical Topology The 'vertical' topology of $\mathcal{V}(\mathcal{H})$ will take into account the usual topology on coset spaces. In fact it is defined as the weak topology associated with the orbits of G on $\mathcal{V}(\mathcal{H})$. Hence its basis open sets are

$$\mathcal{O}(V, N) := \{l_g(V) | g \in N \subseteq G\} \tag{9.20}$$

where $V \in \mathcal{V}(\mathcal{H})$ and N is open in G. Since the action of G is transitive on each orbit by construction, it suffices to let N be a neighbourhood of the identity element, e, of G. The sets $\mathcal{O}(V, N)$ are then a basis of the neighbourhood filter[5] of V. Given this definition we have the following result:

$$\begin{aligned}\mathcal{O}(V, N_1) \cap \mathcal{O}(V, N_2) &= \{l_g(V) | g \in N_1 \subseteq G\} \cap \{l_g(V) | g \in N_2 \subseteq G\} \\ &= \{l_g(V) | g \in N_1 \cap N_2\} \\ &= \mathcal{O}(V, N_1 \cap N_2)\end{aligned} \tag{9.21}$$

[5]Recall that a neighbourhood filter of a point $x \in X$ of a topological space X, is simply the collection of all neighbourhoods of x.

The other requirements for the collection of $\mathcal{O}(V, N)$ to be a basis for the topology are clearly satisfied. Because of the 'vertical' nature of the topology, and the fact that G acts continuously on each orbit, intuitively one would guess that the G-action on $\mathcal{V}(\mathcal{H})$ is continuous in the 'vertical' topology; i.e., the map $G \times \mathcal{V}(\mathcal{H}) \to \mathcal{V}(\mathcal{H})$ is continuous.

Lemma 9.2 *The map $G \times \mathcal{V}(\mathcal{H}) \to \mathcal{V}(\mathcal{H})$ is continuous in the 'vertical' topology.*

In the proof of this lemma we will use a standard result in topology which we will report here for completeness reasons.

Theorem 9.1 *Given a topological space Y and a topological space X whose topology is determined by the family $\{A_\alpha | \alpha \in I\}$ of subsets of X, each with its own topology, then a map $f : X \to Y$ is continuous iff each $f|A_\alpha : A_\alpha \to Y$ is continuous.*

The proof can be found in [95]. We will now give the proof for Lemma 9.2.

Proof Given the nature of the poset $\mathcal{V}(\mathcal{H})$ it follows that it can be written as

$$\mathcal{V}(\mathcal{H}) = \coprod_{w \in \mathcal{V}(\mathcal{H})/G} \mathcal{O}_w \tag{9.22}$$

where $\mathcal{O}_w$ is the orbit associated with the coset w. Thus the group action map is now

$$\begin{aligned} \Phi : G \times \coprod_{w \in \mathcal{V}(\mathcal{H})/G} \mathcal{O}_w &\to \coprod_{w \in \mathcal{V}(\mathcal{H})/G} \mathcal{O}_w \\ \coprod_{w \in \mathcal{V}(\mathcal{H})/G} \mathcal{O}_w \times G &\to \coprod_{w \in \mathcal{V}(\mathcal{H})/G} \mathcal{O}_w \end{aligned} \tag{9.23}$$

where we have used that $G \times \coprod_{w \in \mathcal{V}(\mathcal{H})/G} \mathcal{O}_w \simeq \coprod_{w \in \mathcal{V}(\mathcal{H})/G} G \times \mathcal{O}_w$. Since the G-action is 'vertical', i.e., the G-action acts 'vertically' on each individual fibre/orbit, then according to Theorem 9.1 we have that $\Phi : G \times \coprod_{w \in \mathcal{V}(\mathcal{H})/G} \mathcal{O}_w \to \coprod_{w \in \mathcal{V}(\mathcal{H})/G} \mathcal{O}_w$ is continuous iff $\Phi_{|\mathcal{O}_w} : G \times \mathcal{O}_w \to \coprod_{w \in \mathcal{V}(\mathcal{H})/G} \mathcal{O}_w$ is continuous. However the group action on each orbit is continuous by definition thus Φ is continuous. □

We will denote $\mathcal{V}(\mathcal{H})$ with the 'vertical' topology as $\mathcal{V}(\mathcal{H})_{ver}$.

Bucket Topology An alternative way of defining a topology on $\mathcal{V}(\mathcal{H})$, which renders the action continuous, is to combine the 'vertical' topology with the Alexandroff topology. To do this we define the basis of the 'bucket' topology as all sets of the form

$$\downarrow\mathcal{O}(V, N) := \bigcup_{g \in N} \downarrow l_g(V) \tag{9.24}$$

where $V \in \mathcal{V}(\mathcal{H})$ and $N \subseteq G$ is an open neighbourhood of $e \in G$. These 'buckets' are a basis for the neighbourhood filter of V in the bucket topology. We note that since $\downarrow(V_1 \cap V_2) = \downarrow V_1 \cap \downarrow V_2$, (9.21) shows that

$$\downarrow\mathcal{O}(V, N_1) \cap \downarrow\mathcal{O}(V, N_2) = \downarrow\mathcal{O}(V, N_1 \cap N_2). \tag{9.25}$$

Similarly as above, also in this case all the remaining requirements for the collection of $\downarrow\mathcal{O}(V, N)$ to be a basis for the topology are clearly satisfied.

The following lemma will now be useful; here GV indicates the G orbit through V, i.e. $GV := \{l_g V | g \in G\}$.

Lemma 9.3 *if $V_1 \subseteq V$ then $\downarrow\mathcal{O}(V, N) \cap GV_1$ is open in the 'vertical' topology.*

Proof Given an element $V_0 \in \downarrow\mathcal{O}(V, N) \cap GV_1$, then there exists an element $W \in \mathcal{O}(V, N)$ such that $V_0 \subseteq W$. Since $\mathcal{O}(V, N)$ is open in the 'vertical' topology there exists some N_0 such that $W \in \mathcal{O}(W, N_0) \subseteq \mathcal{O}(V, N)$. We have seen above that the action of each $g \in G$ on $\mathcal{V}(\mathcal{H})$ is order preserving, thus $V_0 \subseteq W$ implies that $l_g V_0 \subseteq l_g W$ for all $g \in G$. Therefore we have

$$V_0 \in \mathcal{O}(V_0, N_0) \subseteq \downarrow\mathcal{O}(W, N_0) \cap GV_1 \subseteq \downarrow\mathcal{O}(V, N) \cap GV_1. \tag{9.26}$$

□

A direct consequence of the above lemma is that the bucket topology induces the 'vertical' topology on each G-orbit. Moreover it is clear from (9.24) that each bucket is the union of Alexandroff open sets. Therefore, every set open in the bucket topology is also open in the Alexandroff topology. The inverse however is not true. If follows that the bucket topology is *weaker* than the Alexandroff topology.

Lemma 9.4 *Let $dim(\mathcal{H}) \geq 3$. Then the bucket topology is not Hausdorff.*

Proof Given any two elements $V_1, V_2 \in \mathcal{V}(\mathcal{H})$ the smallest neighbourhoods of each are respectively $\downarrow\mathcal{O}(V_1, N_1)$ and $\downarrow\mathcal{O}(V_2, N_2)$. However

$$\downarrow\mathcal{O}(V_1, N_1) \cap \downarrow\mathcal{O}(V_2, N_2) \supseteq \downarrow\mathcal{O}(V_1 \cap V_2, N_1 \cap N_2). \tag{9.27}$$

Moreover, we know that there exists V_1 and V_2 such that $V_1 \cap V_2 \neq \mathbb{C}\hat{1}$. Therefore, for all N_1 and N_2, $\downarrow\mathcal{O}(V_1 \cap V_2, N_1 \cap N_2)$ is not empty hence $\downarrow\mathcal{O}(V_1, N_1) \cap \downarrow\mathcal{O}(V_2, N_2)$ is not empty. □

9.2 Topos Analogue of the State Space

We would now like to define the topos analogue of the state space. The way in which we would like to construct such a state space is in analogy with how it is constructed in classical physics. In particular, we would like a state space which allows a definition of physical quantities in terms of maps from the state space to the reals, as it is the case in classical physics.

Since we are in the realm of presheaves on $\mathcal{V}(\mathcal{H})$, the state space will itself be a presheaf, thus it will be defined context wise, i.e. for each abelian von Neumann algebra $V \in \mathcal{V}(\mathcal{H})$. It is precisely of such algebras that we will take advantage of when trying to define the state space. In fact, each algebra V has associated to it its Gel'fand spectrum, which is the topological space $\underline{\Sigma}_V$ of all multiplicative linear functionals of norm 1 on V, i.e. $\underline{\Sigma}_V := \{\lambda : V \to \mathbb{C} | \lambda(\hat{1}) = 1\}$. The property of being multiplicative means that

$$\lambda(\hat{A}\hat{B}) = \lambda(\hat{A})\lambda(\hat{B}); \quad \forall \hat{A}, \hat{B} \in V. \tag{9.28}$$

So the elements λ of the spectrum $\underline{\Sigma}_V$ are, in essence, algebra homomorphisms from V to $\mathbb{C}$. The topology on $\underline{\Sigma}_V$ is that of a compact Hausdorff space in the weak *-topology.

Now, what is interesting is the action of such homomorphisms $\lambda \in \underline{\Sigma}_V$ on self-adjoint operators $\hat{A} \in V$. In fact it turns out that such maps λ actually represent valuations which respect the FUNC principle. To see this consider an operator $\hat{A} \in V$, then, for each element λ of the spectrum $\underline{\Sigma}_V$ we obtain a value $\lambda(\hat{A}) \in sp(\hat{A})$ of A. On the other hand, for each element a of the spectrum of $\hat{A} \in V$, i.e., $a \in sp(\hat{A})$, there exists a corresponding element λ_i such that a is defined as $a = \lambda_i(\hat{A})$.

Moreover, given a Borel function $g : \mathbb{R} \to \mathbb{R}$, then

$$\lambda\big(g(\hat{A})\big) = g\big(\lambda(\hat{A})\big). \tag{9.29}$$

This is precisely the FUNC principle. Therefore the elements of the Gel'fand spectrum can each be interpreted as (different) valuations, i.e. maps which send each self-adjoint operator to an element of its spectrum such that FUNC holds.

Given the topological space $\underline{\Sigma}_V$ it is possible to represent a self-adjoint operator $\hat{A} \in V$ as a map from $\underline{\Sigma}_V$ to $\mathbb{C}$. This is because of the existence of the *Gel'fand representation theorem* which states that each von Neumann algebra V is isomorphic[6] to the algebra of continuous, complex-valued functions (denoted as $C(\underline{\Sigma}_V)$) on its Gel'fand spectrum $\underline{\Sigma}_V$. That is to say the following map is an isomorphisms

$$\begin{aligned} V &\to C(\underline{\Sigma}_V) \\ \hat{A} &\mapsto (\bar{A} : \underline{\Sigma}_V \to \mathbb{C}) \end{aligned} \tag{9.30}$$

where $\bar{A}$ is the Gel'fand transform of the operator $\hat{A}$ and it is defined as $\bar{A}(\lambda) := \lambda(\hat{A})$ for all $\lambda \in \underline{\Sigma}_V$. If $\hat{A}$ is self-adjoint then $\bar{A}$ is real valued and is such that $\bar{A}(\underline{\Sigma}_V) = sp(\hat{A})$.

[6]Definition does not really matter here and we will simply call it isomorphisms.

Thus, for each context $V \in \mathcal{V}(\mathcal{H})$ we have managed to reproduce a situation analogous to classical physics in which self-adjoint operators are identified with functions from a space to the reals. In this sense the topological space $\underline{\Sigma}_V$ can be interpreted as a local state space, one for each $V \in \mathcal{V}(\mathcal{H})$. Obviously the complete quantum picture is only given when we consider the collection of all the local state spaces, since not all operators are contained in a single algebra V. It is precisely such a collection of local state spaces which will define the topos analogue of the state space. This will be called the spectral presheaf and it is defined as follows:

Definition 9.7 The spectral presheaf, $\underline{\Sigma}$, is the covariant functor from the category $\mathcal{V}(\mathcal{H})^{op}$ to **Sets** (equivalently, the contravariant functor from $\mathcal{V}(\mathcal{H})$ to **Sets**) defined by:

- Objects: Given an object V in $\mathcal{V}(\mathcal{H})$, the associated set $\underline{\Sigma}(V) := \underline{\Sigma}_V$ is defined to be the Gel'fand spectrum of the (unital) commutative von Neumann subalgebra V, i.e. the set of all multiplicative linear functionals $\lambda : V \to \mathbb{C}$, such that $\lambda(\hat{\mathbb{1}}) = 1$.
- Morphisms: Given a morphism $i_{V'V} : V' \to V$ $(V' \subseteq V)$ in $\mathcal{V}(\mathcal{H})$, the associated function $\underline{\Sigma}_{VV'} := \underline{\Sigma}(i_{V'V}) : \underline{\Sigma}(V) \to \underline{\Sigma}(V')$ is defined for all $\lambda \in \underline{\Sigma}(V)$ to be the restriction of the functional $\lambda : V \to \mathbb{C}$ to the sub-algebra $V' \subseteq V$, i.e. $\underline{\Sigma}(i_{V'V})(\lambda) := \lambda_{|V'}$.

9.2.1 Example

Given the category $\mathcal{V}(\mathbb{C}^4)$ defined in the previous example we will now define the spectral presheaf. Let us first consider the maximal abelian sub-algebra $V = lin_{\mathbb{C}}(\hat{P}_1, \hat{P}_2, \hat{P}_3, \hat{P}_4)$, the Gel'fand spectrum $\underline{\Sigma}_V$ (which has a discrete topology) of this algebra contains 4 elements

$$\lambda_i(\hat{P}_j) = \delta_{ij} \quad (i = 1, 2, 3, 4). \tag{9.31}$$

We then consider the sub-algebra $V_{\hat{P}_1\hat{P}_2} = lin_{\mathbb{C}}(\hat{P}_1, \hat{P}_2, \hat{P}_3 + \hat{P}_4)$. Its Gel'fand spectrum $\underline{\Sigma}_{V_{\hat{P}_1\hat{P}_2}}$ will contain the elements

$$\underline{\Sigma}_{V_{\hat{P}_1\hat{P}_2}} = \{\lambda'_1, \lambda'_2, \lambda'_3\} \tag{9.32}$$

such that $\lambda'_1(\hat{P}_1) = 1$, $\lambda'_2(\hat{P}_2) = 1$ and $\lambda'_3(\hat{P}_3 + \hat{P}_4) = 1$, while all the rest will be zero.

Since $V_{\hat{P}_1\hat{P}_2} \subseteq V$, there exists a morphisms between the respective spectra as follows (for notational simplicity we will denote $V_{\hat{P}_1\hat{P}_2}$ as V'):

$$\begin{aligned} \underline{\Sigma}_{VV'} : \underline{\Sigma}_V &\to \underline{\Sigma}_{V'} \\ \lambda &\mapsto \lambda_{|V'}. \end{aligned} \tag{9.33}$$

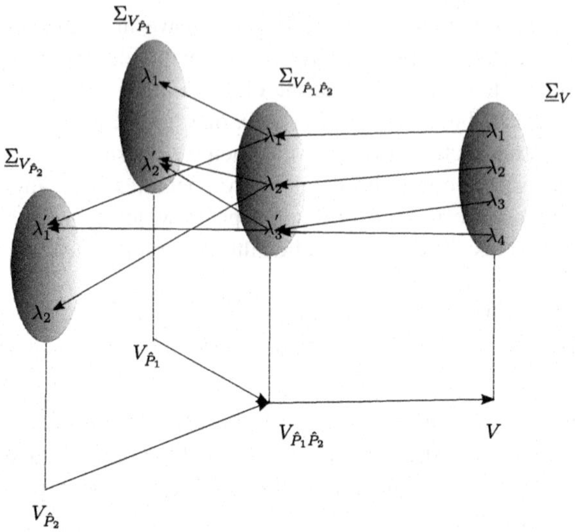

Fig. 9.1 Pictorial representation of the spectral presheaf for contexts V, $V_{\hat{P}_1,\hat{P}_2}$, $V_{\hat{P}_1}$ and $V_{\hat{P}_2}$, which satisfy the following relations $V_{\hat{P}_1} \subset V_{\hat{P}_1,\hat{P}_2} \subset V$, $V_{\hat{P}_2} \subset V_{\hat{P}_1,\hat{P}_2} \subset V$

Such that we obtain the following:

$$\begin{aligned}
\underline{\Sigma}_{VV'}(\lambda_1) &= \lambda_1' \\
\underline{\Sigma}_{VV'}(\lambda_2) &= \lambda_2' \\
\underline{\Sigma}_{VV'}(\lambda_3) &= \lambda_3' \\
\underline{\Sigma}_{VV'}(\lambda_4) &= \lambda_3'.
\end{aligned} \tag{9.34}$$

A worked out representation for contexts V, $V_{\hat{P}_1,\hat{P}_2}$, $V_{\hat{P}_1}$ and $V_{\hat{P}_2}$ is shown in Fig. 9.1.

From the above simple example we can generalise the definition of the spectrum for all sub-algebras $V' \subseteq V$. In particular we get

$$\underline{\Sigma}_{V_{\hat{P}_i,\hat{P}_j}} = \{\lambda_i, \lambda_j, \lambda_{kl}\} \tag{9.35}$$

where

$$\begin{aligned}
\lambda_i(\hat{P}_j) &= \delta_{ij} \\
\lambda_{kl}(\hat{P}_k + \hat{P}_l) &= 1
\end{aligned} \tag{9.36}$$

and all the rest equals zero. On the other hand for contexts $V_{\hat{P}_i}$ we obtain:

$$\underline{\Sigma}_{V_{\hat{P}_i}} = \{\lambda_i, \lambda_{jkl}\} \tag{9.37}$$

where

$$\begin{aligned}
\lambda_i(\hat{P}_i) &= 1 \\
\lambda_{jkl}(\hat{P}_j + \hat{P}_k + \hat{P}_k) &= 1
\end{aligned} \tag{9.38}$$

and all the rest is equal to zero.

9.3 The Spectral Presheaf and the Kochen-Specker Theorem

We will now show how the non existence of global elements of the spectral presheaf $\underline{\Sigma}$ is equivalent to the Kochen-Specker theorem.

Let us consider the topos $\mathbf{Sets}^{\mathcal{V}(\mathcal{H})^{op}}$. In order to define a global element we first of all have to define what a terminal object looks like. This is identified as the presheaf

$$\underline{1}:\mathcal{V}(\mathcal{H})\rightarrow \mathbf{Sets}$$

such that for each $V\in\mathcal{V}(\mathcal{H})$

$$\underline{1}_V:=\{*\}. \tag{9.39}$$

Given a map $i_{V'V}:V'\rightarrow V$ ($V'\subseteq V$), the corresponding morphisms is simply the identity map

$$\underline{1}(i_{V'V}):\{*\}\rightarrow\{*\}. \tag{9.40}$$

We now want to define a global element of the presheaf $\underline{\Sigma}$. Recall that this is defined as a map

$$\gamma:\underline{1}\rightarrow\underline{\Sigma} \tag{9.41}$$

such that for each context we get

$$\begin{aligned}\gamma_V:\underline{1}_V&\rightarrow\underline{\Sigma}_V\\ \{*\}&\rightarrow\underline{\Sigma}_V\\ \{*\}&\mapsto\gamma_V\big(\{*\}\big):=\lambda.\end{aligned} \tag{9.42}$$

Thus at the level of the stalks we retrieve the usual set definition of global element.

The connection between global sections and the Kochen-Specker theorem is given by the following theorem:

Theorem 9.2 *The spectral presheaf $\underline{\Sigma}$ has no global elements iff FUNC does not hold, i.e. $V(f(\hat{A}))\neq f(V(\hat{A}))$ for some Borel function $f:\mathbb{R}\rightarrow\mathbb{R}$, such that $\hat{B}=f(\hat{A})$.*

Proof Let us assume that $\underline{\Sigma}$ did have global sections. This would imply that there existed maps $\gamma:\underline{1}\rightarrow\underline{\Sigma}$, such that to each element $V\in\mathcal{V}(\mathcal{H})$, $\gamma_V:=\gamma_V(\{*\})\in\underline{\Sigma}_V$, i.e. $\gamma_V=\lambda:V\rightarrow\mathbb{C}$. In particular, for each selfadjoint operator $\hat{A}\in V_{sa}$, $\gamma_V(\hat{A})=\lambda(\hat{A})\in\sigma(\hat{A})$, i.e. it is an element of the spectrum of $\hat{A}$. Given a map $i_{V',V} : V' \to V$ ($V' \subseteq V$), then from the properties of global sections we have that

$$\underline{\Sigma}(i_{V',V})\gamma_V = \gamma_{V'}. \tag{9.43}$$

Now consider a self-adjoint operator $\hat{A}$, such that $\hat{A} \in V$ but $\hat{A} \notin V'$. Given the fact that $V' \subseteq V$ it is always possible to find an operator $\hat{B} \in V'$ such that $f(\hat{A}) = \hat{B}$ for some Borel function $f : \mathbb{R} \to \mathbb{R}$. Since $\gamma_V(\hat{A}) \in \sigma(\hat{A})$, by applying (9.43) to $\hat{A}$ we obtain

$$f\big(\gamma_V(\hat{A})\big) = \gamma_{V'}\hat{B} \tag{9.44}$$

which is precisely FUNC. The converse is left as an exercise. □

The above theorem leads immediately to the following statement:

Corollary 9.1 *The Kochen-Specker theorem is equivalent to the statement that, if* $\dim \mathcal{H} > 2$, *the spectral presheaf* $\underline{\Sigma}$ *has no global elements.*

Proof K-S theorem tells us that the FUNC condition can not hold, hence from the above theorem it follows that $\underline{\Sigma}$ has no global sections. □

It is interesting to note that, although the spectral presheaf has no global sections it still has local sections. An explicit construction of a local section of $\underline{\Sigma}$ will be given in Example 13.1.

Chapter 10
Topos Analogue of Propositions

In this chapter we will explain how propositions are represented in topos quantum theory. Such a representation requires the very important notion of daseinisation which, literally translated, means "to bring into existence". Such an operation is used when trying to represent a proposition which is locally defined by a projection operator, in all contexts. In fact it could be the case that, given a context V, the proposition represented by the projection operator $\hat{P}$ does not belong to V, i.e. $\hat{P} \notin V$. It is in such cases that the process of daseinisation is used. What it does is essentially to approximate the proposition $\hat{P}$ as best as possible so that it now belongs to V. This process of approximation is called *coarse graining*.

It is in this sense that the daseinisation brings the proposition 'into existence' in each context V.

10.1 Propositions

We will now describe how propositions are represented in $\mathbf{Sets}^{\mathcal{V}(\mathcal{H})^{op}}$. They are identified with terms of type $P(\Sigma)$, i.e. with sub-objects of the state space object.

In standard quantum theory propositions are represented by projection operators, on the other hand, in topos quantum theory they are identified with clopen (both open and closed) sub-objects of the spectral presheaf. A *clopen* sub-object $\underline{S} \subseteq \underline{\Sigma}$ is an object such that, for each context $V \in \mathcal{V}(\mathcal{H})$, the set $\underline{S}(V)$ is a clopen subset of $\underline{\Sigma}(V)$, where the latter is equipped with the usual compact Hausdorff spectral topology.

In order to understand how propositions are represented by clopen subsets of the state space we need to introduce the concept of 'daseinisation'. Roughly speaking, what daseinisation does is to approximate operators so as to 'fit' into any given context V. In fact, because the formalism defined so far is contextual, any proposition one wants to consider has to be studied within (with respect to) each context $V \in \mathcal{V}(\mathcal{H})$.

To see how this works consider the case in which we would like to analyse the projection operator $\hat{P}$, which corresponds, via the spectral theorem (see Defi-

C. Flori, *A First Course in Topos Quantum Theory*, Lecture Notes in Physics 868,
DOI 10.1007/978-3-642-35713-8_10, © Springer-Verlag Berlin Heidelberg 2013

nition 13.1), to the proposition "$A \in \Delta$".[1] In particular, let us take a context V such that $\hat{P} \notin P(V)$ (the lattice of projection operators in V). We, somehow need to define a projection operator which does belong to V and which is related, in some way, to our original projection operator $\hat{P}$. This can be achieved by approximating $\hat{P}$ from above in V by the 'smallest' projection operator in V, greater than or equal to $\hat{P}$. More precisely, the *outer daseinisation*, $\delta^o(\hat{P})$, of $\hat{P}$ is defined at each context V by[2]

$$\delta^o(\hat{P})_V := \bigwedge\{\hat{R} \in P(V) | \hat{R} \geq \hat{P}\}. \tag{10.1}$$

Since projection operators represent propositions, $\delta^o(\hat{P})_V$ is a coarse graining of the proposition "$A \in \Delta$".

This process of outer daseinisation takes place for all contexts and hence gives, for each projection operator $\hat{P}$, a collection of daseinised projection operators, one for each context V, i.e.,

$$\hat{P} \mapsto \{\delta^o(\hat{P})_V | V \in \mathcal{V}(\mathcal{H})\}. \tag{10.2}$$

Because of the Gel'fand transform, to each operator $\hat{P} \in P(V)$ there is associated the map $\bar{P} : \underline{\Sigma}_V \to \mathbb{C}$, which takes values in $\{0, 1\} \subset \mathbb{R} \subset \mathbb{C}$, since $\hat{P}$ is a projection operator. Thus, $\bar{P}$ is a characteristic function of the subset $S_{\hat{P}} \subseteq \underline{\Sigma}(V)$ defined by

$$S_{\hat{P}} := \{\lambda \in \underline{\Sigma}(V) | \bar{P}(\lambda) := \lambda(\hat{P}) = 1\}. \tag{10.3}$$

Since $\bar{P}$ is continuous with respect to the spectral topology on $\underline{\Sigma}(V)$, then $\bar{P}^{-1}(1) = S_{\hat{P}}$ is a clopen subset of $\underline{\Sigma}(V)$, because both $\{0\}$ and $\{1\}$ are clopen subsets of the Hausdorff space $\mathbb{C}$.

Through the Gel'fand transform it is then possible to define a bijective map between projection operators $\delta^o(\hat{P})_V \in P(V)$ and clopen subsets of[3] $\underline{\Sigma}_V := \underline{\Sigma}(V)$ where, for each *context* V,

$$S_{\delta^o(\hat{P})_V} := \{\lambda \in \underline{\Sigma}_V | \lambda(\delta^o(\hat{P})_V) = 1\}. \tag{10.4}$$

This correspondence between projection operators and clopen sub-objects of the spectral presheaf $\underline{\Sigma}$, which we denote as $Sub_{cl}(\underline{\Sigma})$, implies the existence of a lattice homomorphism for each V

$$\mathfrak{S} : P(V) \to Sub_{cl}(\underline{\Sigma})_V \tag{10.5}$$

---

[1]It should be noted that different propositions correspond to the same projection operator, i.e. the mapping from propositions to projection operators is many to one. Thus, to account for this, one is really associating equivalence class of propositions to each projection operator. The reason why von Neumann algebras were chosen instead of general C* algebras is precisely because all projections representing propositions are contained in the former, but not necessarily in the latter.

[2]Here the ordering $\leq$ is the lattice ordering of the lattice $P(V)$ of projection operators in V. See Appendix for details.

[3]Note that we will use both notations $\underline{\Sigma}_V := \underline{\Sigma}(V)$ interchangeably.

such that

$$\delta^o(\hat{P})_V \mapsto \mathfrak{S}\big(\delta^o(\hat{P})_V\big) := S_{\delta^o(\hat{P})_V}. \tag{10.6}$$

Here $Sub_{cl}(\underline{\Sigma})_V$ is the lattice of clopen subsets of the spectrum $\underline{\Sigma}_V$ with lattice operations given by intersection and union, while the lattice ordering is given by subset inclusion.

It can be shown that the collection of subsets $\{S_{\delta(\hat{P})_V}\}$, $V \in \mathcal{V}(\mathcal{H})$ induces a sub-object of $\underline{\Sigma}$.

In order to understand how this is done let us first give the definition of what a general sub-object of the topos analogue of the state space actually is.

Definition 10.1 A sub-object $\underline{S}$ of the spectral presheaf $\underline{\Sigma}$ is a contravariant functor $\underline{S} : \mathcal{V}(\mathcal{H}) \to$ **Sets** such that:

- $\underline{S}_V := \underline{S}(V)$ is a subset of $\underline{\Sigma}_V$ for all $V \in \mathcal{V}(\mathcal{H})$.
- Given a map $i_{V'V} : V' \subseteq V$, then $\underline{S}(i_{V'V}) : \underline{S}_V \to \underline{S}'_V$ is simply the restriction of the map $\underline{\Sigma}(i_{V'V})$ to the subset $\underline{S}_V \subseteq \underline{\Sigma}_V$, thus it is given by $\lambda \mapsto \lambda_{|V'}$.

Obviously, for clopen sub-objects we simply require that $\underline{S}_V$ be clopen in the above definition.

Theorem 10.1 *For each projection operator $\hat{P} \in P(\mathcal{H})$ the collection*

$$\underline{\delta(\hat{P})} := \big\{S_{\delta(\hat{P})_V} | V \in \mathcal{V}(\mathcal{H})\big\} \tag{10.7}$$

forms a (clopen) sub-object of the spectral presheaf.

Proof We already know that for each $V \in \mathcal{V}(\mathcal{H})$, $S_{\delta(\hat{P})_V} \subseteq \underline{\Sigma}_V$. Therefore, what we need to show is that these clopen subsets get mapped one to another by the presheaf morphisms. To see that this is the case consider an element $\lambda \in S_{\delta(\hat{P})_V}$. Given any $V' \subseteq V$, then by the definition of daseinisation we get $\delta^o(\delta^o(\hat{P})_V)_{V'} = \delta^o(\hat{P})_{V'} = \bigwedge\{\hat{\alpha} \in P(V') | \hat{\alpha} \geq \delta^o(\hat{P})_V\} \geq \delta^o(\hat{P})_V$. Therefore, if $\delta^o(\hat{P})_{V'} - \delta^o(\hat{P})_V = \hat{\beta}$, then $\lambda(\delta(\hat{P})_{V'}) = \lambda(\delta(\hat{P})_V) + \lambda(\hat{\beta}) = 1$ since $\lambda(\delta(\hat{P})_V) = 1$ and $\lambda(\hat{\beta}) \in \{0, 1\}$. Therefore

$$\{\lambda_{|V'} | \lambda \in S_{\delta(\hat{P})_V}\} \subseteq \{\lambda \in S_{\delta(\hat{P})_{V'}}\} \tag{10.8}$$

however $\lambda_{|V'}$ is precisely $\underline{\Sigma}(i_{V'V})\lambda$, therefore

$$\{\lambda_{|V'} | \lambda \in S_{\delta(\hat{P})_V}\} = \underline{\Sigma}(i_{V'V}) S_{\delta(\hat{P})_V}. \tag{10.9}$$

It follows that $\underline{\delta(\hat{P})}$ is a sub-object of $\underline{\Sigma}$. □

We can now define the (outer) daseinisation as a mapping from the projection operators to the sub-object of the spectral presheaf given by

$$\begin{aligned}\delta : P(\mathcal{H}) &\to Sub_{cl}(\underline{\Sigma})\\ \hat{P} &\mapsto \big(\mathfrak{S}\big(\delta^o(\hat{P})_V\big)\big)_{V\in\mathcal{V}(\mathcal{H})} =: \underline{\delta(\hat{P})}.\end{aligned} \tag{10.10}$$

We will sometimes denote $\mathfrak{S}(\delta^o(\hat{P})_V)$ as $\underline{\delta(\hat{P})}_V$.

We know that the sub-objects of any object in a topos form a Heyting algebra. In particular, the sub-objects of the spectral presheaf form a Heyting algebra. However we are only considering the clopen sub-objects. Luckily it turns out that also the collection of all *clopen* sub-objects of $\underline{\Sigma}$ forms a Heyting algebra.

Theorem 10.2 *The collection, $Sub_{cl}(\underline{\Sigma})$, of all clopen sub-objects of $\underline{\Sigma}$ is a Heyting algebra.*

Proof First of all let us consider how the logical connectives are defined.

The '∧'- and '∨'-operations. Given two sub-objects $\underline{T}$ and $\underline{S}$ of $\underline{\Sigma}$, then the '∧'- and '∨'-operations are defined by

$$\begin{aligned}(\underline{S}\wedge\underline{T})_V &:= \underline{S}_V \cap \underline{T}_V\\ (\underline{S}\vee\underline{T})_V &:= \underline{S}_V \cup \underline{T}_V\end{aligned} \tag{10.11}$$

for all contexts $V\in\mathcal{V}(\mathcal{H})$. From the properties of open and closed subsets, it follows that if $\underline{S}_V$ and $\underline{T}_V$ are clopen as subsets then so are $\underline{S}_V\cap\underline{T}_V$ and $\underline{S}_V\cup\underline{T}_V$.

The zero and unit elements. The zero element in $Sub_{cl}(\underline{\Sigma})$ is the empty sub-object

$$\underline{0} := \big\{\emptyset_V \,|\, V\in Ob\big(\mathcal{V}(\mathcal{H})\big)\big\} \tag{10.12}$$

where $\emptyset_V$ is the empty subset of $\underline{\Sigma}_V$ and $Ob(\mathcal{V}(\mathcal{H}))$ simply indicates the objects in the category $\mathcal{V}(\mathcal{H})$.

The unit element in $Sub_{cl}(\underline{\Sigma})$ is the unit sub-object.

$$\underline{\Sigma} := \big\{\underline{\Sigma}_V \,|\, V\in Ob\big(\mathcal{H}(\mathcal{V})\big)\big\}. \tag{10.13}$$

Clearly both $\underline{0}$ and $\underline{\Sigma}$ are clopen sub-objects of $\underline{\Sigma}$.

The '⇒'-operation. We have seen in previous lectures that the negation operation in a Heyting algebra is given by the relative pseudo-complement. In particular $\neg\underline{S} := \underline{S}\Rightarrow\underline{0}$. To understand exactly how such an operation is defined let us first describe $\underline{S}\Rightarrow\underline{T}$. This is defined, for each context $V\in\mathcal{V}(\mathcal{H})$, as

$$\begin{aligned}(\underline{S}\Rightarrow\underline{T})_V &:= \big\{\lambda\in\underline{\Sigma}_V \,|\, \forall V'\subseteq V \text{ if } \underline{\Sigma}(i_{V'V})(\lambda)\in\underline{S}_{V'} \text{ then } \underline{\Sigma}(i_{V'V})(\lambda)\in\underline{T}_{V'}\big\}\\ &= \big\{\lambda\in\underline{\Sigma}_V \,|\, \forall V'\subseteq V \text{ if } \lambda_{|V'}\in\underline{S}_{V'} \text{ then } \lambda_{|V'}\in\underline{T}_{V'}\big\}.\end{aligned} \tag{10.14}$$

From the above it follows that the negation operation in the Heyting algebra $Sub_{cl}(\underline{\Sigma})$ is defined as follows:

$$\begin{aligned}(\neg\underline{S})_V := (\underline{S}\Rightarrow\underline{0})_V &= \big\{\lambda\in\underline{\Sigma}_V \,|\, \forall V'\subseteq V,\, \underline{\Sigma}(i_{V'V})\lambda\notin\underline{S}_{V'}\big\}\\ &= \big\{\lambda\in\underline{\Sigma}_V \,|\, \forall V'\subseteq V,\, \lambda_{|V'}\notin\underline{S}_{V'}\big\}.\end{aligned} \tag{10.15}$$

It is also possible to write the negation in terms of the complement of sets as follows:

$$(\neg \underline{S})_V = \bigcap_{V' \subseteq V} \{\lambda \in \underline{\Sigma}_V | \lambda_{|V'} \in \underline{S}^c_{V'}\} \tag{10.16}$$

where $\underline{S}^c_{V'}$ represents the standard complement of the set $\underline{S}_{V'}$. Since $\underline{S}_{V'}$ is clopen so will $\underline{S}^c_{V'}$. The map $\underline{\Sigma}(i_{V'V}) : \underline{\Sigma}_V \to \underline{\Sigma}'_V$ is continuous and surjective, thus $\underline{\Sigma}(i_{V'V})^{-1}(\underline{S}^c_{V'})$ is clopen. Such a subset is defined as

$$\underline{\Sigma}(i_{V'V})^{-1}(\underline{S}^c_{V'}) = \{\lambda \in \underline{\Sigma}_V | \lambda_{|V'} \in \underline{S}^c_{V'}\}. \tag{10.17}$$

Substituting for the formula of the negation operation we obtain

$$(\neg \underline{S})_V = \bigcap_{V' \subseteq V} \underline{\Sigma}(i_{V'V})^{-1}(\underline{S}^c_{V'}). \tag{10.18}$$

However, the right hand side of the above formula is not guaranteed to be clopen, in fact it is closed and it would only be clopen if the set $\{V'|V' \subseteq V\}$ over which the intersection ranges is actually finite.

Now we know that the collection of all clopen subsets for each $\underline{\Sigma}_V$ is a complete lattice, thus given a family of decreasing subsets there will exist a limiting point of such subsets which will belong to the lattice.

In our case the collection of $\underline{\Sigma}(i_{V'V})^{-1}(\underline{S}^c_{V'})$ for all $\{V'|V' \subseteq V\}$ is a decreasing net of clopen subsets of $\underline{\Sigma}_V$. This means that if $V'' \subseteq V'$ and $\lambda_{V''} \in S^c_{V''}$ then $\lambda_{|V'} \in S^c_{V'}$. That this is the case can be proved by contradiction, in fact if $\lambda_{|V'} \in S_{V'}$ then $\underline{\Sigma}_{V''V'}\lambda_{|V'} = \lambda_{|V''} \in S_{V''}$ which would be a contradiction. Therefore $\underline{\Sigma}(i_{V''V'})^{-1}(\underline{S}^c_{V''}) \subseteq \underline{\Sigma}(i_{V'V})^{-1}(\underline{S}^c_{V'})$. Therefore the right hand side of (10.18) represents a decreasing net of clopen sub-sets of $\underline{\Sigma}_V$. If we now define the limit point of such a net and call it $(\neg S)_V$ we have a definition of the negation of an element as a clopen subset. Thus we define

$$\begin{aligned}(\neg S)_V &:= int \bigcap_{V' \subseteq V} \underline{\Sigma}(i_{V'V})^{-1}(\underline{S}^c_{V'}) \\ &= int \bigcap_{V' \subseteq V} \{\lambda \in \underline{\Sigma}_V | \lambda_{|V'} \in (\underline{S}^c_{V'})\}. \end{aligned} \tag{10.19}$$

□

The fact that $Sub_{cl}(\underline{\Sigma})$ is a Heyting algebra implies that the map (10.10) associates propositions to a distributive lattice. Thus, differently from standard quantum logic, topos quantum logic is distributive. It was recently shown in [99] that $Sub_{cl}(\underline{\Sigma})$ is not only a Heyting algebra, but also a co-Heyting algebra implying that $Sub_{cl}(\underline{\Sigma})$ is actually a bi-Heyting algebra. The definitions of co-Heyting algebra and bi-Heyting algebra are the following

Definition 10.2 A co-Heyting algebra (also called Brouwer algebra J) is a lattice with both bottom and top element: 0 and 1, respectively, such that, given two elements A and B is possible to define the co-Heyting algebra implication $A \Leftarrow B$ as

$$A \Leftarrow B \leq C \quad \text{iff} \quad A \leq B \vee C \quad \forall C \in J. \tag{10.20}$$

It is straightforward to show that the underlying lattice of a co-Heyting algebra is distributive. Moreover, if the underlying lattice is complete then, for all $A \in J$ and all families $(A_i)_{i \in I} \subseteq J$ the following infinite distributivity law holds:

$$A \vee \bigwedge_{i \in I} A_i = \bigwedge_{i \in I} (A \vee A_i). \tag{10.21}$$

The co-Heyting negation is defined in term of the co-Heyting algebra implication as

$$\sim A := (1 \Leftarrow A). \tag{10.22}$$

Therefore $\sim A = \bigvee \{B \in J | A \vee B = 1\}$ is the smallest element in J such that $A \vee \sim A = 1$. From this definition it follows that

$$\text{if } A \leq B \quad \text{then } \sim A \geq \sim B \tag{10.23}$$

$$\sim\sim A \leq A \tag{10.24}$$

$$\sim\sim\sim A = \sim A \tag{10.25}$$

$$\sim A \wedge A \geq 0. \tag{10.26}$$

On the other hand a bi-Heyting algebra is

Definition 10.3 A bi-Heyting algebra K is a lattice which is both a Heyting algebra and a co-Heyting algebra. Therefore, given any two elements $A, B \in K$ there exists both the Heyting algebra implication $A \Rightarrow B$ and the co-Heyting algebra implication $A \Leftarrow B$. A bi-Heyting algebra K is called complete if it is complete as a Heyting algebra and complete as a co-Heyting algebra.

A canonical example of a bi-Heyting algebra is a Boolean algebra B.

10.1.1 Physical Interpretation of Daseinisation

We would like to analyse the significance of daseinisation from a physics perspective [40]. To this end, let us consider a projection $\hat{P}$ which represents the proposition $A \in \Delta$. We now consider a context V such that $\hat{P} \notin V$, thus we approximate this projection so as to be in V, obtaining $\delta^o(\hat{P})_V$. If the projection $\delta^o(\hat{P})_V$ is a spectral projector of the operator, $\hat{A}$, representing the quantity, A, then it represents the proposition $A \in \Gamma$ where $\Delta \subseteq \Gamma$. Therefore, the mapping

$$\begin{aligned} \delta^o_V : P(\mathcal{H}) &\to P(\mathcal{V}) \\ \hat{P} &\mapsto \delta^o(\hat{P})_V \end{aligned} \tag{10.27}$$

is the mathematical implementation of the idea of coarse graining of propositions, i.e. of generalizing a proposition.

If, on the other hand, $\delta^o(\hat{P})_V$ is not a spectral projector of the operator $\hat{A}$ representing the quantity, A, then $\delta^o(\hat{P})_V$ represents the proposition $B \in \Delta'$. The physical quantity B is now represented by the projection operator $\hat{B} \in P(V)$ of which $\delta^o(\hat{P})_V$ is a spectral projector. Given the fact that $\hat{P} \leq \delta^o(\hat{P})_V$, the proposition $B \in \Delta'$ is a coarse graining of $A \in \Delta$, in fact a general form of $B \in \Delta'$ could be $f(A) \in \Delta$, for some Borel function $f : sp(\hat{A}) \to \mathbb{R}$.

Obviously, for many contexts it is the case that $\delta^o(\hat{P})_V = \hat{1}$, which is the most general proposition of all.

From the analysis above we can deduce that, in this framework, there are two types of propositions:

(i) *Global propositions*, which are the propositions we start with and which we want to represent in various contexts, i.e. $(A \in \Delta)$.
(ii) *Local propositions*, which are the individual coarse graining of the global propositions, as referred to individual contexts V.

Thus, for every global proposition we obtain a collection of local propositions

$$\hat{P} \to \big(\delta^o(\hat{P})_V\big)_{V \in \mathcal{V}(\mathcal{H})}. \tag{10.28}$$

In the topos perspective we consider the collection of all these local propositions at the same time, as exemplified by (10.10).

10.2 Properties of the Daseinisation Map

The daseinisation map has certain important properties which are listed in the following theorem [32]:

Theorem 10.3

1. $\underline{\delta(P \vee Q)} = \underline{\delta(P)} \vee \underline{\delta(Q)}$, *i.e. it preserves the "or" operation.*
2. $\underline{\delta(P \wedge Q)} \leq \underline{\delta(P)} \wedge \underline{\delta(Q)}$, *i.e. it does not preserve the "and" operation.*
3. *If* $\hat{P} \leq \hat{Q}$, *then* $\underline{\delta(P)} \leq \underline{\delta(Q)}$.
4. *The daseinisation map is injective but not surjective.*
5. $\underline{\delta(\hat{0})} = \{\emptyset_V | V \in Ob(\mathcal{V}(\mathcal{H}))\}$.
6. $\underline{\delta(\hat{1})} = \{\underline{\Sigma}_V | V \in Ob(\mathcal{V}(\mathcal{H}))\}$.

We will now prove each of them.

Proof 1. For each context $V \in \mathcal{V}(\mathcal{H})$ we compute

$$\delta^o(P \vee Q)_V = \bigwedge\{\alpha \in P(V)|\alpha \geq P \vee Q\}. \tag{10.29}$$

Thus we are looking for the smallest projection operator in V which is 'greater' than $P \vee Q$. This is basically the least upper bound, therefore

$$\begin{aligned}\delta^o(P \vee Q)_V &= \bigwedge\{\alpha \in P(V)|\alpha \geq P \text{ or } \alpha \geq Q\}\\ &= \bigwedge\{\alpha \in P(V)|\alpha \geq P\} \vee \bigwedge\{\alpha \in P(V)|\alpha \geq Q\}\\ &= \delta^o(P)_V \vee \delta^o(Q)_V.\end{aligned} \tag{10.30}$$

In terms of sub-objects of the state space $\underline{\Sigma}$ we have that

$$\begin{aligned}\underline{\delta(P \vee Q)}_V &= S_{\delta^o(P\vee Q)_V} = \{\lambda \in \underline{\Sigma}_V|\lambda(\delta^o(P \vee Q)_V) = 1\}\\ &= \{\lambda \in \underline{\Sigma}_V|\lambda(\delta^o(P) \vee \delta^o(Q)_V) = 1\}\\ &= \{\lambda \in \underline{\Sigma}_V|\lambda(\delta^o(P)) = 1 \vee \lambda(\delta^o(Q)_V) = 1\}\\ &= \{\lambda \in \underline{\Sigma}_V|\lambda(\delta^o(P)) = 1\} \vee \{\lambda \in \underline{\Sigma}_V|\lambda(\delta^o(Q)_V) = 1\}\\ &= S_{\delta^o(P)} \cup S_{\delta^o(Q)_V}\\ &= \underline{\delta(\hat{P})}_V \vee \underline{\delta(\hat{Q})}_V.\end{aligned} \tag{10.31}$$

2. On the other hand, for each context $V \in \mathcal{V}(\mathcal{H})$, we have

$$\delta^o(P \wedge Q)_V = \bigwedge\{\alpha \in P(V)|\alpha \geq P \wedge Q\}. \tag{10.32}$$

Consider the case for which $P \wedge Q = 0$ then $\delta^o(0) = 0$. However, from the definition of daseinisation $\delta^o(P) \geq P$ and $\delta^o(Q) \geq Q$ therefore it can be the case that $\delta^o(P) \wedge \delta^o(Q) \geq 0$. Thus it follows that

$$\underline{\delta(P \wedge Q)} \leq \underline{\delta(P)} \wedge \underline{\delta(Q)} \tag{10.33}$$

$$S_{\delta^o(P\wedge Q)_V} \subseteq S_{\delta^o(P)} \cup S_{\delta^o(Q)_V}. \tag{10.34}$$

3. We want to show that if $\hat{P} \leq \hat{Q}$, then $\underline{\delta(\hat{P})} \subseteq \underline{\delta(\hat{Q})}$. If we consider $\underline{\hat{P}}$ and $\underline{\hat{Q}}$ for all contexts $V \in \mathcal{V}(\mathcal{H})$ we have that (i) for contexts V in which $\hat{P}, \hat{Q} \in V$ (note that if $\hat{Q} \in V$ then $\hat{P} \in V$) then $\hat{P} \leq \hat{Q}$, therefore $S_{\hat{P}} \subseteq S_{\hat{Q}}$; (ii) for contexts V in which neither or only $\hat{P}$ are in $P(V)$, then because $\delta^o(\hat{Q})_V \geq Q$ and $\delta^o(\hat{P})_V \geq P$ are the smallest satisfying such conditions, it follows that $\delta^o(\hat{Q})_V \geq \delta^o(\hat{P})_V$, thus $S_{\delta^o(\hat{P})_V} \subseteq S_{\delta^o(\hat{Q})_V}$. Combining all these results for each $V \in \mathcal{V}(\mathcal{H})$ we obtain that $\underline{\delta(\hat{P})} \subseteq \underline{\delta(\hat{Q})}$.

4. For all $V \in \mathcal{V}(\mathcal{H})$

$$\delta^o(\hat{0})_V = \bigwedge\{\hat{\alpha} \in P(V)|\hat{\alpha} \geq \hat{0}\} = \hat{0}. \tag{10.35}$$

Since $\hat{0} \in P(V)$ for each V, then

$$S_{\delta^o(\hat{0})_V} = \{\lambda \in \underline{\Sigma}_V|\lambda(\hat{0}) = 1\} = \emptyset. \tag{10.36}$$

5. On the other hand for all $V \in \mathcal{V}(\mathcal{H})$

$$\delta^o(\hat{1})_V = \bigwedge\{\hat{\alpha} \in P(V)|\hat{\alpha} \geq \hat{1}\} = \hat{1} \tag{10.37}$$

therefore

$$S_{\delta^o(\hat{1})_V} = \{\lambda \in \underline{\Sigma}_V|\lambda(\hat{1}) = 1\} = \underline{\Sigma}_V := \{\lambda : V \to \mathbb{C}|\lambda(\hat{1}) = 1\}. \tag{10.38}$$

6. The map

$$\delta : P(\mathcal{H}) \to Sub_{cl}(\underline{\Sigma}) \tag{10.39}$$

is injective, but not surjective.

To prove injectivity we notice that there will be a context V such that $\hat{P} \in V$, therefore

$$\hat{P} = \bigwedge_{V \in \mathcal{V}(\mathcal{H})} \delta^o(\hat{P})_V \tag{10.40}$$

thus

$$\hat{P} = \bigwedge_{V \in \mathcal{V}(\mathcal{H})} \delta^o(\hat{P})_V = \bigwedge_{V \in \mathcal{V}(\mathcal{H})} \delta^o(\hat{Q})_V = \hat{Q}. \tag{10.41}$$

Surjectivity We want to show that there are elements in $Sub_{cl}(\underline{\Sigma})$ which are not of the form $\underline{\delta(\hat{P})}$ for some operator $\hat{P}$. We know what the general definition of a sub-object of $\underline{\Sigma}$ is, thus the question is if the clopen sub-objects of the form $\underline{\delta(\hat{P})}$ have any special properties which make them a subset of clopen sub-objects of $\underline{\Sigma}$. To this end, recall that in the definition of a sub-object of $\underline{\Sigma}$ we have the condition that, for each $V' \subseteq V$ the respective presheaf map

$$\underline{\Sigma}(i_{V'V}) : \underline{\Sigma}_V \to \underline{\Sigma}_{V'} \tag{10.42}$$

is such that, for a given subset $\underline{S}_V \subseteq \underline{\Sigma}_V$, then

$$\underline{\Sigma}(i_{V'V})(\underline{S}_V) \subseteq \underline{S}_{V'} \subseteq \underline{\Sigma}_{V'}. \tag{10.43}$$

What about the action of $\underline{\Sigma}(i_{V'V})$ on $\underline{\delta(\hat{P})}_V$? It turns out that, differently from (10.43), we obtain

$$\underline{\Sigma}(i_{V'V})\underline{\delta(\hat{P})}_V = \underline{\delta(\hat{P})}_{V'}. \tag{10.44}$$

To show this we will first define a map between the power-set of $\underline{\Sigma}$ and show that such a map is continuous, closed and open. The map we are interested in is

$$\begin{aligned} r_{V'V} : P(\underline{\Sigma}_V) &\rightarrow P(\underline{\Sigma}_{V'}) \\ S &\mapsto r_{V'V} S := \{\lambda_{|V'} | \lambda \in S\}. \end{aligned} \tag{10.45}$$

Using such a map[4] the analogue of (10.44) is

$$r_{V'V}(S_{\delta^o(\hat{P})_V}) = S_{\underline{\Sigma}(i_{V'V})(\delta^o(\hat{P})_V} = S_{\delta^o(\hat{P})_{V'}}. \tag{10.46}$$

For this to be the case the map r has to be closed, open and continuous.

Let us first show that such a map is continuous. Consider an open basis set[5] $R \in \underline{\Sigma}_{V'}$ we know that $\underline{\Sigma}_V := \{\lambda : V \rightarrow \mathbb{C} | \lambda(\hat{1}) = 1\}$ and, similarly, $\underline{\Sigma}_{V'} := \{\lambda : V' \rightarrow \mathbb{C} | \lambda(\hat{1}) = 1\}$. Moreover if $\lambda \in \underline{\Sigma}_V$ then from the definition of the presheaf maps it follows that $\lambda_{|V'} \in \underline{\Sigma}_{V'}$ when $V' \subseteq V$. We can then define, for any $R \in P(\underline{\Sigma}_{V'})$

$$r^{-1}_{V'V}(R) := R \cap \underline{\Sigma}_V. \tag{10.47}$$

Since the intersection[6] of open sets is open, thus $r^{-1}(S)$ is open.

Next we need to show that r is closed. Consider a closed subset $S \subseteq \underline{\Sigma}_V$. Since $\underline{\Sigma}_V$ is compact so is S and, since r is continuous, then $r(S)$ is compact in $\underline{\Sigma}_{V'}$. But $\underline{\Sigma}_{V'}$ is Hausdorff thus $r(S)$ is closed.

To show that r is open we note that since every $\lambda_{V'} \in \underline{\Sigma}_{V'}$ is of the form $\lambda_V|_{V'}$ for some $\lambda_V \in S \subseteq \underline{\Sigma}_V$, then

$$r(S) = \{\lambda_{V'} | \lambda \in S\} = S \cap \underline{\Sigma}_{V'}. \tag{10.48}$$

If S is open, then $S \cap \underline{\Sigma}_V$ is the intersection of two opens thus it is itself open. What we have proven holds for any pair $V' \subseteq V$.

Given the above properties of r, a clopen subset $S_{\delta^o(\hat{P})_V} \subseteq \underline{\Sigma}_V$ gets mapped to the clopen subset $r(S_{\delta^o(\hat{P})_V}) \in \underline{\Sigma}_{V'}$. Such a subset[7] is

$$r(S_{\delta^o(\hat{P})_V}) = int \bigcap \left\{ S_{\hat{Q}} \in Sub_{cl}(\underline{\Sigma}_{V'}) | r(S_{\delta^o(\hat{P})_V}) \subseteq S_{\hat{Q}} \right\} \tag{10.49}$$

thus $r(S_{\delta^o(\hat{P})_V}) \subseteq S_{\hat{Q}}$.

---

[4]Note that $r_{V'V} S = \underline{\Sigma}(i_{V'V}) S$.

[5]Note that for each $V \in \mathcal{V}(\mathcal{H})$, $\underline{\Sigma}_V$ has the spectral topology (being the spectrum of V) which is compact and Hausdorff. The details of such a topology are not necessary to prove continuity. It is worth saying, though, that it can be shown that a basis for this topology is the collection of clopen subsets. This renders the prof of continuity easier, however we will not use it here. On the other hand, when proving closeness of $r_{V'V}$ we will use the fact that $\underline{\Sigma}_V$ is a Hausdorff compact space.

[6]From now on, for notational simplicity we will write $r_{V'V}$ simply as r.

[7]Note that the *int* operation is needed for the subset to be clopen, otherwise it would only be closed.

We now need to show that $\hat{Q} \geq \delta^o(\hat{P})_V$. We prove this by contradiction. Assume that $\delta^o(\hat{P})_V \geq \hat{Q}$ and define $\hat{R} := \delta(\hat{P})_V - \hat{Q} \in P(V)$, such that $\lambda \in S_{\hat{R}}$. It follows that $\lambda \in S_{\delta^o(\hat{P})_V}$ but $\lambda \notin S_{\hat{Q}}$.

However if $\delta^o(\hat{P})_V = \delta^o(\hat{P})_{V'}$ (which means that $\delta^o(\hat{P})_V \in V' \subseteq V$) then $r(S_{\delta^o(\hat{P})_V}) = S_{\delta^o(\hat{P})_{V'}}$. In fact, given an element $\lambda \in \underline{\delta(\hat{P})}_V = S_{\delta^o(\hat{P})_V}$ by definition $\lambda(\delta^o(\hat{P})_V) = 1$. Since $\delta^o(\hat{P})_{V'} \geq \delta^o(\hat{P})_V$ $(V' \subseteq V)$, then $\lambda_{|V'} \in S_{\delta^o(\hat{P})_{V'}}$. On the other hand, if $\lambda \notin S_{\delta^o(\hat{P})_V}$ then $\lambda(\delta^o(\hat{P})_V) = 0$. Since $\delta^o(\hat{P})_{V'} = \delta^o(\hat{P})_V$ then $\lambda_{|V'} \notin S_{\delta^o(\hat{P})_{V'}}$.

Given the fact that every $\lambda_{V'} \in \underline{\Sigma}_{V'}$ is of the form $\lambda_V|_{V'} = r(\lambda_V)$ for some $\lambda \in S_{\delta^o(\hat{P})_V}$, then $r(S_{\delta^o(\hat{P})_V}) = S_{\delta^o(\hat{P})_{V'}}$ and $r((S_{\delta^o(\hat{P})_V})^c) = S^c_{\delta^o(\hat{P})_{V'}} \in \underline{\Sigma}_{V'}$.

It follows that in our case, since $\hat{Q} \in P(V') \subseteq P(V)$, we have $r(S_{\hat{Q}}) = S_{\hat{Q}}$ and $r((S_{\hat{Q}})^c) = (S_{\hat{Q}})^c$. Then $\lambda_{V'} \notin S_{\hat{Q}} \subseteq \underline{\Sigma}_V$ but $\lambda \in r(S_{\delta^o(\hat{P})_V})$. This means that

$$\hat{Q} \leq \delta^o(\hat{P})_V \Rightarrow r(S_{\delta^o(\hat{P})_V}) \nsubseteq S_{\hat{Q}}, \tag{10.50}$$

which is a contradiction, therefore it must be the case that $\hat{Q} \geq \delta^o(\hat{P})_V$. We can now write $r(S_{\delta^o(\hat{P})_V})$ as

$$\begin{aligned} r(S_{\delta^o(\hat{P})_V}) &= int \bigcap \{S_{\hat{Q}} \in Sub_{cl}(\underline{\Sigma}_{V'}) | \hat{Q} \geq \delta^o(\hat{P})_V\} \\ &= S_{\bigwedge\{\hat{Q} \in P(V') | \hat{Q} \geq \delta^o(\hat{P})_V\}} = S_{\underline{\Sigma}(i_{V'V})\delta^o(\hat{P})_V} \end{aligned} \tag{10.51}$$

therefore

$$\underline{\Sigma}(i_{V'V}) : S_{\delta^o(\hat{P})_V} \to S_{\delta^o(\hat{P})_{V'}}. \tag{10.52}$$

It follows that the clopen sub-objects of the form $\underline{\delta(\hat{P})}$ are such that the presheaf maps are also surjective. □

It is important to understand the significance of conditions 4 and 5 above. In particular the fact that, for all $V \in \mathcal{V}(\mathcal{H})$ we obtain $\delta^o(\hat{0})_V = \hat{0}$ implies that the null projection operator represents propositions of the form $A \in \Delta$, such that $sp(\hat{A}) \cap \Delta = \emptyset$, i.e. it represents false propositions. On the other hand for all $V \in \mathcal{V}(\mathcal{H})$ we obtain $\delta^o(\hat{1})_V = \hat{1}$. Thus the unit projection operator represents propositions of the form $A \in \Delta$, such that $sp(\hat{A}) \cap \Delta = sp(\hat{A})$, i.e. it represents true propositions. Therefore, only in this limiting cases does the topos concept of *true* and *false* resembles the respective classical concepts.

10.3 Example

To illustrate the concept of daseinisation of propositions let us consider a 2 spin system. We are interested in the spin in the z-direction, which is represented by the physical quantity S_z. In particular, we want to consider the following proposition $S_z \in [1.3, 2.3]$. Since the total spin in the z direction can only have values $-2, 0, 2$, the only value in the interval $[1.3, 2.3]$ which S_z can take is 2.

The self-adjoint operator representing S_z is

$$\hat{S}_z = \begin{pmatrix} 2 & 0 & 0 & 0 \\ 0 & 0 & 0 & 0 \\ 0 & 0 & 0 & 0 \\ 0 & 0 & 0 & -2 \end{pmatrix}.$$

The eigenstate with eigenvalue 2 would be $\psi = (1, 0, 0, 0)$, whose associated projector $\hat{P} := \hat{E}[S_z \in [1.3, 2.3]] = |\psi\rangle\langle\psi|$ would be

$$\hat{P}_1 = \begin{pmatrix} 1 & 0 & 0 & 0 \\ 0 & 0 & 0 & 0 \\ 0 & 0 & 0 & 0 \\ 0 & 0 & 0 & 0 \end{pmatrix}.$$

From our definition of $\mathcal{V}(\mathbb{C}^4)$ in Example 9.12, we know that the operator $\hat{S}_z$ is contained in all algebras which contain the projector operators $\hat{P}_1$ and $\hat{P}_4$. These algebras are: (i) the maximal algebra V and (ii) the non maximal sub-algebra $V_{\hat{P}_1\hat{P}_4}$. We will now analyse how the proposition $S_z \in [1.3, 2.3]$, represented by the projection operator $\hat{P}_1$, gets represented in the various abelian von Neumann algebra in $\mathcal{V}(\mathbb{C}^4)$.

1. *Context V and its sub-algebras.*

   Since V, $V_{\hat{P}_1\hat{P}_i}$ $(i \in \{2, 3, 4\})$ and $V_{\hat{P}_1}$ contain the projection operator P_1, then, for all these contexts we have

$$\delta^o(\hat{P}_1)_V = \delta^o(\hat{P}_1)_{V_{\hat{P}_1\hat{P}_i}} = \delta^o(\hat{P}_1)_{V_{\hat{P}_1}} = \hat{P}_1. \tag{10.53}$$

   Instead for context $V_{\hat{P}_i}$ for $i \neq 1$ we have

$$\delta^o(\hat{P}_1)_{V_{\hat{P}_i}} = \hat{P}_1 + \hat{P}_j + \hat{P}_k \quad j \neq i \neq k \in \{2, 3, 4\}. \tag{10.54}$$

   For contexts of the form $V_{\hat{P}_i\hat{P}_j}$, where $i \neq j \neq 1$, we have

$$\delta^o(\hat{P}_1)_{V_{\hat{P}_i}} = \hat{P}_1 + \hat{P}_j + \hat{P}_k \quad \text{for distinct } j, i, k \in \{2, 3, 4\}. \tag{10.55}$$

2. *Other maximal algebras which contain $\hat{P}_1$ and their sub-algebras.*

   Let us consider 4 pairwise orthogonal projection operators $\hat{P}_1, \hat{Q}_2, \hat{Q}_3, \hat{Q}_4$, such that the maximal abelian von Neumann algebra generated by such projections is different from V, i.e.

$$V' = lin_{\mathbb{C}}(\hat{P}_1, \hat{Q}_2, \hat{Q}_3, \hat{Q}_4) \neq V.$$

We then have the following daseinised propositions:

For contexts V' and $V'_{\hat{P}_1}$, as before, we have

$$\delta^o(\hat{P}_1)_{V'} = \delta^o(\hat{P}_1)_{V'_{\hat{P}_1}} = \hat{P}_1 \tag{10.56}$$

for contexts $V_{\hat{Q}_i}$ we have

$$\delta^o(\hat{P}_1)_{V_{\hat{Q}_i}} = \hat{P}_1 + \hat{Q}_j + \hat{Q}_k \quad \text{for distinct } j, i, k \in \{2, 3, 4\}. \tag{10.57}$$

Instead, for context $V_{\hat{Q}_i\hat{Q}_j}$, we have

$$\delta^o(\hat{P}_1)_{V_{\hat{Q}_i\hat{Q}_j}} = \hat{P}_1 + \hat{Q}_k \quad \text{for distinct } j, i, k \in \{2, 3, 4\}. \tag{10.58}$$

3. *Contexts which contain a projection operator which is Implied by* $\hat{P}_1$.
   Let us consider contexts $\tilde{V}$ which contain the projection operator $\hat{Q}$, such that $\hat{Q} \geq \hat{P}_1$, but do not contain $\hat{P}_1$ (if they did contain $\hat{P}_1$, we would be in exactly the same situation as above). In this situation the daseinisated propositions will be

$$\delta^o(\hat{P}_1)_{\tilde{V}} = \hat{Q}. \tag{10.59}$$

4. *Context which neither contain* $\hat{P}_1$ *nor a projection operator implied by it.*
   In these contexts V'' the only coarse grained proposition related to $\hat{P}_1$ is the unity operator, therefore we have

$$\delta^o(\hat{P}_1)_{V''} = \hat{\mathbb{1}}. \tag{10.60}$$

Now that we have defined all the possible coarse grainings of the proposition $\hat{P}_1$, for all possible contexts, we can define the presheaf $\underline{\delta(\hat{P}_1)}$ which is the topos analogue of the proposition $S_z \in [1.3, 2.3]$. As explained in the previous section, in order to obtain the presheaf $\underline{\delta(\hat{P}_1)}$ from the projection $\hat{P}_1$, we must apply the daseinisation map defined in (10.10), so as to obtain

$$\begin{aligned} \delta : P\big(\mathbb{C}^4\big) &\to Sub_{cl}(\underline{\Sigma}) \\ \hat{P}_1 &\mapsto \big(\mathfrak{S}\big(\delta^o(\hat{P}_1)_V\big)\big)_{V \in \mathcal{V}(\mathbb{C}^4)} =: \underline{\delta(\hat{P})} \end{aligned} \tag{10.61}$$

where the map $\mathfrak{S}$ was defined in (10.5), in particular

$$\mathfrak{S}\big(\delta^o(\hat{P}_1)_V\big) := S_{\delta^o(\hat{P}_1)_V} := \big\{\lambda \in \underline{\Sigma}_V \big| \lambda\big(\delta^o(\hat{P}_1)_V\big) = 1\big\}. \tag{10.62}$$

We now need to define the $\underline{\delta(\hat{P}_1)}$-morphisms. In order to do so we will again subdivide our analysis in different cases, as above.

1. *Maximal algebra* V *and its sub-algebras.*

The sub-algebras of V are of two kinds: $V_{\hat{P}_i,\hat{P}_j}$ and $V_{\hat{P}_k}$ for $i,j,k \in \{1,2,3,4\}$, such that in $\mathcal{V}(\mathbb{C}^4)$ we obtain the morphisms $i_{V_{\hat{P}_i\hat{P}_j},V} : V_{\hat{P}_i\hat{P}_j} \subseteq V$ and $i_{V_{\hat{P}_k},V} : V_{\hat{P}_k} \subseteq V$. Correspondingly $\underline{\delta(\hat{P}_1)}$-morphisms with domain $\underline{\delta(\hat{P}_1)}_V$ will be of two kinds. We will analyse one at the time. First we analyse the morphism

$$\underline{\delta(\hat{P}_1)}(i_{V_{\hat{P}_i\hat{P}_j},V}) : \underline{\delta(\hat{P}_1)}_V \to \underline{\delta(\hat{P}_1)}_{V_{\hat{P}_i,\hat{P}_j}}. \tag{10.63}$$

In this context we have

$$\underline{\delta(\hat{P}_1)}_V = \{\lambda \in \underline{\Sigma}_V | \lambda(\delta(\hat{P}_1)_V) = \lambda(\hat{P}_1) = 1\} = \{\lambda_1\}. \tag{10.64}$$

This is the case since, as we saw in the previous chapter $\underline{\Sigma}_V = \{\lambda_1, \lambda_2, \lambda_3, \lambda_4\}$ where $\lambda_i \hat{P}_j = \delta_{ij}$.

On the other hand for the contexts $V_{\hat{P}_i,\hat{P}_j}$ $i,j \in \{1,2,3,4\}$ we have the following:

$$\begin{aligned}\underline{\delta(\hat{P}_1)}_{V_{\hat{P}_1,\hat{P}_j}} &= \{\lambda_1\} \quad \text{where } \lambda_1(\delta(\hat{P}_1)_{V_{\hat{P}_1,\hat{P}_j}} = \hat{P}_1) = 1;\ j \in \{2,3,4\}\\ \underline{\delta(\hat{P}_1)}_{V_{\hat{P}_i,\hat{P}_j}} &= \{\lambda_{1k}\} \quad \text{where } \lambda_{1k}(\delta(\hat{P}_1)_{V_{\hat{P}_i,\hat{P}_j}} = (\hat{P}_1 + \hat{P}_k)) = 1;\ i \neq j \neq k \neq 1.\end{aligned} \tag{10.65}$$

The $\underline{\delta(\hat{P}_1)}$-morphisms for the above contexts would be

$$\begin{aligned}\underline{\delta(\hat{P}_1)}(i_{V_{\hat{P}_1,\hat{P}_j},V})(\lambda_1) &:= \lambda_1\\ \underline{\delta(\hat{P}_1)}(i_{V_{\hat{P}_i,\hat{P}_j},V})(\lambda_1) &:= \lambda_{1k}.\end{aligned} \tag{10.66}$$

The remaining $\underline{\delta(\hat{P}_1)}$-morphisms with domain $\underline{\delta(\hat{P}_1)}_V$ are

$$\underline{\delta(\hat{P}_1))}i_{V_{\hat{P}_i},V} : \underline{\delta(\hat{P}_1)}_V \to \underline{\delta(\hat{P}_1)}_{V_{\hat{P}_i}}. \tag{10.67}$$

In this case the local propositions $\underline{\delta(\hat{P}_1)}_{V_{\hat{P}_i}}$, $i \in \{1,2,3,4\}$ are

$$\begin{aligned}\underline{\delta(\hat{P}_1)}_{V_{\hat{P}_1}} &= \{\lambda_1\}\\ \underline{\delta(\hat{P}_1)}_{V_{\hat{P}_i}} &= \{\lambda_{1jk}\} \quad i,j,k \in \{2,3,4\}.\end{aligned} \tag{10.68}$$

The $\underline{\delta(\hat{P}_1)}$-morphisms are then

$$\begin{aligned}\underline{\delta(\hat{P}_1)}(i_{V_{\hat{P}_1},V}(\lambda_1) &:= \lambda_1\\ \underline{\delta(\hat{P}_1)}(i_{V_{\hat{P}_i},V})(\lambda_1) &:= \lambda_{1kl}.\end{aligned} \tag{10.69}$$

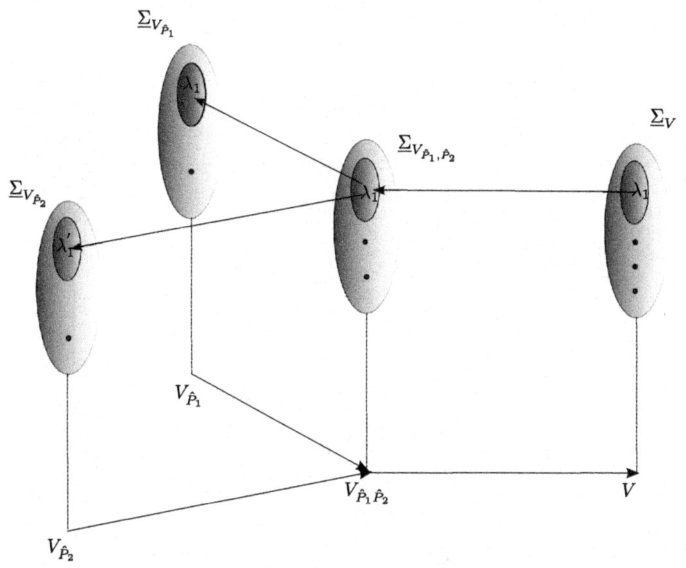

Fig. 10.1 Pictorial representation of the proposition $\underline{\delta \hat{P}_1}$ for contexts V, $V_{\hat{P}_1,\hat{P}_2}$, $V_{\hat{P}_1}$ and $V_{\hat{P}_2}$, which satisfy the following relations $V_{\hat{P}_1} \subset V_{\hat{P}_1,\hat{P}_2} \subset V$ and $V_{\hat{P}_2} \subset V_{\hat{P}_1,\hat{P}_2} \subset V$

A pictorial representation of the proposition $\underline{\delta(\hat{P}_1)}$ for context V and some of its sub-algebras is given in Fig. 10.1. On the other hand Fig. 10.2 is a pictorial representation of the more general proposition $\underline{\delta(\hat{P}_1 + \hat{P}_2)}$.

2. *Other maximal algebras which contain $\hat{P}_1$ and their sub-algebras.*

   As before we consider 4 pairwise orthogonal projection operators $\hat{P}_1$, $\hat{Q}_2$, $\hat{Q}_3$, $\hat{Q}_4$, such that the maximal abelian von Neumann algebra generated by such projections is different from V, i.e. $V' = lin_{\mathbb{C}}(\hat{P}_1, \hat{Q}_2, \hat{Q}_3, \hat{Q}_4) \neq V$.

   We then obtain the following morphisms with domain $\underline{\delta(\hat{P}_1)}_V$:

$$\begin{aligned} \underline{\delta(\hat{P}_1)}(i_{V_{\hat{P}_1},V}) : \underline{\delta(\hat{P}_1)}_V &\to \underline{\delta(\hat{P}_1)}_{V_{\hat{P}_1}} \\ \lambda_1 &\mapsto \lambda_1 \end{aligned} \tag{10.70}$$

$$\begin{aligned} \underline{\delta(\hat{P}_1)}(i_{V_{\hat{Q}_i},V}) : \underline{\delta(\hat{P}_1)}_V &\to \underline{\delta(\hat{P}_1)}_{V_{\hat{Q}_i}} \\ \lambda_1 &\mapsto \rho_{1jk} \end{aligned} \tag{10.71}$$

where $\underline{\Sigma}_{\hat{Q}_i} := \{\rho_i, \rho_{1jk}\}$, such that $\rho_i(\hat{Q}_i) = 1$ and $\rho_{1jk}(\hat{P}_1 + \hat{Q}_j + \hat{Q}_k) = 1$.

$$\begin{aligned} \underline{\delta(\hat{P}_1)}(i_{V_{\hat{Q}_i,\hat{Q}_j},V}) : \underline{\delta(\hat{P}_1)}_V &\to \underline{\delta(\hat{P}_1)}_{V_{\hat{Q}_i},\hat{Q}_j} \\ \lambda_1 &\mapsto \rho_{1k} \end{aligned} \tag{10.72}$$

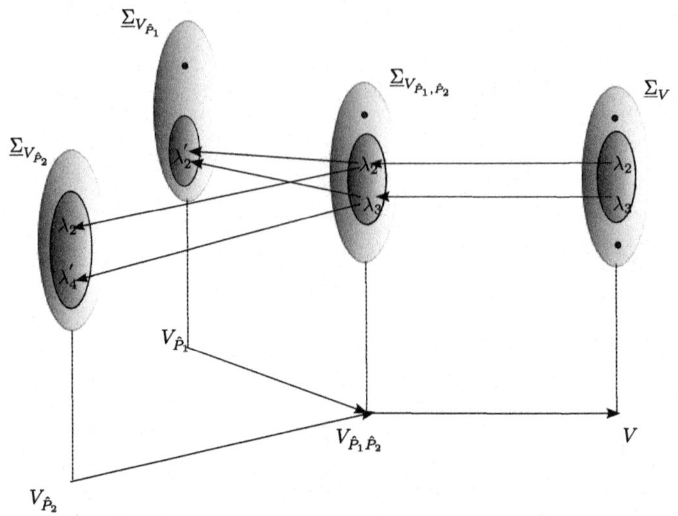

Fig. 10.2 Pictorial representation of the proposition $\underline{\delta(\hat{P}_2+\hat{P}_3)}$ for contexts V, $V_{\hat{P}_1,\hat{P}_2}$, $V_{\hat{P}_1}$ and $V_{\hat{P}_2}$, which satisfy the following relations $V_{\hat{P}_1}\subset V_{\hat{P}_1,\hat{P}_2}\subset V$ and $V_{\hat{P}_2}\subset V_{\hat{P}_1,\hat{P}_2}\subset V$

where $\underline{\Sigma}_{\hat{Q}_i,\hat{Q}_j}:=\{\rho_i,\rho_j,\rho_{1k}\}$, such that $\rho_i(\hat{Q}_i)=1$, $\rho_j(\hat{Q}_j)=1$ and $\rho_{1k}(\hat{P}_1+\hat{Q}_k)=1$. The computation of the remaining maps is left as an exercise.

3. *Contexts which contain a projection operator which is implied by* $\hat{P}_1$.
   We now consider a context $\tilde{V}$ which contains an operator $\hat{Q}$, such that $\hat{Q}\geq\hat{P}$. For such a context we have $\underline{\delta(\hat{P}_1)}_{\tilde{V}}=\{\lambda|\lambda(\hat{Q})=1\}$. Therefore, for sub-algebras which contain the operator $\hat{Q}$ the morphisms will simply map λ to itself. The rest of the maps are easily derivable.
4. *Context which neither contain* $\hat{P}_1$ *or a projection operator implied by it.*
   In such a context V'', whatever its spectrum is, each of the multiplicative linear functionals $\lambda_i\in\underline{\Sigma}_{V''}$ will assign value 1 to $\underline{\delta(\hat{P}_1)}=\hat{\mathbb{1}}$ and so will the elements of the spectrum of the sub-algebras $\bar{V}$ of V''. Thus, all the maps $\underline{\delta(\hat{P}_1)}(i_{\bar{V},V''})$ will simply be equivalent to spectral presheaf maps.

Chapter 11
Topos Analogues of States

In classical physics a pure state, s, is a point in the state space. It is the smallest subset of the state space which has measure one with respect to the Dirac measure δ_s.

Recall that a Dirac measure δ_s on some set S is defined by

$$\delta_s(A) = \begin{cases} 1 & \text{if } s \in A \\ 0 & \text{if } s \notin A \end{cases} \tag{11.1}$$

for any $s \in S$ and any measurable subset $A \subseteq S$.

Identifying states with subsets which have measure one is a consequence of the 1:1 correspondence which subsists between pure states and Dirac measure, i.e. for each pure state, s, there corresponds a unique Dirac measure δ_s. Moreover, propositions which are true in a pure state s are given by subsets of the state space which have measure one, with respect to the Dirac δ_s, i.e. those subsets which contain s. The smallest such subset is the one-element set $\{s\}$. Thus, a pure state can be identified with a single point in the state space.

On the other hand, more general states are represented by more general probability measures on the state space. This is the mathematical framework that underpins classical statistical physics.

However, the spectral presheaf $\underline{\Sigma}$ has *no* points.[1] Indeed, this is equivalent to the Kochen-Specker theorem! Thus the analogue of a pure state must be identified with some other construction. There are two (ultimately equivalent) possibilities: a 'state' can be identified with (i) an element of $P(P(\underline{\Sigma}))$ (the set of all possible sub-objects of the set of all possible sub-objects of $\underline{\Sigma}$); or (ii) an element of $P(\underline{\Sigma})$ (the set of all sub-object of $\underline{\Sigma}$). The first choice is called the *truth-object* option, the second is the *pseudo-state* option. In what follows we will concentrate on the second option. The first option will be described later.

[1] Recall that in a topos τ, a 'point' (or 'global element' or just 'element') of an object O is defined to be a morphism from the terminal object, 1_τ, to O.

C. Flori, *A First Course in Topos Quantum Theory*, Lecture Notes in Physics 868,
DOI 10.1007/978-3-642-35713-8_11, © Springer-Verlag Berlin Heidelberg 2013

11.1 Outer Daseinisation Presheaf

In order to define the truth-object we need to first introduce a new presheaf called the *outer presheaf*, whose definition is as follows:

Definition 11.1 The outer presheaf $\underline{O} : \mathcal{V}(\mathcal{H}) \to Sets$ is defined on

1. Objects: for each $V \in \mathcal{V}(\mathcal{H})$ we obtain $\underline{O}_V := P(V)$, i.e. the collection of all projection operators in V.
2. Morphisms: given a map $i : V' \subseteq V$ in $\mathcal{V}(\mathcal{H})$ the corresponding presheaf map is

$$\begin{aligned} \underline{O}(i_{V'V}) : \underline{O}_V &\to \underline{O}_{V'} \\ \hat{\alpha} &\mapsto \delta^o(\hat{\alpha})_{V'}. \end{aligned} \tag{11.2}$$

The above is a well defined presheaf. To see this all we need to show is that, given another inclusion map $j : V'' \subseteq V'$, then the following holds

$$\underline{O}(i \circ j) = \underline{O}(j) \circ \underline{O}(i). \tag{11.3}$$

Computing the left hand side we get

$$\begin{aligned} \underline{O}(i \circ j) : \underline{O}_V &\to \underline{O}_{V''} \\ \hat{\alpha} &\mapsto \delta^o(\hat{\alpha})_{V''}. \end{aligned} \tag{11.4}$$

Computing the right hand side we get

$$\begin{aligned} &\underline{O}_V \xrightarrow{\underline{O}(i)} \underline{O}_{V'} \xrightarrow{\underline{O}(j)} \underline{O}_{V''} \\ &\hat{\alpha} \mapsto \delta^o(\hat{\alpha})_{V'} \mapsto \delta^o\big(\delta^o(\hat{\alpha})_{V'}\big)_{V''} \end{aligned} \tag{11.5}$$

where $\delta^o(\delta^o(\hat{\alpha})_{V'})_{V''} \geq \delta^o(\hat{\alpha})_{V'}$. In particular, applying the definition of daseinisation recursively it follows trivially that $\delta^o(\delta^o(\hat{\alpha})_{V'})_{V''} = \delta^o(\hat{\alpha})_{V'|V''} = \delta^o(\hat{\alpha})_{V'}$.

Therefore for each projection operator $\hat{P}$, the assignment $V \to \delta^o(\hat{P})_V$ defines a global element of $\underline{O}$. We thus arrive at an alternative, but ultimately equivalent definition of the daseinisation map

$$\begin{aligned} \delta : P(\mathcal{H}) &\to \Gamma(\underline{O}) \\ \hat{P} &\mapsto \big\{\delta^o(\hat{P})_V \big| V \in \mathcal{V}(\mathcal{H})\big\}. \end{aligned} \tag{11.6}$$

11.2 Properties of the Outer-Daseinisation Presheaf

We would like to show that $\underline{O} \subseteq P_{cl}(\underline{\Sigma})$. First of all we introduce the presheaf $P(\underline{\Sigma})$ which is defined as follows:

Definition 11.2 The power object $P\underline{\Sigma}$ of $\underline{\Sigma}$ is the presheaf given by

1. (i) On objects $V \in \mathcal{V}(\mathcal{H})$

$$P\underline{\Sigma}_V := \{\eta_V : \underline{\Sigma}_{\downarrow V} \to \underline{\Omega}_{\downarrow V} | \eta_V \text{ is a natural transformation}\}.$$

Here $\underline{\Sigma}_{\downarrow V}$ is the restriction of $\underline{\Sigma}$ to a smaller poset namely $\downarrow V \subseteq \mathcal{V}(\mathcal{H})$.
2. On morphisms: for $i : V' \subseteq V$ the presheaf maps are

$$\begin{aligned} P\underline{\Sigma}(i_{V'V}) : P\underline{\Sigma}_V &\to P\underline{\Sigma}_{V'} \\ \eta &\mapsto \eta_{|V'} \end{aligned} \tag{11.7}$$

where here $\eta_{|V'} : \underline{\Sigma}_{\downarrow V'} \to \underline{\Omega}_{\downarrow V'}$.

It follows that $P_{cl}(\underline{\Sigma})$ is simply $P(\underline{\Sigma})$ restricted to clopen sub-presheaves of $\underline{\Sigma}$. To prove that $\underline{\mathcal{O}} \subseteq P_{cl}(\underline{\Sigma})$ we need to show that there exists a monic arrow $i : \underline{\mathcal{O}} \to P_{cl}(\underline{\Sigma})$. This will be done by first defining a monic arrow $i : \underline{\mathcal{O}} \to P(\underline{\Sigma})$ and then restricting it to $P_{cl}(\underline{\Sigma})$. First of all we recall, from the definition of exponential (7.1) that there exists, in any topos τ, a bijection

$$Hom_\tau(A, C^B) \to Hom_\tau(A \times B, C). \tag{11.8}$$

We would like to utilise this bijection to define the map $i : \mathcal{O} \to P(\underline{\Sigma})$. To do so we need to utilise another result of topos theory which states that sub-objects of a given object are in bijective correspondence with maps from the object in question to the sub-object classifier.[2] Thus for the case at hand $P(\underline{\Sigma}) \simeq \underline{\Omega}^{\underline{\Sigma}}$. Substituting this in (11.8) with $A = \underline{O}$ and $\tau = Sets^{\mathcal{V}(\mathcal{H})^{op}}$ we get

$$Hom_{Sets^{\mathcal{V}(\mathcal{H})^{op}}}\big(\underline{\mathcal{O}}, P(\underline{\Sigma})\big) \to Hom_{Sets^{\mathcal{V}(\mathcal{H})^{op}}}(\underline{\mathcal{O}} \times \underline{\Sigma}, \underline{\Omega}). \tag{11.9}$$

Consider a map $j \in Hom_{Sets^{\mathcal{V}(\mathcal{H})^{op}}}(\underline{\mathcal{O}} \times \underline{\Sigma}, \underline{\Omega})$ which, for each $V \in \mathcal{V}(\mathcal{H})$, we define as follows:

$$\begin{aligned} j_V : (\underline{\mathcal{O}} \times \underline{\Sigma})_V &\to \underline{\Omega}_V \\ (\hat{\alpha}, \lambda) &\mapsto j_V(\hat{\alpha}, \lambda) := \big\{V' \subseteq V | \underline{\Sigma}(i_{V'V})\lambda \in S_{\underline{\mathcal{O}}(i_{V'V})\hat{\alpha}}\big\} \end{aligned} \tag{11.10}$$

where $\hat{\alpha} \in P(V)$, $\underline{\mathcal{O}}(i_{V'V})\hat{\alpha} = \delta^o(\hat{\alpha})_{V'}$ and $S_{\underline{\mathcal{O}}(i_{V'V})\hat{\alpha}} := \{\lambda \in \underline{\Sigma}_{V'} | \langle \lambda, \delta^o(\hat{\alpha})_{V'} \rangle = 1\}$.

Since for all $V \in \mathcal{V}(\mathcal{H})$ and $\hat{\alpha} \in \underline{\mathcal{O}}_V$

$$S_{\underline{\Sigma}(i_{V'V})\hat{\alpha}} = \underline{\Sigma}(i_{V'V}) S_{\hat{\alpha}} \tag{11.11}$$

it follows that we can write $j_V(\hat{\alpha}, \lambda)$ as follows:

$$j_V(\hat{\alpha}, \lambda) := \big\{V' \subseteq V | \underline{\Sigma}(i_{V'V})\lambda \in \underline{\Sigma}(i_{V'V}) S_{\hat{\alpha}}\big\} \tag{11.12}$$

for all $(\hat{\alpha}, \lambda) \in \underline{\mathcal{O}}(V) \times \underline{\Sigma}_V$.

---

[2] This was shown when we defined the sub-object classifier (see Definition 8.1).

Lemma 11.1 *$j_V(\hat{\alpha}, \lambda)$ is a sieve in $\underline{\Omega}_V$.*

Proof We want to show that

$$j_V(\hat{\alpha}, \lambda) := \big\{V' \subseteq V \,|\, \underline{\Sigma}(i_{V'V})\lambda \in \underline{\Sigma}(i_{V'V})S_{\hat{\alpha}}\big\} \tag{11.13}$$

is a sieve.

To this end we need to show that if $V' \in j_V(\hat{\alpha}, \lambda)$ then for all $V'' \subseteq V'$, $V'' \in j_V(\hat{\alpha}, \lambda)$. So let us assume that $V' \in j_V(\hat{\alpha}, \lambda)$ then, $\lambda_{|V'} \in S_{\delta^o(\hat{\alpha})_{V'}}$ and $\lambda_{|V'}(\delta^o(\hat{\alpha})_{V'}) = 1$. Moreover, for any $V'' \subseteq V'$, $\underline{\Sigma}(i_{V''V'})\lambda_{|V'} \in \underline{\Sigma}_{V''}$ and $\underline{\Sigma}(i_{V''V'})S_{\delta^o(\hat{\alpha})_{V'}} = S_{\delta^o(\hat{\alpha})_{V''}}$.

However, since $\delta^o(\hat{\alpha})_{V''} \geq \delta^o(\hat{\alpha})_{V'}$ then $\lambda_{|V'}(\delta^o(\hat{\alpha})_{V''}) = 1$, in particular $(\lambda_{|V'})_{|V''}(\delta^o(\hat{\alpha})_{V''}) = 1$. Therefore $\underline{\Sigma}(i_{V''V'})\lambda_{|V'} \in \underline{\Sigma}(i_{V''V'})S_{\delta^o(\hat{\alpha})_{V'}} = S_{\delta^o(\hat{\alpha})_{V''}}$, i.e. $V'' \in j_V(\hat{\alpha}, \lambda)$. □

We now need to show that the collection of maps $j_V : (\underline{O} \times \underline{\Sigma})_V \to \underline{\Omega}_V$, for each $V \in \mathcal{V}(\mathcal{H})$ as defined above, combine together to form a natural transformation $j : \underline{O} \times \underline{\Sigma} \to \underline{\Omega}$.

Proof We want to prove that $j : \underline{O} \times \underline{\Sigma} \to \underline{\Omega}$ is indeed a natural transformation. Thus we need to show that for all pairs $V' \subseteq V$ the following diagram commutes:

$$\begin{array}{ccc}
\underline{O}_V \times \underline{\Sigma}_V & \xrightarrow{\;j_V\;} & \underline{\Omega}_V \\
\Big\downarrow{\scriptstyle f(i_{V'V})} & & \Big\downarrow{\scriptstyle \underline{\Omega}(i_{V'V})} \\
\underline{O}_{V'} \times \underline{\Sigma}_{V'} & \xrightarrow{\;j_{V'}\;} & \underline{\Omega}_{V'}
\end{array}$$

If we chase the diagram around we obtain

$$\begin{aligned}
\underline{\Omega}(i_{V'V}) \circ j_V(\hat{\alpha}, \lambda) &= \big(j_V(\hat{\alpha}, \lambda)\big)_{|V'} = j_V(\hat{\alpha}, \lambda) \cap \downarrow V' \\
&= \big\{V'' \subseteq V' \,|\, V'' \in j_V(\hat{\alpha}, \lambda)\big\} \\
&= \big\{V'' \subseteq V' \,|\, \underline{\Sigma}(i_{V''V})\lambda \in \underline{\Sigma}(i_{V'V})(S_{\hat{\alpha}})\big\}.
\end{aligned} \tag{11.14}$$

In order to define the left path of the diagram we need to define what the maps $f(i_{V'V})$ are.

$$\begin{aligned}
f(i_{V'V}) : (\underline{O}_V \times \underline{\Sigma}_V) &\to (\underline{O}_{V'} \times \underline{\Sigma}_{V'}) \\
(\hat{\alpha}, \lambda) &\mapsto f(i_{V'V})(\hat{\alpha}, \lambda) := \big(\underline{O}(i_{V'V})\hat{\alpha}, \underline{\Sigma}(i_{V''V})\lambda\big).
\end{aligned} \tag{11.15}$$

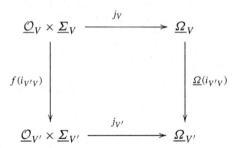

Therefore going around the diagram we obtain

$$\begin{aligned} j_{V'} \circ f(i_{V'V})(\hat{\alpha}, \lambda) &= j_{V'}\big(\underline{O}(i_{V'V})\hat{\alpha}, \underline{\Sigma}(i_{V'V})\lambda\big) \\ &= j_{V'}\big(\delta^o(\hat{\alpha})_{V'}, \lambda_{|V'}\big) \\ &= \big\{V'' \subseteq V' \big| \underline{\Sigma}(i_{V''V'})(\lambda_{|V'}) \in \underline{\Sigma}(i_{V''V'})S_{\delta^o(\hat{\alpha})_{V'}}\big\}. \end{aligned} \tag{11.16}$$

Now since

$$\begin{aligned} \underline{\Sigma}(i_{V''V})\lambda = \lambda_{V''} = \underline{\Sigma}(i_{V''V'})(\lambda_{|V'}) \\ \underline{\Sigma}(i_{V''V})(S_{\hat{\alpha}}) = S_{\delta^o(\hat{\alpha})_{V''}} = \underline{\Sigma}(i_{V''V'})S_{\delta^o(\hat{\alpha})_{V'}}. \end{aligned} \tag{11.17}$$

It follows that the above diagram commutes. □

Because of the equivalence in (11.9), to the map j there corresponds a map $i : \underline{O} \to P(\underline{\Sigma})$. However we are interested in $i : \underline{O} \to P_{cl}(\underline{\Sigma})$, but this restriction poses no problems.

In fact for each context $V \in \mathcal{V}(\mathcal{H})$ we have $\underline{O}_V = P(V)$, which is a lattice of operator with the usual lattice ordering. Now it is possible to put some topology on this set. Whatever topology we choose the entire set $\underline{O}_V$ will be both open and closed. Thus, for each clopen sub-objects $\underline{A} \subseteq \underline{\Sigma}$ we now form a clopen sub-object of $\underline{O} \times \underline{\Sigma}$ as follows:

$$\underline{O} \times \underline{A} \subseteq \underline{O} \times \underline{\Sigma}. \tag{11.18}$$

It follows that a sub-object of $\underline{O} \times \underline{\Sigma}$ is a clopen sub-object iff for each $V \in \mathcal{V}(\mathcal{H})$, $\underline{O}_V \times \underline{A}_V$ is a clopen subset of $\underline{O}_V \times \underline{\Sigma}_V$. We then have that $Sub_{cl}(\underline{O} \times \underline{\Sigma}) \subseteq Sub(\underline{O} \times \underline{\Sigma})$.

However we know that

$$Sub(X) \simeq Hom(X, \underline{\Omega}) \tag{11.19}$$

therefore

$$Sub_{cl}(\underline{O} \times \underline{\Sigma}) \subseteq Sub(\underline{O} \times \underline{\Sigma}) \simeq Hom(\underline{O} \times \underline{\Sigma}, \underline{\Omega}) \simeq Hom\big(\underline{O}, P(\underline{\Sigma})\big) \tag{11.20}$$

but[3]

$$Sub_{cl}(\underline{O} \times \underline{\Sigma}) \simeq Hom_{cl}(\underline{O} \times \underline{\Sigma}, \underline{\Omega}) \subseteq Hom(\underline{O} \times \underline{\Sigma}, \underline{\Omega}). \tag{11.21}$$

A moment of thought reveals that

$$Hom_{cl}(\underline{O} \times \underline{\Sigma}, \underline{\Omega}) \simeq Hom\big(\underline{O}, P_{cl}(\underline{\Sigma})\big). \tag{11.22}$$

The map $i : \underline{O} \to P_{cl}(\underline{\Sigma})$ is then the desired map.

[3] Here $Hom_{cl}(\underline{O} \times \underline{\Sigma}, \underline{\Omega})$ indicates $Hom(\underline{O} \times \underline{\Sigma}, \underline{\Omega})$ but restricted to clopen sub-objects of $\underline{O} \times \underline{\Sigma}$.

Theorem 11.1 $i : \underline{\mathcal{O}} \to P_{cl}(\underline{\Sigma})$ *is monic: for each* $V \in \mathcal{V}(\mathcal{H})$, $i_V : \underline{\mathcal{O}}_V \to P_{cl}(\underline{\Sigma})_V$ *is injective.*

Proof From the isomorphism

$$Sub(X) \simeq Hom(X, \Omega)$$

it follows that

$$P(\underline{\Sigma})_V := Sub(\underline{\Sigma}_{\downarrow V}).$$

If we then restrict to clopen sub-objects we get

$$P_{cl}(\underline{\Sigma})_V := Sub_{cl}(\underline{\Sigma}_{\downarrow V}) \tag{11.23}$$

therefore, for each context $V \in \mathcal{V}(\mathcal{H})$ we have that

$$\begin{aligned} i_V : \underline{\mathcal{O}}_V &\to P_{cl}(\underline{\Sigma}_V) \\ P(V) &\to Sub_{cl}(\underline{\Sigma}_{\downarrow V}) \\ \hat{P} &\mapsto (\underline{S}_{\hat{P}})_{\downarrow V}. \end{aligned} \tag{11.24}$$

Thus the map is clearly monic since $\underline{S}_{\hat{P}} = \mathfrak{S}(\hat{P})$ where $\mathfrak{S}$ is the lattice homomorpshim defined in (11.59). □

It is interesting now to compare the two definitions of daseinisation given so far:

$$\begin{aligned} \delta &: P(\mathcal{H}) \to \Gamma(\underline{\mathcal{O}}) \\ \delta &: P(\mathcal{H}) \to Sub_{cl}(\underline{\Sigma}). \end{aligned} \tag{11.25}$$

These two definitions, although seemingly different are exactly the same. In fact as we have just shown, $\underline{\mathcal{O}} \subseteq P_{cl}(\underline{\Sigma})$, thus a global element γ of $\underline{\mathcal{O}}$ will pick out, for each $V \in \mathcal{V}(\mathcal{H})$, an element $\gamma_V \in \underline{\mathcal{O}}_V \subseteq P_{cl}(\underline{\Sigma})$. Clearly, considering all the contexts together, γ will give rise to a (clopen due to how $\underline{\mathcal{O}}$ is defined) sub-object of $\underline{\Sigma}$, hence $\Gamma(\underline{\mathcal{O}}) \in P_{cl}(\underline{\Sigma})$.

11.3 Truth Object Option

The truth object is an object in the topos $\mathbf{Sets}^{\mathcal{V}(\mathcal{H})^{op}}$ which enables us to represent both pure states and density matrices. This is not possible if we were to use the pseudo-state option (defined in Sect. 11.4) which allows only a topos representation for pure states.

In order to understand precisely what a truth object is we will first analyse it in the context of classical physics and, subsequently, for quantum theory.

11.3.1 Example of Truth Object in Classical Physics

Let us consider a proposition $(A \in \Delta)$ meaning that the value of the quantity A lies in Δ. We want to define the truth value of such a proposition with respect to a given state $s \in X$ where X is the state space. In classical theory, the truth value of the above proposition in the state s is given by

$$v(A \in \Delta; s) := \begin{cases} 1 & \text{iff } s \in f^{-1}_{A\in\Delta}(\Delta) \\ 0 & \text{otherwise} \end{cases}$$

where $f^{-1}_{A\in\Delta}(\Delta) \subseteq X$ is the subset of the state space for which the proposition $(A \in \Delta)$ is true.

Another way of defining truth values is through the truth object $\mathbb{T}^s$, which is state dependent. The definition of the truth object is as follows: for each state s we define the set

$$\mathbb{T}^s := \{S \subseteq X | s \in S\}.$$

Since $(s \in f^{-1}_{A\in\Delta}(\Delta))$ iff $f^{-1}_{A\in\Delta}(\Delta) \in \mathbb{T}^s$, we can now write the above truth value in the following equivalent way:

$$v(A \in \Delta; \mathbb{T}) := \begin{cases} 1 & \text{iff } f^{-1}_{A\in\Delta}(\Delta) \in \mathbb{T}^s \\ 0 & \text{otherwise.} \end{cases}$$

In classical physics propositions $A \in \Delta$ are identified with subsets $S : \{s | A(s) \in \Delta\}$ of the state space X, for which that proposition is true.

11.3.2 Truth Object in Quantum Theory

We now want to define the state dependent truth object $\mathbb{T}^{|\psi\rangle}$ for quantum theory, i.e. we need to define an object $\underline{\mathbb{T}}^{|\psi\rangle}$ of type $P(P(\underline{\Sigma}))$ in $\mathbf{Sets}^{\mathcal{V}(\mathcal{H})^{op}}$. However, since propositions are represented by clopen sub-objects we actually need to restrict our attention to an element of type $P(P_{cl}(\underline{\Sigma}))$. Thus $\underline{\mathbb{T}}^{|\psi\rangle}$ has to be a subpresheaf of $P_{cl}(\underline{\Sigma})$, i.e. $\underline{\mathbb{T}}^{|\psi\rangle} \subseteq P_{cl}(\underline{\Sigma})$. Given a state ψ, the precise way in which this presheaf is defined is as follows:

Definition 11.3 The presheaf $\underline{\mathbb{T}}^{|\psi\rangle}$ has as:

(i) *Objects*: For each context V we get the set

$$\underline{\mathbb{T}}^{|\psi\rangle}_V := \{\hat{\alpha} \in \underline{O}_V | Prob(\hat{\alpha}, |\psi\rangle) = 1\} \tag{11.26}$$

$$= \{\hat{\alpha} \in \underline{O}_V | \langle\psi|\hat{\alpha}|\psi\rangle = 1\}. \tag{11.27}$$

(ii) *Morphisms*: Given two contexts $i : V' \subseteq V$ the associated morphisms is

$$\begin{aligned} \underline{\mathbb{T}}^{|\psi\rangle}(i_{V',V}) : \underline{\mathbb{T}}^{|\psi\rangle}_V &\to \underline{\mathbb{T}}^{|\psi\rangle}_{V'} \\ \hat{\alpha} &\mapsto \delta(\hat{\alpha})_{V'}. \end{aligned} \tag{11.28}$$

What about the truth object as defined for a density matrix? In that case the definition if as follows:

Definition 11.4 Given a density matrix ρ, the truth-object $\underline{\mathbb{T}}^{\rho}$ associated to it is defined on

1. Objects: for each $V \in \mathcal{V}(\mathcal{H})$ we obtain

$$\begin{aligned} \underline{\mathbb{T}}^{\rho}_V &:= \{\hat{\alpha} \in \underline{O}_V | Prob(\hat{\alpha}, \rho) = 1\} \\ &= \{\hat{\alpha} \in \underline{O}_V | \operatorname{tr}(\hat{\rho}\hat{\alpha}) = 1\}. \end{aligned} \tag{11.29}$$

2. Morphisms: Given two contexts $i : V' \subseteq V$ the associated morphisms is

$$\begin{aligned} \underline{\mathbb{T}}^{\rho}(i_{V',V}) : \underline{\mathbb{T}}^{|\psi\rangle}_V &\to \underline{\mathbb{T}}^{\rho}_{V'} \\ \hat{\alpha} &\mapsto \delta(\hat{\alpha})_{V'}. \end{aligned} \tag{11.30}$$

From the definition it follows that $\mathbb{T}^{|\psi\rangle}$ is a sub-object of $\underline{\Omega}$. However in the previous section we showed that $\underline{\Omega} \subseteq P_{cl}(\underline{\Sigma})$ it follows that $\mathbb{T}^{|\psi\rangle} \in P(P_{cl}(\underline{\Sigma}))$.

11.3.3 Example

We will now construct the truth object for the 4 dimensional Hilbert space $\mathcal{H}(\mathbb{C}^4)$ associated to the state $|\psi\rangle = (1, 0, 0, 0)$. We will then define the truth object associated to the density matrix $\rho = \sum_i p_i |\psi\rangle\langle\psi|$.

Pure State In our analysis we will only consider the maximal algebra $V := lin_{\mathbb{C}}(\hat{P}_1, \hat{P}_2, \hat{P}_3, \hat{P}_4)$ and all its subalgebras. We then have:

$$\underline{\mathbb{T}}^{|\psi\rangle}_V = \{\hat{\alpha} \in \underline{O}_V | \langle\psi|\hat{\alpha}|\psi\rangle = 1\} \tag{11.31}$$

$$\begin{aligned} &= \{\hat{P}_1, \hat{P}_1 + \hat{P}_{j\in\{2,3,4\}}, \hat{P}_1 + \hat{P}_{l\in\{2,3\}} + \hat{P}_{k\in\{3,4\}}, \\ &\quad \hat{P}_1 + \hat{P}_2 + \hat{P}_3 + \hat{P}_4\} \end{aligned} \tag{11.32}$$

$$\mathbb{T}^{|\psi\rangle}_{V_{\hat{P}_1}} = \{\hat{P}_1, \hat{P}_1 + \hat{P}_2 + \hat{P}_3 + \hat{P}_4\} \tag{11.33}$$

$$\mathbb{T}^{|\psi\rangle}_{V_{\hat{P}_2}} = \{\hat{P}_1 + \hat{P}_3 + \hat{P}_4, \hat{P}_1 + \hat{P}_2 + \hat{P}_3 + \hat{P}_4\} \tag{11.34}$$

$$\mathbb{T}^{|\psi\rangle}_{V_{\hat{P}_3}} = \{\hat{P}_1 + \hat{P}_2 + \hat{P}_4, \hat{P}_1 + \hat{P}_2 + \hat{P}_3 + \hat{P}_4\} \tag{11.35}$$

$$\mathbb{T}^{|\psi\rangle}_{V_{\hat{P}_4}} = \{\hat{P}_1 + \hat{P}_2 + \hat{P}_3, \hat{P}_1 + \hat{P}_2 + \hat{P}_3 + \hat{P}_4\} \tag{11.36}$$

$$\mathbb{T}^{|\psi\rangle}_{V_{\hat{P}_1,\hat{P}_2}} = \{\hat{P}_1, \hat{P}_1 + \hat{P}_2, \hat{P}_1 + \hat{P}_3 + \hat{P}_4, \hat{P}_1 + \hat{P}_2 + \hat{P}_3 + \hat{P}_4\} \tag{11.37}$$

$$\mathbb{T}^{|\psi\rangle}_{V_{\hat{P}_1,\hat{P}_3}} = \{\hat{P}_1, \hat{P}_1 + \hat{P}_3, \hat{P}_1 + \hat{P}_2 + \hat{P}_4, \hat{P}_1 + \hat{P}_2 + \hat{P}_3 + \hat{P}_4\} \tag{11.38}$$

$$\mathbb{T}^{|\psi\rangle}_{V_{\hat{P}_1,\hat{P}_4}} = \{\hat{P}_1, \hat{P}_1 + \hat{P}_4, \hat{P}_1 + \hat{P}_2 + \hat{P}_3, \hat{P}_1 + \hat{P}_2 + \hat{P}_3 + \hat{P}_4\} \tag{11.39}$$

$$\mathbb{T}^{|\psi\rangle}_{V_{\hat{P}_2,\hat{P}_3}} = \{\hat{P}_1 + \hat{P}_4, \hat{P}_1 + \hat{P}_2 + \hat{P}_4, \hat{P}_1 + \hat{P}_3 + \hat{P}_4, \hat{P}_1 + \hat{P}_2 + \hat{P}_3 + \hat{P}_4\} \tag{11.40}$$

$$\mathbb{T}^{|\psi\rangle}_{V_{\hat{P}_2,\hat{P}_4}} = \{\hat{P}_1 + \hat{P}_3, \hat{P}_1 + \hat{P}_2 + \hat{P}_3, \hat{P}_1 + \hat{P}_3 + \hat{P}_4, \hat{P}_1 + \hat{P}_2 + \hat{P}_3 + \hat{P}_4\} \tag{11.41}$$

$$\mathbb{T}^{|\psi\rangle}_{V_{\hat{P}_3,\hat{P}_4}} = \{\hat{P}_1 + \hat{P}_2, \hat{P}_1 + \hat{P}_2 + \hat{P}_3, \hat{P}_1 + \hat{P}_2 + \hat{P}_4, \hat{P}_1 + \hat{P}_2 + \hat{P}_3 + \hat{P}_4\}. \tag{11.42}$$

The maps between the different truth objects are:

$$\underline{\mathbb{T}}^{|\psi\rangle}(i_{V,V_1})(\hat{P}_1) = \hat{P}_1 \tag{11.43}$$

$$\underline{\mathbb{T}}^{|\psi\rangle}(i_{V,V_1})(\hat{P}_1 + \hat{P}_{j\in\{2,3,4\}}) = \hat{P}_1 + \hat{P}_2 + \hat{P}_3 + \hat{P}_4 \tag{11.44}$$

$$\underline{\mathbb{T}}^{|\psi\rangle}(i_{V,V_1})(\hat{P}_1 + \hat{P}_{l\in\{2,3\}} + \hat{P}_{k\in\{2,4\}}) = \hat{P}_1 + \hat{P}_2 + \hat{P}_3 + \hat{P}_4 \tag{11.45}$$

$$\underline{\mathbb{T}}^{|\psi\rangle}(i_{V,V_1})(\hat{P}_1 + \hat{P}_2 + \hat{P}_3 + \hat{P}_4) = \hat{P}_1 + \hat{P}_2 + \hat{P}_3 + \hat{P}_4. \tag{11.46}$$

The remaining maps are left as an exercise.

Density Matrix Let us consider again a 4 dimensional Hilbert space $\mathbb{C}^4$ representing our 2-spin system. We would like to consider a density matrix

$$\rho = \sum_i p_i |\psi\rangle\langle\psi|. \tag{11.47}$$

In particular, we consider a situation in which $1/2$ get $S_z = 2$ while $1/2$ get $S_z = -2$, thus our density matrix is

$$\hat{\rho} = 1/2 \begin{pmatrix} 1 & 0 & 0 & 0 \\ 0 & 0 & 0 & 0 \\ 0 & 0 & 0 & 0 \\ 0 & 0 & 0 & 0 \end{pmatrix} + 1/2 \begin{pmatrix} 0 & 0 & 0 & 0 \\ 0 & 0 & 0 & 0 \\ 0 & 0 & 0 & 0 \\ 0 & 0 & 0 & 1 \end{pmatrix} = 1/2\hat{P}_1 + 1/2\hat{P}_4.$$

We now consider the context $V = lin_{\mathbb{C}}(\hat{P}_1, \hat{P}_2, \hat{P}_3, \hat{P}_4)$ and compute

$$\underline{\mathbb{T}}^{\rho}_V = \{\hat{\alpha} \in \underline{O}_V \,|\, \mathrm{tr}(\hat{\rho}\hat{\alpha}) = 1\}. \tag{11.48}$$

By considering all possible operators in V we obtain

$$\underline{\mathbb{T}}^{\rho}_{V}=\big\{(\hat{P}_1+\hat{P}_4),(\hat{P}_1+\hat{P}_2+\hat{P}_4),(\hat{P}_1+\hat{P}_3+\hat{P}_4),(\hat{P}_1+\hat{P}_2+\hat{P}_3+\hat{P}_4)\big\}. \tag{11.49}$$

For context $V_{\hat{P}_1,\hat{P}_2}=lin_{\mathbb{C}}(\hat{P}_1,\hat{P}_2,\hat{P}_3+\hat{P}_4)$ we obtain

$$\underline{\mathbb{T}}^{\rho}_{V_{\hat{P}_1,\hat{P}_2}}=\big\{(\hat{P}_1+\hat{P}_3+\hat{P}_4),(\hat{P}_1+\hat{P}_2+\hat{P}_3+\hat{P}_4)\big\}. \tag{11.50}$$

For context $V_{\hat{P}_2,\hat{P}_3}=lin_{\mathbb{C}}(\hat{P}_2,\hat{P}_3,\hat{P}_1+\hat{P}_4)$ we obtain

$$\underline{\mathbb{T}}^{\rho}_{V_{\hat{P}_2,\hat{P}_3}}=\big\{(\hat{P}_1+\hat{P}_4),(\hat{P}_1+\hat{P}_2+\hat{P}_4),(\hat{P}_1+\hat{P}_3+\hat{P}_4),(\hat{P}_1+\hat{P}_2+\hat{P}_3+\hat{P}_4)\big\}. \tag{11.51}$$

For context $V_{\hat{P}_3,\hat{P}_4}=lin_{\mathbb{C}}(\hat{P}_3,\hat{P}_4,\hat{P}_1+\hat{P}_2)$ we obtain

$$\underline{\mathbb{T}}^{\rho}_{V_{\hat{P}_3,\hat{P}_4}}=\big\{(\hat{P}_1+\hat{P}_2+\hat{P}_4),(\hat{P}_1+\hat{P}_2+\hat{P}_3+\hat{P}_4)\big\}. \tag{11.52}$$

For context $V_{\hat{P}_1}=lin_{\mathbb{C}}(\hat{P}_1,\hat{P}_2+\hat{P}_3+\hat{P}_4)$ we obtain

$$\underline{\mathbb{T}}^{\rho}_{V_{\hat{P}_1}}=\big\{(\hat{P}_1+\hat{P}_2+\hat{P}_3+\hat{P}_4)\big\}. \tag{11.53}$$

For context $V_{\hat{P}_2}=lin_{\mathbb{C}}(\hat{P}_2,\hat{P}_1+\hat{P}_3+\hat{P}_4)$ we obtain

$$\underline{\mathbb{T}}^{\rho}_{V_{\hat{P}_2}}=\big\{(\hat{P}_1+\hat{P}_3+\hat{P}_4),(\hat{P}_1+\hat{P}_2+\hat{P}_3+\hat{P}_4)\big\}. \tag{11.54}$$

Morphisms:

$$\begin{aligned}
\underline{\mathbb{T}}^{\rho}_{V}&\to\underline{\mathbb{T}}^{\rho}_{V_{\hat{P}_1,\hat{P}_2}}\\
(\hat{P}_1+\hat{P}_4)&\mapsto(\hat{P}_1+\hat{P}_3+\hat{P}_4)\\
(\hat{P}_1+\hat{P}_2+\hat{P}_4)&\mapsto(\hat{P}_1+\hat{P}_2+\hat{P}_3+\hat{P}_4)\\
(\hat{P}_1+\hat{P}_3+\hat{P}_4)&\mapsto(\hat{P}_1+\hat{P}_2+\hat{P}_3+\hat{P}_4)\\
(\hat{P}_1+\hat{P}_2+\hat{P}_3+\hat{P}_4)&\mapsto(\hat{P}_1+\hat{P}_2+\hat{P}_3+\hat{P}_4)
\end{aligned} \tag{11.55}$$

$$\begin{aligned}
\underline{\mathbb{T}}^{\rho}_{V}&\to\underline{\mathbb{T}}^{\rho}_{V_{\hat{P}_3,\hat{P}_4}}\\
(\hat{P}_1+\hat{P}_4)&\mapsto(\hat{P}_1+\hat{P}_2+\hat{P}_4)\\
(\hat{P}_1+\hat{P}_2+\hat{P}_4)&\mapsto(\hat{P}_1+\hat{P}_2+\hat{P}_4)\\
(\hat{P}_1+\hat{P}_3+\hat{P}_4)&\mapsto(\hat{P}_1+\hat{P}_2+\hat{P}_3+\hat{P}_4)\\
(\hat{P}_1+\hat{P}_2+\hat{P}_3+\hat{P}_4)&\mapsto(\hat{P}_1+\hat{P}_2+\hat{P}_3+\hat{P}_4).
\end{aligned} \tag{11.56}$$

The rest are left as an exercise.

11.4 Pseudo-state Option

A pseudo-state is the object in our topos which most resembles the notion of a point state since it represents the smallest sub-object of the state space $\underline{\Sigma}$. Since $\underline{\Sigma}$ is a presheaf, its sub-objects will be them self presheafs, thus the *pseudo-state* is a presheaf, i.e. an object in $\mathbf{Sets}^{\mathcal{V}(\mathcal{H})^{op}}$.

Specifically, given a pure quantum state $\psi \in \mathcal{H}$, we define the presheaf

$$\underline{\mathfrak{w}}^{|\psi\rangle} := \underline{\delta(|\psi\rangle\langle\psi|)} \tag{11.57}$$

such that for each context V we have

$$\underline{\delta(|\psi\rangle\langle\psi|)}_V := \mathfrak{S}\Big(\bigwedge\{\hat{\alpha} \in P(V) \mid |\psi\rangle\langle\psi| \leq \hat{\alpha}\}\Big) = \mathfrak{S}\big(\delta^o(|\psi\rangle\langle\psi|)\big) \subseteq \underline{\Sigma}(V). \tag{11.58}$$

The map $\mathfrak{S}$ was defined in (11.59) of Chap. 10 however we will report it below for the sake of completeness:

$$\mathfrak{S} : P(V) \to Sub_{cl}(\underline{\Sigma})_V \tag{11.59}$$

is such that

$$\delta^o(|\psi\rangle\langle\psi|)_V \mapsto \mathfrak{S}\big(\delta^o(|\psi\rangle\langle\psi|)_V\big) := \underline{S}_{\delta^o(|\psi\rangle\langle\psi|)_V}. \tag{11.60}$$

Thus, for each context $V \in \mathcal{V}(\mathcal{H})$, the projection operator $\underline{\delta(|\psi\rangle\langle\psi|)}_V$ is the smallest projection operator implied by $|\psi\rangle\langle\psi|$. Since $|\psi\rangle\langle\psi|$ projects on a 1-dimensional sub-space of the Hilbert space, i.e. it projects on a state, $\underline{\delta(|\psi\rangle\langle\psi|)}_V$ identifies the smallest sub-space of $\mathcal{H}$ equal or bigger than the one dimensional sub-space $|\psi\rangle$.

The collection $(\underline{\delta(|\psi\rangle\langle\psi|)}_V)_{V\in\mathcal{V}(\mathcal{H})} =: (\underline{\mathfrak{w}}_V)_{V\in\mathcal{V}(\mathcal{H})}$ forms a sub-presheaf of $\underline{\Sigma}$ which is defined as follows:

Definition 11.5 For each state $|\psi\rangle \in \mathcal{H}$ we obtain the pseudo-state $\underline{\mathfrak{w}}^{|\psi\rangle} \in \mathbf{Sets}^{\mathcal{V}(\mathcal{H})^{op}}$ which is defined on

- Objects: for each context $V \in \mathcal{V}(\mathcal{H})$ we obtain

$$\underline{\delta(|\psi\rangle\langle\psi|)}_V := \{\lambda \in \underline{\Sigma}_V | \lambda(\delta^o(|\psi\rangle\langle\psi|)_V) = 1\}. \tag{11.61}$$

- Morphisms: for each $i_{V'V} : V' \subseteq V$ the corresponding map is simply the spectral presheaf map restricted to $\underline{\mathfrak{w}}^{|\psi\rangle}$, i.e.

$$\begin{aligned} \underline{\mathfrak{w}}^{|\psi\rangle}(i_{V'V}) : \underline{\mathfrak{w}}^{|\psi\rangle}_V &\to \underline{\mathfrak{w}}^{|\psi\rangle}_{V'} \\ \lambda &\mapsto \lambda_{|V'}. \end{aligned} \tag{11.62}$$

We then have the interesting result:

Theorem 11.2 *The map* $|\psi\rangle \to \underline{\mathfrak{w}}^{|\psi\rangle}$ *is injective.*

Proof We first show that if $\underline{\mathbb{T}}^{|\psi\rangle} = \underline{\mathbb{T}}^{|\phi\rangle}$ then $|\phi\rangle = e^{i\lambda}|\psi\rangle$. Applying the definitions for each $V \in \mathcal{V}(\mathcal{H})$ we have

$$\begin{aligned}
\underline{\mathbb{T}}_V^{|\psi\rangle} &= \{\hat{P} \in P(V) | \hat{P} \geq |\psi\rangle\langle\psi|\} \\
&= \{\hat{P} \in P(V) | \langle\psi|\hat{P}|\psi\rangle = 1\} \\
&= \underline{\mathbb{T}}_V^{|\phi\rangle} \\
&= \{\hat{P} \in P(V) | \langle\phi|\hat{P}|\phi\rangle = 1\}.
\end{aligned} \tag{11.63}$$

However if $\langle\psi|\hat{P}|\psi\rangle = 1$ then $\langle\psi e^{-i\lambda}|\hat{P}|e^{i\lambda}\psi\rangle = 1 = \langle\phi|\hat{P}|\phi\rangle = 1$ for some λ. Moreover $e^{i\lambda}\psi$ will be the only state which would satisfy this for all $\hat{P} \in \underline{\mathbb{T}}_V^{|\psi\rangle} = \underline{\mathbb{T}}_V^{|\phi\rangle}$, therefore $|\phi\rangle = e^{i\lambda}|\psi\rangle$, i.e. $|\psi\rangle \to \underline{\mathbb{T}}^{|\psi\rangle}$ is injective. Since the association $\underline{\mathbb{T}}^{|\psi\rangle} \to \underline{\mathfrak{w}}^{|\psi\rangle}$ is injective, as shown in Sect. 11.5, the result follows. □

Thus, for each state $|\psi\rangle$, there is associated a topos pseudo-state, $\underline{\mathfrak{w}}^{|\psi\rangle}$, which is defined as a sub-object of the spectral presheaf $\underline{\Sigma}$.

This presheaf $\underline{\mathfrak{w}}^{|\psi\rangle}$ is interpreted as the smallest clopen sub-object of $\underline{\Sigma}$, which represents the proposition[4] which is totally true in the state $|\psi\rangle$, namely the proposition $\underline{\delta(|\psi\rangle\langle\psi|)}$. Roughly speaking, it is the closest one can get to defining a point in $\underline{\Sigma}$.

11.4.1 Example

We will now give an example of how to define pseudo-states in our 4 dimensional Hilbert space $\mathbb{C}^4$. This is very similar to the example for propositions, since also for pseudo-states the concept of daseinisation is utilised. However, for pedagogical reasons we will, nonetheless, report it below.

Let us consider a state $\psi = (0, 1, 0, 0)$. The respective projection operator is

$$\hat{P}_2 = |\psi\rangle\langle\psi| = \begin{pmatrix} 0 & 0 & 0 & 0 \\ 0 & 1 & 0 & 0 \\ 0 & 0 & 0 & 0 \\ 0 & 0 & 0 & 0 \end{pmatrix}.$$

We now want to compute the outer daseinisation of such a projection operator for various contexts $V \in \mathcal{V}(\mathbb{C}^4)$. As it was done for the proposition, we will subdivide our analysis in different cases:

[4]Recall that in the topos framework, propositions are identified with clopen sub-objects of the state space.

1. *Context V and its sub-algebras.*
   Since the maximal algebra V is such that $|\psi\rangle\langle\psi| \in P(V)$, it follows that:

$$\delta^o\big(|\psi\rangle\langle\psi|\big)_V = |\psi\rangle\langle\psi|. \tag{11.64}$$

   This also holds for any sub-algebra of V containing $|\psi\rangle\langle\psi|$, i.e. $V_{\hat{P}_2,\hat{P}_i}$ for $i = \{1,3,4\}$ and $V_{\hat{P}_2}$.
   Instead, for the algebras $V_{\hat{P}_i,\hat{P}_j}$, where $i, j \in \{1,3,4\}$, we have

$$\delta^o\big(|\psi\rangle\langle\psi|\big)_{V_{\hat{P}_i,\hat{P}_j}} = \hat{P}_2 + \hat{P}_k \quad \text{for } k \neq i \neq j. \tag{11.65}$$

   On the other hand, for contexts $V_{\hat{P}_i}$ for $i \in \{1,3,4\}$ we have

$$\delta^o\big(|\psi\rangle\langle\psi|\big)_{V_{\hat{P}_i}} = \hat{P}_2 + \hat{P}_k + \hat{P}_j \quad \text{for } k \neq i \neq j. \tag{11.66}$$

2. *Other maximal algebras which contain $|\psi\rangle\langle\psi|$ and their sub-algebras.*
   Let us consider the 4 pairwise orthogonal operators $(\hat{Q}_1, \hat{P}_2, \hat{Q}_3, \hat{Q}_4)$, such that $V' := lin_{\mathbb{C}}(\hat{Q}_1, \hat{P}_2, \hat{Q}_3, \hat{Q}_4) \neq V$. For these contexts we obtain

$$\delta^o\big(|\psi\rangle\langle\psi|\big)_{V'} = \hat{P}_2 \tag{11.67}$$

$$\delta^o\big(|\psi\rangle\langle\psi|\big)_{V_{\hat{Q}_i}} = \hat{P}_2 + \hat{Q}_j + \hat{Q}_k$$

$$\delta^o\big(|\psi\rangle\langle\psi|\big)_{V_{\hat{Q}_i,\hat{Q}_j}} = \hat{P}_2 + \hat{Q}_k.$$

3. *Contexts that contain a projection operator implied by $|\psi\rangle\langle\psi|$.*
   If V'' contains $\hat{Q} \geq |\psi\rangle\langle\psi|$ then

$$\delta^o\big(|\psi\rangle\langle\psi|\big)_{V'} = \hat{Q}. \tag{11.68}$$

4. *Contexts contain neither $|\psi\rangle\langle\psi|$ nor a projection operator implied by it.*

$$\delta^o\big(|\psi\rangle\langle\psi|\big)_{V'} = \hat{1}. \tag{11.69}$$

We leave the definition of the $\underline{\mathfrak{w}}^{|\psi\rangle}$-morphisms as an exercise.

11.5 Relation Between Pseudo-state Object and Truth Object

We are now interested in understanding the relation between the two distinct ways in which a pure state is defined in the topos formulation of quantum theory. We expect that the two definitions turn out to be equivalent. We will first analyse the relation in classical physics then turn our attention to quantum theory.

Recall that in classical physics $\mathbb{T}^s : \{X \subseteq S | s \in X\}$. It then follows trivially that:

$$\{s\} = \bigcap \{X \in S | s \in X\} = \bigcap \{X \in \mathbb{T}^s\}. \tag{11.70}$$

Following the example of classical physics we will now try to define $\underline{\mathfrak{w}}^{|\psi\rangle}$ in terms of $\underline{\mathbb{T}}^{|\psi\rangle}$ and vice versa.

To this end, consider the assignment

$$V \mapsto \bigwedge \{\hat{\alpha} \in \underline{\mathbb{T}}_V^{|\psi\rangle}\} = \bigwedge \{\hat{\alpha} \in \underline{O}_V | |\psi\rangle\langle\psi| \leq \hat{\alpha}\} \tag{11.71}$$

where the last equality follows from the application of the definition of $\underline{\mathbb{T}}_V^{|\psi\rangle}$. But

$$\underline{\mathfrak{w}}^{|\psi\rangle} : V \mapsto \delta^o(|\psi\rangle\langle\psi|)_V = \bigwedge \{\hat{\alpha} \in \underline{O}_V | |\psi\rangle\langle\psi| \leq \hat{\alpha}\} \tag{11.72}$$

such that $\underline{\mathfrak{w}}_V^{|\psi\rangle}$ is the smallest projection operator in V for which $\underline{\mathfrak{w}}_V^{|\psi\rangle} \geq |\psi\rangle\langle\psi|$, therefore

$$\underline{\mathfrak{w}}_V^{|\psi\rangle} = \bigwedge \{\hat{\alpha} \in \underline{\mathbb{T}}_V^{|\psi\rangle}\}. \tag{11.73}$$

On the other hand

$$\underline{\mathbb{T}}_V^{|\psi\rangle} = \{\hat{\alpha} \in \underline{O}_V | \hat{\alpha} \geq |\psi\rangle\langle\psi|\}. \tag{11.74}$$

However, as said above, $\underline{\mathfrak{w}}_V^{|\psi\rangle}$ is the smallest projection operator in V such that $\underline{\mathfrak{w}}_V^{|\psi\rangle} \geq |\psi\rangle\langle\psi|$, therefore

$$\underline{\mathbb{T}}_V^{|\psi\rangle} = \{\hat{\alpha} \in \underline{O}_V | \hat{\alpha} \geq \underline{\mathfrak{w}}_V^{|\psi\rangle}\}. \tag{11.75}$$

It follows that there is a one to one correspondence between $\underline{\mathfrak{w}}^{|\psi\rangle}$ and $\underline{\mathbb{T}}^{|\psi\rangle}$.

Chapter 12
Truth Values

We now arrive at the very important concept of truth values in a topos and how they are assigned to quantum propositions. The important feature of these new truth values is that, given any set of quantum propositions, even incompatible ones, it is always possible to assess their truthfulness simultaneously. Moreover the set of truth values forms a Heyting algebra thus, we obtain an intuitionistic logic. What this implies is that a more realist picture of quantum theory emerges. In fact, it now makes sense to say that quantities possess values since, at each moment in time, we can assess whether any statement regarding properties of a physical quantity is true or not. However, since the set of truth values is bigger than the classical set $\{0, 1\}$ we do not obtain a strictly realist interpretation, but rather a broader definition of realism, which is called *neo-realism*.

12.1 Representation of Sub-object Classifier

We will now describe how the sub-object classifier is defined in the topos $\mathbf{Sets}^{\mathcal{V}(\mathcal{H})^{op}}$. Such an object represents the truth value object whose elements (global sections) are truth values, which get assigned to propositions (clopen sub-objects of $\underline{\Sigma}$). As we will see, we end up with a multi valued logic.

In the topos $\mathbf{Sets}^{\mathcal{V}(\mathcal{H})^{op}}$ the sub-object classifier Ω is identified with the following presheaf:

Definition 12.1 The presheaf $\underline{\Omega} \in \mathbf{Sets}^{\mathcal{V}(\mathcal{H})^{op}}$ is defined on

1. Objects. For any $V \in \mathcal{V}(\mathcal{H})$, the set $\underline{\Omega}_V$ is defined as the set of all sieves on V.
2. Morphisms. Given a morphism $i_{V'V} : V' \to V$ $(V' \subseteq V)$, the associated morphisms in $\underline{\Omega}$ is

$$\underline{\Omega}(i_{V'V}) : \underline{\Omega}_V \to \underline{\Omega}_{V'} \tag{12.1}$$

$$S \mapsto \underline{\Omega}\big((i_{V'V})\big)(S) := \big\{V'' \subseteq V' \big| V'' \in S\big\} = S \cap \downarrow V'. \tag{12.2}$$

C. Flori, *A First Course in Topos Quantum Theory*, Lecture Notes in Physics 868,
DOI 10.1007/978-3-642-35713-8_12, © Springer-Verlag Berlin Heidelberg 2013

In Chap. 7 we explain what a sieve is, however, for the particular case in which we are interested, namely sieves defined on the poset $\mathcal{V}(\mathcal{H})$, the definition of a sieve can be simplified as follows:

Definition 12.2 For all $V \in \mathcal{V}(\mathcal{H})$, a sieve S on V is a collection of sub-algebras $V' \subseteq V$ such that, if $V' \in S$ and $V'' \subseteq V'$, then $V'' \in S$. Thus S is a downward closed set.

In this case a maximal sieve on V is

$$\downarrow V := \{V' \in \mathcal{V}(\mathcal{H}) | V' \subseteq V\}. \tag{12.3}$$

In order for $\underline{\Omega}$ to be a presheaf, we need to show that indeed $\underline{\Omega}((i_{V'V}))(S) := \{V'' \subseteq V' | V'' \in S\}$ defines a sieve on V'. Thus we need to show that $\underline{\Omega}((i_{V'V}))(S) := \{V'' \subseteq V' | V'' \in S\}$ is a downward closed set with respect to V'. It is straightforward to deduce this from the definition.

Given the sub-object classifier $\underline{\Omega}$, truth values are defined as elements of $\underline{\Omega}$, thus as global sections. In particular, a global element γ of $\underline{\Omega}$ is a map $\gamma : \underline{1} \rightarrow \underline{\Omega}$, such that for each context $V \in \mathcal{V}(\mathcal{H})$ we have

$$\begin{aligned} \gamma_V : \underline{1}_V &\rightarrow \underline{\Omega}_V \\ \{*\} &\mapsto \gamma_V(\{*\}) = S \in \underline{\Omega}_V. \end{aligned} \tag{12.4}$$

Moreover, for any morphisms $i_{V'V} : V' \subseteq V$, the fact that γ is a global section implies that the following diagram commutes:

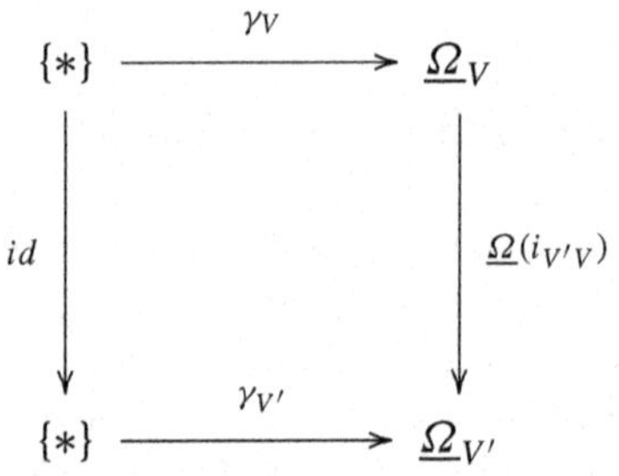

Chasing the diagram around we obtain

$$\underline{\Omega}(i_{V'V}) \circ \gamma_V(\{*\}) = \underline{\Omega}(i_{V'V})(S) = S \cap \downarrow V'. \tag{12.5}$$

On the other hand

$$id \circ \gamma_{V'}\{*\} = S'. \tag{12.6}$$

Commutativity is precisely the condition for γ to be a global section namely:

$$\underline{\Omega}(i_{V'V}) \circ \gamma_V = \gamma_{V'}. \tag{12.7}$$

The significance of the local components γ_V of the global section of the presheaf $\underline{\Omega}$ is that of 'local' truth values. In particular, each proposition and each state is defined as a collection of 'local' representations, one for each $V \in \mathcal{V}(\mathcal{H})$. Such 'local' representations, are glued together by the categorical structure of $\mathcal{V}(\mathcal{H})$. Now for each context we obtain a 'local' truth value of the 'local' proposition given the 'local' state. Such 'local' truth values are represented by the global element computed at that particular context, i.e. γ_V. All such 'local' truth values are 'glued' together by the structure of global sections (relation (12.7)) which, in turn, follows the categorical structure of the base category $\mathcal{V}(\mathcal{H})$. Thus, again, we obtain the quantum picture by considering a collection of 'local' representatives. However, it is only the collection that corresponds to a well defined object in our theory, each local representative on its own is meaningless.

Coming back to truth values, the global section that consists entirely of principal sieves is interpreted as representing the value 'totally true'. In classical Boolean logic this is just 'true'. Similarly, the global section that consists of empty sieves is interpreted as 'totally false'. In classical Boolean logic this is just 'false'.

A very important property of sieves is that the set $\underline{\Omega}_V$ of sieves on V has the structure of a Heyting algebra, where the unit element $\underline{1}_{\underline{\Omega}_V} \in \underline{\Omega}_V$ is represented by the principal sieve $\downarrow V$ and, the null element $\underline{0}_{\underline{\Omega}_V} \in \underline{\Omega}_V$, is represented by the empty set $\emptyset$.

Moreover $\underline{\Omega}_V$ is equipped with a partial ordering given by subset inclusion, such that $S_i \leq S_j$ iff $S_i \subseteq S_j$. In this context the logical connectives are given by

$$\begin{aligned} S_i \wedge S_j &:= S_i \cap S_j \\ S_i \vee S_j &:= S_i \cup S_j \\ S_i \Rightarrow S_j &:= \{V' \subseteq V \mid \forall V'' \subseteq V' \text{ if } V'' \in S_i \text{ then } V'' \in S_j\}. \end{aligned} \tag{12.8}$$

Being a Heyting algebra, the negation is given by the pseudo-complement. In particular, given an element S, its pseudo-complement (negation) is the element

$$\begin{aligned} \neg S &:= S \Rightarrow 0 \\ \neg S_V &:= \{V' \subseteq V \mid \forall V'' \subseteq V', V'' \notin S\}. \end{aligned} \tag{12.9}$$

From the fact that for each context $V \in \mathcal{V}(\mathcal{H})$, $\underline{\Omega}_V$ is a Heyting algebra, it is possible to show that the collection of global elements $\Gamma(\underline{\Omega})$ of $\underline{\Omega}$ is itself a Heyting algebra.

Theorem 12.1 *The collection $\Gamma(\underline{\Omega})$ is a Heyting algebra.*

In order to prove the above theorem we first of all need to define a partial ordering on $\Gamma\underline{\Omega}$, so the operations of $\vee$ and $\wedge$ can be defined.

Definition 12.3 A partial ordering on $\Gamma\underline{\Omega}$ can be constructed in a 'local' way by defining

$$\gamma_1 \leq \gamma_2 \quad \text{iff} \quad \forall V \in \mathcal{V}(\mathcal{H}), \quad \gamma_1(V) \subseteq \gamma_2(V) \tag{12.10}$$

where the ordering on the right hand side of the above equation is the subset ordering in $\underline{\Omega}_V$.

Given such an ordering we can now define the logical connective $\vee, \wedge, \Rightarrow, \neg$ and check whether they respect a Heyting algebra structure. Thus we have the following:

Theorem 12.2 *Given two sections γ_1 and γ_2 the 'or' is defined as the global element $\gamma_1 \vee \gamma_2$ such that for each context $V \in \mathcal{V}(\mathcal{H})$*

$$(\gamma_1 \vee \gamma_2)(V) := \gamma_1(V) \vee \gamma_2(V). \tag{12.11}$$

The right hand side is the usual ordering in $\underline{\Omega}_V$.

Proof We need to show that indeed $\gamma_1 \vee \gamma_2 \in \Gamma(\underline{\Omega})$, i.e. that $\underline{\Omega}(i_{V'V})((\gamma_1 \vee \gamma_2)(V)) = (\gamma_1 \vee \gamma_2)(V')$ for all $V' \subseteq V$. In particular

$$\begin{aligned}
&\underline{\Omega}(i_{V'V})\big((\gamma_1 \vee \gamma_2)(V)\big) \\
&\quad = \underline{\Omega}(i_{V'V})\big(\gamma_1(V) \vee \gamma_2(V)\big) = \big(\gamma_1(V) \vee \gamma_2(V)\big) \wedge \downarrow V' \\
&\quad = \big(\gamma_1(V) \wedge \downarrow V'\big) \vee \big(\gamma_2(V) \wedge \downarrow V'\big) = \gamma_1\big(V'\big) \vee \gamma_2\big(V'\big) \\
&\quad = (\gamma_1 \vee \gamma_2)\big(V'\big).
\end{aligned} \tag{12.12}$$

□

Similarly we have:

Theorem 12.3 *Given two sections γ_1 and γ_2, the 'and' is defined as the global element $\gamma_1 \wedge \gamma_2$ such that for each context $V \in \mathcal{V}(\mathcal{H})$*

$$(\gamma_1 \wedge \gamma_2)(V) := \gamma_1(V) \wedge \gamma_2(V). \tag{12.13}$$

The right hand side is the usual ordering in $\underline{\Omega}_V$.

The proof of this is similar to one for $\vee$. All that remains to define is the implication operation and hence the negation operation.

Theorem 12.4 *Given two sections γ_1 and γ_2 the 'implication' is defined as the global element $\gamma_1 \Rightarrow \gamma_2$, such that for each context $V \in \mathcal{V}(\mathcal{H})$*

$$(\gamma_1 \Rightarrow \gamma_2)(V) := \big(\gamma_1(V) \Rightarrow \gamma_2(V)\big). \tag{12.14}$$

Proof Again we need to show that in order for this to be a well defined definition, for all $V' \subseteq V$ the following must hold:

$$\underline{\Omega}(i_{V'V})(\gamma_1 \Rightarrow \gamma_2)(V) = (\gamma_1 \Rightarrow \gamma_2)\big(V'\big). \tag{12.15}$$

In particular,

$$
\begin{aligned}
&\underline{\Omega}(i_{V'V})(\gamma_1 \Rightarrow \gamma_2)(V) \\
&= \underline{\Omega}(i_{V'V})\big(\gamma_1(V) \Rightarrow \gamma_2(V)\big) && (12.16)\\
&= \big(\gamma_1(V) \Rightarrow \gamma_2(V)\big) \wedge \downarrow V' && (12.17)\\
&= \big\{V_i \subseteq V \big| \forall V_j \subseteq V_i \text{ if } V_j \in \gamma_1(V) \text{ then } V_j \in \gamma_2(V)\big\} \wedge \downarrow V' \\
&= \big\{V_i \subseteq V \big| \forall V_j \subseteq V_i \text{ if } V_j \in \gamma_1(V) \wedge \downarrow V' \text{ then } V_j \in \gamma_2(V) \wedge \downarrow V'\big\} \\
&= \big\{V_i \subseteq V' \big| \forall V_j \subseteq V_i \text{ if } V_j \in \gamma_1\big(V'\big) \text{ then } V_j \in \gamma_2\big(V'\big)\big\} \\
&= \gamma_1\big(V'\big) \Rightarrow \gamma_2\big(V'\big) \\
&= (\gamma_1 \Rightarrow \gamma_2)\big(V'\big). && (12.18)
\end{aligned}
$$

We then have that for each context $V \in \mathcal{V}(\mathcal{H})$, $(\gamma_1 \Rightarrow \gamma_2)(V)$ is the relative pseudo-complement of $\gamma_1(V)$ relative to $\gamma_2(V)$. Therefore, $\gamma_1 \Rightarrow \gamma_2$ is the relative pseudo-complement of γ_1 relative to γ_2. The definition of negation follows trivially. □

We have now all the ingredients to prove Theorem 12.1.

Proof Theorems 12.2, 12.3 and 12.4 imply that it is possible to define, on the set $\Gamma(\underline{\Omega})$, all the logical connective. These are defined locally in terms of the logical connective in $\underline{\Omega}_V$, for each $V \in \mathcal{V}(\mathcal{H})$. Moreover, the presheaf maps preserve the logical connectives thus defined. Since global sections satisfy the presheaf maps, it follows that $\Gamma(\underline{\Omega})$ acquires the Heyting algebra structure from the Heyting algebra structure on each $\underline{\Omega}_V$, $V \in \mathcal{V}(\mathcal{H})$. □

12.1.1 Example

We will now describe an example of the sub-object classifier for the case of our 4 dimensional Hilbert space $\mathcal{H} = \mathbb{C}^4$. Let us start with the *maximal algebras* $V = lin_{\mathbb{C}}(\hat{P}_1, \hat{P}_2, \hat{P}_3, \hat{P}_4)$. What follows can be generalised to any maximal sub-algebra, not just V.

The collection of sieves on V will be

$$\underline{\Omega}_V := \{0_{\underline{\Omega}_V}, S, S_{12}, S_{13}, S_{14}, S_{23}, S_{24}, S_{34}, S_1, S_2, S_3, S_4, \ldots\} \quad (12.19)$$

where the sieves in $\underline{\Omega}_V$ are defined as follows:

$$S = \big\{V, V_{\hat{P}_i,\hat{P}_j}, V_{\hat{P}_k}, \big| i, j, k \in \{1, 2, 3, 4, \}\big\} = \text{principal sieve on } V \quad (12.20)$$

$$S_{ij} = \{V_{\hat{P}_i,\hat{P}_j}, V_{\hat{P}_i}, V_{\hat{P}_j}\} = \text{principal sieve on } V_{\hat{P}_i,\hat{P}_j} \quad (12.21)$$

$$S_i = \{V_{\hat{P}_i}\} \quad (12.22)$$

$$\underline{0}_{\underline{\Omega}_V} = \{\emptyset\}. \tag{12.23}$$

We now consider a non maximal algebra $V_{\hat{P}_i,\hat{P}_j}$. The collection of sieves on such an algebras is[1]

$$\underline{\Omega}_{V_{\hat{P}_i,\hat{P}_j}} := \{\underline{0}_{\underline{\Omega}_{V_{\hat{P}_i,\hat{P}_j}}}, S_{ij}, S_i, S_j, \cdots\} \tag{12.24}$$

where the definitions of the individual sieves are the same as before.

Similarly, for the context $V_{\hat{P}_i}$ we have

$$\underline{\Omega}_{V_{\hat{P}_i}} := \{S_i\}. \tag{12.25}$$

We now want to define the $\underline{\Omega}$-morphisms. To this end, let us first consider the $\underline{\Omega}$-morphism with domain $\underline{\Omega}_V$. There are various such morphisms, one for each pair $i, j \in \{1, 2, 3, 4\}$, as follows:

$$\underline{\Omega}(i_{V_{\hat{P}_i,\hat{P}_j},V}) : \underline{\Omega}_V \to \underline{\Omega}_{V_{\hat{P}_i,\hat{P}_j}} \tag{12.26}$$

where $i_{V_{\hat{P}_i,\hat{P}_j},V} : V_{\hat{P}_i,\hat{P}_j} \subseteq V$. $\underline{\Omega}(i_{V_{\hat{P}_i,\hat{P}_j},V})$ is defined component-wise as follows:

$$S \mapsto S_{ij} \qquad S_{ij} \mapsto S_{ij} \tag{12.27}$$

$$S_{ik} \mapsto S_i \qquad S_{kj} \mapsto S_j \tag{12.28}$$

$$S_i \mapsto S_i \qquad S_{kl} \mapsto \underline{0}_{\underline{\Omega}_{V_{\hat{P}_i,\hat{P}_j}}} \tag{12.29}$$

$$S_j \mapsto S_j \qquad S_k \mapsto \underline{0}_{\underline{\Omega}_{V_{\hat{P}_i,\hat{P}_j}}}.$$

Moreover, for each $k \in \{1, 2, 3, 4\}$, such that $i_{V_{\hat{P}_k},V} : V_{\hat{P}_k} \subseteq V$ we have the following $\underline{\Omega}$-morphisms:

$$\underline{\Omega}_{V,V_{\hat{P}_k}} : \underline{\Omega}_V \to \underline{\Omega}_{V_{\hat{P}_k}} \tag{12.30}$$

which component-wise are defined as follows:

$$\begin{aligned} S &\mapsto S_k & S_{ij} &\mapsto \underline{0}_{\underline{\Omega}_{V_{\hat{P}_k}}} \\ S_{ik} &\mapsto S_k & S_l &\mapsto \underline{0}_{\underline{\Omega}_{V_{\hat{P}_k}}} \\ S_k &\mapsto S_k & S_j &\mapsto \underline{0}_{\underline{\Omega}_{V_{\hat{P}_k}}} \\ & & S_i &\mapsto \underline{0}_{\underline{\Omega}_{V_{\hat{P}_k}}}. \end{aligned} \tag{12.31}$$

[1]Note that $\underline{0}_{\underline{\Omega}_V} = \underline{0}_{\underline{\Omega}_{V_{\hat{P}_i,\hat{P}_j}}} = \underline{0}_{\underline{\Omega}_{V_{\hat{P}_i}}} = \{\emptyset\}$.

It is straightforward to extend the definition of $\underline{\Omega}$-morphisms for all contexts $V_i \in \mathcal{V}(\mathbb{C}^4)$.

12.2 Truth Values Using the Pseudo-state Object

We are now ready to turn to the question of how truth values are assigned to propositions which, in this case, are represented by sub-presheaves $\underline{\delta(\hat{P})} \subseteq \underline{\Sigma}$. For this purpose it is worth thinking again about classical physics. There, as previously stated, we know that a proposition $\hat{A} \in \Delta$ is true for a given state s if $s \in f_{\hat{A}}^{-1}(\Delta)$, i.e. if s belongs to those subsets $f_{\hat{A}}^{-1}(\Delta)$ of the state space for which the proposition $\hat{A} \in \Delta$ is true. Therefore, given a state s, all true propositions of s are represented by those measurable subsets which contain s, i.e. those subsets which have measure 1 with respect to the Dirac measure δ_s.

As discussed in Chap. 10, in the quantum case a proposition of the form "$A \in \Delta$" is represented by the presheaf $\underline{\delta(\hat{E}[A \in \Delta])}$, where $\hat{E}[A \in \Delta]$ is the spectral projector for the self-adjoint operator[2] $\hat{A}$, which projects onto the subset Δ of the spectrum of $\hat{A}$. On the other hand, states are represented by the presheaves $\underline{\mathfrak{w}}^{|\psi\rangle}$. As described above, these identifications are obtained using the maps $\mathfrak{S} : P(V) \rightarrow \mathrm{Sub_{cl}}(\underline{\Sigma}_V)$, $V \in \mathcal{V}(\mathcal{H})$, and the daseinisation map $\delta : P(\mathcal{H}) \rightarrow \mathrm{Sub_{cl}}(\underline{\Sigma})$, with the properties that

$$\begin{aligned} \big\{\mathfrak{S}\big(\delta(\hat{P})_V\big) \,\big|\, V \in \mathcal{V}(\mathcal{H})\big\} &:= \underline{\delta(\hat{P})} \subseteq \underline{\Sigma} \\ \big\{\mathfrak{S}\big(\mathfrak{w}_V^{|\psi\rangle}\big) \,\big|\, V \in \mathcal{V}(\mathcal{H})\big\} &:= \underline{\mathfrak{w}}^{|\psi\rangle} \subseteq \underline{\Sigma}. \end{aligned} \tag{12.32}$$

As a consequence, within the structure of formal typed languages (see Appendix), both presheaves $\underline{\mathfrak{w}}^{|\psi\rangle}$ and $\underline{\delta(\hat{P})}$ are terms of type $P\underline{\Sigma}$, i.e. they are sub-objects of the spectral presheaf.

We now want to define the condition by which, for each context V, the proposition $(\underline{\delta(\hat{P})})_V$ is true given $\underline{\mathfrak{w}}_V^{|\psi\rangle}$.[3] To this end we recall that, for each context V, the projection operator $\mathfrak{w}_V^{|\psi\rangle}$ can be written as follows:

$$\begin{aligned} \mathfrak{w}_V^{|\psi\rangle} &= \bigwedge\big\{\hat{\alpha} \in P(V) \big| \,|\psi\rangle\langle\psi| \leq \hat{\alpha}\big\} \\ &= \bigwedge\big\{\hat{\alpha} \in P(V) \big| \langle\psi|\hat{\alpha}\,|\psi\rangle = 1\big\} \\ &= \delta^o\big(\,|\psi\rangle\langle\psi|\big)_V. \end{aligned} \tag{12.33}$$

---

[2]Note that $\hat{E}[A \in \Delta] = \hat{P}_{A\in\Delta}$.

[3]Recall that $\mathfrak{w}_V^{|\psi\rangle} := \delta^o(|\psi\rangle\langle\psi|)_V$ represents the projection operator while $\underline{\mathfrak{w}}_V^{|\psi\rangle}$ indicates the subset of $\underline{\Sigma}_V$. Although ultimately they are equivalent, it is always worth highlighting what specific role $\mathfrak{w}^{|\psi\rangle}$ has.

This represents the smallest projection in $P(V)$ which has expectation value equal to one with respect to the state $|\psi\rangle$. The associated subset of the Gel'fand spectrum is defined as

$$\underline{\mathfrak{w}}_V^{|\psi\rangle} = \mathfrak{S}\Big(\bigwedge\{\hat{\alpha} \in P(V) | \langle\psi|\hat{\alpha}|\psi\rangle = 1\}\Big)$$
$$= \big\{\lambda \in \underline{\Sigma}_V | \lambda\big(\delta^o\big(|\psi\rangle\langle\psi|\big)_V\big) = 1\big\}. \qquad (12.34)$$

It follows that $\underline{\mathfrak{w}}^{|\psi\rangle} = \{\underline{\mathfrak{w}}_V^{|\psi\rangle} \mid V \in \mathcal{V}(\mathcal{H})\}$ is the sub-object of the spectral presheaf $\underline{\Sigma}$ such that, at each context $V \in \mathcal{V}(\mathcal{H})$, it identifies those subsets of the Gel'fand spectrum which correspond (through the map $\mathfrak{S}$) to the smallest projections of that context, which have expectation value equal to one with respect to the state $|\psi\rangle$, i.e. which are true in $|\psi\rangle$.

On the other hand, as previously defined, at a given context V, the operator $\delta(\hat{P})_V$ is:

$$\delta^o(\hat{P})_V := \bigwedge\{\hat{\alpha} \in P(V) | \hat{P} \leq \hat{\alpha}\}. \qquad (12.35)$$

Thus the sub-presheaf $\underline{\delta(\hat{P})}$ is defined as the sub-object of $\underline{\Sigma}$, such that at each context V it defines the subset $\underline{\delta(\hat{P})}_V$ of the Gel'fand spectrum $\underline{\Sigma}_V$, which represents (through the map $\mathfrak{S}$) the projection operator $\delta(\hat{P})_V$.

We are interested in defining the condition by which the proposition represented by the sub-object $\underline{\delta(\hat{P})}$ is true, given the state $\underline{\mathfrak{w}}^{|\psi\rangle}$. Let us analyse this condition for each context V. In this case, we need to define the condition by which the projection operator $\delta(\hat{P})_V$, associated to the proposition $\underline{\delta(\hat{P})}$ is true, given the local pseudostate $\mathfrak{w}^{|\psi\rangle}$. Since at each context V the pseudo-state defines the smallest projection in that context which is true with probability one, i.e. $\mathfrak{w}_V^{|\psi\rangle}$, for any other projection to be true, given this pseudo-state, this projection must be a coarse-graining of $\mathfrak{w}_V^{|\psi\rangle}$, i.e. it must be implied by $\mathfrak{w}_V^{|\psi\rangle}$. Thus, if $\mathfrak{w}_V^{|\psi\rangle}$ is the smallest projection in $P(V)$, which is true with probability one, then the projector $\delta^o(\hat{P})_V$ will be true if and only if $\delta^o(\hat{P})_V \geq \mathfrak{w}_V^{|\psi\rangle}$. This condition is a consequence of the fact that, if $\langle\psi|\hat{\alpha}|\psi\rangle = 1$, then for all $\hat{\beta} \geq \hat{\alpha}$ it follows that $\langle\psi|\hat{\beta}|\psi\rangle = 1$.

So far we have defined a 'truthfulness' relation at the level of projection operators, namely $\delta^o(\hat{P})_V \geq \mathfrak{w}_V^{|\psi\rangle}$. Through the map $\mathfrak{S}$ it is possible to shift this relation to the level of sub-objects of the Gel'fand spectrum:

$$\mathfrak{S}\big(\mathfrak{w}_V^{|\psi\rangle}\big) \subseteq \mathfrak{S}\big(\delta^o(\hat{P})_V\big) \qquad (12.36)$$

$$\underline{\mathfrak{w}}_V^{|\psi\rangle} \subseteq \underline{\delta(\hat{P})}_V$$
$$\big\{\lambda \in \underline{\Sigma}(V) | \lambda\big(\delta^o\big(|\psi\rangle\langle\psi|\big)_V\big) = 1\big\} \subseteq \big\{\lambda \in \underline{\Sigma}(V) | \lambda\big(\big(\delta^o(\hat{P})\big)_V\big) = 1\big\}. \qquad (12.37)$$

What the above equation reveals is that, at the level of sub-objects of the Gel'fand spectrum, for each context V, a 'proposition' can be said to be (totally) true for a given pseudo-state if, and only if, the subset of the Gel'fand spectrum, associated to the pseudo-state, is a subsets of the corresponding subsets of the Gel'fand spectrum associated to the proposition. It is straightforward to see that if $\delta(\hat{P})_V \geq \mathfrak{w}_V^{|\psi\rangle}$ then $\mathfrak{S}(\mathfrak{w}_V^{|\psi\rangle}) \subseteq \mathfrak{S}(\delta(\hat{P})_V)$ since, for projection operators, the map λ takes only the values 0, 1.

We still need a further abstraction in order to work directly with the presheaves $\underline{\mathfrak{w}}^{|\psi\rangle}$ and $\underline{\delta(\hat{P})}$. Thus we want the analogue of (12.36) at the level of sub-objects of the spectral presheaf, $\underline{\Sigma}$. This relation is easily derived to be

$$\underline{\mathfrak{w}}^{|\psi\rangle} \subseteq \underline{\delta(\hat{P})}. \tag{12.38}$$

Equation (12.38) shows that, whether or not a proposition $\underline{\delta(\hat{P})}$ is 'totally true' given a pseudo-state $\underline{\mathfrak{w}}^{|\psi\rangle}$, is determined by whether or not the pseudo-state is a sub-presheaf of the presheaf $\underline{\delta(\hat{P})}$. With this motivation, we can now define the *local truth value* of the proposition "$A \in \Delta$" at stage V, given the state $|\psi\rangle$, as:

$$\begin{aligned} v\big(A \in \Delta;\ |\psi\rangle\big)_V &= v\big(\underline{\mathfrak{w}}^{|\psi\rangle} \subseteq \underline{\delta\big(\hat{E}[A \in \Delta]\big)}\big)_V \\ &:= \big\{V' \subseteq V \big| \underline{\mathfrak{w}}_{V'}^{|\psi\rangle} \subseteq \underline{\delta\big(\hat{E}[A \in \Delta]\big)}_{V'}\big\} \\ &= \big\{V' \subseteq V \big| \langle\psi|\delta\big(\hat{E}[A \in \Delta]\big)_V |\psi\rangle = 1\big\}. \end{aligned} \tag{12.39}$$

The last equality is derived by the fact that $\underline{\mathfrak{w}}_V^{|\psi\rangle} \subseteq \underline{\delta(\hat{P})}_V$ which is a consequence, at the level of projection operators, of $\delta^o(\hat{P})_V \geq (\mathfrak{w}^{|\psi\rangle})_V$. But, since $\mathfrak{w}_V^{|\psi\rangle}$ is the smallest projection operator such that $\langle\psi|\mathfrak{w}_V^{|\psi\rangle}|\psi\rangle = 1$, then $\delta^o(\hat{P})_V \geq \mathfrak{w}_V^{|\psi\rangle}$ implies that $\langle\psi|\delta^o(\hat{P}_V)|\psi\rangle = 1$.

The right hand side of (12.39) means that the truth value, defined at V of the proposition "$A \in \Delta$", given the state $\underline{\mathfrak{w}}^{|\psi\rangle}$, is given in terms of all those sub-contexts $V' \subseteq V$ for which the projection operator $\delta(\hat{E}[A \in \Delta])_V$ has expectation value equal to one with respect to the state $|\psi\rangle$. In other words, this *local* truth value is defined to be the set of all those sub-contexts for which the proposition is totally true.

So we can see how a local truth value can be seen as a measure of how far the proposition is from being true in the classical sense. In fact, what the sieve $v(A \in \Delta;\ |\psi\rangle)_V$ tells you is how much you have to generalise your original proposition for it to be true.

We now need to show that indeed $v(A \in \Delta;\ |\psi\rangle)_V = S$ is a sieve. To this end we simply need to show that it is closed under left composition.

Proof Consider an algebra $V' \in S$ then, given any other algebra $V'' \subseteq V$ we want to show that $V'' \in S$. Now since $V' \in S$, then $\delta^o(\hat{P})_{V'} \geq (\underline{\mathfrak{w}}^{|\psi\rangle})_{V'}$, however from the definition of daseinisation we have that $\delta^o(\hat{P})_{V''} \geq \delta^o(\hat{P})_{V'}$ therefore $\delta^o(\hat{P})_{V''} \geq (\underline{\mathfrak{w}}^{|\psi\rangle})_{V'}$, which implies that $V'' \in S$. □

When considering all the contexts $V \in \mathcal{V}(\mathcal{H})$ together, we obtain the notion of *global truth value*:

$$v\big(A \in \Delta;\, |\psi\rangle\big) = v\big(\underline{\mathfrak{w}}^{|\psi\rangle} \subseteq \underline{\delta\big(\hat{E}[A \in \Delta]\big)}\big). \tag{12.40}$$

This represents a global section of $\underline{\Omega}$, i.e. an element of the Heyting algebra $\Gamma\underline{\Omega}$. Pictorially we have the following situation regarding truth values:

$$\begin{array}{ccc}
\underline{\mathfrak{w}}^{|\psi\rangle} & \hookrightarrow & \underline{\delta(\hat{P})} \\
\downarrow & & \downarrow \\
\underline{1} & \xrightarrow{\ \gamma\ } & \underline{\Omega}
\end{array}$$

where to each sub-object relation which identifies the mathematical concept of valuation we associate a global element $\gamma : 1 \to \underline{\Omega}$ which represents the global truth value. Such a global truth value will have local components defined as follows:

$$\begin{array}{ccc}
(\underline{\mathfrak{w}}^{|\psi\rangle})_V & \hookrightarrow & \delta^o(\hat{P})_V \\
\downarrow & & \downarrow \\
\underline{1}_V & \xrightarrow{\ \gamma_V\ } & \underline{\Omega}_V
\end{array}$$

which pick out a particular sieve for each context.

The reason why all this works is that local truth values defined in this way form a *sieve* on V; and the set of all of these is a Heyting algebra.

Specifically: $v(\underline{\mathfrak{w}}^{|\psi\rangle} \subseteq \underline{\delta(\hat{P})})$ is a global element defined at stage V of the sub-object classifier $\underline{\Omega} := (\underline{\Omega}_V)_{V \in \mathcal{V}(\mathcal{H})}$ where, for all $V \in \mathcal{V}(\mathcal{H})$, $\underline{\Omega}_V$ represents the set of all sieves defined at stage V. In turn, the set of all global truth values($\Gamma(\underline{\Omega})$) forms a Heyting algebra.

12.3 Example

Let us consider again a 4 dimensional Hilbert space $\mathbb{C}^4$ whose category of abelian von Neumann sub-algebras is $\mathcal{V}(\mathbb{C}^4)$. In Example 9.1.2 we have already defined the state space $\underline{\Sigma}$. We now want to define the truth values of the proposition $S_z \in [-3, -1]$ which has corresponding projector operator $\hat{P} := \hat{E}[S_z \in$

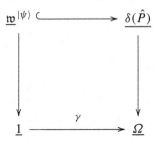

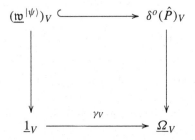

$[-3,-1]] = |\phi\rangle\langle\phi|$. This proposition is equivalent to the projection operator $\hat{P}_4 :=$ diag$(0,0,0,1)$. The reason being that $\hat{S}_z :=$ diag$(2,0,0,-2)$ thus the only value in the set $[-3,-1]$ which $\hat{S}_z$ can take is -2.

The state we will consider will be $\psi = (1,0,0,0)$ with respective operator $|\psi\rangle\langle\psi| = \hat{P}_1$. First of all we will analyse the context $V = lin_{\mathbb{C}}(\hat{P}_1, \hat{P}_2, \hat{P}_3, \hat{P}_4)$. Since $\hat{P}_4 \in V$ it follows that $\delta^o(\hat{P}_4)_V = \hat{P}_4$, but $\langle\psi|\hat{P}_4|\psi\rangle = 0$, thus we need to go to smaller contexts. In particular, only contexts in which $\delta^o(\hat{P}_4)$ is an operator which is implied by $\hat{P}_1$ will contribute to the truth value. We thus obtain:

$$v\big(\underline{\mathfrak{w}}^{|\psi\rangle} \subseteq \underline{\delta(\hat{P}_4)}\big)_V := \big\{V' \subseteq V \big| \langle\psi|\delta\big(\hat{E}[A \in \Delta]\big)|\psi\rangle = 1\big\} \tag{12.41}$$

$$= \{V_{\hat{P}_{i\in\{2,3\}}}, V_{\hat{P}_2\hat{P}_3}\}. \tag{12.42}$$

On the other hand for other sub-algebras we obtain

$$v\big(\underline{\mathfrak{w}}^{|\psi\rangle} \subseteq \underline{\delta(\hat{P}_4)}\big)_{V_{\hat{P}_1}} := \{\emptyset\} \tag{12.43}$$

$$v\big(\underline{\mathfrak{w}}^{|\psi\rangle} \subseteq \underline{\delta(\hat{P}_4)}\big)_{V_{\hat{P}_2}} := \{V_{\hat{P}_2}\} \tag{12.44}$$

$$v\big(\underline{\mathfrak{w}}^{|\psi\rangle} \subseteq \underline{\delta(\hat{P}_4)}\big)_{V_{\hat{P}_3}} := \{V_{\hat{P}_3}\} \tag{12.45}$$

$$v\big(\underline{\mathfrak{w}}^{|\psi\rangle} \subseteq \underline{\delta(\hat{P}_4)}\big)_{V_{\hat{P}_4}} := \{\emptyset\} \tag{12.46}$$

$$v\big(\underline{\mathfrak{w}}^{|\psi\rangle} \subseteq \underline{\delta(\hat{P}_4)}\big)_{V_{\hat{P}_1,\hat{P}_2}} := \{V_{\hat{P}_2}\} \tag{12.47}$$

$$v\big(\underline{\mathfrak{w}}^{|\psi\rangle} \subseteq \underline{\delta(\hat{P}_4)}\big)_{V_{\hat{P}_1,\hat{P}_3}} := \{V_{\hat{P}_3}\} \tag{12.48}$$

$$v\big(\underline{\mathfrak{w}}^{|\psi\rangle} \subseteq \underline{\delta(\hat{P}_4)}\big)_{V_{\hat{P}_1,\hat{P}_4}} := \{\emptyset\} \tag{12.49}$$

$$v\big(\underline{\mathfrak{w}}^{|\psi\rangle} \subseteq \underline{\delta(\hat{P}_4)}\big)_{V_{\hat{P}_2,\hat{P}_3}} := \{V_{\hat{P}_2,\hat{P}_3}, V_{\hat{P}_2}, V_{\hat{P}_3}\} \tag{12.50}$$

$$v\big(\underline{\mathfrak{w}}^{|\psi\rangle} \subseteq \underline{\delta(\hat{P}_4)}\big)_{V_{\hat{P}_2,\hat{P}_4}} := \{V_{\hat{P}_2}\} \tag{12.51}$$

$$v\big(\underline{\mathfrak{w}}^{|\psi\rangle} \subseteq \underline{\delta(\hat{P}_4)}\big)_{V_{\hat{P}_3,\hat{P}_4}} := \{V_{\hat{P}_3}\}. \tag{12.52}$$

As it can be seen, the truth values obtained by using the pseudo-state object coincide with the truth values obtained using the truth object.

12.4 Truth Values Using the Truth-Object

We have shown in Sect. 11.5 of Chap. 11 that the truth object $\underline{\mathbb{T}}^{|\psi\rangle}$ can be written in terms of the respective pseudo-state as follows:

$$\underline{\mathbb{T}}_V^{|\psi\rangle} = \big\{\hat{\alpha} \in \underline{O}_V \big| \hat{\alpha} \geq \underline{\mathfrak{w}}_V^{|\psi\rangle}\big\}. \tag{12.53}$$

This implies that the truth value of a given proposition, at a given context $V \in \mathcal{V}(\mathcal{H})$, can be defined as

$$\begin{aligned} v\big(\underline{\delta(\hat{P})} \in \underline{\mathbb{T}}^{|\psi\rangle}\big)_V &:= \big\{V' \subseteq V \big| \underline{\delta(\hat{P})}_{V'} \in \underline{\mathbb{T}}^{|\psi\rangle}_{V'}\big\} \\ &= \big\{V' \subseteq V \big| \langle\psi|\underline{\delta(\hat{P})}_{V'}|\psi\rangle = 1\big\}. \end{aligned} \tag{12.54}$$

Mimicking the above expression we define the truth value in terms of the density matrix truth object as:

$$\begin{aligned} v\big(\underline{\delta(\hat{P})} \in \underline{\mathbb{T}}^{\rho}\big)_V &:= \big\{V' \subseteq V \big| \underline{\delta(\hat{P})}_{V'} \in \underline{\mathbb{T}}^{\rho}_{V'}\big\} \\ &= \big\{V' \subseteq V \big| \mathrm{tr}\big(\rho\underline{\delta(\hat{P})}_{V'}\big) = 1\big\}. \end{aligned} \tag{12.55}$$

12.4.1 Example

We would now like to define the truth values of the proposition $S_z \in [-3, -1]$, which has corresponding projector operator $\hat{P} := \hat{E}[S_z \in [-3, -1]] = |\phi\rangle\langle\phi|$ in terms of the truth object. We will first consider a pure state and then a density matrix.

Pure State The proposition $S_z \in [-3, -1]$ is equivalent to the projection operator $\hat{P}_4$. We now consider the state $|\psi\rangle = (1, 0, 0, 0)$, whose associated truth object was defined in Example 11.3.3 of Chap. 11. Consequently we obtain the following truth values:

$$v\big(\underline{\delta(\hat{P}_4)} \in \underline{\mathbb{T}}^{|\psi\rangle}\big)_V := \{V_{\hat{P}_{i\in\{2,3\}}}, V_{\hat{P}_2\hat{P}_3}\} \tag{12.56}$$

$$v\big(\underline{\delta(\hat{P}_4)} \in \underline{\mathbb{T}}^{|\psi\rangle}\big)_{V_{\hat{P}_{i\in\{2,3\}}}} := \{V_{\hat{P}_i}\} \tag{12.57}$$

$$v\big(\underline{\delta(\hat{P}_4)} \in \underline{\mathbb{T}}^{|\psi\rangle}\big)_{V_{\hat{P}_{i\in\{1,4\}}}} := \{\emptyset\} \tag{12.58}$$

$$v\big(\underline{\delta(\hat{P}_4)} \in \underline{\mathbb{T}}^{|\psi\rangle}\big)_{V_{\hat{P}_1,\hat{P}_2}} := \{V_{\hat{P}_2}\} \tag{12.59}$$

$$v\big(\underline{\delta(\hat{P}_4)} \in \underline{\mathbb{T}}^{|\psi\rangle}\big)_{V_{\hat{P}_1,\hat{P}_3}} := \{V_{\hat{P}_3}\} \tag{12.60}$$

$$v\big(\underline{\delta(\hat{P}_4)} \in \underline{\mathbb{T}}^{|\psi\rangle}\big)_{V_{\hat{P}_1,\hat{P}_4}} := \{\emptyset\} \tag{12.61}$$

$$v\big(\underline{\delta(\hat{P}_4)} \in \underline{\mathbb{T}}^{|\psi\rangle}\big)_{V_{\hat{P}_2,\hat{P}_3}} := \{V_{\hat{P}_2}, V_{\hat{P}_3}, V_{\hat{P}_2,\hat{P}_3}\} \tag{12.62}$$

$$v\big(\underline{\delta(\hat{P}_4)} \in \underline{\mathbb{T}}^{|\psi\rangle}\big)_{V_{\hat{P}_2,\hat{P}_4}} := \{V_{\hat{P}_2}\} \tag{12.63}$$

$$v\big(\underline{\delta(\hat{P}_4)} \in \underline{\mathbb{T}}^{|\psi\rangle}\big)_{V_{\hat{P}_3,\hat{P}_4}} := \{V_{\hat{P}_3}\}. \tag{12.64}$$

As it can be seen, the truth values obtained by using the truth object coincide with the truth values obtained using the pseudo-state object.

Density Matrix Give the proposition $S_z \in [1.3, 2.3]$ represented by the projection operator $\hat{P}_1$, we would like to compute its truthfulness with respect to the truth object associated to the density matrix

$$\hat{\rho} = 1/2 \begin{pmatrix} 1 & 0 & 0 & 0 \\ 0 & 0 & 0 & 0 \\ 0 & 0 & 0 & 0 \\ 0 & 0 & 0 & 0 \end{pmatrix} + 1/2 \begin{pmatrix} 0 & 0 & 0 & 0 \\ 0 & 0 & 0 & 0 \\ 0 & 0 & 0 & 0 \\ 0 & 0 & 0 & 1 \end{pmatrix}.$$

The corresponding truth object was defined in Example 11.3.3 of Chap. 11.

We can now compute the truth values which are defined by

$$\begin{aligned} v\big(\underline{\delta(\hat{P})} \in \underline{\mathbb{T}}^{\rho}\big)_V &:= \big\{V' \subseteq V \big| \underline{\delta(\hat{P})}_{V'} \in \underline{\mathbb{T}}^{\rho}_{V'}\big\} \\ &= \big\{V' \subseteq V \big| \mathrm{tr}\big(\rho \underline{\delta(\hat{P})}_{V'}\big) = 1\big\}. \end{aligned} \tag{12.65}$$

We consider context V, here the daseinised proposition is simply itself: $\delta^o(\hat{P}_1) = \hat{P}_1$ but $\hat{P}_1 \notin \underline{\mathbb{T}}^{\rho}_V$ since $\mathrm{tr}(\hat{\rho}\hat{P}_1) \neq 1$ thus we need to go to a smaller algebra, i.e. we need to go to a smaller algebra, i.e. we need to generalise our proposition.

Let us consider $V_{\hat{P}_3,\hat{P}_4}$, in this context we get $\delta^o(\hat{P}_1) = \hat{P}_1 + \hat{P}_2$ but again $\hat{P}_1 + \hat{P}_2 \notin \underline{\mathbb{T}}^{\rho}_{V_{\hat{P}_3,\hat{P}_4}}$ since $\mathrm{tr}(\hat{\rho}(\hat{P}_1 + \hat{P}_2)) \neq 1$.

On the other hand for $V_{\hat{P}_2,\hat{P}_3}$ we obtain $\delta^o(\hat{P}_1) = \hat{P}_1 + \hat{P}_4$. In this case $\hat{P}_1 + \hat{P}_4 \in \underline{\mathbb{T}}^{\rho}_{V_{\hat{P}_2,\hat{P}_3}}$ since $\mathrm{tr}(\hat{\rho}(\hat{P}_1 + \hat{P}_4)) = 1$. Similarly, for context $V_{\hat{P}_2}$, since $\delta^o(\hat{P}_1) = \hat{P}_1 + \hat{P}_4 + \hat{P}_3$ then $\mathrm{tr}(\hat{\rho}(\hat{P}_1 + \hat{P}_4\hat{P}_3)) = 1$ and $V_{\hat{P}_3} \in \underline{\mathbb{T}}^{\rho}_{V_{\hat{P}_2,\hat{P}_3}}$. The same reasoning holds for $V_{\hat{P}_4}$.

It follows that:

$$v\big(\underline{\delta(\hat{P})} \in \underline{\mathbb{T}}^{\rho}\big)_V = \{V_{\hat{P}_2,\hat{P}_3}, V_{\hat{P}_2}, V_{\hat{P}_3}\}. \tag{12.66}$$

Applying similar reasoning one can compute truth values for all the remaining contexts.

12.5 Relation Between the Truth Values

Let us reiterate the relation between $\underline{\mathbb{T}}^{|\psi\rangle}$ and $\underline{\mathfrak{w}}^{|\psi\rangle}$.

1. Since for each $V \in \mathcal{V}(\mathcal{H})$, $\mathfrak{w}_V^{|\psi\rangle}$ is the smallest projection operator in V, such that $\mathfrak{w}_V^{|\psi\rangle} \geq |\psi\rangle\langle\psi|$, then if $\hat{\alpha} \in \underline{\mathbb{T}}_V^{|\psi\rangle}$ it follows that $\hat{\alpha} \geq \mathfrak{w}_V^{|\psi\rangle}$.
2. If $\hat{\alpha} \geq \underline{\mathfrak{w}}_V^{|\psi\rangle}$, then from the definition of $\underline{\mathbb{T}}^{|\psi\rangle}$ it follows that $\hat{\alpha} \in \underline{\mathbb{T}}_V^{|\psi\rangle}$.

The above relations imply the following:

$$\hat{\alpha} \in \underline{\mathbb{T}}_V^{|\psi\rangle} \quad \text{iff} \quad \hat{\alpha} \geq \mathfrak{w}_V^{|\psi\rangle} \tag{12.67}$$

which, when applied to a daseinised proposition becomes

$$\delta^o(\hat{P})_V \in \underline{\mathbb{T}}^{|\psi\rangle}_V \quad \text{iff} \quad \delta^o(\hat{P})_V \geq \mathfrak{w}^{|\psi\rangle}_V. \tag{12.68}$$

Alternatively, we can express this relation in terms of subsets of the state space $\underline{\Sigma}_V$ as follows:

$$S_{\delta^o(\hat{P})_V} \in \underline{\mathbb{T}}^{|\psi\rangle}_V \quad \text{iff} \quad S_{\delta^o(\hat{P})_V} \supseteq S_{\mathfrak{w}^{|\psi\rangle}_V} \tag{12.69}$$

therefore (12.54) can be written as

$$v\big(\underline{\delta(\hat{P})} \in \underline{\mathbb{T}}^{|\psi\rangle}\big)_V = \big\{V' \subseteq V \big| \delta^o(\hat{P})_V \geq \mathfrak{w}^{|\psi\rangle}_V\big\} \tag{12.70}$$

but this is precisely $v(\underline{\mathfrak{w}}^{|\psi\rangle} \subseteq \underline{\delta(\hat{P})})_V$, thus as presheaves

$$\underline{\delta(\hat{P})} \in \underline{\mathbb{T}}^{|\psi\rangle} \quad \text{is equivalent to} \quad \underline{\mathfrak{w}}^{|\psi\rangle} \subseteq \underline{\delta(\hat{P})}. \tag{12.71}$$

Therefore the truth values as computed with respect to the pseudo-state or with respect to the truth object are exactly the same:

$$v\big(\underline{\mathfrak{w}}^{|\psi\rangle} \subseteq \underline{\delta(\hat{P})}\big) = v\big(\underline{\delta(\hat{P})} \in \underline{\mathbb{T}}^{|\psi\rangle}\big). \tag{12.72}$$

Chapter 13
Quantity Value Object and Physical Quantities

In this chapter we will introduce the quantity value object. As the name suggests such an object will be used to assign values to quantities. In classical theory the quantity value object is simply the real numbers since each quantity has, as its value, an element of the Reals. Similarly, in standard quantum theory we have the reals as the quantity value object.

As we will see, in topos quantum theory the quantity value object will be an object which has the same role as the reals have in standard quantum theory, but its elements will not be numbers. Clearly each element in the quantity value object will be related to real numbers in some way, but they themselves will not be real numbers.

It should be noted that in any topos there is already an object which represents the real numbers, in fact there are few of them, however the quantity value object we use for topos quantum theory will not be one of them. As we will see, the motivations for the particular choice of quantity value object comes from physics requirements.

In fact, this object will then be used to define the topos analogues of physical quantities. In particular we will define physical quantities in the same way as they are defined in classical physics, namely as maps from the state space to the quantity value object.

This feature is in accordance with our aim to make quantum theory in a topos 'look like' classical physics in **Sets**.

13.1 Topos Representation of the Quantity Value Object

In the topos $\mathbf{Sets}^{\mathcal{V}(\mathcal{H})}$ the representation of the quantity value object is given by the following presheaf [32]:

Definition 13.1 The presheaf $\underline{\mathbb{R}}^{\leftrightarrow}$ has as

C. Flori, *A First Course in Topos Quantum Theory*, Lecture Notes in Physics 868,
DOI 10.1007/978-3-642-35713-8_13, © Springer-Verlag Berlin Heidelberg 2013

(i) Objects:[1]

$$\underline{\mathbb{R}}^{\leftrightarrow}_V := \{(\mu, \nu)|\mu, \nu : \downarrow V \to \mathbb{R}|\mu \text{ is order preserving, } \nu \text{ is order reversing;} \\ \mu \leq \nu\}. \tag{13.1}$$

(ii) Morphisms: given two contexts $V' \subseteq V$ the corresponding morphism is

$$\begin{aligned} \underline{\mathbb{R}}^{\leftrightarrow}_{V,V'} : \underline{\mathbb{R}}^{\leftrightarrow}_V &\to \underline{\mathbb{R}}^{\leftrightarrow}_{V'} \\ (\mu, \nu) &\mapsto (\mu_{|V'}, \nu_{|V'}). \end{aligned} \tag{13.2}$$

This presheaf is where physical quantities take their values, thus it has the same role as the reals in classical physics.

The reason why the quantity value object is defined in terms of order reversing and order preserving functions is because, in general, in quantum theory one can only give approximate values to the quantities. In particular, in most cases, the best approximation to the value of a physical quantity one can give is the smallest interval of possible values of that quantity.

Let us analyse the presheaf $\underline{\mathbb{R}}^{\leftrightarrow}$ in more details. To this end we assume that we want to define the value of a physical quantity A given a state ψ. If ψ is an eigenstate of A, then we would get a sharp value of the quantity A say a. If ψ is not an eigenstate, then we would get a certain range Δ of values for A, where $\Delta \in sp(\hat{A})$.

Let us assume that $\Delta = [a, b]$, then what the presheaf $\underline{\mathbb{R}}^{\leftrightarrow}$ does is to single out these extreme points a and b, so as to give a range (unsharp) of values for the physical quantity A. Obviously, since we are in the topos of presheaves, we have to define each object contextually, i.e. for each context $V \in \mathcal{V}(\mathcal{H})$. It is precisely to accommodate this fact that the pair of order reversing and order preserving functions were chosen to define the extreme values of our intervals.

To understand this let us consider a context V, such that the self-adjoint operator $\hat{A}$, which represents the physical quantity A, does belong to V. The range of values of A at V is $[a, b]$. If we then consider the context $V' \subseteq V$, such that $\hat{A} \notin V$, we will have to approximate $\hat{A}$ so as to fit V'. The precise way in which self-adjoint operators are approximated will be described later on, however, such an approximation will inevitably coarse-grain $\hat{A}$, i.e. it will make it more general.

It follows that the range of possible values of such an approximated operator, call it $\hat{A}_1$, will be bigger. Therefore the range of values of $\hat{A}_1$ at V' will be $[c, d] \supseteq [a, b]$, where $c \leq a$ and $d \geq b$. These relations between the extremal points can be achieved by the presheaf $\underline{\mathbb{R}}^{\leftrightarrow}$ through the order reversing and order preserving functions. Specifically, given that $a := \mu(V)$, $b := \nu(V)$ since $V' \subseteq V$, it follows that $c := \mu(V') \leq \mu(V)$ (μ being order preserving) and $d := \nu(V') \geq \nu(V)$ (ν being order reversing). By definition $\mu(V) \leq \nu(V)$, this implies that as one goes to smaller and smaller contexts the intervals $(\mu(V)_i, \nu(V)_i)$ keep getting bigger or stay the same.

---

[1]A map $\mu : \downarrow V \to \mathbb{R}$ is said to be order preserving if $V' \subseteq V$ implies that $\mu(V') \leq \mu(V)$. On the other hand, a map $\nu : \downarrow V \to \mathbb{R}$ is order reversing if $V' \subseteq V$ implies that $\nu(V') \supseteq \nu(V)$.

13.2 Inner Daseinisation

We will now introduce a different kind of daseinisation called inner daseinisation. The role of such daseinisation is to approximate projection operators but from below. In fact, while outer daseinisation would pick the smallest projection operator implied by the original projection operator (hence approximation from above), inner daseinisation picks the biggest projection operators which implies the original one (hence approximation form below). The precise definition of inner daseinisation is the following:

Definition 13.2 Given a projection operator $\hat{P}$, for each context $V \in \mathcal{V}(\mathcal{H})$, inner daseinisation is defined as:

$$\delta^i(\hat{P})_V = \bigvee \{\hat{\alpha} \in P(V) | \hat{\alpha} \leq \hat{P}\}. \tag{13.3}$$

It follows that $\delta^i(\hat{P})_V$ is the best approximation in V of $\hat{P}$ obtained by taking the 'largest' projection operator in V which implies $\hat{P}$. From the definition, given $V' \subseteq V$ then

$$\delta^i\big(\delta^i(\hat{P}_V)\big)_{V'} = \delta^i(\hat{P})_{V'} \leq \delta^i(\hat{P})_V \quad \text{and} \quad \delta^i(\hat{P})_V \leq \hat{P}. \tag{13.4}$$

Since $\hat{P} \leq \delta^o(\hat{P})_V$ we then have

$$\delta^i(\hat{P})_V \leq \delta^o(\hat{P})_V. \tag{13.5}$$

Given this definition we can construct the inner presheaf, which is the analogue of the outer presheaf, as follows

Definition 13.3 The inner presheaf $\underline{I}$ is defined over the category $\mathcal{V}(\mathcal{H})$ on:

1. Objects: for each $V \in \mathcal{V}(\mathcal{H})$ we obtain $\underline{I}_V := P(V)$.
2. Morphisms: for each $i_{V'V} : V' \subseteq V$ the corresponding presheaf map is

$$\begin{aligned} \underline{I}(i_{V'V}) : \underline{I}_V &\rightarrow \underline{I}_{V'} \\ \hat{\alpha} &\mapsto \underline{I}(i_{V'V})(\hat{\alpha}) := \delta^i(\hat{\alpha})_{V'} \end{aligned} \tag{13.6}$$

for all $\hat{\alpha} \in P(V)$.

Similarly, as was the case for the outer presheaf, the assignment

$$\hat{P} \mapsto \{\delta^i(\hat{P})_V \big| V \in \mathcal{V}(\mathcal{H})\} \tag{13.7}$$

defines a global element of $\underline{I}$. Thus we can write inner daseinisation as

$$\begin{aligned} \delta^i : P(\mathcal{H}) &\rightarrow \Gamma(\underline{I}) \\ \hat{P} &\mapsto \{\delta^i(\hat{P})_V \big| V \in \mathcal{V}(\mathcal{H})\}. \end{aligned} \tag{13.8}$$

Equivalently, we can define inner daseinisation as a mapping from projection operators to sub-objects of the spectral presheaf

$$\begin{aligned}\delta^i : P(\mathcal{H}) &\to Sub_{cl}(\underline{\Sigma}) \\ \hat{P} &\mapsto \{\underline{T}_{\delta^i(\hat{P})_V} | V \in \mathcal{V}(\mathcal{H})\}\end{aligned} \tag{13.9}$$

where, for each $V \in \mathcal{V}(\mathcal{H})$, $\underline{T}_{\delta^i(\hat{P})_V} := \{\lambda \in \underline{\Sigma}_V | \lambda(\delta^i(\hat{P})_V) = 1\}$. The collection of all these clopen subsets forms a clopen sub-object $\underline{\delta^i(\hat{P})} \subseteq \underline{\Sigma}$.

It is interesting to see how the inner daseinisation helps in the definition of the negation operation. In particular we would like to understand the presheaf

$$\underline{\delta(\neg\hat{P})} \quad \text{where } \neg\hat{P} = \hat{\mathbb{1}} - \hat{P} \text{ in the lattice } P(\mathcal{H}). \tag{13.10}$$

In order to define $\underline{\delta(\neg\hat{P})}$ one makes use of both inner and outer daseinisation obtaining for each context $V \in \mathcal{V}(\mathcal{H})$

$$\underline{O}(i_{V'V})(\neg\hat{P}) = \neg\underline{I}(i_{V'V})(\hat{\alpha}). \tag{13.11}$$

Proof Computing the left hand side of (13.11) we obtain

$$\underline{O}(i_{V'V})(\neg\hat{P}) = \delta^o(\neg\hat{P})_{V'} = \delta^o(\hat{\mathbb{1}} - \hat{P})_{V'} = \bigwedge\{\hat{\beta} \in P(V') | \hat{\beta} \geq \hat{\mathbb{1}} - \hat{P}\}. \tag{13.12}$$

Let us assume that the right hand side of the above equation is the projection operator $\hat{\beta}$. Then such projection operator is the smallest projection operator in V' which is bigger than $\hat{1} - \hat{P}$, i.e. $\hat{\beta} \geq \hat{1} - \hat{P}$. Therefore $\hat{\mathbb{1}} - \hat{\beta} = \hat{\beta}^c$ is the biggest projection operator in $P(V')$ such that $\hat{\mathbb{1}} - \hat{\beta} = \hat{\beta}^c \leq \hat{P}$. Thus $\hat{\beta}^c = \delta^i(\hat{P})_{V'}$.

If we now consider the right hand side of (13.11) we obtain

$$\neg\underline{I}(i_{V'V})(\hat{P}) = \hat{\mathbb{1}} - \delta^i(\hat{P})_{V'} = \hat{\mathbb{1}} - \hat{\beta}^c = \hat{\mathbb{1}} - \hat{\mathbb{1}} + \hat{\beta} = \hat{\beta} \tag{13.13}$$

hence (13.11) is proved. □

It follows that for each projection operator $\hat{P} \in P(\mathcal{H})$ and each context $V \in \mathcal{V}(\mathcal{H})$ we obtain

$$\delta^o(\neg\hat{P})_V = \hat{\mathbb{1}} - \delta^i(\hat{P})_V. \tag{13.14}$$

13.3 Spectral Decomposition

In quantum theory, observables, i.e. things that we measure, are identified with self-adjoint operators. Thus, the study of their eigenvalues and how these eigenvalues are measured is very important. Here is where the spectral theorem and consequently the spectral decomposition come into the picture.

Given a self-adjoint operator $\hat{A}$, the spectral theorem essentially tells us that it is possible to write $\hat{A}$ as

$$\hat{A} = \int_{\sigma(A)} \lambda d\hat{E}^{\hat{A}}_{\lambda}. \tag{13.15}$$

Such an expression is called the spectral decomposition of $\hat{A}$. $\sigma(A) \subseteq \mathbb{R}$ represents the spectrum of the operator $\hat{A}$ and $\{\hat{E}^{\hat{A}}_{\lambda} | \lambda \in \sigma(\hat{A})\}$ is the spectral family of $\hat{A}$. Such family determines the set of spectral projection operators of $\hat{A}$, namely $\hat{E}[A \in \Delta]$ as follows:

$$\hat{E}[A \in \Delta] := \int_{\Delta} d\hat{E}^{\hat{A}}_{\lambda} \tag{13.16}$$

where Δ is a Borel subset of the spectrum $\sigma(\hat{A})$ of A. What such projection operators represent are subspaces of the Hilbert space for which the states $|\psi\rangle$ have a value of A which lies in the interval Δ. Therefore, if a is a value in the discrete spectrum of $\hat{A}$, then

$$\hat{E}\big[A \in \{a\}\big] \tag{13.17}$$

projects onto the eigenstates of $\hat{A}$ with eigenvalue a.

In this setting, given a bounded Borel function $f : \mathbb{R} \to \mathbb{R}$, then the 'transformed' operator $f(\hat{A})$ has spectral decomposition given by

$$f(\hat{A}) = \int_{\sigma(A)} f(\lambda) d\hat{E}^{\hat{A}}_{\lambda}. \tag{13.18}$$

The formal theorem for the spectral decomposition for bounded operators is as follows:

Theorem 13.1 [75] *Given a bounded self-adjoint operator $\hat{A}$ on $\mathcal{H}$ there exists a family of projection operators $\{\hat{E}_{\lambda} | \lambda \in \mathbb{R}\}$ called the spectral family of $\hat{A}$, such that the following conditions are satisfied*:

1. $\hat{E}_{\lambda} \leq \hat{E}_{\lambda'}$ *for* $\lambda \leq \lambda'$.
2. *The net* $\lambda \to \hat{E}_{\lambda}$ *of projection operators in the lattice* $P(\mathcal{H})$ *is bounded above by* $\hat{1}$, *and below by* $\hat{0}$, *i.e.*

$$\lim_{\lambda \to \infty} \hat{E}_{\lambda} = \hat{1}$$
$$\lim_{\lambda \to -\infty} \hat{E}_{\lambda} = \hat{0}. \tag{13.19}$$

3. $\hat{E}_{\lambda+0} = \hat{E}_{\lambda}$.
4. $\hat{A} = \int_{\sigma(A) \subseteq \mathbb{R}} \lambda d\hat{E}^{\hat{A}}_{\lambda}$.

5. *The map* $\lambda \to \hat{E}_\lambda$ *is right-continuous:*[2]

$$\bigwedge_{\epsilon \downarrow 0} \hat{E}_{\lambda+\epsilon} = \hat{E}_\lambda \tag{13.20}$$

for all $\lambda \in \mathbb{R}$.

Given the spectral decomposition it is possible to define a different type of ordering[3] on operators, called *spectral ordering*. The reason why this order was chosen rather than the standard operator ordering is because the former preserves the relation between the spectrums of the operator, i.e. if $\hat{A} \leq_s \hat{B}$, then $sp(\hat{A}) \subseteq sp(\hat{B})$. This feature will reveal itself very important when defining the values for physical quantitates.

We will now define what the spectral order is. Consider two self-adjoint operators $\hat{A}$ and $\hat{B}$ with spectral families $(\hat{E}_r^{\hat{A}})_{r\in\mathbb{R}}$ and $(\hat{E}_r^{\hat{B}})_{r\in\mathbb{R}}$, respectively. The spectral order is then defined as follows:[4]

$$\hat{A} \leq_s \hat{B} \quad \text{iff} \quad \forall r \in \mathbb{R} \quad \hat{E}_r^{\hat{A}} \geq \hat{E}_r^{\hat{B}}. \tag{13.21}$$

From the definition it follows that the spectral order implies the usual order between operators, i.e. if $\hat{A} \leq_s \hat{B}$ then $\hat{A} \leq \hat{B}$, but the converse is not true. Thus the spectral order is a partial order on $B(\mathcal{H})_{sa}$ (the self-adjoint operators in $B(\mathcal{H})$) that is coarser than the usual one.

It is easy to see that the spectral ordering defines a genuine partial ordering on $B(\mathcal{H})_{sa}$. In fact, each bounded set S of self-adjoint operators has a minimum $\bigwedge S \in B(\mathcal{H})_{sa}$ and a maximum $\bigvee S \in B(\mathcal{H})_{sa}$ with respect to the spectral order, i.e. $B(\mathcal{H})_{sa}$ is a 'boundedly complete' lattice with respect to the spectral order.

If we define the spectral order for projection operators then we will obtain exactly the usual partial ordering. In fact if we consider two projection operators $\hat{P}$ and $\hat{Q}$ then their spectral decomposition is

$$\hat{E}_\lambda^{\hat{P}} = \begin{cases} \hat{0} & \text{if } \lambda < 0 \\ \hat{1} - \hat{P} & \text{if } 0 \leq \lambda < 1 \\ \hat{P} & \text{if } 1 \leq \lambda. \end{cases} \tag{13.22}$$

It follows that

$$\hat{P} \leq_s \hat{Q} \quad \text{iff} \quad \hat{P} \leq \hat{Q}. \tag{13.23}$$

[2]Could equivalently require left continuity.

[3]Recall that the standard operator ordering is given as follows: $\hat{A} \leq \hat{B}$ iff $\langle\psi|\hat{A}|\psi\rangle \leq \langle\psi|\hat{B}|\psi\rangle$ for all $\langle\psi| - |\psi\rangle : N_{sa} \to \mathbb{C}$, where N_{sa} represents the collection of self-adjoint operators on the Hilbert space $\mathcal{H}$.

[4]Note that the spectral order of self-adjoint operators is defined in terms of the usual ordering of the respective projection operators in the spectral decomposition.

Thus the spectral order coincides with the usual partial order on $P(\mathcal{H})$.

Moreover, if $\hat{A}$ and $\hat{B}$ are self-adjoint operators such that (i) either $\hat{A}$ or $\hat{B}$ is a projection, or (ii) $[\hat{A}, \hat{B}] = 0$, then $\hat{A} \leq_s \hat{B}$ iff $\hat{A} \leq \hat{B}$.

13.3.1 Example of Spectral Decomposition

Let us consider again the 4 dimensional Hilbert space $\mathbb{C}^4$. We would like to define the spectral decomposition of the self-adjoint operator

$$\hat{S}_z = \begin{pmatrix} 2 & 0 & 0 & 0 \\ 0 & 0 & 0 & 0 \\ 0 & 0 & 0 & 0 \\ 0 & 0 & 0 & -2 \end{pmatrix}.$$

For such an operator the spectral family is

$$\hat{E}_\lambda^{\hat{S}_z} = \begin{cases} \hat{0} & \text{if } \lambda < -2 \\ \hat{P}_4 & \text{if } -2 \leq \lambda < 0 \\ \hat{P}_4 + \hat{P}_3 + \hat{P}_2 & \text{if } 0 \leq \lambda < 2 \\ \hat{P}_4 + \hat{P}_3 + \hat{P}_2 + \hat{P}_1 & \text{if } 2 \leq \lambda. \end{cases} \tag{13.24}$$

If we now consider the coarse grained $\hat{S}_z^2$ whose matrix representation is

$$\hat{S}_z^2 = \begin{pmatrix} 4 & 0 & 0 & 0 \\ 0 & 0 & 0 & 0 \\ 0 & 0 & 0 & 0 \\ 0 & 0 & 0 & 4 \end{pmatrix}$$

then the corresponding spectral decomposition will be

$$\hat{E}_\lambda^{\hat{S}_z^2} = \begin{cases} \hat{0} & \text{if } \lambda < 0 \\ \hat{P}_2 + \hat{P}_3 & \text{if } 0 \leq \lambda < 4 \\ \hat{P}_4 + \hat{P}_3 + \hat{P}_2 + \hat{P}_1 & \text{if } 4 \leq \lambda. \end{cases} \tag{13.25}$$

Utilising the spectral ordering to determine which one is 'bigger' we would obtain the following:

$$\begin{cases} \hat{E}_\lambda^{\hat{S}_z^2} = \hat{E}_\lambda^{\hat{S}_z} = \hat{0} & \text{for } \lambda < -2 \\ \hat{E}_\lambda^{\hat{S}_z^2} \leq \hat{E}_\lambda^{\hat{S}_z} & \text{for } -2 \leq \lambda < 0 \\ \hat{E}_\lambda^{\hat{S}_z^2} \leq \hat{E}_\lambda^{\hat{S}_z} & \text{for } 0 \leq \lambda < 2 \\ \hat{E}_\lambda^{\hat{S}_z^2} \leq \hat{E}_\lambda^{\hat{S}_z} & \text{for } 2 \leq \lambda < 4 \\ \hat{E}_\lambda^{\hat{S}_z^2} = \hat{E}_\lambda^{\hat{S}_z} & \text{for } 4 \leq \lambda. \end{cases} \tag{13.26}$$

Thus, as expected,

$$\hat{S}_z^2 \geq_s \hat{S}_z. \tag{13.27}$$

13.4 Daseinisation of Self-adjoint Operators

Given the discussion above regarding the spectral order we are now ready to define the concept of both inner and outer daseinisation of self-adjoint operators. To this end let us consider a self-adjoint operator $\hat{A}$ and a context V, such that $\hat{A} \notin V_{sa}$ (V_{sa} denotes the collection of self-adjoint operators in V). We then need to approximate $\hat{A}$, so as to be in V. However, since we eventually want to define an interval of possible values of $\hat{A}$ at V, we will approximate $\hat{A}$ both from above and from below. In particular, we will consider the pair of operators

$$\begin{aligned} \delta^o(\hat{A})_V &:= \bigwedge\{\hat{B} \in V_{sa} | \hat{A} \leq_s \hat{B}\} \\ \delta^i(\hat{A})_V &:= \bigvee\{\hat{B} \in V_{sa} | \hat{A} \geq_s \hat{B}\}. \end{aligned} \tag{13.28}$$

In the above equation $\delta^o(\hat{A})_V$ represents the smallest self-adjoint operator in V, which is spectrally larger or equal to $\hat{A}$, while $\delta^i(\hat{A})_V$ represents the biggest self-adjoint operator in V_{sa}, that is spectrally smaller or equal to $\hat{A}$. The process represented by δ^i is what we have defined as *inner daseinisation*, while δ^o represents the *outer daseinisation*.

From the definition of $\delta^i(\hat{A})_V$ it follows that if $V' \subseteq V$ then $\delta^i(\hat{A})_{V'} \leq_s \delta^i(\hat{A})_V$. Moreover, from (13.28) it follows that:

$$sp(\hat{A}) \subseteq sp\big(\delta^i(\hat{A})_V\big), \qquad sp\big(\delta^o(\hat{A})_V\big) \subseteq sp(\hat{A}) \tag{13.29}$$

which, as mentioned above, is precisely the reason why the spectral order was chosen.

Given the spectral decomposition of $\hat{A}$: $\hat{A} = \int_{\mathbb{R}} \lambda d(\hat{E}_\lambda^{\hat{A}})$, if we apply the definition of spectral order to inner and outer daseinisation we obtain:

$$\delta^i(\hat{A})_V \leq_s \delta^o(\hat{A})_V \quad \text{iff} \quad \forall r \in \mathbb{R} \quad \hat{E}_r^{\delta^i(\hat{A})_V} \geq \hat{E}_r^{\delta^o(\hat{A})_V}. \tag{13.30}$$

Since $\delta^i(\hat{A})_V \leq \delta^o(\hat{A})_V$, it follows that for all $r \in \mathbb{R}$

$$\begin{aligned} \hat{E}_r^{\delta^i(\hat{A})_V} &:= \delta^o\big(\hat{E}_r^{\hat{A}}\big)_V \\ \hat{E}_r^{\delta^o(\hat{A})_V} &:= \delta^i\big(\hat{E}_r^{\hat{A}}\big)_V. \end{aligned} \tag{13.31}$$

The spectral family described by the second equation is right-continuous, while the first is not. To overcome this problem we define the following:

$$\hat{E}_r^{\delta^i(\hat{A})_V} := \bigwedge_{s>r} \delta^o\big(\hat{E}_s^{\hat{A}}\big)_V. \tag{13.32}$$

In this way, given a spectral family $\lambda \to \hat{E}_\lambda$, it is then possible to construct, for each $V \in \mathcal{V}(\mathcal{H})$ two other spectral families

$$\begin{aligned} &\lambda \to \bigwedge_{s>r} \delta^o\big(\hat{E}_s^{\hat{A}}\big) \\ &\lambda \to \delta^i(\hat{E}_\lambda). \end{aligned} \tag{13.33}$$

These will be precisely the spectral families we will utilise to define the inner and outer daseinisation of self-adjoint operators. In particular, we can now write inner and outer daseinisation for self-adjoint operators as follows:

$$\begin{aligned} \delta^o(\hat{A})_V &:= \int_{\mathbb{R}} \lambda d\big(\delta^i\big(\hat{E}_\lambda^{\hat{A}}\big)\big) \\ \delta^i(\hat{A})_V &:= \int_{\mathbb{R}} \lambda d\Big(\bigwedge_{\mu>\lambda} \delta^o\big(\hat{E}_\mu^{\hat{A}}\big)\Big). \end{aligned} \tag{13.34}$$

Note that the above integrals should be interpreted as Riemann Stieltjes integrals, i.e.

$$\hat{A} = \int_{\mathbb{R}} \lambda d\hat{P}_\lambda \quad \text{means} \quad \langle\psi|\hat{A}|\psi\rangle = \int_{\mathbb{R}} \lambda\langle\psi|d\hat{P}_\lambda|\psi\rangle. \tag{13.35}$$

This explains the condition of right-continuity.

We can now define the analogues of the inner and outer presheaves but for self-adjoint operators rather than projection operators.

Definition 13.4 [32] The outer de Groote presheaf $\underline{\mathbb{O}} : \mathcal{V}(\mathcal{H}) \to$ **Sets** is defined on:

1. Objects: for each $V \in \mathcal{V}(\mathcal{H})$ we define $\underline{\mathbb{O}}_V := V_{sa}$ (the collection of self adjoin operators in V).
2. Morphisms: given a map $i : V' \subseteq V$ the corresponding presheaf map is $\underline{\mathbb{O}}(i_{V'V}) : \underline{\mathbb{O}}_V \to \underline{\mathbb{O}}_{V'}$ which is defined as follows:

$$\begin{aligned}\underline{\mathbb{O}}(i_{V'V})(\hat{A}) &:= \delta^o(\hat{A})_{V'} \\ &= \int_{\mathbb{R}} \lambda d\big(\delta^i(\hat{E}^{\hat{A}}_{\lambda})_{V'}\big) \\ &= \int_{\mathbb{R}} \lambda d\big(\underline{I}(i_{V'V})(\hat{E}^{\hat{A}}_{\lambda})\big)\end{aligned} \tag{13.36}$$

for all $\hat{A} \in \underline{\mathbb{O}}_V$.

In the above definition, $\underline{I}$ is the inner presheaf for projection operators defined in Sect. 13.2. On the other hand the inner de Groote presheaf is:

Definition 13.5 [32] The inner de Groote presheaf $\underline{\mathbb{I}} : \mathcal{V}(\mathcal{H}) \to$ **Sets** is defined on:

1. Objects: for each $V \in \mathcal{V}(\mathcal{H})$ we obtain $\underline{\mathbb{I}}_V := V_{sa}$.
2. Morphisms: given a map $i : V' \subseteq V$ then the corresponding presheaf map is $\underline{\mathbb{I}}(i_{V'V}) : \underline{\mathbb{I}}_V \to \underline{\mathbb{I}}_{V'}$ such that

$$\begin{aligned}\underline{\mathbb{I}}(i_{V'V})(\hat{A}) &:= \delta^i(\hat{A})_{V'} \\ &= \int_{\mathbb{R}} \lambda d\Big(\bigwedge_{\mu>\lambda} \big(\delta^o(\hat{E}^{\hat{A}}_{\mu})_{V'}\big)\Big) \\ &= \int_{\mathbb{R}} \lambda d\Big(\bigwedge_{\mu>\lambda} \big(\underline{O}(i_{V'V})(\hat{E}^{\hat{A}}_{\mu})\big)\Big).\end{aligned} \tag{13.37}$$

For all $\hat{A} \in \underline{\mathbb{I}}_V$. Here $\underline{O}$ is the outer presheaf for projection operators defined in Sect. 11.1.

As it can be expected, the inner and outer daseinisation map give rise to global elements of the inner and outer de Groote presheaves:

$$\begin{aligned}\delta^o(\hat{A}) &: V \mapsto \delta^o(\hat{A})_V \\ \delta^i(\hat{A}) &: V \mapsto \delta^i(\hat{A})_V.\end{aligned} \tag{13.38}$$

We then reach the following theorem:

Theorem 13.2 *The maps*

$$\begin{aligned}\delta^i : B(\mathcal{H})_{sa} &\to \Gamma\underline{\mathbb{I}} \\ \hat{A} &\mapsto \delta^i(\hat{A})\end{aligned} \tag{13.39}$$

and

$$\begin{aligned}\delta^o : B(\mathcal{H})_{sa} &\to \Gamma\underline{\mathbb{O}} \\ \hat{A} &\mapsto \delta^o(\hat{A})\end{aligned} \tag{13.40}$$

are injective.

Proof Given the definition of inner daseinisation we know that $\hat{A} \geq_s \delta^i(\hat{A})_V$ for each $V \in \mathcal{V}(\mathcal{H})$. Therefore, since there exists a V such that $\hat{A} \in V_{sa}$ it follows that

$$\hat{A} = \bigvee_{V \in \mathcal{V}(\mathcal{H})} \delta^i(\hat{A})_V. \tag{13.41}$$

Therefore, if $\delta^i(\hat{A}) = \delta^i(\hat{B})$ it follows that:

$$\hat{A} = \bigvee_{V \in \mathcal{V}(\mathcal{H})} \delta^i(\hat{A})_V = \bigvee_{V \in \mathcal{V}(\mathcal{H})} \delta^i(\hat{B})_V = \hat{B}. \tag{13.42}$$

Similarly for outer daseinisation, if $\delta^o(\hat{A}) = \delta^o(\hat{B})$ then

$$\hat{A} = \bigwedge_{V \in \mathcal{V}(\mathcal{H})} \delta^i(\hat{A})_V = \hat{B} = \bigwedge_{V \in \mathcal{V}(\mathcal{H})} \delta^o(\hat{B})_V. \tag{13.43}$$

$\square$

13.4.1 Example

Let us consider the 2 spin system in $\mathbb{C}^4$ defined in previous examples. We are interested in the spin in the z-direction, which is represented by the physical quantity S_z. The self-adjoint operator representing S_z is

$$\hat{S}_z = \begin{pmatrix} 2 & 0 & 0 & 0 \\ 0 & 0 & 0 & 0 \\ 0 & 0 & 0 & 0 \\ 0 & 0 & 0 & -2 \end{pmatrix}.$$

We now would like to define both inner and outer daseinisation of $\hat{S}_z$. The spectral family of $\hat{S}_z$ is

$$\hat{E}_\lambda^{\hat{S}_z} = \begin{cases} \hat{0} & \text{if } \lambda < -2 \\ \hat{P}_4 & \text{if } -2 \leq \lambda < 0 \\ \hat{P}_4 + \hat{P}_3 + \hat{P}_2 & \text{if } 0 \leq \lambda < 2 \\ \hat{P}_4 + \hat{P}_3 + \hat{P}_2 + \hat{P}_1 & \text{if } 2 \leq \lambda. \end{cases} \tag{13.44}$$

However, since $\hat{S}_z \in V$ then for context V we obtain $\delta^o(\hat{S}_z)_V = \hat{S}_z$ and $\delta^i(\hat{S}_z)_V = \hat{S}_z$. If we then go to smaller sub-algebras we obtain the following daseinised spectral decomposition:

1. For $j = \{2, 4\}$ we obtain

$$\delta^i\big(\hat{E}_\lambda^{\hat{S}_z}\big)_{V_{\hat{P}_j\hat{P}_3}} = \begin{cases} \hat{0} & \text{if } \lambda < -2 \\ \hat{0} & \text{if } -2 \leq \lambda < 0 \\ \hat{P}_j + \hat{P}_3 & \text{if } 0 \leq \lambda < 2 \\ \hat{\mathbb{1}} & \text{if } 2 \leq \lambda. \end{cases} \tag{13.45}$$

2. For $j = \{2, 3\}$

$$\delta^i\big(\hat{E}_\lambda^{\hat{S}_z}\big)_{V_{\hat{P}_1\hat{P}_j}} = \begin{cases} \hat{0} & \text{if } \lambda < -2 \\ \hat{0} & \text{if } -2 \leq \lambda < 0 \\ \hat{P}_2 + \hat{P}_3 + \hat{P}_4 & \text{if } 0 \leq \lambda < 2 \\ \hat{\mathbb{1}} & \text{if } 2 \leq \lambda. \end{cases} \tag{13.46}$$

3. For $V_{\hat{P}_1,\hat{P}_4}$

$$\delta^i\big(\hat{E}_\lambda^{\hat{S}_z}\big)_{V_{\hat{P}_1\hat{P}_4}} = \begin{cases} \hat{0} & \text{if } \lambda < -2 \\ \hat{P}_4 & \text{if } -2 \leq \lambda < 0 \\ \hat{P}_2 + \hat{P}_3 + \hat{P}_4 & \text{if } 0 \leq \lambda < 2 \\ \hat{\mathbb{1}} & \text{if } 2 \leq \lambda. \end{cases} \tag{13.47}$$

4. For $V_{\hat{P}_2,\hat{P}_4}$

$$\delta^i\big(\hat{E}_\lambda^{\hat{S}_z}\big)_{V_{\hat{P}_2\hat{P}_4}} = \begin{cases} \hat{0} & \text{if } \lambda < -2 \\ \hat{P}_4 & \text{if } -2 \leq \lambda < 0 \\ \hat{P}_2 + \hat{P}_4 & \text{if } 0 \leq \lambda < 2 \\ \hat{\mathbb{1}} & \text{if } 2 \leq \lambda. \end{cases} \tag{13.48}$$

5. For $j = \{2, 3\}$

$$\delta^i\big(\hat{E}_\lambda^{\hat{S}_z}\big)_{V_{\hat{P}_j}} = \begin{cases} \hat{0} & \text{if } \lambda < -2 \\ \hat{0} & \text{if } -2 \leq \lambda < 0 \\ \hat{P}_j & \text{if } 0 \leq \lambda < 2 \\ \hat{\mathbb{1}} & \text{if } 2 \leq \lambda. \end{cases} \tag{13.49}$$

6. For $V_{\hat{P}_1}$

$$\delta^i\big(\hat{E}_\lambda^{\hat{S}_z}\big)_{V_{\hat{P}_1}} = \begin{cases} \hat{0} & \text{if } \lambda < -2 \\ \hat{0} & \text{if } -2 \leq \lambda < 0 \\ \hat{P}_2 + \hat{P}_3 + \hat{P}_4 & \text{if } 0 \leq \lambda < 2 \\ \hat{1} & \text{if } 2 \leq \lambda. \end{cases} \tag{13.50}$$

7. Finally for $V_{\hat{P}_4}$

$$\delta^i\big(\hat{E}_\lambda^{\hat{S}_z}\big)_{V_{\hat{P}_4}} = \begin{cases} \hat{0} & \text{if } \lambda < -2 \\ \hat{P}_4 & \text{if } -2 \leq \lambda < 0 \\ \hat{P}_4 & \text{if } 0 \leq \lambda < 2 \\ \hat{1} & \text{if } 2 \leq \lambda. \end{cases} \tag{13.51}$$

Similarly for outer daseinisation we obtain the following:

1. For distinct $i, j = \{2, 3\}$

$$\delta^o\big(\hat{E}_\lambda^{\hat{S}_z}\big)_{V_{\hat{P}_j\hat{P}_1}} = \begin{cases} \hat{0} & \text{if } \lambda < -2 \\ \hat{P}_4 + \hat{P}_i & \text{if } -2 \leq \lambda < 0 \\ \hat{P}_j + \hat{P}_4 + \hat{P}_i & \text{if } 0 \leq \lambda < 2 \\ \hat{1} & \text{if } 2 \leq \lambda. \end{cases} \tag{13.52}$$

2. For $j = \{1, 2, 3\}$

$$\delta^o\big(\hat{E}_\lambda^{\hat{S}_z}\big)_{V_{\hat{P}_j}} = \begin{cases} \hat{0} & \text{if } \lambda < -2 \\ \hat{1} - \hat{P}_j & \text{if } -2 \leq \lambda < 0 \\ \hat{1}(-\hat{P}_j \text{ if } j = 1) & \text{if } 0 \leq \lambda < 2 \\ \hat{1} & \text{if } 2 \leq \lambda. \end{cases} \tag{13.53}$$

3. For $j = \{2, 3\}$ and $V_{\hat{P}_4}$

$$\delta^o\big(\hat{E}_\lambda^{\hat{S}_z}\big)_{V_{\hat{P}_4}} = \delta^o\big(\hat{E}_\lambda^{\hat{S}_z}\big)_{V_{\hat{P}_j\hat{P}_4}} = \begin{cases} \hat{0} & \text{if } \lambda < -2 \\ \hat{P}_4 & \text{if } -2 \leq \lambda < 0 \\ \hat{1} & \text{if } 0 \leq \lambda < 2 \\ \hat{1} & \text{if } 2 \leq \lambda. \end{cases} \tag{13.54}$$

4. For $V_{\hat{P}_1,\hat{P}_4}$

$$\delta^o(\hat{E}_\lambda^{\hat{S}_z})_{V_{\hat{P}_4\hat{P}_1}} = \begin{cases} \hat{0} & \text{if } \lambda < -2 \\ \hat{P}_4 & \text{if } -2 \leq \lambda < 0 \\ \hat{P}_2 + \hat{P}_3 + \hat{P}_4 & \text{if } 0 \leq \lambda < 2 \\ \hat{1} & \text{if } 2 \leq \lambda. \end{cases} \tag{13.55}$$

5. For $V_{\hat{P}_2,\hat{P}_3}$

$$\delta^o(\hat{E}_\lambda^{\hat{S}_z})_{V_{\hat{P}_2\hat{P}_3}} = \begin{cases} \hat{0} & \text{if } \lambda < -2 \\ \hat{P}_4 + \hat{P}_1 & \text{if } -2 \leq \lambda < 0 \\ \hat{1} & \text{if } 0 \leq \lambda < 2 \\ \hat{1} & \text{if } 2 \leq \lambda. \end{cases} \tag{13.56}$$

We can now compute the daseinisation of $\hat{S}_z$ for each contexts. These are:

1. For $V_{\hat{P}_4}$

$$\delta^o(\hat{S}_z)_{V_{\hat{P}_4}} = \begin{pmatrix} 2 & 0 & 0 & 0 \\ 0 & 2 & 0 & 0 \\ 0 & 0 & 2 & 0 \\ 0 & 0 & 0 & -2 \end{pmatrix} \quad \text{and} \quad \delta^i(\hat{S}_z)_{V_{\hat{P}_4}} = \begin{pmatrix} 0 & 0 & 0 & 0 \\ 0 & 0 & 0 & 0 \\ 0 & 0 & 0 & 0 \\ 0 & 0 & 0 & -2 \end{pmatrix}. \tag{13.57}$$

2. For $V_{\hat{P}_1}$

$$\delta^o(\hat{S}_z)_{V_{\hat{P}_1}} = \begin{pmatrix} 2 & 0 & 0 & 0 \\ 0 & 0 & 0 & 0 \\ 0 & 0 & 0 & 0 \\ 0 & 0 & 0 & 0 \end{pmatrix} \quad \text{and} \quad \delta^i(\hat{S}_z)_{V_{\hat{P}_1}} = \begin{pmatrix} 2 & 0 & 0 & 0 \\ 0 & -2 & 0 & 0 \\ 0 & 0 & -2 & 0 \\ 0 & 0 & 0 & -2 \end{pmatrix}. \tag{13.58}$$

3. For $V_{\hat{P}_2}$

$$\delta^o(\hat{S}_z)_{V_{\hat{P}_2}} = \begin{pmatrix} 2 & 0 & 0 & 0 \\ 0 & 0 & 0 & 0 \\ 0 & 0 & 2 & 0 \\ 0 & 0 & 0 & 2 \end{pmatrix} \quad \text{and} \quad \delta^i(\hat{S}_z)_{V_{\hat{P}_2}} = \begin{pmatrix} 2 & 0 & 0 & 0 \\ 0 & 0 & 0 & 0 \\ 0 & 0 & -2 & 0 \\ 0 & 0 & 0 & -2 \end{pmatrix}. \tag{13.59}$$

4. For $V_{\hat{P}_3}$

$$\delta^o(\hat{S}_z)_{V_{\hat{P}_3}} = \begin{pmatrix} 2 & 0 & 0 & 0 \\ 0 & 2 & 0 & 0 \\ 0 & 0 & 0 & 0 \\ 0 & 0 & 0 & 2 \end{pmatrix} \quad \text{and} \quad \delta^i(\hat{S}_z)_{V_{\hat{P}_3}} = \begin{pmatrix} -2 & 0 & 0 & 0 \\ 0 & -2 & 0 & 0 \\ 0 & 0 & 0 & 0 \\ 0 & 0 & 0 & -2 \end{pmatrix}. \tag{13.60}$$

5. For $V_{\hat{P}_1,\hat{P}_2}$

$$\delta^o(\hat{S}_z)_{V_{\hat{P}_1,\hat{P}_2}} = \begin{pmatrix} 2 & 0 & 0 & 0 \\ 0 & 0 & 0 & 0 \\ 0 & 0 & 0 & 0 \\ 0 & 0 & 0 & 0 \end{pmatrix} \quad \text{and} \quad \delta^i(\hat{S}_Z)_{V_{\hat{P}_1,\hat{P}_2}} = \begin{pmatrix} 2 & 0 & 0 & 0 \\ 0 & 0 & 0 & 0 \\ 0 & 0 & -2 & 0 \\ 0 & 0 & 0 & -2 \end{pmatrix}. \tag{13.61}$$

6. For $V_{\hat{P}_1,\hat{P}_3}$

$$\delta^o(\hat{S}_z)_{V_{\hat{P}_1,\hat{P}_3}} = \begin{pmatrix} 2 & 0 & 0 & 0 \\ 0 & 0 & 0 & 0 \\ 0 & 0 & 0 & 0 \\ 0 & 0 & 0 & 0 \end{pmatrix} \quad \text{and} \quad \delta^i(\hat{S}_z)_{V_{\hat{P}_1,\hat{P}_3}} = \begin{pmatrix} 2 & 0 & 0 & 0 \\ 0 & -2 & 0 & 0 \\ 0 & 0 & 0 & 0 \\ 0 & 0 & 0 & -2 \end{pmatrix}. \tag{13.62}$$

7. For $V_{\hat{P}_1,\hat{P}_4}$

$$\delta^o(\hat{S}_z)_{V_{\hat{P}_1,\hat{P}_4}} = \begin{pmatrix} 2 & 0 & 0 & 0 \\ 0 & 0 & 0 & 0 \\ 0 & 0 & 0 & 0 \\ 0 & 0 & 0 & -2 \end{pmatrix} \quad \text{and} \quad \delta^i(\hat{S}_z)_{V_{\hat{P}_1,\hat{P}_4}} = \begin{pmatrix} 2 & 0 & 0 & 0 \\ 0 & 0 & 0 & 0 \\ 0 & 0 & 0 & 0 \\ 0 & 0 & 0 & -2 \end{pmatrix}. \tag{13.63}$$

8. For $V_{\hat{P}_2,\hat{P}_3}$

$$\delta^o(\hat{S}_z)_{V_{\hat{P}_2,\hat{P}_3}} = \begin{pmatrix} 2 & 0 & 0 & 0 \\ 0 & 0 & 0 & 0 \\ 0 & 0 & 0 & 0 \\ 0 & 0 & 0 & 2 \end{pmatrix} \quad \text{and} \quad \delta^i(\hat{S}_z)_{V_{\hat{P}_2,\hat{P}_3}} = \begin{pmatrix} -2 & 0 & 0 & 0 \\ 0 & 0 & 0 & 0 \\ 0 & 0 & 0 & 0 \\ 0 & 0 & 0 & -2 \end{pmatrix}. \tag{13.64}$$

9. For $V_{\hat{P}_2,\hat{P}_4}$

$$\delta^o(\hat{S}_z)_{V_{\hat{P}_2,\hat{P}_4}} = \begin{pmatrix} 2 & 0 & 0 & 0 \\ 0 & 0 & 0 & 0 \\ 0 & 0 & 2 & 0 \\ 0 & 0 & 0 & -2 \end{pmatrix} \quad \text{and} \quad \delta^i(\hat{S}_z)_{V_{\hat{P}_2,\hat{P}_4}} = \begin{pmatrix} 0 & 0 & 0 & 0 \\ 0 & 0 & 0 & 0 \\ 0 & 0 & 0 & 0 \\ 0 & 0 & 0 & -2 \end{pmatrix}. \tag{13.65}$$

10. For $V_{\hat{P}_3,\hat{P}_4}$

$$\delta^o(\hat{S}_z)_{V_{\hat{P}_3,\hat{P}_4}} = \begin{pmatrix} 2 & 0 & 0 & 0 \\ 0 & 2 & 0 & 0 \\ 0 & 0 & 0 & 0 \\ 0 & 0 & 0 & 0 \end{pmatrix} \quad \text{and} \quad \delta^i(\hat{S}_z)_{V_{\hat{P}_3,\hat{P}_4}} = \begin{pmatrix} 0 & 0 & 0 & 0 \\ 0 & 0 & 0 & 0 \\ 0 & 0 & 0 & 0 \\ 0 & 0 & 0 & -2 \end{pmatrix}. \tag{13.66}$$

It is straightforward to read off from the matrices that, for each context $V \in \mathcal{V}(\mathcal{H})$

$$\delta^o(\hat{S}_z)_V \geq_s \hat{S}_z \quad \text{and} \quad \delta^i(\hat{S}_z)_V \leq_s \hat{S}_z. \tag{13.67}$$

On the other hand

$$sp\big(\delta^o(\hat{S}_z)_V\big) \subseteq sp(\hat{S}_z) \quad \text{and} \quad sp\big(\delta^i(\hat{S}_z)_V\big) \subseteq sp(\hat{S}_z). \tag{13.68}$$

13.5 Topos Representation of Physical Quantities

We will now define the topos analogue of a physical quantity. Since we are trying to render quantum theory more realist we will mimic, in the context of the topos $\mathbf{Sets}^{\mathcal{V}(\mathcal{H})^{op}}$, the way in which physical quantities are defined in classical theory. To this end we recall that in classical theory, physical quantities are represented by functions from the state space to the reals, i.e. each physical quantity, A, is represented by a map $f_A : S \to \mathbb{R}$. Similarly we want to define, in topos quantum theory, physical quantities as a functor $A : \underline{\Sigma} \to \underline{\mathbb{R}}^{\leftrightarrow}$. This can be achieved with the aid of both inner and outer daseinisation. In particular, in the topos framework of quantum theory a physical quantity A, with associated self-adjoint operator $\hat{A}$, is represented by the map [32]

$$\breve{\delta}(\hat{A}) : \underline{\Sigma} \to \underline{\mathbb{R}}^{\leftrightarrow} \tag{13.69}$$

which, at each context V, is defined as

$$\begin{aligned} \breve{\delta}(\hat{A})_V : \underline{\Sigma}_V &\to \underline{\mathbb{R}}^{\leftrightarrow}_V \\ \lambda &\mapsto \breve{\delta}(\hat{A})_V(\lambda) := \big(\breve{\delta}^i(\hat{A})_V(\lambda), \breve{\delta}^o(\hat{A})_V(\lambda)\big). \end{aligned} \tag{13.70}$$

Where $\breve{\delta}^o(\hat{A})_V$ is the order reversing function defined by:

$$\breve{\delta}^o(\hat{A})_V(\lambda) : \downarrow V \to sp(\hat{A}) \tag{13.71}$$

such that

$$\begin{aligned} \big(\breve{\delta}^o(\hat{A})_V(\lambda)\big)(V') &:= \overline{\delta^o(\hat{A})_{V'}}\big(\underline{\Sigma}(i_{V'V})(\lambda)\big) \\ &= \overline{\delta^o(\hat{A})_{V'}}(\lambda_{|V'}) \end{aligned}$$

$$= \langle \lambda_{|V'}, \delta^o(\hat{A})_{V'} \rangle$$
$$= \langle \lambda, \delta^o(\hat{A})_{V'} \rangle$$
$$= \lambda\big(\delta^o(\hat{A})_{V'}\big). \tag{13.72}$$

Here we have used the Gel'fand transform $\overline{\delta^o(\hat{A})_V} : \underline{\Sigma}_V \to \mathbb{R}$.

The choice of order reversing functions was determined by the fact that, for all $V' \subseteq V$ since $\delta^o(\hat{A})_{V'} \geq \delta^o(\hat{A})_V$ then

$$\overline{\delta^o(\hat{A})_{V'}}(\lambda_{|V'}) = \overline{\delta^o(\hat{A})_{V'}}\big(\underline{\Sigma}(i_{V'V})(\lambda)\big) \geq \overline{\delta^o(\hat{A})_V}(\lambda). \tag{13.73}$$

On the other hand, the order preserving function is defined by:

$$\breve{\delta}^i(\hat{A})_V(\lambda) : \downarrow\!V \to sp(\hat{A}) \tag{13.74}$$

such that

$$\big(\breve{\delta}^i(\hat{A})_V(\lambda)\big)(V') := \overline{\delta^i(\hat{A})_{V'}}\big(\underline{\Sigma}(i_{V'V})(\lambda)\big)$$
$$= \overline{\delta^i(\hat{A})_{V'}}(\lambda_{|V'})$$
$$= \langle \lambda_{|V'}, \delta^i(\hat{A})_{V'} \rangle$$
$$= \langle \lambda, \delta^i(\hat{A})_{V'} \rangle$$
$$= \lambda\big(\delta^i(\hat{A})_{V'}\big). \tag{13.75}$$

In this case the appropriate Gel'fand transform to use is $\overline{\delta^i(\hat{A})_{V'}} : \underline{\Sigma}_V \to \mathbb{R}$. Here the choice of such an order preserving function is because, for $i : V' \subseteq V$, since $\delta^i(\hat{A})_{V'} \leq \delta^i(\hat{A})_V$ then

$$\overline{\delta^i(\hat{A})_{V'}}(\lambda_{|V'}) = \overline{\delta^o(\hat{A})_{V'}}\big(\underline{\Sigma}(i_{V'V})(\lambda)\big) \leq \overline{\delta^i(\hat{A})_V}(\lambda). \tag{13.76}$$

The definition of $\breve{\delta}(\hat{A}) : \underline{\Sigma}_V \to \mathbb{R}$ represents the mathematical implementation of the idea explained in Sect. 13.1. There we have examined that when going to smaller contexts the coarse graining of the self-adjoint operators imply/induce an equivalent coarse graining of the interval of possible values for that operator. Therefore such an interval either becomes bigger or stays the same. This enlarging of the interval is precisely what is achieved by $\breve{\delta}(\hat{A})$. In fact as we go to smaller context $V' \subseteq V$ the interval of possible values of the operator $\hat{A}$, which gets picked by $\breve{\delta}(\hat{A})$, becomes bigger or stays the same:

$$\big[\delta^i(\hat{A})_{V'}(\lambda), \delta^o(\hat{A})_{V'}(\lambda)\big] \geq \big[\delta^i(\hat{A})_V(\lambda), \delta^o(\hat{A})_V(\lambda)\big]. \tag{13.77}$$

We now need to show that indeed the map $\breve{\delta}(\hat{A}) : \underline{\Sigma} \to \underline{\mathbb{R}}^{\leftrightarrow}$ is a well defined natural transformation between the presheaves $\underline{\Sigma}$ and $\underline{\mathbb{R}}^{\leftrightarrow}$. Therefore, given any

$V' \subseteq V$ we have to show that the following diagram commutes:

$$\begin{array}{ccc}
\underline{\Sigma}_V & \xrightarrow{\breve{\delta}(\hat{A})_V} & \underline{\mathbb{R}}^{\leftrightarrow}_V \\
{\scriptstyle \underline{\Sigma}(i_{V'V})}\downarrow & & \downarrow{\scriptstyle \underline{\mathbb{R}}^{\leftrightarrow}(i_{V'V})} \\
\underline{\Sigma}_{V'} & \xrightarrow{\breve{\delta}(\hat{A})_{V'}} & \underline{\mathbb{R}}^{\leftrightarrow}_{V'}
\end{array}$$

Going one way round the diagram we obtain

$$\begin{aligned}
\underline{\mathbb{R}}^{\leftrightarrow}(i_{V'V})\big(\breve{\delta}^i(\hat{A})_V(\lambda), \breve{\delta}^o(\hat{A})_V(\lambda)\big) &= \big(\breve{\delta}^i(\hat{A})_V(\lambda), \breve{\delta}^o(\hat{A})_V(\lambda)\big)_{V'} \\
&= \big(\big(\breve{\delta}^i(\hat{A})_V(\lambda)\big)_{|V'}, \big(\breve{\delta}^o(\hat{A})_V(\lambda)\big)_{|V'}\big) \\
&= \big(\breve{\delta}^i(\hat{A})_{V'}(\lambda_{|V'}), \breve{\delta}^o(\hat{A})_{V'}(\lambda_{|V'})\big). \qquad (13.78)
\end{aligned}$$

On the other hand

$$\begin{aligned}
\big(\breve{\delta}^i(\hat{A})_V(\cdot), \breve{\delta}^o(\hat{A})_V(\cdot)\big)\underline{\Sigma}(i_{V'V})(\lambda) &= \big(\breve{\delta}^i(\hat{A})_V(\cdot), \breve{\delta}^o(\hat{A})_V(\cdot)\big)(\lambda_{|V'}) \\
&= \big(\breve{\delta}^i(\hat{A})_V(\lambda_{|V'}), \breve{\delta}^o(\hat{A})_V(\lambda_{|V'})\big) \\
&= \big(\breve{\delta}^i(\hat{A})_{V'}(\lambda_{|V'}), \breve{\delta}^o(\hat{A})_{V'}(\lambda_{|V'})\big). \qquad (13.79)
\end{aligned}$$

Thus $\breve{\delta}(\hat{A})$ is a well defined natural transformation.

13.6 Interpreting the Map Representing Physical Quantities

In order to really understand the map $\breve{\delta}(\hat{A}) : \underline{\Sigma} \to \underline{\mathbb{R}}^{\leftrightarrow}$ and how exactly one can use it in topos quantum theory, we need to analyse how expectation values are computed. To this end consider a vector $|\psi\rangle$ in a Hilbert space $\mathcal{H}$. We are interested in computing the expectation value of the self-adjoint operator $\hat{A}$. This is defined (in standard quantum theory) as follows:

$$\langle\psi|\hat{A}|\psi\rangle = \int_{\|\hat{A}\|}^{\|\hat{A}\|} \lambda d\langle\psi|\hat{E}^{\hat{A}}_{\lambda}|\psi\rangle. \qquad (13.80)$$

We can now re-write the above expression using the map $\breve{\delta}(\hat{A})$. To do so we note that for each $V \in \mathcal{V}(\mathcal{H})$ the Gel'fand spectrum $\underline{\Sigma}_V$ of that algebra will contain a special element $\lambda^{|\psi\rangle}$ defined by $\lambda^{|\psi\rangle}(\hat{A}) := \langle\psi|\hat{A}|\psi\rangle$ for all $\hat{A} \in V$. Such an element of the spectrum is characterised by the properties (i) $\lambda^{|\psi\rangle}(|\psi\rangle\langle\psi|) = 1$ and (ii) $\lambda^{|\psi\rangle}(\hat{P}) = 0$ for all $\hat{P} \in P(V)$ such that $\hat{P}|\psi\rangle\langle\psi| = 0$.

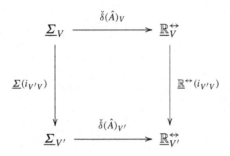

Given such a definition, it follows that:

$$\lambda^{|\psi\rangle} \in \underline{\Sigma}_V \quad \text{iff} \quad |\psi\rangle\langle\psi| \in P(V). \tag{13.81}$$

We can now re-write (13.80) as follows: for each context V such that $|\psi\rangle\langle\psi| \in P(V)$

$$\begin{aligned}\langle\psi|\hat{A}|\psi\rangle &= \int_{\delta^i(\hat{A})_V\lambda^{|\psi\rangle}}^{\delta^o(\hat{A})_V\lambda^{|\psi\rangle}} \lambda d\langle\psi|\hat{E}^{\hat{A}}_\lambda|\psi\rangle \\ &= \int_{\langle\psi|\delta^i(\hat{A})_V|\psi\rangle}^{\langle\psi|\delta^o(\hat{A})_V|\psi\rangle} \lambda d\langle\psi|\hat{E}^{\hat{A}}_\lambda|\psi\rangle.\end{aligned} \tag{13.82}$$

Therefore it is possible to interpret (in the language of canonical quantum theory) $\delta^o(\hat{A})_V\lambda^{|\psi\rangle}$ and $\delta^i(\hat{A})_V\lambda^{|\psi\rangle}$ as the largest, respectively smallest result of measurements of a physical quantity $\hat{A}$ given the state $|\psi\rangle$. Obviously if $|\psi\rangle$ is an eigenstate of $\hat{A}$ then

$$\langle\psi|\hat{A}|\psi\rangle = \langle\psi|\delta^o(\hat{A})_V|\psi\rangle = \langle\psi|\delta^i(\hat{A})_V|\psi\rangle. \tag{13.83}$$

However, if it is not an eigenstate then

$$\langle\psi|\delta^o(\hat{A})_V|\psi\rangle \geq \langle\psi|\hat{A}|\psi\rangle \geq \langle\psi|\delta^i(\hat{A})_V|\psi\rangle. \tag{13.84}$$

If we now go to smaller context $i : V' \subseteq V$, then the properties of inner and outer daseinisations imply that $\delta^o(\hat{A})_{V'} \geq \delta^o(\hat{A})_V$ while $\delta^i(\hat{A})_{V'} \leq \delta^i(\hat{A})_V$. It follows that

$$\begin{aligned}\delta^o(\hat{A})_V\lambda^{|\psi\rangle} &= \langle\psi|\delta^o(\hat{A})_V|\psi\rangle \leq \delta^o(\hat{A})_{V'}\lambda^{|\psi\rangle} = \langle\psi|\delta^o(\hat{A})_{V'}|\psi\rangle \\ \delta^i(\hat{A})_V\lambda^{|\psi\rangle} &= \langle\psi|\delta^i(\hat{A})_V|\psi\rangle \geq \delta^i(\hat{A})_{V'}\lambda^{|\psi\rangle} = \langle\psi|\delta^i(\hat{A})_{V'}|\psi\rangle.\end{aligned} \tag{13.85}$$

Therefore the interval between the smallest and largest possible value for $\hat{A}$ becomes bigger and bigger as we go to smaller and smaller algebras. This is because we approximate our self-adjoint operator more and more.

In this setting, for each context $V \in \mathcal{V}(\mathcal{H})$, the map

$$\begin{aligned}\breve{\delta}(\hat{A})_V : \underline{\Sigma}_V &\to \underline{\mathbb{R}}^{\leftrightarrow}_V \\ \lambda &\mapsto \breve{\delta}(\hat{A})_V(\lambda) = \big(\breve{\delta}^i(\hat{A})_V(\lambda), \breve{\delta}^o(\hat{A})_V(\lambda)\big)\end{aligned} \tag{13.86}$$

defines the interval or range of possible values of the quantity A at stages $V' \subseteq V$. In particular, for each $\lambda \in \underline{\Sigma}_V$ we obtain a map $\breve{\delta}(\hat{A})_V(\lambda) : \downarrow V \to sp(\hat{A}) \times sp(\hat{A})$. The role of this map is to pick an interval of values, namely, for all $V' \in \downarrow V$, $\breve{\delta}(\hat{A})_V(\lambda)(V') = (\breve{\delta}^i(\hat{A})_V(\lambda)(V'), \breve{\delta}^o(\hat{A})_V(\lambda)(V'))$ is such that it picks the range of values

$$\big[\breve{\delta}^i(\hat{A})_V(\lambda), \breve{\delta}^o(\hat{A})_V(\lambda)\big] \cap sp(\hat{A}). \tag{13.87}$$

As we go to smaller and smaller contexts $V'' \subseteq V'$ the range of values becomes bigger[5]

$$\big[\breve{\delta}^i(\hat{A})_{V''}(\lambda), \breve{\delta}^o(\hat{A})_{V''}(\lambda)\big] \cap sp(\hat{A}). \tag{13.88}$$

Thus reiterating, the map $\breve{\delta}(\hat{A}) : \underline{\Sigma} \to \underline{\mathbb{R}}^{\leftrightarrow}$ is defined such that as we go down to smaller sub-algebras $V' \subseteq V$, the range of possible values of our physical quantity becomes bigger or stays the same. This is because $\hat{A}$ gets approximated both from above, through the process of outer daseinisation and from below, through the process of inner daseinisation. Such an approximation gets coarser as V' gets smaller, which basically means that V' contains less and less projections, i.e. less and less information. However, such an interpretation can only be local since the state space $\underline{\Sigma}$ has no global elements.

13.7 Computing Values of Quantities Given a State

We would now like to define the value of a physical quantity $\hat{A}$ given a state $\underline{\mathfrak{w}}^{|\psi\rangle}$. Again, as for all the constructs we have defined so far, we would like to mimic classical physics, i.e. we would like to define values for quantities given a state, in the same way as it is done in classical physics. To this end, recall that in classical physics, given a state s, and a physical quantity f_A, the value of the latter given the former is $f_A(s)$.

Similarly we would like to define the value of $\breve{\delta}(\hat{A})$ given $\underline{\mathfrak{w}}^{|\psi\rangle}$ as something like $\breve{\delta}(\hat{A})(\underline{\mathfrak{w}}^{|\psi\rangle})$.

To this end let us recall how the pseudo-state $\underline{\mathfrak{w}}^{|\psi\rangle}$ is defined. For each context V we first define the approximate operator

$$\delta^o\big(|\psi\rangle\langle\psi|\big)_V := \bigwedge\big\{\hat{\alpha} \in \underline{O}_V \,|\, |\psi\rangle\langle\psi| \leq \hat{\alpha}\big\}. \tag{13.89}$$

We then associate a subset of the state space $\underline{\Sigma}_V$:

$$\underline{\mathfrak{w}}^{|\psi\rangle}_V := \big\{\lambda \in \underline{\Sigma}_V \big| \lambda\big(\delta^o\big(|\psi\rangle\langle\psi|\big)_V\big) = 1\big\}. \tag{13.90}$$

Thus for each state $|\psi\rangle$, we get the presheaf $\underline{\mathfrak{w}}^{|\psi\rangle} \subseteq \underline{\Sigma}_V$. Now since the physical quantity $\breve{\delta}(\hat{A})$ is defined in terms of a map $\breve{\delta}(\hat{A}) : \underline{\Sigma} \to \underline{\mathbb{R}}^{\leftrightarrow}$ whose codomain is $\underline{\Sigma}$, such a map has to be defined also on any sub-object of $\underline{\Sigma}$. Thus we obtain the composite

$$\underline{\mathfrak{w}}^{|\psi\rangle} \hookrightarrow \underline{\Sigma} \xrightarrow{\breve{\delta}(\hat{A})} \underline{\mathbb{R}}^{\leftrightarrow}. \tag{13.91}$$

[5]Recall that $\delta^o(\hat{A})_V \geq \hat{A}$ and $\delta^i(\hat{A})_V \leq \hat{A}$ imply that $sp(\delta^i(\hat{A})_V) \subseteq sp(\hat{A})$ and $sp(\hat{A}) \subseteq sp(\delta^o(\hat{A})_V)$ respectively.

In this setting one can indeed write the value of a physical quantity given a state as

$$\breve{\delta}(\hat{A})\big(\underline{\mathfrak{w}}^{|\psi\rangle}\big) \tag{13.92}$$

such that for each context $V \in \mathcal{V}(\mathcal{H})$ we obtain

$$\begin{aligned} \breve{\delta}(\hat{A})_V : \underline{\mathfrak{w}}_V^{|\psi\rangle} &\to \underline{\mathbb{R}}_V^{\leftrightarrow} \\ \lambda &\mapsto \breve{\delta}(\hat{A})_V(\lambda) := \big(\breve{\delta}^i(\hat{A})_V(\lambda), \breve{\delta}^o(\hat{A})_V(\lambda)\big). \end{aligned} \tag{13.93}$$

Theorem 13.3 *$\breve{\delta}(\hat{A})(\underline{\mathfrak{w}}^{|\psi\rangle})$ is a sub-object of $\underline{\mathbb{R}}^{\leftrightarrow}$.*

Proof We first need to show that $\breve{\delta}(\hat{A})_V(\underline{\mathfrak{w}}^{|\psi\rangle})_V \subseteq \underline{\mathbb{R}}_V^{\leftrightarrow}$ for each $V \in \mathcal{V}(\mathcal{H})$. This follows trivially from the definition in (13.93). Thus what remains to show is that for each $V' \subseteq V$

$$\underline{\mathbb{R}}^{\leftrightarrow}(i_{V'V})\big(\breve{\delta}(\hat{A})_V\big(\underline{\mathfrak{w}}_V^{|\psi\rangle}\big)\big) \subseteq \breve{\delta}(\hat{A})_{V'}\big(\underline{\mathfrak{w}}^{|\psi\rangle}\big)_{V'}. \tag{13.94}$$

Given $\lambda \in \underline{\mathfrak{w}}_V^{|\psi\rangle}$, then

$$\underline{\mathbb{R}}^{\leftrightarrow}(i_{V'V})\big(\breve{\delta}(\hat{A})_V(\lambda)\big) = \big(\breve{\delta}(\hat{A})_V(\lambda)\big)_{|V'} = \breve{\delta}(\hat{A})_{V'}(\lambda_{|V'}). \tag{13.95}$$

However form the definition of pseudo-state we have

$$\underline{\mathfrak{w}}_{V'}^{|\psi\rangle} = \underline{\Sigma}(i_{V'V})\big(\underline{\mathfrak{w}}_V^{|\psi\rangle}\big) = \big\{\lambda_{|V'} | \lambda \in \underline{\mathfrak{w}}_V^{|\psi\rangle}\big\} \tag{13.96}$$

therefore each element in $\underline{\mathfrak{w}}_{V'}^{|\psi\rangle}$ comes from restricting an element in $\underline{\mathfrak{w}}_V^{|\psi\rangle}$. It follows that

$$\underline{\mathbb{R}}^{\leftrightarrow}(i_{V'V})\big(\breve{\delta}(\hat{A})_V\big(\underline{\mathfrak{w}}_V^{|\psi\rangle}\big)\big) = \breve{\delta}(\hat{A})_{V'}\big(\underline{\mathfrak{w}}^{|\psi\rangle}\big)_{V'}. \tag{13.97}$$

We thus obtain a commuting diagram

$$\begin{array}{ccc} \underline{\mathfrak{w}}_V^{|\psi\rangle} & \xrightarrow{\breve{\delta}(\hat{A})_V} & \underline{\mathbb{R}}_V^{\leftrightarrow} \\ {\scriptstyle \underline{\Sigma}(i_{V'V})}\Big\downarrow & & \Big\downarrow{\scriptstyle \underline{\mathbb{R}}^{\leftrightarrow}(i_{V'V})} \\ \underline{\mathfrak{w}}_{V'}^{|\psi\rangle} & \xrightarrow{\breve{\delta}(\hat{A})_{V'}} & \underline{\mathbb{R}}_{V'}^{\leftrightarrow}. \end{array}$$

□

13.7.1 Examples

We will now give two examples to show how the map representing self-adjoint operators actually works [24].

Eigenvalue-Eigenstate Link We will first consider the case in which the state $|\psi\rangle$ is an eigenstate of $\hat{A}$. We then consider abelian sub-algebras for which $\hat{A} \in V$, such that $\delta^o(\hat{A}) = \delta^i(\hat{A}) = \hat{A}$. The condition $\hat{A} \in V$ also implies that $|\psi\rangle\langle\psi| \in P(V)$, thus $\underline{\mathfrak{w}}^{|\psi\rangle}_V$ will contain the single element $\lambda_{|\psi\rangle\langle\psi|} \in \underline{\Sigma}_V$, such that $\lambda_{|\psi\rangle\langle\psi|}(|\psi\rangle\langle\psi|) = 1$ while $\lambda_{|\psi\rangle\langle\psi|}(\hat{Q}) = 0$ for all $\hat{Q}|\psi\rangle\langle\psi| = 0$. It follows that:

$$\breve{\delta}(\hat{A})_V\big(\underline{\mathfrak{w}}^{|\psi\rangle}_V\big) = \big(\breve{\delta}^i(\hat{A})_V(\lambda_{|\psi\rangle\langle\psi|}), \breve{\delta}^o(\hat{A})_V(\lambda_{|\psi\rangle\langle\psi|})\big). \tag{13.98}$$

Since $\hat{A} \in V$, then $\breve{\delta}^i(\hat{A})_V(\lambda_{|\psi\rangle\langle\psi|}) = \breve{\delta}^o(\hat{A})_V(\lambda_{|\psi\rangle\langle\psi|}) = \lambda_{|\psi\rangle\langle\psi|}(\hat{A})$, which is the eigenvalue of $\hat{A}$ given the state $|\psi\rangle$. Thus we get the usual eigenvalue eigenstate link.

Interval We now give both a simple example and a more complicated example on how to define values for quantities.

Simple Example Let us consider the simple self-adjoint projection operator $|\psi\rangle\langle\psi|$. Such an operator has $sp(|\psi\rangle\langle\psi|) = \{0, 1\}$. For each context $V \in \mathcal{V}(\mathcal{H})$ the map $\breve{\delta}(|\psi\rangle\langle\psi|) : \underline{\Sigma} \to \underline{\mathbb{R}}^{\leftrightarrow}$ is

$$\begin{aligned} \breve{\delta}\big(|\psi\rangle\langle\psi|\big)_V : \underline{\Sigma}_V &\to \underline{\mathbb{R}}^{\leftrightarrow}_V \\ \lambda &\mapsto \big(\breve{\delta}^i\big(|\psi\rangle\langle\psi|\big)_V(\lambda), \breve{\delta}^o\big(|\psi\rangle\langle\psi|\big)_V(\lambda)\big). \end{aligned} \tag{13.99}$$

However, given the spectrum of $|\psi\rangle\langle\psi|$, we obtain

$$\begin{aligned} \breve{\delta}^i\big(|\psi\rangle\langle\psi|\big)_V(\lambda) &: \downarrow V \to \{0, 1\} \\ \breve{\delta}^o\big(|\psi\rangle\langle\psi|\big)_V(\lambda) &: \downarrow V \to \{0, 1\} \end{aligned} \tag{13.100}$$

such that for all $V' \subseteq V$ we have

$$\begin{aligned} \breve{\delta}^i\big(|\psi\rangle\langle\psi|\big)_V(\lambda)\big(V'\big) &:= \big\langle\lambda, \delta^i\big(|\psi\rangle\langle\psi|\big)_{V'}\big\rangle \\ \breve{\delta}^o\big(|\psi\rangle\langle\psi|\big)_V(\lambda)\big(V'\big) &:= \big\langle\lambda, \delta^o\big(|\psi\rangle\langle\psi|\big)_{V'}\big\rangle. \end{aligned} \tag{13.101}$$

We then consider the pseudo-state $\underline{\mathfrak{w}}^{|\psi\rangle}$ and want to evaluate $\breve{\delta}(|\psi\rangle\langle\psi|)(\underline{\mathfrak{w}}^{|\psi\rangle})$. From the definition of pseudo-state, given a context V, it follows that for all $\lambda \in \underline{\mathfrak{w}}^{|\psi\rangle}_V$ and for all $V' \subseteq V$

$$\breve{\delta}^o\big(|\psi\rangle\langle\psi|\big)_V(\lambda)\big(V'\big) = \big\langle\lambda, \delta^o\big(|\psi\rangle\langle\psi|\big)_{V'}\big\rangle = 1. \tag{13.102}$$

We call the constant function with value 1 on all $V' \subseteq V$ as $1_{\downarrow V}$.

On the other hand, $\breve{\delta}^i(|\psi\rangle\langle\psi|)_V(\lambda)$ is such that for all $V' \subseteq V$ we obtain

$$\breve{\delta}^i\big(|\psi\rangle\langle\psi|\big)_V(\lambda)(V') = \begin{cases} 1 & \text{if } |\psi\rangle\langle\psi| \in V' \\ 0 & \text{if } |\psi\rangle\langle\psi| \notin V'. \end{cases} \tag{13.103}$$

We then arrive at a complete description of $\breve{\delta}(|\psi\rangle\langle\psi|)(\underline{\mathfrak{w}}^{|\psi\rangle})$ as follows: given any context $V \in \mathcal{V}(\mathcal{H})$ if $\lambda \in \underline{\mathfrak{w}}_V^{|\psi\rangle}$ then we obtain

$$\breve{\delta}\big(|\psi\rangle\langle\psi|\big)_V(\lambda) = \big(\breve{\delta}^i\big(|\psi\rangle\langle\psi|\big)_V(\lambda), 1_{\downarrow V}\big). \tag{13.104}$$

A More Complicated Example Let us consider again the 2 spin system in $\mathbb{C}^4$ defined in Example 13.4.1. We want to compute $\breve{\delta}(\hat{S}_z) : \underline{\Sigma} \to \mathbb{R}^{\leftrightarrow}$. Since we are dealing with presheaves we need to define this expression context-wise, for each V, i.e.

$$\begin{aligned} \breve{\delta}(\hat{S}_z)_V : \underline{\Sigma}_V &\to \mathbb{R}^{\leftrightarrow}_V \\ \lambda &\mapsto \breve{\delta}(\hat{S}_z)(\lambda) = \big(\breve{\delta}^i(\hat{S}_z)_V(\lambda), \breve{\delta}^o(\hat{S}_z)_V(\lambda)\big) \end{aligned} \tag{13.105}$$

where

$$\begin{aligned} \breve{\delta}^i(\hat{S}_z)_V(\lambda) : \downarrow V &\to \mathbb{R} \\ V' &\mapsto \lambda\big(\delta^i(\hat{S}_z)_{V'}\big) \end{aligned} \tag{13.106}$$

and similar for $\breve{\delta}^o(\hat{S}_z)_V(\lambda)$. Thus, in order to compute $\breve{\delta}^i(\hat{S}_z)_V$ and $\breve{\delta}^o(\hat{S}_z)_V$ we need to find the inner and outer daseinisation of the spectral family of $\hat{S}_z$ since we want to apply the formulas

$$\begin{aligned} \delta^o(\hat{S}_z)_V &= \int_{\mathbb{R}} \lambda d\big(\delta^i\big(\hat{E}^{\hat{S}_z}_\lambda\big)\big) \\ \delta^i(\hat{S}_z)_V &= \int_{\mathbb{R}} \lambda d\bigg(\bigwedge_{\mu>\lambda} \delta^o\big(\hat{E}^{\hat{S}_z}_\lambda\big)\bigg). \end{aligned} \tag{13.107}$$

Both the inner and outer daseinisation of the spectral family of $\hat{S}_z$ are computed in Sect. 13.4.1.

Now given a state $|\psi\rangle = (1, 0, 0, 0)$ we want to compute the value of the physical quantity $\breve{\delta}(\hat{S}_z)$. Thus for each context V we need to define the pair $(\breve{\delta}^i(\hat{S}_z)_V(\cdot), \breve{\delta}^o(\hat{S}_z)_V(\cdot))$ which will then act on all $\lambda \in \underline{\mathfrak{w}}_V^{|\psi\rangle}$. In particular, for each $\lambda \in \underline{\mathfrak{w}}_V^{|\psi\rangle}$ we obtain

$$\begin{aligned} \breve{\delta}^i(\hat{S}_z)_V(\lambda) &: \downarrow V \to \mathbb{R} \\ \breve{\delta}^o(\hat{S}_z)_V(\lambda) &: \downarrow V \to \mathbb{R}. \end{aligned} \tag{13.108}$$

Let us consider the context $V_{\hat{P}_4}$, here $\underline{\mathfrak{w}}^{|\psi\rangle}_{V_{\hat{P}_4}} = \{\lambda\}$, such that $\lambda(\delta^o(|\psi\rangle\langle\psi|)_{V_{\hat{P}_4}}) = 1$ where

$$\delta^o\big(|\psi\rangle\langle\psi|\big)_{V_{\hat{P}_4}} = \begin{pmatrix} 1 & 0 & 0 & 0 \\ 0 & 1 & 0 & 0 \\ 0 & 0 & 1 & 0 \\ 0 & 0 & 0 & 0 \end{pmatrix} \tag{13.109}$$

hence

$$\begin{aligned} \breve{\delta}^o(\hat{S}_z)_{V_{\hat{P}_4}}(\lambda)(V_{\hat{P}_4}) &= \lambda\big(\breve{\delta}^o(\hat{S}_z)_{V_{\hat{P}_4}}\big) = \lambda\big(\text{diag}(2,2,2,-2)\big) = 2 \\ \breve{\delta}^i(\hat{S}_z)_{V_{\hat{P}_4}}(\lambda)(V_{\hat{P}_4}) &= \lambda\big(\breve{\delta}^i(\hat{S}_z)_{V_{\hat{P}_4}}\big) = \lambda\big(\text{diag}(0,0,0,-2)\big) = 0. \end{aligned} \tag{13.110}$$

Note that this is equivalent to

$$\langle\psi|\delta^o(\hat{S}_z)_{V_{\hat{P}_4}}|\psi\rangle = \begin{pmatrix} 1 & 0 & 0 & 0 \end{pmatrix} \cdot \begin{pmatrix} 2 & 0 & 0 & 0 \\ 0 & 2 & 0 & 0 \\ 0 & 0 & 2 & 0 \\ 0 & 0 & 0 & -2 \end{pmatrix} \begin{pmatrix} 1 \\ 0 \\ 0 \\ 0 \end{pmatrix} = 2 \tag{13.111}$$

and

$$\langle\psi|\delta^i(\hat{S}_z)_{V_{\hat{P}_4}}|\psi\rangle = \begin{pmatrix} 1 & 0 & 0 & 0 \end{pmatrix} \cdot \begin{pmatrix} 0 & 0 & 0 & 0 \\ 0 & 0 & 0 & 0 \\ 0 & 0 & 0 & 0 \\ 0 & 0 & 0 & -2 \end{pmatrix} \begin{pmatrix} 1 \\ 0 \\ 0 \\ 0 \end{pmatrix} = 0. \tag{13.112}$$

Similarly, for context $V_{\hat{P}_2,\hat{P}_4}$ we obtain

$$\langle\psi|\delta^o(\hat{S}_z)_{V_{\hat{P}_2,\hat{P}_4}}|\psi\rangle = \begin{pmatrix} 1 & 0 & 0 & 0 \end{pmatrix} \cdot \begin{pmatrix} 2 & 0 & 0 & 0 \\ 0 & 0 & 0 & 0 \\ 0 & 0 & 2 & 0 \\ 0 & 0 & 0 & -2 \end{pmatrix} \begin{pmatrix} 1 \\ 0 \\ 0 \\ 0 \end{pmatrix} = 2 \tag{13.113}$$

and

$$\langle\psi|\delta^i(\hat{S}_z)_{V_{\hat{P}_2}}|\psi\rangle = \begin{pmatrix} 1 & 0 & 0 & 0 \end{pmatrix} \cdot \begin{pmatrix} 0 & 0 & 0 & 0 \\ 0 & 0 & 0 & 0 \\ 0 & 0 & 0 & 0 \\ 0 & 0 & 0 & -2 \end{pmatrix} \cdot \begin{pmatrix} 1 \\ 0 \\ 0 \\ 0 \end{pmatrix} = 0. \tag{13.114}$$

For $V_{\hat{P}_2}$ we obtain

$$\langle\psi|\delta^o(\hat{S}_z)_{V_{\hat{P}_2}}|\psi\rangle = \begin{pmatrix} 1 & 0 & 0 & 0 \end{pmatrix} \cdot \begin{pmatrix} 2 & 0 & 0 & 0 \\ 0 & 0 & 0 & 0 \\ 0 & 0 & 2 & 0 \\ 0 & 0 & 0 & 2 \end{pmatrix} \begin{pmatrix} 1 \\ 0 \\ 0 \\ 0 \end{pmatrix} = 2. \tag{13.115}$$

and

$$\langle\psi|\delta^i(\hat{S}_z)_{V_{\hat{P}_2}}|\psi\rangle = \begin{pmatrix}1 & 0 & 0 & 0\end{pmatrix} \cdot \begin{pmatrix} -2 & 0 & 0 & 0 \\ 0 & 0 & 0 & 0 \\ 0 & 0 & -2 & 0 \\ 0 & 0 & 0 & -2 \end{pmatrix} \cdot \begin{pmatrix}1\\0\\0\\0\end{pmatrix} = -2. \quad (13.116)$$

Given the above results we have that for $V_{\hat{P}_2,\hat{P}_4}$ then the pair of order preserving, order reversing functions for the physical quantity $\breve{\delta}(\hat{S}_z)$ is $(\breve{\delta}^i(\hat{S}_z)_{V_{\hat{P}_2,\hat{P}_4}}(\lambda)$, $\breve{\delta}^o(\hat{S}_z)_{V_{\hat{P}_2,\hat{P}_4}}(\lambda))$ where

$$\begin{aligned} \breve{\delta}^i(\hat{S}_z)_{V_{\hat{P}_2,\hat{P}_4}}(\lambda) &: \downarrow V_{\hat{P}_2,\hat{P}_4} \rightarrow sp(\hat{S}_z) \\ \breve{\delta}^o(\hat{S}_z)_{V_{\hat{P}_2,\hat{P}_4}}(\lambda) &: \downarrow V_{\hat{P}_2,\hat{P}_4} \rightarrow sp(\hat{S}_z). \end{aligned} \quad (13.117)$$

Since the latter, as we have seen, for all $V' \subseteq V$ has constant value 2 we can write

$$\big(\breve{\delta}^i(\hat{S}_z)_{V_{\hat{P}_2,\hat{P}_4}}(\lambda), \breve{\delta}^o(\hat{S}_z)_{V_{\hat{P}_2,\hat{P}_4}}(\lambda)\big) = \big(\breve{\delta}^i(\hat{S}_z)_{V_{\hat{P}_2,\hat{P}_4}}(\lambda), 2_{\downarrow V_{\hat{P}_2,\hat{P}_4}}\big). \quad (13.118)$$

This represents the interval of the value of the physical quantity S_z for $\downarrow V_{\hat{P}_2,\hat{P}_1}$. Putting all the results together for all contexts $V \in \mathcal{V}(\mathcal{H})$ we obtain the desired physical quantity $\breve{\delta}(\hat{S}_z)$. Similar analysis can be performed for all the remaining contexts.

We will now give an example of a local section of the presheaf $\underline{\Sigma}$. The reason we postponed this example to the end of this Chapter is because we will utilise the concept of daseinisation of self adjoint operators.

Example 13.1 Consider the abelian von Neuman category $\mathcal{V}(\mathbb{C}^4)$ defined in Section 9.1.2 of this chapter. Given two maxima algebras $V = lin_{\mathbb{C}}(\hat{P}_1, \hat{P}_2, \hat{P}_3, \hat{P}_4)$, and $V' = lin_{\mathbb{C}}(\hat{Q}_1, \hat{Q}_2.\hat{P}_3, \hat{P}_4)$ and all their sub-algebras, we would like to define the local section of the spectral presheaf $\underline{\Sigma}$. The local section ρ of $\underline{\Sigma}$ are maps $\rho : \underline{U} \rightarrow \underline{\Sigma}$ where $\underline{U} \subseteq \underline{1}$ is a sub-object of the terminal object. Thus for each algebra $V \in \mathcal{V}(\mathcal{H})$ we have

$$\rho_V : \underline{U}_V \rightarrow \underline{\Sigma}_V. \quad (13.119)$$

We know that $\underline{U}_V$ takes values in $\{\{*\}, \{\emptyset\}\}$. For $\{\emptyset\}$ there is no map to $\underline{\Sigma}_V$ thus, for each ρ we will only consider the elements $V \in \mathcal{V}(\mathcal{H})$ which comprise its domain.

Let us consider the maximal algebra V and all its sub-algebras. We would like to consider local sections defined on them. Each local section is closed downwards, i.e. if $V \in dom(\rho)$ then for all $V' \subseteq V$ $V' \in dom(\rho)$. Moreover the condition on a local section is

$$\rho_{V'}\big(\{*\}\big) = \underline{\Sigma}(i_{V'V})\rho_V\big(\{*\}\big). \quad (13.120)$$

Therefore since $\underline{\Sigma}_V := \{\lambda_1, \lambda_2, \lambda_3, \lambda_4\}$ we will have $\rho^i_V(\{*\}) := \lambda_i$, $i \in \{1, 2, 3, 4\}$. When going to context then one is simply following the presheaf maps.[6]

We are now interested in defining the action of these sections on the operators $\hat{A} = a_1\hat{P}_1 + a_2\hat{P}_2 + a_3\hat{P}_3 + a_4\hat{P}_4$, $\hat{B} = b_1(\hat{P}_1 + \hat{P}_2) + b_3\hat{P}_3 + b_4\hat{P}_4$ and $\hat{C} = c_1(\hat{P}_1 + \hat{P}_2 + \hat{P}_3) + c_4\hat{P}_4$. Such operators can be related as follows:

$$\hat{B} = f(\hat{A}) \tag{13.121}$$

where

$$\begin{aligned}
f(\hat{P}_1) &= \hat{P}_1 + \hat{P}_2, \Rightarrow f(a_1) = b_1\\
f(\hat{P}_2) &= \hat{P}_1 + \hat{P}_2, \Rightarrow f(a_2) = b_1\\
f(\hat{P}_3) &= \hat{P}_3, \Rightarrow f(a_3) = b_3\\
f(\hat{P}_4) &= \hat{P}_4, \Rightarrow f(a_4) = b_4.
\end{aligned} \tag{13.122}$$

Similarly one can define

$$\hat{C} = g(\hat{A}) = h(\hat{B}). \tag{13.123}$$

We now consider the context V to which all the above operators belong. Then for each $\rho^i_V(*) = \lambda_i$ we will have

$$\begin{aligned}
&\lambda_i(\hat{A}) = a_i\\
&\lambda_1(\hat{B}) = \lambda_2(\hat{B}) = b_1\\
&\lambda_i(\hat{B}) = b_i\\
&\vdots
\end{aligned} \tag{13.124}$$

On the other hand, for the context $V_{\hat{P}_3,\hat{P}_4}$, then $\hat{A} \notin V_{\hat{P}_3,\hat{P}_4}$ while $\hat{B}, \hat{C} \in V_{\hat{P}_3,\hat{P}_4}$. If we consider the spectral decomposition of $\hat{A}$ and define $\delta^o(\hat{A})_{V_{\hat{P}_3,\hat{P}_4}}$ this will obviously belong to $V_{\hat{P}_3,\hat{P}_4}$ and it turns out that for certain relations between the elements of the spectra $sp(\hat{A})$ then $f(\delta^o(\hat{A})_{V_{\hat{P}_3,\hat{P}_4}}) = \hat{B}$. To see this first we have to define a relation among the elements $sp(\hat{A})$ and we will simply choose $a_1 \leq a_2 \leq a_3 \leq a_4$. We then have

$$\delta^i(\hat{E}^{\hat{A}}_\lambda)_{V_{\hat{P}_3,\hat{P}_4}} = \begin{cases} \hat{0} & \text{if } \lambda < a_1\\ \hat{P}_1 + \hat{P}_2 & \text{if } a_1 \leq \lambda < a_2\\ \hat{P}_1 + \hat{P}_2 & \text{if } a_2 \leq \lambda < a_3\\ \hat{P}_1 + \hat{P}_2 + \hat{P}_3 & \text{if } a_3 \leq \lambda < a_4\\ \hat{1} & \text{if } a_4 \leq \lambda. \end{cases}$$

[6]For notational simplicity we will, from now on, write $\rho(\{*\})$ as ρ.

It then follows that $\hat{B} = f(\delta^o(\hat{A})_{V_{\hat{P}_3,\hat{P}_4}})$ obtaining

$$f(\rho^i_V(A)) = \underline{\Sigma}(i_{V_{\hat{P}_3,\hat{P}_4}V})\rho^i_V(f(\hat{A}_V)) = \rho^i_{V_{\hat{P}_3,\hat{P}_4}} f(\delta^o(\hat{A})_{V_{\hat{P}_3,\hat{P}_4}}) = \rho^i_{V_{\hat{P}_3,\hat{P}_4}}(\hat{B}) \tag{13.125}$$

which is FUNC.

By going to the context $V_{\hat{P}_4}$ and daseinising both $\hat{A}$ and $\hat{B}$ we obtain, for an appropriate choice of relations between the respective spectra, that $h(\delta^o(\hat{A})_{V_{\hat{P}_4}}) = g(\delta^o(\hat{B})_{V_{\hat{P}_4}}) = \hat{C}$. Given the maps $i : V \subseteq V_{\hat{P}_4}$ and $j : V_{\hat{P}_3,\hat{P}_4} \subseteq V_{\hat{P}_4}$ we then obtain

$$\begin{aligned} h(\rho^i_V(\hat{A}_V)) &= \underline{\Sigma}(i_{V_{\hat{P}_4}V})\rho^i_V(h(\hat{A}_V)) = \rho^i_{V_{\hat{P}_4}} h(\delta^o(\hat{A})_{V_{\hat{P}_4}}) = \rho^i_{V_{\hat{P}_4}}(\hat{C}) \\ g(\rho^i_{V_{\hat{P}_3,\hat{P}_4}}(\hat{B}_V)) &= \underline{\Sigma}(i_{V_{\hat{P}_4}V_{\hat{P}_3,\hat{P}_4}})\rho^i_V(g(\hat{B}_V)) = \rho^i_{V_{\hat{P}_4}} g(\delta^o(\hat{B})_{V_{\hat{P}_4}}) = \rho^i_{V_{\hat{P}_4}}(\hat{C}). \end{aligned} \tag{13.126}$$

Chapter 14
Sheaves

In this chapter we will describe the concept of a sheaf and its relation to presheaves and certain bundles[1] called étale bundles. The need for introducing sheaves is because we will have to change the topos used to describe quantum theory from the topos of presheaves over $\mathcal{V}(\mathcal{H})$ to the topos of sheave over $\mathcal{V}(\mathcal{H})$, but now seen not merely as a category but as a topological space. This change is required in order to introduce the concept of probabilities (Chap. 15) and of group transformations (Chap. 16) in topos quantum theory.

We will also introduce a very useful way to define maps between topoi, in particular topoi of sheaves over some topological space. Such maps are called *geometric morphisms*.

14.1 Sheaves

We will now describe what a sheaf is. In order to do this we will first give the bundle theoretical definition and then give the categorical definition. This equivalence of descriptions is possible since there is a 1:1 correspondence between sheaves and a special type of bundles, namely: *étale bundles* which are defined as follows:

Definition 14.1 Given a topological space X, a bundle $p_E : E \to X$ is said to be étale iff p_A is a local homeomorphism. By this we mean that, for each $e \in E$ there exists an open set V with $e \in V \subseteq E$, such that pV is open in X and $p_{|V}$ is a homeomorphism $V \to pV$.

If for example $X = \mathbb{R}^2$, then for each point of a fibre there will be an open disk isomorphic to an open disk in $\mathbb{R}^2$, not necessarily of the same size. Such a collection of open disks on each fibre are glued together by the topology on E.

[1] Recall that given two topological spaces X and Y a bundle is simply a continuos map $p : X \to Y$. In this case X is a bundle over Y. Moreover, for each $y \in Y$ then $p^{-1}(y) =: X_y$ is called the fibre of the bundle at y. Bundles over Y can also be though of as objects in the comma category $Top \downarrow Y$.

C. Flori, *A First Course in Topos Quantum Theory*, Lecture Notes in Physics 868, DOI 10.1007/978-3-642-35713-8_14,

Another example of étale bundles are covering spaces [96]. However, although all covering spaces are étale, it is not the case that all étale bundles are covering spaces.

Given an étale bundle $p_E : E \to X$ and an open subset $U \subseteq X$, then the pullback of p_E via $i : U \subseteq X$ is étale:

$$\begin{array}{ccc} E_U & \longrightarrow & E \\ \mu \downarrow & & \downarrow p_E \\ U & \xrightarrow{\quad i \quad} & X \end{array}$$

i.e. $E_U \to U$ is étale.

This result generalises as follows:

Lemma 14.1 *Given any continuous map $f : X \to Y$ and an étale bundle $p_E : E \to Y$ then $g : f^*E \to X$ is étale over X.*

Proof We want to show that g is a local homeomorphism. From the definition of pullback $f^*E := \{(x, e) | f(x) = p_E(e)\} \subseteq X \times E$. Therefore, given an element $(x, e) \in f^*E$, we want to show that there exists an open neighbourhood $(V, U) \ni (x, e)$ which is mapped homeomorphically onto V via g. Since p_E is étale, there exists an open neighbourhood $U \ni e$ which is mapped homeomorphically into an open set $p_E(U)$ in Y. Then, since f is continuous, it follows that $f^{-1}(p_E(U)) \times U$ is open in $X \times E$ and is a neighbourhood of the given point (x, e). If we then define $(f^{-1}(p_E(U)) \times U) \cap f^*E$ this will be an open by the definition of subspace topology and it will be a neighbourhood of (x, e) in f^*E. Given the definition of pullback, then $g(f^{-1}(p_E(U)) \times U) = f^{-1}(p_E(U))$, i.e. $f^{-1}(p_E(U)) \times U$ will be mapped homeomorphically onto $f^{-1}(p_E(U))$. Thus g is étale. □

Each étale bundle is equipped with an étale topology on the fiber space. Such a topology is defined in terms of sections of the bundle as follows:

Definition 14.2 Given an étale bundle $p_E : E \to X$, both p_E and any section $s : X \to E$ of p_E are open maps. Through every point $e \in E$ there is at least one local section $s : U \to E$ where U is an open neighbourhood of e. The images of $s(U)$ for all sections form a base for the topology of E. If s and t are two sections, the set $W = \{x | s(x) = t(x)\}$ where both sections are defined and agree on, is open in X.

From the definition it follows that each fibre has a discrete topology since, by definition of a section, s will pick an element in each stalk $p^{-1}(x)$ for all $x \in U$.

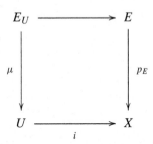

In this setting a sheaf can essentially be thought of as an étale bundle, i.e. a bundle with some extra topological properties. In particular, given a topological space X, a sheaf over X is a pair (E, p_E) consisting of a topological space E and a continuous map $p_E : E \to X$, which is a local homeomorphism.

Thus, pictorially, one can imagine that to each point, in each fibre, one associates an open disk (each of which will have a different size) thus obtaining a stack of open disks for each fibre. These different open disks are then glued together by the topology on E.

The above is the more intuitive definition of what a sheaf is. Now we come to the technical definition which is the following:

Definition 14.3 A sheaf of sets F on a topological space X is a functor[2] $F : \mathcal{O}(X)^{op} \to Sets$, such that each open covering $U = \bigcup_i U_i$, $i \in I$ of an open set U of X determines an equaliser

$$F(U) \overset{e}{\rightarrowtail} \prod_i F(U_i) \underset{q}{\overset{p}{\rightrightarrows}} \prod_{i,j} F(U_i \cap U_j)$$

where for $t \in F(U)$ we have $e(t) = \{t_{|U_i} | i \in I\}$ and for a family $t_i \in F(U_i)$ we obtain

$$p\{t_i\} = \{t_i|_{U_i \cap U_j}\}, \qquad q\{t_i\} = \{t_j|_{U_i \cap U_j}\}. \tag{14.1}$$

Given the definition of product, it follows that the maps e, p, and q above are determined through the diagram

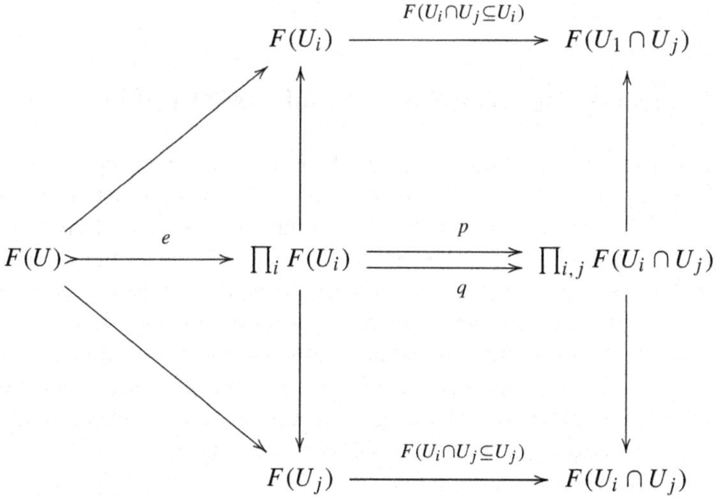

[2]Here $\mathcal{O}(X)$ indicates the category of open sets of X ordered by inclusions.

It is clear from the above definition that a sheaf is a special type of presheaf. In fact, given a topological space X, $Sh(X)$ is a full subcategory of $Sets^{\mathcal{O}(X)^{op}}$. Similarly as $Sets^{\mathcal{O}(X)^{op}}$, also $Sh(X)$ forms a topos.

14.1.1 Simple Example

A very simple example of a sheaf is the following: consider the presheaf

$$\begin{aligned} C : \mathcal{O}(X) :\to Sets \\ U \mapsto C(U) := \{f \mid f : U \to \mathbb{R} \text{ if continuous}\}. \end{aligned} \tag{14.2}$$

This is a well-defined presheaf. In fact, for $U' \subseteq U$, the presheaf maps are defined through restriction

$$\begin{aligned} C(U) \to C(U') \\ f \mapsto f_{|U'}. \end{aligned} \tag{14.3}$$

In fact, if we consider a covering U_i, $i \in I$ of U, such that we have continuous functions $f_i : U_i \to \mathbb{R}$ for all $i \in I$, it follows that there exists at most one map $f : U \to \mathbb{R}$ such that $f_{|U_i} = f_i$. Moreover, such a map exists iff

$$f_i(x) = f_j(x) \quad \forall x \in U_i \cap U_j \ (i, j \in I). \tag{14.4}$$

It follows that the requirement of the map $e : C(U) \to \coprod_{i \in I} C(U_i)$ being an equaliser is satisfied, thus C is a sheaf.

14.2 Connection Between Sheaves and Étale Bundles

From the definitions given above, it seems hard to fully understand what the connection between sheaves and étale bundles is. In order to understand this connection we need to introduce the notion of *germ* of a function. Once we have introduced this notion, it can be shown that each sheaf is a sheaf of local sections of a suitable bundle. All this will become clear as we proceed. First of all: what is a *germ*? *Germs* represent constructions which define local properties of functions. In particular they indicate how similar two functions are locally. Because of this locality requirement, *germs* are generally defined on functions acting on topological spaces such that the word local acquires meaning. For example one can consider the measure of 'locality' to be a power series expansion of a function around some fixed point. Thus, one can say that two holomorphic functions $f, g : U \to \mathbb{C}$ have the same germ at a point $a \in U$ iff the power series expansions around that point are the same. Thus f, g agree on some neighbourhood of a, i.e., with respect to that neighbourhood they "look" the same.

This definition obviously holds only if a power series expansion exists, however it is possible to generalise such a definition in a way that it only requires topological properties of the spaces involved. For example two functions $f, g : X \to E$ have the same germ at $x \in X$ if there exists some neighbourhood of x on which they agree. In this case we write[3] $germ_x f = germ_x g$ which implies that $f(x) = g(x)$. However the converse is not true.

How do we generalise such a definition of germs in the case of presheaves? Let us consider a presheaf $P : \mathcal{O}(X) \to Sets \in Sets^{\mathcal{O}(X)^{op}}$ where X is a topological space and $\mathcal{O}(X)^{op}$ is the category of open sets with reverse ordering to the inclusion ordering. Given a point $x \in X$ and two neighbourhoods U and V of x, the presheaf P assigns two sets $P(U)$ and $P(V)$. Now consider two elements $t \in P(V)$ and $s \in P(U)$, we then say that t and s have the same germ at x iff there exists some open $W \subseteq U \cap V$ such that $x \in W$ and $s_{|W} = t_{|W} \in P(W)$.

The condition of having the same germ at x defines an equivalence relation which is denoted as $germ_x s$. Thus $t \in germ_x s$ iff, given two opens $U, V \ni x$ on which s, respectively t, are defined, then there exists some $W \subseteq U \cap V$ such that $x \in W$ and $t_{|W} = s_{|W} \in P(W)$. It follows that the set of all elements obtained through the P presheaf get 'quotiened' through the equivalence relation of "*belonging to the same germ*". Therefore, for each point $x \in X$, there will exist a collection of germs at x, i.e., a collection of equivalence classes:

$$P_x := \{germ_x s | s \in P(U), x \in U \text{ open in } X\}. \tag{14.5}$$

We can now collect all these sets of germs for all points $x \in X$ defining

$$\Lambda_P = \coprod_{x \in X} P_x = \{\text{all } germ_x s | x \in X, s \in P(U)\}. \tag{14.6}$$

What we have done so far is, basically, to divide the presheaf space into equivalence classes. We can now define the map

$$\begin{aligned} p : \Lambda_P &\to X \\ germ_x s &\mapsto x \end{aligned} \tag{14.7}$$

which sends each germ to the point in which it is taken. It follows that each $s \in P(U)$ defines a function

$$\begin{aligned} \dot{s} : U &\to \Lambda_P \\ x &\mapsto germ_x s. \end{aligned} \tag{14.8}$$

It is straightforward to see that $\dot{s}$ is a section of $p : \Lambda_P \to X$. Since the assignment $s \mapsto \dot{s}$ is unique, it is possible to replace each element s in the original presheaf with a section $\dot{s}$ on the set of germs Λ_P.

---

[3]This should be read as: the germ of f at x is the same as the germ of g at x.

We now define a topology on Λ_P by considering as basis of open sets all the image sets $\dot{s}(U) \subseteq \Lambda_P$ for U open in X, i.e. open sets are unions of images of sections. Such a topology obviously makes p continuous. In fact, given an open set $U \subseteq X$ then $p^{-1}(U)$ is open by definition of the topology on Λ_P, since $p^{-1}(U) = \bigcup_{s_i \in P(U)} \dot{s}_i(U)$.

On the other hand it is also possible to show that the sections $\dot{s}$ as defined above are continuous with respect to the topology on Λ_P. To understand this, consider two elements $t \in P(V)$ and $s \in P(U)$ such that $W \subseteq V \cap U$ and $s_{|W} = t_{|W}$. If we consider the set of all points $y \in V \cap U \subseteq X$ for which $\dot{s}(y) = \dot{t}(y)$ this set contains the open set $W \subseteq V \cap U$. We want to show that given an open $O \in \Lambda_P$, then $\dot{s}^{-1}(O)$ is open in X. Without loss of generality we can choose O to be a basis set, i.e.

$$\dot{s}(W) = \{germ_x(s) | \forall x \in W\}. \tag{14.9}$$

Thus $\dot{s}^{-1}\dot{s}(W) = W$ consists of all those points x such that $\dot{s}(x) = \dot{t}(x)$ for $t, s \in germs_x(s)$.

One can also show that $\dot{s}$ is open and an injection. The property of being open follows directly from the definition of the topology on Λ_p since the basis of open sets are all the image sets $\dot{s}(U) \subseteq \Lambda_P$ for U open in X. To show that it is injective we need to show that if $germ_x s = germ_y s$ then $x = y$. This follows from the definition of germs at a point. Putting all these results together we have show that $\dot{s} : U \to \dot{s}(U)$ is a homeomorphism. So we have managed to construct a bundle $p : \Lambda_P \to X$ which is a local homeomorphism since each point $germ_x(s) \in \Lambda_P$ has an open neighbourhood $\dot{s}(U)$ such that p, restricted to $\dot{s}(U)$, $p : \dot{s}(U) \to X$ has a two sided inverse $\dot{s} : U \to \dot{s}(U)$:

$$p \circ \dot{s} = id_U; \qquad \dot{s} \circ p = id_{\dot{s}(U)} \tag{14.10}$$

hence p is a local homeomorphism.

The above reasoning shows how, given a presheaf P, it is possible to construct an étale bundle $p : \Lambda_P \to X$ out of it. Given such a bundle, it is then possible to construct a sheaf in terms of it. In particular we have the following theorem:

Theorem 14.1 *The presheaf*

$$\begin{aligned} \Gamma(\Lambda_P) : \mathcal{O}^{op} &\to Sets \\ U &\mapsto \{\dot{s} | s \in P(U)\} \end{aligned} \tag{14.11}$$

is a sheaf.

Proof In the presheaf

$$\begin{aligned} \Gamma(\Lambda_P) : \mathcal{O}(X)^{op} &\to Sets \\ U &\mapsto \{\dot{s} | s \in P(U)\} \end{aligned} \tag{14.12}$$

the maps are defined by restriction, i.e. given $U_i \subseteq U$ then

$$\begin{aligned} \Gamma(\Lambda_P) : \mathcal{O}(X)^{op} &\to Sets \\ U_i &\mapsto \{\dot{s}_i | s_i \in P(U_i)\} \end{aligned} \tag{14.13}$$

where $\dot{s} \mapsto \dot{s}_i$ is defined via $\dot{s}_i = P(i_{U_i U})\dot{s}$. Now since

$$\begin{aligned} \dot{s} : U &\to \Lambda_p(U) \\ x &\mapsto germ_x s \end{aligned} \tag{14.14}$$

while

$$\begin{aligned} \dot{s}_i : U_i &\to \Lambda_p(U_i) \\ y &\mapsto germ_y s_i . \end{aligned} \tag{14.15}$$

Since $U_i \subseteq U$ then

$$\begin{aligned} \dot{s} : U_i &\to \Lambda_p(U_i) \\ y &\mapsto germ_y s \end{aligned} \tag{14.16}$$

it follows that we can write $\dot{s}_i = \dot{s}_{|U_i}$.

In order to show that the above is indeed a sheaf we need to show that the diagram

$$\Gamma(\Lambda_p(U)) \xrightarrow{e} \prod_i \Gamma(\Lambda_p(U_i)) \underset{q}{\overset{p}{\rightrightarrows}} \prod_{i,j} \Gamma(\Lambda_p(U_i \cap U_j))$$

is an equaliser. By applying the definition of the sheaf maps we obtain

$$\begin{aligned} e : \Gamma(\Lambda_p(U)) &\to \prod_i \Gamma(\Lambda_p(U_i)) \\ \dot{s} &\to e(\dot{s}) = \{\dot{s}_{U_i} | i \in I\} = \{\dot{s}_i | i \in I\}. \end{aligned} \tag{14.17}$$

On the other hand

$$p(\dot{s}_i) = \{s_i|_{U_i \cap U_j}\} = \{s_{|U_i \cap U_j}\} \tag{14.18}$$

while

$$q(\dot{s}_j) = \{s_j|_{U_i \cap U_j}\} = \{s_{|U_i \cap U_j}\}. \tag{14.19}$$

Therefore $p \circ e = q \circ e$. To show that e is an equaliser, we need to show that it is universal among the arrows with that property. However, since we are in **Sets**, from (4.42), we only need to show that $\Gamma(\Lambda_p(U)) \subseteq \prod_i \Gamma(\Lambda_P(U_i))$ is the subset containing all $\{\dot{s}_i\}$ such that $q(\{\dot{s}_i\}) = p(\{\dot{s}_i\})$. The first thing is to show that e is injective which means that a section $\dot{s} \in \Gamma(\Lambda_P(U))$ is uniquely determined by its restrictions on the cover of U. Therefore assuming that $e(\dot{s}) = e(\dot{t})$ we get

$\{\dot{s}_i\} = \{\dot{t}_i\}$, hence for each i we get $\dot{s}_i(y) = \dot{t}_i(y) = germ_y(s_i) = germ_y(t_i)$ for all $y \in U_i$. From the definition of Λ_P (14.18) it follows that $\dot{s} = \dot{t}$.

Let us now assume that $q(\{\dot{s}_i\}) = p(\{\dot{s}_i\})$ holds and define $\dot{s} : U \to \Lambda_P$, $x \mapsto germ_x(s_i)$ where i is such that $x \in U_i$. Assuming that $x \in U_j$ then since $q(\{\dot{s}_i\}) = p(\{\dot{s}_i\})$ it follows that $germ_x(s_i) = germ_x(s_j)$, so $\dot{s}$ is well-defined. By construction $e(\dot{s}) = \{\dot{s}_i\}$. □

$\Gamma(\Lambda_P)$ is called the sheaf of cross sections of the bundle $p : \Lambda_P \to X$. We can now define a natural transformation

$$\eta : P \to \Gamma \circ \Lambda_P \tag{14.20}$$

such that for each open set $U \in \mathcal{O}(X)^{op}$ we obtain

$$\begin{aligned} \eta_U : P_U &\to \Gamma(\Lambda_P)(U) \\ s &\mapsto \dot{s}. \end{aligned} \tag{14.21}$$

Theorem 14.2 *If P is a sheaf then η is an isomorphism.*

Proof We need to show that η is 1:1 and onto.

1. One-to-one:

We want to show that if $\dot{s} = \dot{t}$ then $t = s$. Given $t, s \in P(U)$, $\dot{s} = \dot{t}$ means that $germ_x(s) = germ_x(t)$ for all $x \in U$. Therefore there exist opens $V_x \subseteq U$ such that $x \in V_x$ and $t_{|V_x} = s_{|V_x}$. The collection of these opens V_x for all $x \in U$ forms a cover of U such that $s_{|V_x} = t_{|V_x}$. This implies that s, t agree on the map $P(U) \to \coprod_{x \in U} P(V_x)$. From the sheaf requirements it follows that $t = s$.

2. Onto:

We want to show that any section $h : U \to \Lambda_p$ is of the form $\eta_U(s) = \dot{s}$ for some $s \in P(U)$. Let us consider a section $h : U \to \Lambda_p$, this will pick for each $x \in U$ an element, say $h(x) = germ_x(s_x)$. Therefore for each $x \in U$ there will exist an open $U_x \ni x$ such that $s_x \in P(U_x)$. Since $\dot{s}_x(U_x)$ is open by the definition of the topology on Λ_P, continuity of h implies that $h^{-1}(\dot{s}_x(U_x)) = \{z \in U_x | h(z) = germ_z(s_x)\}$ is open. Let us denote this set by W_x, this is an open neighbourhood of x. Then for all $z \in W_x \cap W_y$,

$$germ_z(s_x) = h(z) = germ_z(s_y). \tag{14.22}$$

By the sheaf condition, this implies $s_x = s_y$ on $W_x \cap W_y$. We thus obtain a family of elements $s_x \in P(W_x)$ such that they agree on the two maps $\prod_{x \in U} P(W_x) \rightrightarrows \prod_{y,x \in U} P(W_x \cap W_y)$. From the condition of being a sheaf it follows that there exists an $s \in P(U)$ such that $s_{|W_x} = s_{x|W_x}$. Then at each $x \in U$ we have $h(x) = germ_x(s_x) = germ_x(s) = \dot{s}(x)$. Therefore $h = \dot{s}$. □

It follows that all sheaves are sheaves of cross sections of some bundle.

Moreover it is possible to generalise the above process and define a pair of functors

$$Sets^{\mathcal{O}(X)^{op}} \xrightarrow{\Lambda} Bund(X) \xrightarrow{\Gamma} Sh(X). \tag{14.23}$$

Which if we combine together we get the so called *sheafification functor*:

$$\Gamma\Lambda : Sets^{\mathcal{O}(X)^{op}} \rightarrow Sh(X). \tag{14.24}$$

Such a functor sends each presheaf P on X to the "best approximation" $\Gamma\Lambda_P$ of P by a sheaf.

In the case of étale bundles we then obtain the following equivalence of categories

$$\text{Etalé}(X) \underset{\Gamma}{\overset{\Lambda}{\leftrightarrows}} Sh(X)$$

The pair of functors Γ and Λ are an adjoint pair (see Sect. 14.4). Here we have restricted the functors to act on $Sh(X) \subseteq Sets^{\mathcal{O}(X)^{op}}$.

14.3 Sheaves on a Partially Ordered Set

In the case at hand, since our base category $\mathcal{V}(\mathcal{H})$ is a poset, we have an interesting result. In particular, each poset P is equipped with an Alexandroff topology whose basis is given by the collection of all lower sets in the poset P, i.e., the basis can be taken to consist of sets of the form $\downarrow p := \{p' \in P | p' \leq p\}$, $p \in P$. Note that a function $\alpha : P_1 \rightarrow P_2$ between posets P_1 and P_2 is continuous with respect to the Alexandroff topologies on each poset, if and only if it is order preserving. In fact we have the following lemma:

Lemma 14.2 *Let $\alpha : P_1 \rightarrow P_2$ be a map between posets P_1 and P_2. Then α is order preserving if and only if for each lower set $L \subseteq P_2$, we have that $\alpha^{-1}(L)$ is a lower subset of P_1.*

Proof Let us assume that α is order preserving and let $L \subseteq P_2$ be lower. Now let $z \in \alpha^{-1}(L) \in P_1$, i.e., $\alpha(z) = l$ for some $l \in L$, and suppose $y \in P_1$ is such that $y \leq z$. Since α is order preserving, we have $\alpha(y) \leq \alpha(z) = l \in L$, which, since L is lower, means that $\alpha(y) \in L$, i.e., $y \in \alpha^{-1}(L)$. Hence $\alpha^{-1}(L)$ is lower.

Conversely, suppose that for any lower set $L \in P_2$, we have that $\alpha^{-1}(L) \in P_1$ is lower, and consider a pair $x, y \in P_1$ such that $x \leq y$. Now $\downarrow\alpha(y)$ is lower in P_2 and hence $\alpha^{-1}(\downarrow\alpha(y))$ is a lower subset of P_1. However $\alpha(y) \in \downarrow\alpha(y)$ and hence $y \in \alpha^{-1}(\downarrow\alpha(y))$. Therefore, the fact that $x \leq y$ implies that $x \in \alpha^{-1}(\downarrow\alpha(y))$, i.e., $\alpha(x) \in \downarrow\alpha(y)$, which means that $\alpha(x) \leq \alpha(y)$. Therefore α is order preserving. □

The dual of a lower set topology is the topology of upper sets, i.e. the topology generated by the sets $\uparrow p := \{p' \in P | p' \geq p\}$. Given such a topology it is a standard result [17, 18] that, for any poset P,

$$\mathbf{Sets}^P \simeq Sh(P^+) \tag{14.25}$$

where P^+ denotes the complete Heyting algebra of upper sets, which are the duals of lower sets. It follows that

$$\mathbf{Sets}^{P^{op}} \simeq Sh\big((P^{op})^+\big) \simeq Sh\big(P^-\big) \tag{14.26}$$

where P^- denotes the set of all lower sets in P. In particular, for the poset $\mathcal{V}(\mathcal{H})$ we have

$$\mathbf{Sets}^{\mathcal{V}(\mathcal{H})^{op}} \simeq Sh\big(\mathcal{V}(\mathcal{H})^-\big). \tag{14.27}$$

Thus every presheaf in our theory is in fact a sheaf with respect to the topology $\mathcal{V}(\mathcal{H})^-$. We will denote by $\bar{\underline{A}}$ the sheaves over $\mathcal{V}(\mathcal{H})$, while the respective presheaf will be denoted by $\underline{A}$. Moreover, in order to simplify the notation we will write $Sh(\mathcal{V}(\mathcal{H})^-)$ as just $Sh(\mathcal{V}(\mathcal{H}))$.

We shall frequently use the particular class of lower sets in $\mathcal{V}(\mathcal{H})$ of the form

$$\downarrow V := \{V' | V' \subseteq V\} \tag{14.28}$$

where $V \in Ob(\mathcal{V}(\mathcal{H}))$. It is easy to see that the set of all of these is a basis for the topology $\mathcal{V}(\mathcal{H})^-$. Moreover

$$\downarrow V_1 \cap \downarrow V_2 = \downarrow (V_1 \cap V_2) \tag{14.29}$$

i.e., these basis elements are closed under finite intersections.

It should be noted that $\downarrow V$ is the 'smallest' open set containing V, i.e., the intersection of all open neighbourhoods of V is $\downarrow V$. The existence of such a smallest open neighbourhood is typical of an Alexandroff space.

If we were to include the minimal algebra $\mathbb{C}(\hat{\mathbb{1}})$ in $\mathcal{V}(\mathcal{H})$ then, for any V_1, V_2 the intersection $V_1 \cap V_2$ would be non-empty. This would imply that $\mathcal{V}(\mathcal{H})$ is non-Hausdorff. To avoid this, we will exclude the minimal algebra from $\mathcal{V}(\mathcal{H})$. This means that, when $V_1 \cap V_2$ equals $\mathbb{C}(\hat{\mathbb{1}})$ we will not consider it.

More precisely, the semi-lattice operation $V_1, V_2 \to V_1 \cap V_2$ becomes a partial operation which is defined as $V_1 \cap V_2$ only if $V_1 \cap V_2 \neq \mathbb{C}(\hat{\mathbb{1}})$, otherwise it is undefined.

This restriction implies that when considering the topology on the poset $\mathcal{V}(\mathcal{H}) - \mathbb{C}(\hat{\mathbb{1}})$ we obtain

$$\downarrow V_1 \cap \downarrow V_2 = \begin{cases} \downarrow (V_1 \cap V_2) & \text{if } V_1 \cap V_2 \neq \mathbb{C}(\hat{\mathbb{1}}); \\ \emptyset & \text{otherwise.} \end{cases} \tag{14.30}$$

There are a few properties regarding sheaves on a poset worth mentioning:

1. When constructing sheaves it suffices to restrict attention to the basis elements of the form $\downarrow V$, $V \in Ob(\mathcal{V}(\mathcal{H}))$. For a given presheaf $\underline{A}$, we define the associated sheaf $\bar{\underline{A}}$ by defining it on the basis as

$$\bar{\underline{A}}(\downarrow V) := \underline{A}_V \tag{14.31}$$

where the left hand side is the sheaf using the topology $\mathcal{V}(\mathcal{H})^-$ and the right hand side is the presheaf on $\mathcal{V}(\mathcal{H})$. Given a presheaf map there is an associated restriction map for sheaves. In particular, given $i_{V_1 V} : V_1 \to V$ with associated presheaf map $\underline{A}(i_{V_1 V}) : \underline{A}_V \to \underline{A}_{V_1}$, then the restriction map $\rho_{V_1 V} : \bar{\underline{A}}(\downarrow V) \to \bar{\underline{A}}(\downarrow V_1)$ for the sheaf $\bar{\underline{A}}$ is defined as

$$a_{|\downarrow V_1} = \rho_{V_1 V}(a) := \underline{A}(i_{V_1 V})(a) \tag{14.32}$$

for all $a \in \bar{\underline{A}}(\downarrow V) = \underline{A}_V$.

In the following we will explain how this extends uniquely to a sheaf defined on all open sets in $\mathcal{V}(\mathcal{H})$.

Proof. We want to show that the sheaf condition holds on the basis sets for the sheaf defined in (14.31). To this end consider a covering $\downarrow V = \bigcup_i \downarrow V_i$ we then verify that the following is an equaliser diagram:

$$\bar{\underline{A}}(\downarrow V) \overset{e}{\rightarrowtail} \prod_i \bar{\underline{A}}(\downarrow V_i) \underset{q}{\overset{p}{\rightrightarrows}} \prod_{i,j} \bar{\underline{A}}(\downarrow (V_i \cap V_j))$$

Indeed we have that

$$\begin{aligned} &\underline{A}_V \to \underline{A}_{V_i} \to \underline{A}_{(V_i \cap V_j)} \\ &a \mapsto \underline{A}(i_{V_i V})(a) \mapsto \underline{A}(i_{V_i, V_i \cap V_j})\big(\underline{A}(i_{V_i V})(a)\big) = \underline{A}(i_{V, V_i \cap V_j})a. \end{aligned} \tag{14.33}$$

On the other hand

$$\begin{aligned} &\underline{A}_V \to \underline{A}_{V_j} \to \underline{A}_{(V_i \cap V_j)} \\ &a \mapsto \underline{A}(i_{V_j V})(a) \mapsto \underline{A}(i_{V_i, V_i \cap V_j})\big(\underline{A}(i_{V_j V})(a)\big) = \underline{A}(i_{V, V_i \cap V_j})a. \end{aligned} \tag{14.34}$$

Thus $p \circ e = q \circ e$. The next step is to show that e is injective. That this is the case follows from the fact that for $\downarrow V = \bigcup_i \downarrow V_i$ to hold then then there exists an i such that $V = V_i$. We then need to show that for any element $\{a_i\} \in \prod_i \bar{\underline{A}}(\downarrow V_i)$ such that $p(\{a_i\}) = q(\{a_i\})$ then $\{a_i\} = e(a)$ for some $a \in \underline{A}_V$. However since $V = V_i$ for some i then the collection $\{a_i\}$ contains some $a \in \underline{A}_V$. For such an a we have $a_{|V \cap V_j} = a_{|V_j} = a_{j|V_j}$ for all j. Hence $\{a_i\} = e(a)$. □

2. We will now extend the above definition of sheaf to all open sets in $\mathcal{V}(\mathcal{H})^-$. Given an open set $O = \bigcup_{V \in O} \downarrow V$ in $\mathcal{V}(\mathcal{H})^-$, the unique way of extending a sheaf defined on each $\downarrow V$ to a sheaf defined on all open sets O is as follows:

$$\bar{\underline{A}}(O) = \lim_{\longleftarrow V \in O} \bar{\underline{A}}(\downarrow V) = \lim_{\longleftarrow V \subseteq O} \underline{A}_V. \tag{14.35}$$

Where $\lim_{\longleftarrow}$ indicates the inverse limit. A consequence of the above is that

$$\bar{\underline{A}}(O) = \Gamma \underline{A}_{|O}. \tag{14.36}$$

Where the presheaf $\underline{A}_{|O}$ has as objects $\underline{A}_{V_i}$ for each $V_i \in O$ and morphisms $\underline{A}_V \to \underline{A}_{V_j}; a \mapsto \underline{A}(i_{V,V_j})a$ where $V_j \in \downarrow V$. We then obtain following:

$$\bar{\underline{A}}\left(\bigcup_i \downarrow(V_i)\right) = \lim_{\longleftarrow V \subseteq O} \underline{A}_{V_i}$$

$$= \left\{\{a_i\} \in \prod_i \underline{A}_{V_i} | a_{i|V_i \cap V_j} = a_{j|V_i \cap V_j}\right\}$$

$$= \Gamma \underline{A}|_O. \tag{14.37}$$

In fact, given a global section $\gamma \in \Gamma \underline{A}|_O$, by definition this will be such that given any pair $V_i, V_j \in O$ then $\gamma_{V_i}(\{*\}) = a_i \in \underline{A}_{V_i}$ and $\gamma_{V_j}(\{*\}) = a_j \in \underline{A}_{V_j}$. However the condition of global section is that $\underline{A}(i_{V'V})\gamma_V = \gamma_{V'}$ for all $V' \subseteq V$, therefore in our case we have that

$$\begin{aligned}\underline{A}(i_{V_i \cap V_j, V_i})\gamma_{V_i}\left(\{*\}\right) &= \gamma_{V_i \cap V_j}\left(\{*\}\right) = a_{i|V_i \cap V_j} \\ &= \underline{A}(i_{V_i \cap V_j, V_j})\gamma_{V_j}\left(\{*\}\right) \\ &= \gamma_{V_i \cap V_j}\left(\{*\}\right) = a_{j|V_i \cap V_j}.\end{aligned} \tag{14.38}$$

The connection with (14.31) is given by the fact that $\Gamma \bar{\underline{A}}_{|\downarrow V} \simeq \underline{A}_V$.

3. For presheaves on partially ordered sets the sub-object classifier $\underline{\Omega}^{\mathcal{V}(\mathcal{H})}$ has some interesting properties. In particular, given the set $\underline{\Omega}_V^{\mathcal{V}(\mathcal{H})}$ of sieves on V, there exists a bijection between sieves in $\underline{\Omega}_V^{\mathcal{V}(\mathcal{H})}$ and lower sets of V. To understand this let us consider any sieve S on V, we can then define the lower set of V

$$L_S := \bigcup_{V_1 \in S} \downarrow V_1. \tag{14.39}$$

Conversely, given a lower set L of V we can construct a sieve on V

$$S_L := \{V_2 \subseteq V | \downarrow V_2 \subseteq L\}. \tag{14.40}$$

However if $\downarrow V_2 \subseteq L_S$ then $\downarrow V_2 \subseteq \bigcup_{V_1 \in S} \downarrow V_1$, therefore $V_2 \in S$. On the other hand if $V_2 \in S_L$ (S_L sieve on V), then $V_2 \subseteq V$ and $\downarrow V_2 \subseteq L$, therefore $V_2 \in \bigcup_{V_1 \in S} \downarrow V_1$, i.e. $V_2 \in L_S$. This implies that the above operations are inverse of each other, and therefore

$$\bar{\underline{\Omega}}^{\mathcal{V}(\mathcal{H})}(\downarrow V) := \underline{\Omega}_V^{\mathcal{V}(\mathcal{H})} \simeq \Theta(V) \tag{14.41}$$

where $\Theta(V)$ is the collection of lower subsets (i.e. open subsets in $\mathcal{V}(\mathcal{H})$) of V. This is equivalent to the fact that, in a topological space X, we have that $\Omega^X(O)$ is the set of all open subsets of $O \subseteq X$.

14.4 Adjunctions

Consider two categories $\mathcal{C}$ and $\mathcal{D}$ and two functors between them going in opposite directions

$$F:\mathcal{D}\rightarrow\mathcal{C} \quad \text{and} \quad G:\mathcal{C}\rightarrow\mathcal{D}. \tag{14.42}$$

F is said to be *left adjoint* to G (or G is *right adjoint* to F) iff given any two objects $A\in\mathcal{C}$ and $X\in\mathcal{D}$ there exists a natural bijection between morphisms, i.e.

$$\frac{X\xrightarrow{f}G(A)}{F(X)\xrightarrow{h}A}. \tag{14.43}$$

What this means is that there is an exact correspondence between certain types of any maps, i.e. to each map from X to $G(A)$ there uniquely corresponds a map from $F(X)$ to A. In other words, f uniquely determines h and vice versa. Therefore we can write the following bijection

$$\theta: Hom_{\mathcal{D}}\big(X,G(A)\big)\xrightarrow{\sim}Hom_{\mathcal{C}}\big(F(X),A\big). \tag{14.44}$$

Such a bijection is said to be natural in the sense that, given any morphisms $\alpha: A\rightarrow A'$ in $\mathcal{C}$ and $\beta: X'\rightarrow X$ in $\mathcal{D}$, then the composition between these arrows and f and h above creates, yet, another correspondence:

$$\frac{X'\xrightarrow{\beta}X\xrightarrow{f}G(A)\xrightarrow{G(\alpha)}G(A')}{F(X')\xrightarrow{f(\beta)}F(X)\xrightarrow{h}A\xrightarrow{\alpha}A'}. \tag{14.45}$$

The symbol to indicate an adjunction relation between functors is $F\dashv G$ (F is left adjoint of G).

An important consequence of adjunctions is the existence of unit and co-unit morphisms. These are defined as follows

Definition 14.4 Given an adjunction $F\dashv G$ with corresponding bijection (14.44), taking $h=id_{F(X)}$ then we obtain a map

$$\eta_X: X\rightarrow GF(X) \tag{14.46}$$

such that $\theta(\eta_X)=id_{F(X)}$. Such a map is called the unit of the adjunction. Moreover, given a map f, then h is uniquely determined such that the following diagram

commutes

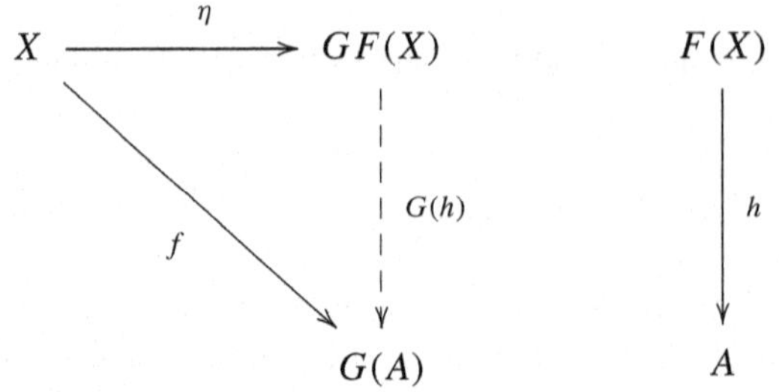

η is universal among the arrows which make the above diagram commute (i.e. any other such arrow uniquely factors through η).

Similarly we also have the notion of *co-unit* of the adjunction which is defined as follows:

Definition 14.5 By taking $G(A) = X$ in (14.44) and f the identity on $G(A)$ then h becomes the map

$$\epsilon_A : FG(A) \to (A). \tag{14.47}$$

Therefore $\theta^{-1}(\epsilon_A) = id_{G(A)}$. Such a map is called the co-unit of the adjunction. Moreover, given any $g : F(X) \to A$ there exists a unique f, such that ϵ is universal among the arrows which make the following diagram commute:

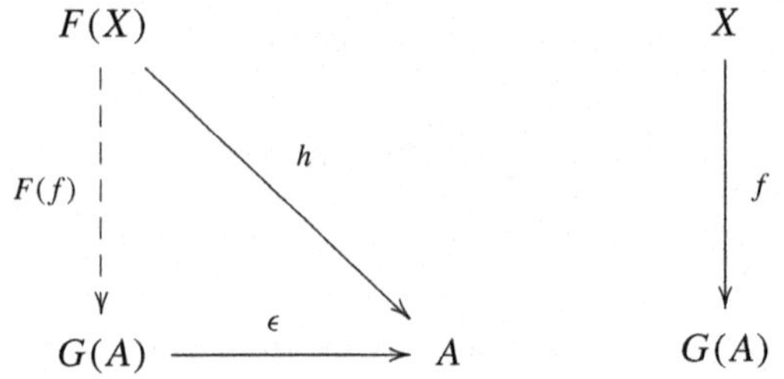

14.4.1 Example

Posets In a poset we can define an adjunction as follows. Suppose $g : Q \to P$ is a monotone map between posets. Then, given any element $x \in P$ we call a g-approximation of x (from above) an element $y \in Q$, such that $x \leq g(y)$. Moreover among all such approximations there might be the best one (it is a bit similar to how one defines the greatest lower bound): a best g-approximation of x is an element $y \in Q$ such that

$$x \leq g(y) \quad \text{and} \quad \forall z \in Q \quad \big(x \leq g(z) \Rightarrow y \leq z\big). \tag{14.48}$$

If a best g-approximation exists then it is clearly unique since we are in a poset (at most one arrow between any two elements). Moreover if it does exists for all $x \in P$, then we have a function $f : P \to Q$ such that, for all $x \in P$, $z \in Q$:

$$x \leq g(z) \quad \Leftrightarrow \quad f(x) \leq z. \tag{14.49}$$

We say that f is the left adjoint of g, and g is the right adjoint of f. Again it is trivial to see that the left adjoint of g, if it exists, is uniquely determined by g.

Exponential Consider a category $\mathcal{C}$ in which products are defined. For a given object $A \in \mathcal{C}$ one can define the functor

$$\begin{aligned} A \times - : \mathcal{C} &\to \mathcal{C} \\ B &\mapsto A \times B. \end{aligned} \tag{14.50}$$

It is possible to define the right adjoint of such a functor, namely:

$$\begin{aligned} (-)^A : \mathcal{C} &\to \mathcal{C} \\ B &\mapsto B^A \end{aligned} \tag{14.51}$$

which is simply the *exponential* as defined in Sect. 7.1. We then obtain the adjunction

$$(A \times -) \dashv (-)^A. \tag{14.52}$$

The property of being an adjoint pair implies that there exists the bijection

$$\frac{C \to B^A}{A \times C \to B}. \tag{14.53}$$

In this context the *co-unit* map is

$$\epsilon : A \times B^A \to B \tag{14.54}$$

such that, given any map $h : A \times C \to B$ there exists a unique $f : C \to B^A$ such that $\epsilon \circ (1 \times f) = h$, i.e.

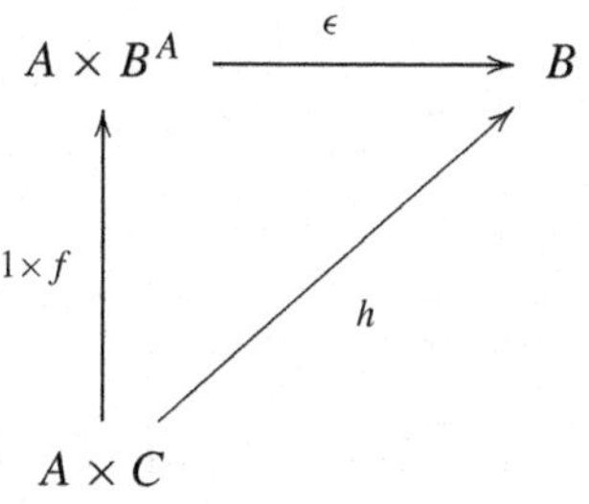

commutes. Thus $\epsilon = ev$ (where ev was defined in Definition 7.1) and we get the usual definition of exponentiation.

14.5 Geometric Morphisms

We will now introduce a very important concept in topos theory, namely the idea of geometric morphisms. Such objects are very important because they allow to define maps between topoi in a way that a lot of internal relations are preserved.

Definition 14.6 A *geometric morphism* $\phi : \tau_1 \to \tau_2$ between topoi τ_1 and τ_2 is defined to be a pair of functors $\phi_* : \tau_1 \to \tau_2$ and $\phi^* : \tau_2 \to \tau_1$ called, respectively, the *direct image* and the *inverse image* part of the geometric morphism, such that

1. $\phi^* \dashv \phi_*$ i.e., ϕ^* is the left adjoint of ϕ_*.
2. ϕ^* is left exact, i.e., it preserves all finite limits.

In the case of presheaf topoi, an important source of such geometric morphisms arises from functors between the base categories, according to the following theorem:

Theorem 14.3 *A functor $\theta : A \to B$ between two categories A and B induces a geometric morphism (also denoted θ)*

$$\theta : \mathbf{Sets}^{A^{op}} \to \mathbf{Sets}^{B^{op}} \tag{14.55}$$

of which the inverse image part $\theta^* : \mathbf{Sets}^{B^{op}} \to \mathbf{Sets}^{A^{op}}$ is such that

$$F \mapsto \theta^*(F) := F \circ \theta. \tag{14.56}$$

Proof To prove the above theorem we need to show that θ^* is left exact and that it has a right adjoint θ_*. Left exactness means that θ^* preserves all finite limits. To show that this is indeed the case, all that needs to be done is to show that it preserves terminal object and pullback. Let us consider first the terminal object $\underline{1}^B$. We recall that in $\mathbf{Sets}^{B^{op}}$ the terminal object simply assigns to each element $b \in B$ the singleton $\{*\}$. Similarly for the terminal object in $\mathbf{Sets}^{A^{op}}$.

$$\underline{1}^B \mapsto \theta^*\big(\underline{1}^B\big) := \underline{1} \circ \theta \tag{14.57}$$

such that for all $a \in A$

$$\underline{1}^B \circ \theta(a) = \underline{1}^B(b) = \{*\}. \tag{14.58}$$

Hence by uniqueness of the terminal object it follows that $\theta^*(\underline{1}^B) = \underline{1}^A$.

We now need to show that θ^* preserves pullbacks. To this end we recall that the definition of pullbacks in $\mathbf{Sets}^{A^{op}}$ (see Sect. 7.2) relies on the existence of pullbacks in **Sets**. Thus to show that θ^* preserves pullbacks all that needs to be done is to show that given $X, Y, Z, P \in \mathbf{Sets}^{B^{op}}$, if the following diagram is a pullback in **Sets**

$$\begin{array}{ccc} X(b) & \longrightarrow & Y(b) \\ \downarrow & & \downarrow \\ Z(b) & \longrightarrow & P(b) \end{array}$$

for any $b \in B$, then so will be

$$\begin{array}{ccc} \theta^* X(a) & \longrightarrow & \theta^* Y(a) \\ \downarrow & & \downarrow \\ \theta^* Z(a) & \longrightarrow & \theta^* P(a) \end{array}$$

for any $a \in A$.

However, this is always the case since $\theta^* X(a) = X \circ \theta(a) = X(b')$ thus the above diagram becomes

$$\begin{array}{ccc} X(b') & \longrightarrow & Y(b') \\ \downarrow & & \downarrow \\ Z(b') & \longrightarrow & P(b') \end{array}$$

and by assumption this is a pullback.

What remains to be shown is that there exists the direct image part θ_* such that $\theta^* \dashv \theta_*$. This means that we want to construct $\theta_* F$ such that $\forall G \in \mathbf{Sets}^{A^{op}}$

$$\frac{G \xrightarrow{\tau} \theta_* F}{\theta^* G \xrightarrow{\sigma} F}. \tag{14.59}$$

The way in which θ_* is defined is through a Kan extension. For simplicity of exposition we will, from now on, consider diagrams as belonging to A^{op} and B^{op} instead of A and B so that any functor is now covariant rather than contravariant.

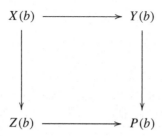

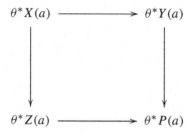

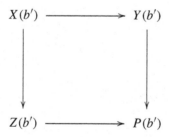

On an object $b \in B$, $\theta_* F(b)$ is defined to be the limit of the composite functor

$$(\theta \downarrow b)^{op} \xrightarrow{U} A^{op} \xrightarrow{F} \textbf{Sets}$$

where U is the forgetful functor from the comma category (see Sect. 4.2.1) $(\theta \downarrow b)^{op}$ to A^{op}. This means that for any two objects $g, g' \in (\theta \downarrow b)$ and any morphism $f : g \to g'$ such that

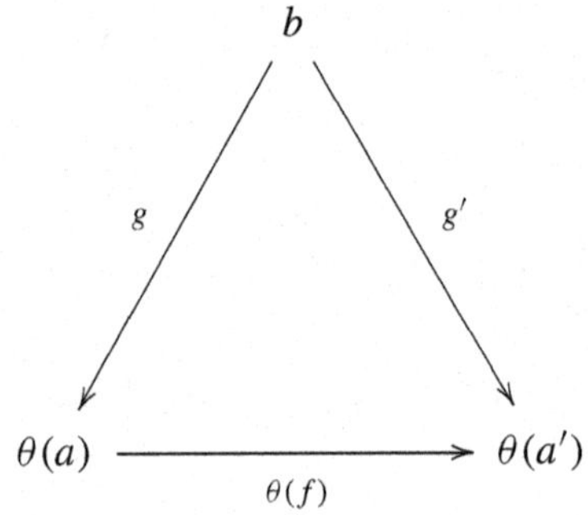

commutes we obtain

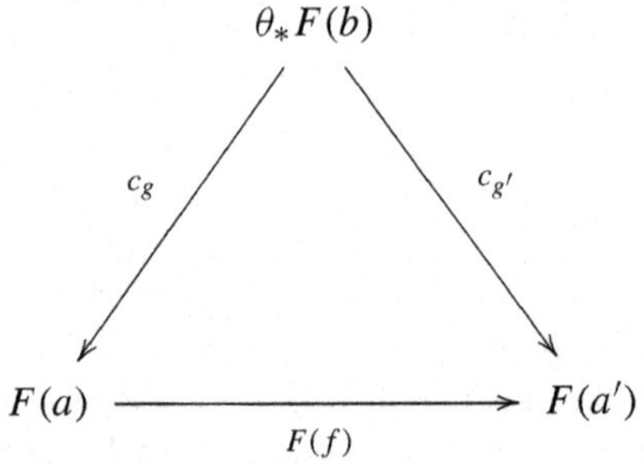

which is a tiny part of the limit diagram defining $\theta_* F(b)$, where c_g and $c_{g'}$ are components of the limiting cone. It follows from the universal property of limits that $\theta_* F(b)$ is covariant in $b \in B^{op}$.

Given $\sigma : G\theta \to F$ we construct $\tau : G \to \theta_* F$ in terms of its components $\tau_b : G(b) \to \theta_* F(b)$. These are defined from the fact that, given the cone

Diagram 14.1

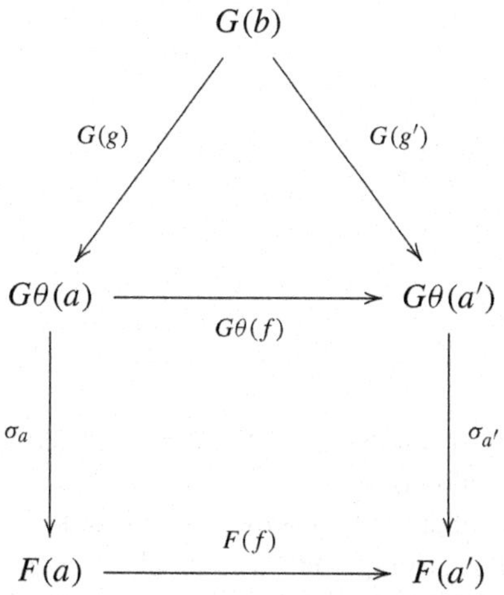

by the universal property of limits, this uniquely factors through $\theta_* F(b)$. We now need to check that all the components τ_b form a natural transformation, hence we need to check naturality on $h : b \to b'$:

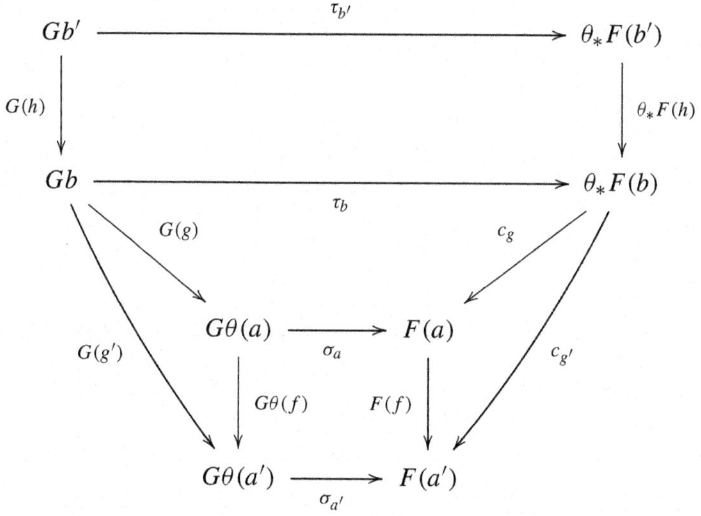

Conversely, given $\tau : G \to \theta_* F$ we can construct σ componentwise as the composition

$$\sigma_a : G\theta(a) \xrightarrow{\tau_{\theta(a)}} \theta_* F(\theta(a)) \xrightarrow{c_{id_{\theta(a)}}} F(a) \tag{14.60}$$

Naturality can be checked through the following commuting diagram

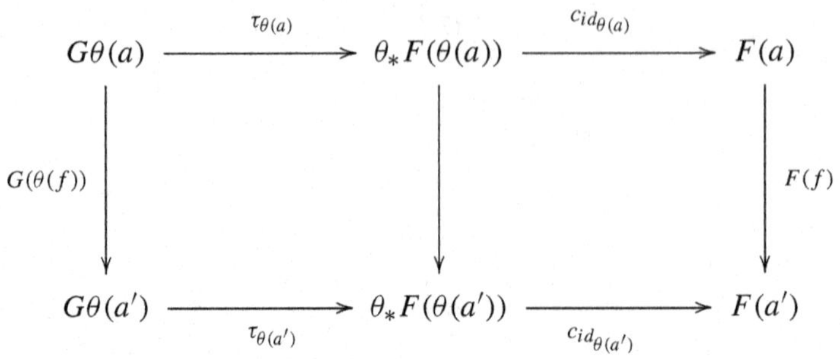

The left square commutes from the naturality of τ while commutativity of the right square expresses functoriality of $\theta_* F$.

We now need to show that the two above constructions of τ from σ and σ from τ are inverse of each other. Going one way, namely $\sigma \mapsto \tau \mapsto \sigma'$ then it is easy to show that indeed $\sigma = \sigma'$. On the other hand going the other way $\tau \mapsto \sigma \mapsto \tau'$ we first substitute the definition of σ in the Diagram 14.1 obtaining

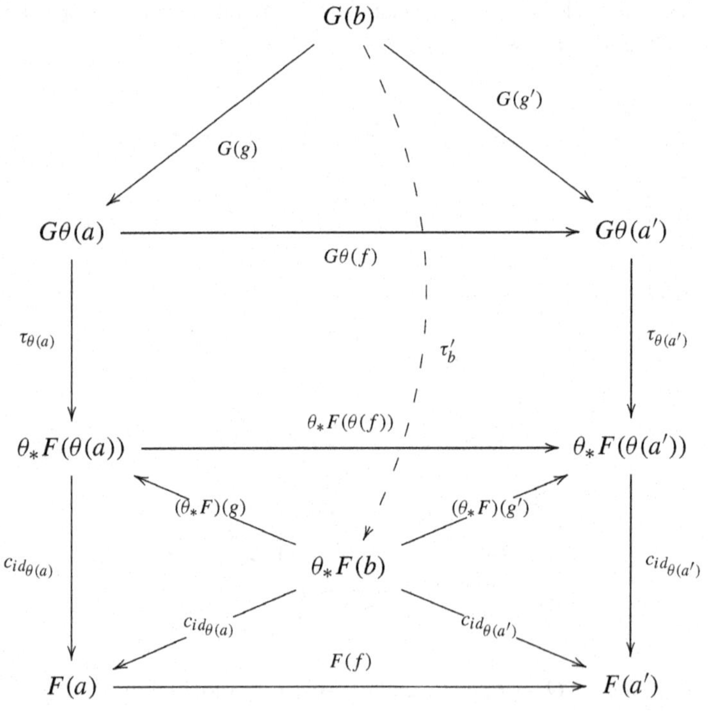

Since in this diagram, τ'_b can be chosen also to be τ_b, by uniqueness it follows that $\tau'_b = \tau_b$. □

There is also a standard way of defining geometric morphisms between categories of sheaves.

Theorem 14.4 *Given a continuous map $f : X \to Y$ between topological spaces, this induces a geometric morphism*

$$f : Sh(X) \to Sh(Y) \tag{14.61}$$

such that

$$\begin{aligned} f_* : Sh(X) &\to Sh(Y) \\ F &\mapsto F \circ f^{-1}. \end{aligned} \tag{14.62}$$

and

$$f^* : Sh(Y) \to Sh(X). \tag{14.63}$$

Here f^* is defined as the composition $Sh(Y) \xrightarrow{\Lambda} \text{Étale}(Y) \xrightarrow{\hat{f}} \text{Étale}(X) \xrightarrow{\Gamma} Sh(X)$, where $\hat{f}$ is the pullback of an étale bundle over Y along $f : X \to Y$.

Proof We will first show that f^* is left exact. Since both Γ and Λ are equivalences all that needs to be shown is that $\hat{f}$ is left exact, i.e. preserves terminal objects and pullbacks. Thus given the terminal object $1_{\text{Étale}(Y)} = id_Y : Y \to Y$ in Étale(Y), the pullback is $\hat{f}(1_{\text{Étale}(Y)}) = id_X : X \to X$ which is $1_{\text{Étale}(X)}$ since

$$\begin{array}{ccc} X & \xrightarrow{f} & Y \\ {\scriptstyle id_X}\downarrow & & \downarrow{\scriptstyle id_Y} \\ X & \xrightarrow{f} & Y \end{array}$$

is a pullback diagram. On the other hand given a pullback in Étale(Y).

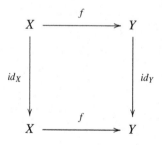

Diagram 14.2

$$\begin{array}{ccc} E_4 & \xrightarrow{j_{43}} & E_3 \\ {\scriptstyle j_{41}}\downarrow & & \downarrow{\scriptstyle j_{32}} \\ E_1 & \xrightarrow{j_{12}} & E_2 \end{array}$$

the functoriality of $\hat{f}$ gives the commutative diagram

Diagram 14.3

$$\begin{array}{ccc} \hat{f}(E_4) & \xrightarrow{\hat{f}(j_{43})} & \hat{f}(E_3) \\ {\scriptstyle \hat{f}(j_{41})}\downarrow & & \downarrow{\scriptstyle \hat{f}(j_{32})} \\ \hat{f}(E_1) & \xrightarrow{\hat{f}(j_{12})} & \hat{f}(E_2) \end{array}$$

where for each $i \in \{1, 2, 3, 4\}$ we have

$$\begin{array}{ccc} \hat{f}(E_i) & \xrightarrow{k_i} & E_i \\ \downarrow & & \downarrow \\ X & \xrightarrow{f} & Y \end{array}$$

We now want to prove the universal property of pullbacks for the Diagram 14.3, i.e. we want to show that given some object P with arrows $h : P \to \hat{f}(E_3)$ and

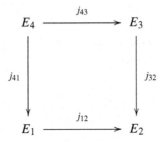

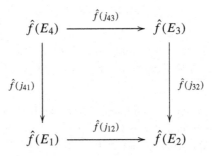

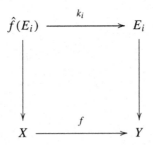

$i : P \to \hat{f}(E_1)$ such that $\hat{f}(j_{32}) \circ h = \hat{f}(j_{12}) \circ i$, then there exists a unique arrow $l : P \to \hat{f}(E_4)$ such that $\hat{f}(j_{43}) \circ l = h$ and $\hat{f}(j_{41}) \circ l = i$. However, since Diagram 14.2 is a pullback, we have the unique arrow $k_4 \circ l$ such that $j_{43} \circ k_4 \circ l = k_3 \circ h$ and $j_{41} \circ k_4 \circ l = k_4 \circ i$. On the other hand from the universality property of the pullback defining $\hat{f}(E_4)$, we obtain the unique $l : P \to \hat{f}(E_4)$.

It now remains to show that $f^* \dashv f_*$. To this end we will use the equivalence between étale bundles and sheaves discussed in Sect. 14.2 and show the existence of a natural bijection

$$\frac{\hat{f}E \overset{\alpha}{\to} \Lambda A \text{ in Étale}(X)}{\Gamma E \overset{\beta}{\to} f_* A \text{ in } Sh(Y)}. \tag{14.64}$$

As a first step we construct β given α by defining its components β_V for all $V \subseteq Y$:

$$\beta_V : \Gamma E(V) \to f_* A(V) = A \circ f^{-1}(V) = A\big(f^{-1}(V)\big) \tag{14.65}$$

$$s \mapsto \beta_V(s). \tag{14.66}$$

To define $\beta_V(s)$ we will utilise the universal property of $\hat{f}E$ to construct $k_V(s)$ as follows

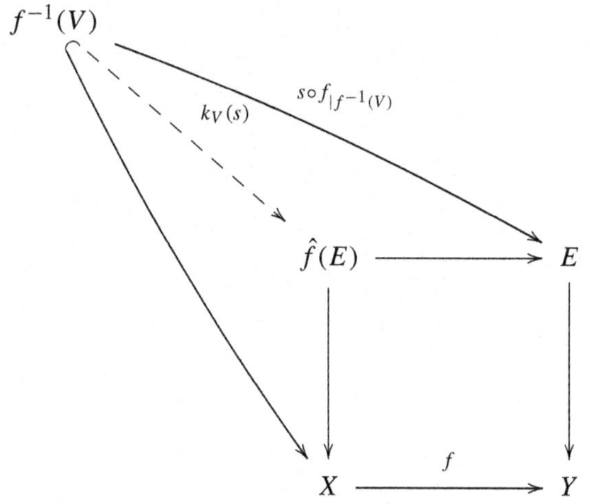

We then define

$$\beta_V(s) := \alpha \circ k_V(s) : f^{-1}(V) \to \Lambda A. \tag{14.67}$$

This is a continuous local section of the bundle ΛA and hence can be considered an element $\beta_V(s) \in (\Gamma \Lambda A)(f^{-1}(V)) = A(f^{-1}(V))$. Having defined β_V we need to verify that it is a natural transformation. Thus we consider $i_{V'V} : V' \subseteq V$ and construct

Diagram 14.4

$$\begin{array}{ccccc}
(\Gamma E)(V) & \xrightarrow{k_V} & (\Gamma \hat{f}(E))(f^{-1}(V)) & \xrightarrow{\Gamma\alpha_{f^{-1}(V)}} & A(f^{-1}(V)) \\
\Big\downarrow {\scriptstyle (\Gamma E)(i_{V'V})} & & \Big\downarrow {\scriptstyle (\Gamma \hat{f}(E))(i_{f^{-1}(V')f^{-1}(V)})} & & \Big\downarrow {\scriptstyle A(i_{f^{-1}(V')f^{-1}(V)})} \\
(\Gamma E)(V') & \xrightarrow{k_{V'}} & (\Gamma \hat{f} E)(f^{-1}(V')) & \xrightarrow{\Gamma\alpha_{f^{-1}(V')}} & A(f^{-1}(V'))
\end{array}$$

The commutativity of the right square follows from the property of local sections. On the other hand the commutativity of the left square expresses naturalily of k which can be seen from the following diagram

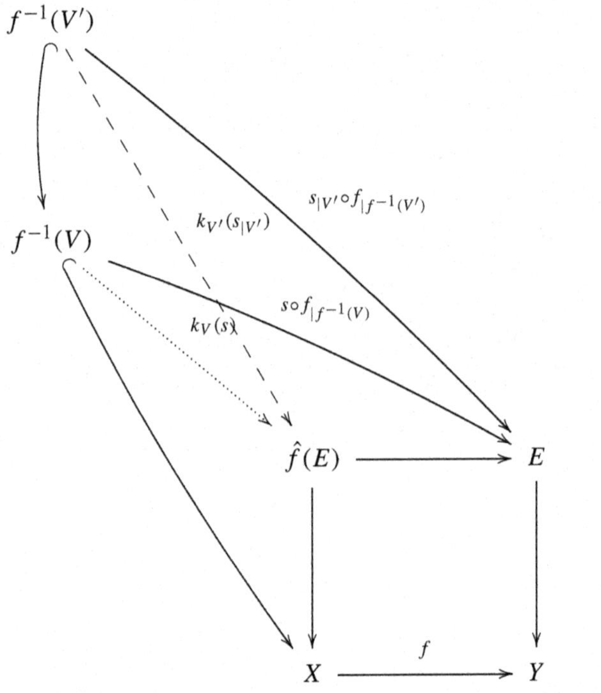

Uniqueness of the dashed arrow implies that the triangle comprising the dashed and dotted arrows commutes:

$$k_V(s)_{|f^{-1}(V')} = k_{V'}(s_{|V'}). \tag{14.68}$$

This equation implies the commutativity of the left square in the Diagram 14.4.

We now need to construct a map α given β. To this end, given the bundle $p_A : \Lambda A \to X$ we define $\alpha(e, x) \in p_A^{-1}(x) = A_x$ where $p_E(e) = f(x)$ and $p_E : E \to Y$.

We begin by choosing some neighbourhood $W_e \subseteq E$ of $e \in E$ such that $p_E(W_e)$ is open in Y and

$$p_{E|W_e} : W_e \to p_E(W_e) \tag{14.69}$$

is a homeomorphism. Its inverse is a local section $s_e \in (\Gamma E)(p_E(W_e))$. Since $\beta_{p_E(W_e)}(s_e) \in A(f^{-1}(p_E(W_e)))$ and $x \inf^{-1}(p_E(W_e))$ we can define $\alpha(e,x) := germ_x \beta_{p_E(W_e)}(s_e)$. Next we need to show that i) $\alpha(e,x)$ is well-defined and ii) α is continuous. For the first requirement we need to show that the definition is independent of the choice of W_e. In particular considering another W_e' then

$$p_{E|W_e \cap W_e'} : W_e \cap W_e' \to p_E\big(W_e \cap W_e'\big) \tag{14.70}$$

whose inverse is $s_{e|p_E(W_e \cap W_e')}$. In the stork A_x the corresponding elements

$$germ_x \beta_{p_E(W_e)}(s_e) = germ_x \beta_{p_E(W_e \cap W_e')}(s_{e|p_E(W_e \cap W_e')}). \tag{14.71}$$

The same argument shows that this also equals $germ_x \beta_{p_E(W_e')}(s_e')$.

With regards to continuity we take a basis open set in ΛA which is defined in terms of some $t \in A(U)$ where $U \subseteq X$ is open and given by $\{germ_x t | x \in U\}$. Then we prove the set

$$\alpha^{-1}\big(\{germ_x t | x \in U\}\big) = \big\{(e,x) \in \hat{f} E | x \in U \text{ and } germ_x \beta_{P_E(W_e)}(s_e) = germ_x t\big\} \tag{14.72}$$

to be open by showing that $(e,x) \in \alpha^{-1}(\{germ_x t | x \in U\})$ implies that there exists some neighbourhood of (e,x) also contained in $\alpha^{-1}(\{germ_x t | x \in U\})$. In particular, the assumption $germ_x \beta_{P_E(W_e)}(s_e) = germ_x t$ means that there exists an open $U' \subseteq U$ with $x \in U'$ on which

$$\beta_{P_E(W_e)}(s_e)_{|U'} = t_{|U'}. \tag{14.73}$$

If $(e', x') \in \hat{f} E$ is any other point with $e' \in W_e$ and $x' \in U'$ then we can choose $s_{e'} = s_e$ so that $\alpha(e', x') = germ_{x'} \beta_{P_E(W_e)}(s_e)$ which is equal to $germ_{x'} t$ by (14.73).

As a last step we need to show that the two construction for α and β defined above are inverse of each other. To this end, given a β we construct the associated α in terms of which we construct β'. We then show that $\beta = \beta'$. In order to show that $\beta_V(s) = \beta_V'(s)$ it is sufficient to show that the equality holds for each germ. Hence we compute

$$\begin{aligned} germ_x \beta_V'(s) &= germ_x\big(\alpha \circ k_V(s)\big) = \alpha\big(k_V(s)(x)\big) = \alpha\big(s\big(f(x)\big), x\big) \\ &= germ_x \beta_{p_E(W_e)}(s) = germ_x \beta_V(s). \end{aligned} \tag{14.74}$$

We then go the other direction and show that, given an α we compute the corresponding β and show that the associated α' is equal to α. In particular

$$\begin{aligned} \alpha'(e,x) &= germ_x \beta_{P_E(W_e)}(s_e) = germ_x(\alpha \circ k_{p_E(W_e)}(s_e) \\ &= \alpha\big(k_{p_E(W_e)}(s_e)(x)\big) = \alpha(e,x). \end{aligned} \tag{14.75}$$

□

14.6 Group Action and Twisted Presheaves

In this section we will briefly analyse the problem of twisted presheaves.

Given a group G, its action on the base category $\mathcal{V}(\mathcal{H})$ is defined as $l_g(V) := \hat{U}_g V \hat{U}_g^{-1} := \{\hat{U}_g \hat{A} \hat{U}_g^{-1} | \hat{A} \in V\}$, $g \in G$ for a unitary representation $g \mapsto \hat{U}_g$. When considering the topos $\mathbf{Sets}^{\mathcal{V}(\mathcal{H})^{op}}$, for each g we obtain the functor $l_{\hat{U}_g} : \mathcal{V}(\mathcal{H}) \to \mathcal{V}(\mathcal{H})$ which induces a geometric morphism

$$l_{\hat{U}_g} : \mathbf{Sets}^{\mathcal{V}(\mathcal{H})^{op}} \to \mathbf{Sets}^{\mathcal{V}(\mathcal{H})^{op}} \tag{14.76}$$

whose inverse image part is

$$\begin{aligned} l^*_{\hat{U}_g} : \mathbf{Sets}^{\mathcal{V}(\mathcal{H})^{op}} &\to \mathbf{Sets}^{\mathcal{V}(\mathcal{H})^{op}} \\ \underline{F} &\mapsto l^*_{\hat{U}_g}(\underline{F}) = \underline{F} \circ l_{\hat{U}_g}. \end{aligned} \tag{14.77}$$

The above geometric morphism acts on the spectral presheaf $\underline{\Sigma}^{\mathcal{V}(\mathcal{H})}$, the quantity value object $\underline{\mathbb{R}}^{\leftrightarrow}$, truth values and daseinisation. Let us analyse each of these actions in detail.

14.6.1 Spectral Presheaf

Given the speactral presheaf $\underline{\Sigma} \in \mathbf{Sets}^{\mathcal{V}(\mathcal{H})^{op}}$, the respective twisted presheaf $\underline{\Sigma}^{\hat{U}}$ associated to the unitary operator $\hat{U}$ is:

Definition 14.7 The twisted presheaf $\underline{\Sigma}^{\hat{U}}$ has as:

- Objects: for each $V \in \mathcal{V}(\mathcal{H})$ it assigns the Gel'fand spectrum of the algebra $\hat{U} V \hat{U}^{-1}$, i.e., $\underline{\Sigma}^{\hat{U}}_V := \{\lambda : \hat{U} V \hat{U}^{-1} \to \mathbb{C} | \lambda(\hat{1}) = 1\}$..
- Morphisms: for each $i_{V'V} : V' \to V$ $(V' \subseteq V)$ it assigns the presheaf maps

$$\begin{aligned} \underline{\Sigma}^{\hat{U}}(i_{V'V}) : \underline{\Sigma}^{\hat{U}}_V &\to \underline{\Sigma}^{\hat{U}}_{V'} \\ \lambda &\mapsto \lambda_{|\hat{U} V' \hat{U}^{-1}}. \end{aligned} \tag{14.78}$$

The action of each element of the group on $\underline{\Sigma}$ is given by the following theorem:

Theorem 14.5 *For each $\hat{U} \in \mathcal{U}(\mathcal{H})$, there is a natural isomorphism $\iota^{\hat{U}} : \underline{\Sigma} \to \underline{\Sigma}^{\hat{U}}$ which is defined through commutativity of the following diagram:*

$$
\begin{array}{ccc}
\underline{\Sigma}_V & \xrightarrow{\;\iota^{\hat{U}}_V\;} & \underline{\Sigma}^{\hat{U}}_V \\
\Big\downarrow {\scriptstyle \underline{\Sigma}_V(i_{V'V})} & & \Big\downarrow {\scriptstyle \underline{\Sigma}^{\hat{U}}_V(i_{V'V})} \\
\underline{\Sigma}_{V'} & \xrightarrow[\;\iota^{\hat{U}}_{V'}\;]{} & \underline{\Sigma}^{\hat{U}}_{V'}
\end{array}
$$

where, at each stage V

$$\big(\iota^{\hat{U}}_V(\lambda)\big)(\hat{A}) := \big\langle \lambda, \hat{U}\hat{A}\hat{U}^{-1} \big\rangle \tag{14.79}$$

for all $\lambda \in \underline{\Sigma}_V$ and $\hat{A} \in V_{sa}$.

Proof To show that the above is indeed a natural isomorphisms we simply need to show that for all $V \in \mathcal{V}(\mathcal{H})$ then

$$\iota^{\hat{U}}_{V'} \circ \underline{\Sigma}(i_{V'V})(\lambda) = \underline{\Sigma}^{\hat{U}}(i_{V'V})\iota^{\hat{U}}_{V'}(\lambda). \tag{14.80}$$

Now applying the definitions we obtain from the right hand side

$$\underline{\Sigma}^{\hat{U}}(i_{V'V})\iota^{\hat{U}}_{V'}(\lambda)(\hat{A}) = \underline{\Sigma}^{\hat{U}}(i_{V'V})\big\langle \lambda, \hat{U}\hat{A}\hat{U}^{-1} \big\rangle = \big\langle \lambda_{|V'}, \hat{U}\hat{A}\hat{U}^{-1} \big\rangle. \tag{14.81}$$

On the other hand the left hand side gives

$$\iota^{\hat{U}}_{V'} \circ \underline{\Sigma}(i_{V'V})(\lambda)(\hat{A}) = \iota^{\hat{U}}_{V'}\big(\lambda_{|V'}(\hat{A})\big) = \big\langle \lambda_{|V'}, \hat{U}\hat{A}\hat{U}^{-1} \big\rangle. \tag{14.82}$$

□

14.6.2 Quantity Value Object

Similarly, for the quantity value object we obtain the following theorem:

Theorem 14.6 *For each $\hat{U} \in \mathcal{U}(\mathcal{H})$, there exists a natural isomorphism $k^{\hat{U}} : \underline{\mathbb{R}}^{\leftrightarrow} \to (\underline{\mathbb{R}}^{\leftrightarrow})^{\hat{U}}$, such that for each $V \in \mathcal{V}(\mathcal{H})$ we obtain the individual components $k^{\hat{U}} : \underline{\mathbb{R}}^{\leftrightarrow}_V \to (\underline{\mathbb{R}}^{\leftrightarrow})^{\hat{U}}_V$ defined as*

$$k^{\hat{U}}_V(\mu, \nu)\big(l^{\hat{U}}(V')\big) := \big(\mu(V'), \nu(V')\big) \tag{14.83}$$

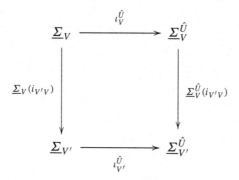

for all $V' \subseteq V$.

Here, $\mu \in \underline{\mathbb{R}}^{\leftrightarrow}_V$ is an order preserving function $\mu : \downarrow V \to \mathbb{R}$ such that, if $V_2 \subseteq V_1 \subseteq V$, then $\mu(V_2) \geq \mu(V_1) \geq \mu(V)$, while ν is an order reversing function $\nu : \downarrow V \to \mathbb{R}$ such that, if $V_2 \subseteq V_1 \subseteq V$, then $\nu(V_2) \leq \nu(V_1) \leq \nu(V)$.

In (14.83) we have used the bijection between the sets $\downarrow l^{\hat{U}}(V)$ and $\downarrow V$.

14.6.3 Daseinisation

We recall the concept of daseinisation: given a projection operator $\hat{P}$ its daseinisation with respect to each context V is

$$\delta^o(\hat{P})_V := \bigwedge\{\hat{Q} \in \mathcal{P}(V) | \hat{Q} \geq \hat{P}\} \tag{14.84}$$

where $P(V)$ represents the collection of projection operators in V.

If we then act upon it by any $\hat{U}$ we obtain

$$\begin{aligned} \hat{U}\delta^o(\hat{P})_V\hat{U}^{-1} &:= \hat{U} \bigwedge\{\hat{Q} \in \mathcal{P}(V) | \hat{Q} \geq \hat{P}\}\hat{U}^{-1} \\ &= \bigwedge\{\hat{U}\hat{Q}\hat{U}^{-1} \in \mathcal{P}(l_{\hat{U}}(V)) | \hat{Q} \geq \hat{P}\} \\ &= \bigwedge\{\hat{U}\hat{Q}\hat{U}^{-1} \in \mathcal{P}(l_{\hat{U}}(V)) | \hat{U}\hat{Q}\hat{U}^{-1} \geq \hat{U}\hat{P}\hat{U}^{-1}\} \\ &= \delta^o(\hat{U}\hat{P}\hat{U}^{-1})_{l_{\hat{U}}(V)} \end{aligned} \tag{14.85}$$

where the second and third equation hold since the map $\hat{Q} \to \hat{U}\hat{Q}\hat{U}^{-1}$ is weakly continuous.

What this implies is that the clopen sub-objects which represent propositions, i.e., $\underline{\delta(\hat{P})}$, get mapped to one another by the action of the group.

14.6.4 Truth Values

Now that we have defined the group action on daseinisation we can define the group action on the truth values. We recall that for a pure state $|\psi\rangle$ the truth object at each stage V is defined as

$$\begin{aligned} \underline{\mathbb{T}}^{|\psi\rangle}_V &:= \{\hat{\alpha} \in \mathcal{P}(V) | Prob(\hat{\alpha}; |\psi\rangle) = 1\} \\ &= \{\hat{\alpha} \in \mathcal{P}(V) | \langle\psi|\hat{\alpha}|\psi\rangle = 1\}. \end{aligned} \tag{14.86}$$

For each context $V \in \mathcal{V}(\mathcal{H})$ the truth value is

$$v(\underline{\delta(\hat{P})} \in \underline{\mathbb{T}}^{|\psi\rangle})_V := \{V' \subseteq V | \delta^o(\hat{P})_{V'} \in \underline{\mathbb{T}}^{|\psi\rangle}_{V'}\}$$

$$= \big\{V' \subseteq V \big| \langle\psi|\delta^o(\hat{P})_{V'}|\psi\rangle = 1\big\}.$$

We now act upon it with a group element $\hat{U}$ obtaining

$$\begin{aligned} l_{\hat{U}}\big(v\big(\delta^o(\hat{P}) \in \underline{\mathbb{T}}^{|\psi\rangle}\big)_V\big) &:= l_{\hat{U}}\big\{V' \subseteq V \big| \langle\psi|\delta^o(\hat{P})_{V'}|\psi\rangle = 1\big\} \\ &= \big\{l_{\hat{U}}V' \subseteq l_{\hat{U}}V \big| \langle\psi|\delta^o(\hat{P})_{V'}|\psi\rangle = 1\big\} \\ &= \big\{l_{\hat{U}}V' \subseteq l_{\hat{U}}V \big| \langle\psi|\hat{U}^{-1}\hat{U}\delta^o(\hat{P})_{V'}\hat{U}^{-1}\hat{U}|\psi\rangle = 1\big\} \\ &= \big\{l_{\hat{U}}V' \subseteq l_{\hat{U}}V \big| \langle\psi|\hat{U}^{-1}\delta^o\big(\hat{U}\hat{P}\hat{U}^{-1}\big)_{l_{\hat{U}}(V')}\hat{U}|\psi\rangle = 1\big\}s \\ &= v\big(\delta^o\big(\hat{U}\hat{P}\hat{U}^{-1}\big) \in \underline{\mathbb{T}}^{\hat{U}|\psi\rangle}\big)_{l_{\hat{U}}(V)}. \end{aligned} \tag{14.87}$$

We then obtain the following equality:

$$l_{\hat{U}}\big(v\big(\delta^o(\hat{P}) \in \underline{\mathbb{T}}^{|\psi\rangle}\big)_V\big) = v\big(\delta^o\big(\hat{U}\hat{P}\hat{U}^{-1}\big) \in \underline{\mathbb{T}}^{\hat{U}|\psi\rangle}\big)_{l_{\hat{U}}(V)}. \tag{14.88}$$

Thus truth values are invariant under the group transformations. This is the topos analogue of Dirac covariance, i.e., given a state $|\psi\rangle$ and a physical quantity $\hat{A}$, we would obtain the same predictions if we replaced the state by $\hat{U}|\psi\rangle$ and the quantity by $\hat{U}\hat{A}\hat{U}^{-1}$.

Chapter 15
Probabilities in Topos Quantum Theory

The main idea in the topos formulation of quantum theory it to have logic as a fundamental concept and try and derive other concepts in terms of it. This is what is done in the case of probabilities.

In fact, in the topos approach, probabilities acquire a logical interpretation rather than a relative frequency interpretation. In this setting probabilities are described in terms of truth values.

This implies that for the concept of probabilities to be well defined one does not require the notions of external observer, observed system and measurement. All that is required is an internal logic in terms of which truth values can be assigned. Probabilities, are then derived by these truth values. Thus truth values are seen as fundamental concepts while probabilities as derived concepts.

The importance of this lies in the fact that in the topos approach truth values can be assigned to any quantum proposition and, since probabilities are defined in terms of truth values, probabilities can always be assigned even in the context of closed systems. For the interested reader, the full analysis of these ideas can be found in [24, 25, 39].

15.1 General Definition of Probabilities in the Language of Topos Theory

In this section we will outline the general way in which probabilities can be described in a topos. How this general definition will apply to classical physics and quantum physics will be described in subsequent sections.

First of all we need to define a possible topos in which probabilities can be expressed in terms of truth values in that topos. In particular we are looking for a way to combine, on the one hand probabilities which are described by numbers in the interval $[0, 1]$ and, on the other hand, truth values expressed in a topos τ, which are defined as global elements of the sub-object classifier Ω^{τ}.

C. Flori, *A First Course in Topos Quantum Theory*, Lecture Notes in Physics 868, DOI 10.1007/978-3-642-35713-8_15, © Springer-Verlag Berlin Heidelberg 2013

What this amounts to is to find a topos τ such that the following holds:

$$\mathcal{O}(X) \simeq \Gamma\Omega^{Sh(X)}. \tag{15.1}$$

If we are able to achieve this, it will mean that we have found a bijective correspondence between probabilities and truth values in a topos τ.

The first step in achieving the equivalence in (15.1) is to define the topological space $(0, 1)$, whose open sets are the intervals $(0, r)$ for $0 \leq r \leq 1$. This topological space is denoted as $(0, 1)_L$ and the collection of open sets as $\mathcal{O}((0, 1)_L)$. This forms a category under inclusion. One can then define a bijection

$$\begin{aligned} \beta : [0, 1] &\to \mathcal{O}\big((0, 1)_L\big) \\ r &\mapsto (0, r). \end{aligned} \tag{15.2}$$

The strategy is then to associate the open sets $(0, r)$ with truth values in a certain topos. Such open sets will then, in turn, be associated with the probability r.

In this way we will not be loosing anything by considering open sets $(0, r)$ which don't actually contain r as opposed to closed sets $(0, r]$ or $[0, r]$. Moreover, if we were to consider either $(0, r]$ or $[0, r]$ we would run into troubles. In fact, if we had chosen $[0, r]$ we would have obtained situations in which all propositions are totally true with probability zero, even the totally false proposition (more details later on). If instead we had chosen $(0, r]$ we wouldn't have obtained a topology, since these sets do not close under arbitrary unions.

Having chosen the topology on our space, we then know from topos theory that, for any topological space X, there is the following isomorphisms of Heyting algebras

$$\mathcal{O}(X) \simeq \Gamma\Omega^{Sh(X)} \tag{15.3}$$

where $Sh(X)$ identifies the topos of sheaves over the topological space X. In order to prove the above isomorphism we need to define the sub-object classifier for sheaves.

Definition 15.1 Given a topological space X, the sub-object classifier $\Omega^{Sh(X)} \in Sh(X)$ is defined on

1. Objects: for a given open $U \subseteq X$ it assigns the set of all open subsets of U, i.e.

$$\Omega_U = \{V | V \subseteq U, V \text{ open in } X\}. \tag{15.4}$$

2. Morphisms: if $W \subseteq U$ then

$$\begin{aligned} \Omega_U &\to \Omega_W \\ V &\mapsto V \cap W. \end{aligned} \tag{15.5}$$

We can now prove equation (15.1).

Proof Let us consider the map

$$\begin{aligned} i : \mathcal{O}(X) &\to \Gamma\Omega^{Sh(X)} \\ U &\mapsto i(U) := \gamma^U. \end{aligned} \tag{15.6}$$

Here γ^U is such that for any context $W \subseteq X$ then $\gamma^U_W(\{*\}) := \downarrow(U \cap W)$. The first thing to show is that i is a bijection.

1. 1:1. Given $i(U) = i(V)$ then for each context $W \subseteq X$ we have

$$\gamma^U_W = \gamma^V_W = \downarrow(U \cap W =) \downarrow(V \cap W). \tag{15.7}$$

In particular if we choose the context U then

$$\gamma^U_U = \gamma^V_U = \downarrow U = \downarrow(V \cap U). \tag{15.8}$$

Hence $V = U$.

2. Onto. We need to show that any section γ is of the form $i(U) = \gamma^U$. Let us consider a general section $\gamma : 1 \to \Omega$ then, for any open $W \subseteq X$, $\gamma_W(\{*\}) = S \in \Omega_W$. In particular for $W = X$ we have $\gamma_X(\{*\}) = S = \downarrow V$ for some $V \subseteq X$. Moreover, given $j : W' \subseteq X$ then

$$\Omega(j_{W'X})\gamma_X\big(\{*\}\big) = \Omega(j_{W'X})(\downarrow V) = \downarrow\big(V \cap W'\big) = \gamma^V_{W'}\big(\{*\}\big) \tag{15.9}$$

hence $\gamma = \gamma^V$.

$$\Omega(j_{W'W})\gamma_W\big(\{*\}\big) = \Omega(j_{W'W})(\downarrow V) = \downarrow V \cap W' = \gamma^V_{W'}\big(\{*\}\big) \tag{15.10}$$

We now need to show that i is a homomorphism. We will study each logical connective separately.

1. *And*: let us consider $i(U \cap W) = \gamma^{U \cap W}$ such that for all $W \subseteq X$ we have

$$\gamma^{U\cap W}_V = \downarrow(U \cap W \cap V) = \downarrow U \cap \downarrow W \cap \downarrow V = \downarrow(U \cap V) \cap \downarrow(W \cap V) = \gamma^U_V \cap \gamma^W_V \tag{15.11}$$

hence $\gamma^{U\cap W} = \gamma^U \wedge \gamma^W$.

2. *Or*: given $i(U \cup W) = \gamma^{U\cup W}$, then for all $W \subseteq X$ we have

$$\gamma^{U\cup W}_V = \downarrow(U \cup W) \cap \downarrow V = \downarrow(U \cup W \cap V) = \downarrow\big((U \cap V) \cup (W \cap V)\big) = \gamma^U_V \cup \gamma^W_V \tag{15.12}$$

hence $\gamma^{U\cup W} = \gamma^U \vee \gamma^W$.

3. *Implication*: let us consider $i(U \Rightarrow W) = \gamma^{U\Rightarrow W}$ such that for all $V \subseteq X$ we

$$\gamma^{U\Rightarrow W}_V = \downarrow(U \Rightarrow W) \cap V. \tag{15.13}$$

First of all we need to understand what is $\downarrow(U \Rightarrow W)$. We know from the general definition of implication that

$$Z \leq U \Rightarrow W \quad \text{iff} \quad Z \cap U \subseteq W \tag{15.14}$$

or, in other words, $U \Rightarrow W$ is the greatest element of the set $\{Z \subseteq X | Z \cap U \subseteq W\}$ which for simplicity of notation we call $\tilde{V}$.[1] Thus

$$\gamma_V^{U \Rightarrow W} = \downarrow(\tilde{V} \cap V). \tag{15.15}$$

On the other hand $\gamma_V^U \Rightarrow \gamma_V^W = \downarrow(U \cap V) \Rightarrow \downarrow(W \cap V)$ is equal to $\downarrow V'$ where V' is the greatest element of the set $\{V'' \subseteq X | \downarrow V'' \cap \downarrow(U \cap V) \subseteq \downarrow(W \cap V)\}$.

We need to show that $\downarrow(\tilde{V} \cap V) = \downarrow V'$ for all $V \in \mathcal{V}(\mathcal{H})$. In particular we have that $\downarrow V' \cap \downarrow(U \cap V) \subseteq \downarrow(W \cap V)$ iff $V' \cap U \cap V \subseteq W \cap V$. In this case $V' \cap V \cap U \subseteq W$ and $V' \cap V \in \{Z \subseteq X | Z \cap U \subseteq W\}$. On the other hand for all Z with $Z \cap U \subseteq W$, then $Z \cap U \cap V \subseteq W \cap V$, hence $Z \cap V \in \{V'' \subseteq X | \downarrow V'' \cap \downarrow(U \cap V) \subseteq \downarrow(W \cap V)\}$. It follows that

$$\{Z \cap V \subseteq X | Z \cap V \cap U \subseteq W\} = \big\{V'' \subseteq X | \downarrow V'' \cap \downarrow(U \cap V) \subseteq \downarrow(W \cap V)\big\}. \tag{15.16}$$

Which implies that

$$\gamma_V^{U \Rightarrow W} = \gamma_V^U \Rightarrow \gamma_V^W. \tag{15.17}$$

□

In order to obtain (15.1), a possible topos which could be used is $\tau = Sh((0,1)_L)$. In such a topos the isomorphisms we are after is

$$\begin{aligned} \sigma : \mathcal{O}\big((0,1)_L\big) &\to \Gamma \Omega^{Sh((0,1)_L)} \\ (0,p) &\mapsto \sigma(0,p) := \big(l(p) : \underline{1} \to \Omega^{Sh((0,1)_L)}\big) \end{aligned} \tag{15.18}$$

such that for each stage/context $(0,r) \in \mathcal{O}((0,1)_L)$ we have the truth value

$$l(p)_{(0,r)} = \begin{cases} \{(0,r') \in \mathcal{O}((0,1)_L) | r' \leq r\} = \Omega_{(0,r)}^{Sh((0,1)_L)} & \text{if } p \geq r \\ \{(0,r') \in \mathcal{O}((0,1)_L) | r' \leq p\} & \text{if } 0 < p < r \\ \emptyset & \text{if } p = 0. \end{cases} \tag{15.19}$$

If we use the isomorphisms $\{(0,r') \in \mathcal{O}(0,1)_L | r' \leq r\} \simeq [0,r]$ then we can write the above in the following way

$$l(p)_{(0,r)} = \begin{cases} [0,r] = \Omega_{(0,r)}^{Sh((0,1)_L)} & \text{if } p \geq r \\ [0,p] & \text{if } 0 < p < r \\ \emptyset & \text{if } p = 0. \end{cases} \tag{15.20}$$

As it can be seen from the definition, l is nothing but the combination of β (15.2) and σ (15.18). Thus $l : [0,1] \to \Gamma \Omega^{Sh((0,1)_L)}$ is a bijection between probabilities

[1] Note that $\tilde{V} = \bigcup_i V_i$ such that $V_i \cap U \subseteq W$.

and truth values:

$$\begin{aligned} [0,1] &\xrightarrow{\beta} \mathcal{O}\big((0,1)_L\big) \xrightarrow{\sigma} \Gamma\Omega^{Sh((0,1)_L)} \\ p &\mapsto (0,p) \mapsto l(p). \end{aligned} \tag{15.21}$$

Theorem 15.1 *The map l is an order preserving isomorphisms.*

To prove the above theorem we first of all need to define an ordering on $\Gamma\Omega^{Sh((0,1)_L)}$. This is defined as follows:

Definition 15.2 Given any two global elements $\gamma, \beta \in \Gamma\Omega^{Sh((0,1)_L)}$, then

$$\gamma \geq \beta \quad \text{if } \forall (0,r) \in (0,1)_L \quad \gamma_{(0,r)}\big(\{*\}\big) \geq \alpha_{(0,r)}\big(\{*\}\big). \tag{15.22}$$

Given this definition we can now prove the theorem.

Proof We want to show that given $p' \geq p$ then $l(p') \geq l(p)$. Clearly $p' \geq p$ implies that $\beta(p') = (0, p') \geq \beta(p) = (0, p)$. We then apply the map σ obtaining $l(p')$ and $l(p)$. From the definition of the l map we know that for each context $(0, r)$ we obtain the following cases:

1. If $p' \geq p \geq r$ then $l(p')_{(0,r)} = l(p)(0, r) = [0, r]$ implying that $l(p') = l(p)$.
2. If $p' \geq r$ but $0 < p < r$ then $l(p')_{(0,r)} = [0, r] > l(p)_{(0,r)} = [0, p]$ implying that $l(p') > l(p)$.
3. If $0 < p \geq p' < r$ then $l(p')_{(0,r)} = [0, p'] \geq l(p)_{(0,r)} = [0, p]$ implying that $l(p') \geq l(p)$.
4. If $0 < p' < r$ while $p = 0$ then $l(p')_{(0,r)} = [0, p'] > l(p)_{(0,r)} = 0$ implying that $l(p') > l(p)$.
5. If $p' = p = 0$ then $l(p')_{(0,r)} = 0 = l(p)_{(0,r)} = 0$ implying that $l(p') = l(p)$.

Therefore $l(p') \geq l(p)$ when $p' \geq p$. □

This implies that probabilities are faithfully represented as truth values in the topos $Sh((0, 1)_L)$.

15.2 Example for Classical Probability Theory

We will now apply the topos theoretic description of probabilities defined in the previous section to a classical system. As a first step we will define truth values of propositions regarding a classical system and, then, show how such truth values are related to classical probabilities.

Let us consider a proposition $(A \in \Delta)$ meaning that the value of the quantity A lies in Δ. We want to define the truth value of such a proposition with respect to a given state $s \in X$, where X is the state space. Recall that in classical theory the truth value of the above proposition in the state s is given by

$$v(A \in \Delta; s) := \begin{cases} 1 & \text{iff } s \in A^{-1}(\Delta) \\ 0 & \text{otherwise} \end{cases}$$

where $A^{-1}(\Delta) \subseteq X$ is the subset of the state space for which the proposition $(A \in \Delta)$ is true.

Another way of defining truth values is through the truth object $\mathbb{T}^s$ which is state dependent. The definition of the truth object was defined in Chap. 11, but we will report it here for clarity reasons:

for each state s, we define the set

$$\mathbb{T}^s := \{S \subseteq X | s \in S\}.$$

Since $(s \in A^{-1}(\Delta))$ iff $A^{-1}(\Delta) \in \mathbb{T}^s$ we can now write the truth value above in the following equivalent way:

$$v(A \in \Delta; s) := \begin{cases} 1 & \text{iff } A^{-1}(\Delta) \in \mathbb{T}^s \\ 0 & \text{otherwise.} \end{cases}$$

It follows that the truth value $v(A \in \Delta; s)$ is equivalent to the truth value of the mathematical statement $A^{-1}(\Delta) \in \mathbb{T}^s$. We denote the truth value of such a statement as $[[A^{-1}(\Delta) \in \mathbb{T}^s]]$.

We would now like to relate the above defined truth values to probability measures. To this end we associate to the space X the probability measure

$$\mu : Sub(X) \to [0, 1] \tag{15.23}$$

where $Sub(X)$ denotes the measurable subsets of X.

It is now possible to define a measure dependent truth object as follows:

$$\mathbb{T}^{\mu}_r := \{S \subseteq X | \mu(S) \geq r\} \tag{15.24}$$

for all $r \in (0, 1]$.[2] What the above truth object defines are all those propositions which are true with probability equal or greater than r.

So far, the objects we have defined are simply sets. However, we would like to find their analogues in the topos $Sh((0, 1)_L)$, which was used to define probabilities, so that truth values and probabilities have a common ground in which to be compared. Thus what we are looking for is a way of expressing both truth object $\mathbb{T}^{\mu}$ and the truth values $v(A \in \Delta; s)$ as objects in $Sh((0, 1)_L)$.

To this end we perform the following "translations".

[2]Note that we will not include the value $r = 0$. The reason, as it will be explained later on, is to avoid obtaining situations in which all propositions are totally true with probability zero.

First of all we map the state space X to the constant sheaf

$$X \to \underline{X}. \tag{15.25}$$

What this means is that for all $(0,r) \in \mathcal{O}((0,1)_L)$

$$\underline{X}(0,r) = X \tag{15.26}$$

and the maps are simply the identity maps.

For any measurable set $S \in Sub(X)$ we define the constant sheaf $S \to \underline{S}$; $\underline{S}(0,r) = S$, thus obtaining the map

$$\begin{aligned} \Delta : Sub(X) &\to Sub_{Sh(((0,1)_L))}(\underline{X}) \\ S &\mapsto \underline{S}. \end{aligned} \tag{15.27}$$

The analogue of the truth object in $Sh(((0,1)_L))$ is then, $\underline{\mathbb{T}}^{\mu}_{(0,r)} := \{\underline{S} \subseteq \underline{X} | \mu(\underline{S}) \geq r\}$ for all $(0,r) \in \mathcal{O}((0,1)_L)$.

Now that we have defined the analogue of the relevant objects in the topos $Sh(((0,1)_L))$ we can define the truth value of a proposition $(A \in \Delta := S) \subseteq X$ as a global section of $\underline{\Omega}^{Sh(((0,1)_L))}$. In particular, for all contexts $(0,r) \in \mathcal{O}((0,1)_L)$ we obtain:

$$\begin{aligned} [[\underline{S} \in \underline{\mathbb{T}}^{\mu}]](0,r) &:= \{(0,r') \leq (0,r) | \underline{S}_{(0,r')} \in \underline{\mathbb{T}}^{\mu}_{(0,r')}\} \\ &= \{(0,r') \leq (0,r) | \mu(S) \geq r'\} \\ &= [0, \mu(S)] \cap (0,r] \\ &= [0, \min\{\mu(S), r\}] \end{aligned} \tag{15.28}$$

which is an element of $\underline{\Omega}^{Sh((0,1)_L)}(0,r)$. Thus, globally, we get

$$[[\underline{S} \in \underline{\mathbb{T}}^{\mu}]] \in \Gamma(\underline{\Omega}^{Sh((0,1)_L)}).$$

It is easy to see that for any set $S \subseteq X$ the value $\mu(S)$ can be recovered by the valuation $[[\underline{S} \in \underline{\mathbb{T}}^{\mu}]] \in \Gamma(\underline{\Omega}^{Sh((0,1)_L)})$. This means that probabilities can be replaced by truth values without any information being lost.

In particular, in the topos $Sh((0,1)_L)$ the probability measure $\mu : Sub(X) \to [0,1]$ can be uniquely expressed through the map ϵ^{μ} (defined below), in the sense that there exists a bijective correspondence between μ and ϵ^{μ}, i.e. for each μ we get:

$$\begin{aligned} \epsilon^{\mu} : Sub_{Sh((0,1)_L)}(\underline{X}) &\to \Gamma(\underline{\Omega}^{Sh((0,1)_L)}) \\ \underline{S} &\mapsto [[\underline{S} \in \underline{\mathbb{T}}^{\mu}]]. \end{aligned} \tag{15.29}$$

What this means is that we have effectively replaced the probability measure $\mu : Sub(X) \to [0,1]$ with the collection of truth values $\Gamma(\underline{\Omega}^{Sh((0,1)_L)})$.

Since $\Gamma(\underline{\Omega}^{Sh((0,1)_L)})$ is a Heyting algebra, probabilities are now interpreted in the context of intuitionistic logic.

What has been done so far can be summarised by the following commutative diagram:

$$\begin{array}{ccc} Sub(X) & \xrightarrow{\quad\mu\quad} & [0,1] \\ \Big\downarrow{\scriptstyle \Delta} & & \Big\downarrow{\scriptstyle l} \\ Sub_{Sh((0,1)_L)}(\underline{X}) & \xrightarrow{\quad\epsilon^{\mu}\quad} & \Gamma\underline{\Omega}^{Sh((0,1)_L)} \end{array}$$

By chasing the diagram around we now have the following equalities:

$$\begin{aligned} \big[l \circ \mu(S)\big](r) &= l\big(\mu(S)\big)(r) = \big\{(0,r') \subseteq (0,r) | \mu(S) \geq r'\big\} \\ &= [[\underline{S} \in \underline{T}^{\mu}]](r) = \epsilon^{\mu}\big(\Delta(S)\big)(r). \end{aligned} \tag{15.30}$$

It was also shown in [24] that it is possible to define the logical analogue of the σ-additivity of the measure μ. We recall that the σ-additivity of a measure μ is defined as follows: for any countable family $(S_i)_{i\in N}$ of pairwise disjoint, measurable subsets of X (the space where the measure is defined), we have

$$\mu\left(\bigcup_i S_i\right) = \sum_i \mu(S_i). \tag{15.31}$$

Let us now consider a countable increasing family $(\tilde{S}_i)_{i\in N}$ of measurable subsets of X defined as follows:

$$\tilde{S}_i := \tilde{S}_{i-1} \cup \tilde{S}_i \quad \text{for } i > 0 \text{ while } \tilde{S}_0 := S_0. \tag{15.32}$$

Then (15.31) becomes

$$\mu\left(\bigvee_i \tilde{S}_i\right) = \mu\left(\bigcup_i \tilde{S}_i\right) = \sup_i \mu(\tilde{S}_i) = \bigvee_i \mu(\tilde{S}_i) \tag{15.33}$$

i.e., μ preserves countable joins (suprema).

The logical analogue of this is as follows:

given a countable increasing family of measurable subsets $(\tilde{S}_i)_{i\in N}$ of X we then have

$$(\epsilon^{\mu} \circ \Delta)\left(\bigvee_i \tilde{S}_i\right) = (l \circ \mu)\left(\bigvee_i (\tilde{S}_i)\right). \tag{15.34}$$

We know that μ preserves countable joins, we now want to show that l does as well. To this end we will explicitly compute, for each context $(0,r)$, how l acts on countable unions. Since $\mu : sub(X) \to [0,1]$ then $\mu(\bigvee_i(\tilde{S}_i)) \overset{(15.33)}{=} \bigvee_i(\mu(\tilde{S}_i)) = \bigvee_i p_i$ where $p_{i\in I}$ is some family of real numbers in the interval $[0,1]$. Thus the question is: what is $l(\bigvee_i p_i)$ as computed for each context $(0,r)$?

By applying the definition we obtain

$$l\Big(\bigvee_i p_i\Big)_{(0,r)} := \begin{cases} \underline{\Omega}^{Sh((0,1)_L)}_{(0,r)} & \text{if } \bigvee_i p_i \geq r \\ \{(0,r') \in \mathcal{O}((0,1)_L) | r' \leq \bigvee_i p_i\} & \text{if } 0 < \bigvee_i p_i < r \\ \emptyset & \text{if } \bigvee_i p_i = 0. \end{cases} \tag{15.35}$$

Or, equivalently,

$$l\Big(\bigvee_i p_i\Big)_{(0,r)} := \begin{cases} [0,r] = \underline{\Omega}^{Sh((0,1)_L)}_{(0,r)} & \text{if } \bigvee_i p_i \geq r \\ [0, \bigvee_i p_i] & \text{if } 0 < \bigvee_i p_i < r \\ \emptyset & \text{if } \bigvee_i p_i = 0. \end{cases} \tag{15.36}$$

On the other hand if we computed $\bigvee_i l(p_i)_{(0,r)}$ for all contexts $(0,r) \in \mathcal{O}((0,1)_L)$ we would obtain

$$\bigvee_i l(p_i)_{(0,r)} = \bigvee_i \begin{cases} \underline{\Omega}^{Sh((0,1)_L)}_{(0,r)} & \text{if } p_i \geq r \\ \{(0,r') \in \mathcal{O}((0,1)_L) | r' \leq p_i\} & \text{if } 0 < p_i < r \\ \emptyset & \text{if } p_i = 0. \end{cases} \tag{15.37}$$

Or equivalently

$$\bigvee_i l(p_i)_{(0,r)} = \bigvee_i \begin{cases} [0,r] = \underline{\Omega}^{Sh((0,1)_L)}_{(0,r)} & \text{if } p_i \geq r \\ [0, p_i] & \text{if } 0 < p_i < r \\ \emptyset & \text{if } p_i = 0. \end{cases} \tag{15.38}$$

Thus it follows that $\bigvee_i l(p_i)_{(0,r)} = l(\bigvee_i p_i)_{(0,r)}$ for all contexts $(0,r) \in \mathcal{O}((0,1)_L)$ hence

$$(l \circ \mu)\Big(\bigvee_i (\tilde{S}_i)\Big) = \bigvee_i (l \circ \mu)(\tilde{S}_i). \tag{15.39}$$

It follows that the logical description of the σ-additivity is

$$(\epsilon^{\mu} \circ \Delta)\Big(\bigvee_i \tilde{S}_i\Big) = \bigvee_i (l \circ \mu)(\tilde{S}_i). \tag{15.40}$$

So far we have only described the classical aspects of the topos interpretation of probability. However, it is possible to extend such ideas to the quantum case.

15.3 Quantum Probabilities

We now would like to show how the interpretation of probabilities as truth values in an appropriate topos can be applied to the quantum case. So far, in the literature it was shown how, within the topos $Sh(\mathcal{V}(\mathcal{H}))$[3] (the topos formed by the collection of sheaves on the poset $\mathcal{V}(\mathcal{H})$ of all abelian von Neumann algebras of a given Hilbert space $\mathcal{H}$) the truth value of a proposition $A \in \Delta$, given a state ψ, is defined as[4]

$$\begin{aligned} v(A \in \Delta, \psi)(V) &= \big[\big[\delta\big(\hat{E}[A \in \Delta]\big) \in {}^{org}\underline{\mathbb{T}}^{|\psi\rangle}\big]\big]_V \\ &:= \big\{V' \subseteq V \big| \langle\psi|\delta\big(\hat{E}[A \in \Delta]\big)_{V'}|\psi\rangle = 1\big\} \in \underline{\Omega}_V^{Sh(\mathcal{V}(\mathcal{H}))} \end{aligned} \tag{15.41}$$

where, for the quantum case, the truth object is the sheaf ${}^{org}\underline{\mathbb{T}}^{|\psi\rangle}$ which is defined, at each context $V \in \mathcal{V}(\mathcal{H})$, as the set

$${}^{org}\underline{\mathbb{T}}_V^{|\psi\rangle} := \big\{\hat{\alpha} \in P(V) | \langle\psi|\hat{\alpha}|\psi\rangle = 1\big\}. \tag{15.42}$$

This definition of truth value works perfectly well when we consider a pure state $|\psi\rangle$. However, if we consider mixed states with associate density matrix, say $\rho = \sum_{i=1}^{N} r_i|\psi\rangle\langle\psi|$, then the obvious analogue of (15.41), namely

$$\begin{aligned} v(A \in \Delta, \rho)(V) &= \big[\big[\delta\big(\hat{E}[A \in \Delta]\big) \in {}^{org}\underline{\mathbb{T}}^{\rho}\big]\big]_V \\ &:= \big\{V' \subseteq V \big| \mathrm{tr}\big(\rho\hat{E}[A \in \Delta]\big)_{V'} = 1\big\} \in \underline{\Omega}_V^{Sh(\mathcal{V}(\mathcal{H}))} \end{aligned} \tag{15.43}$$

where truth object is

$${}^{org}\underline{\mathbb{T}}_V^{\rho} := \big\{\hat{\alpha} \in P(V) | \mathrm{tr}(\rho\hat{\alpha}) = 1\big\} \tag{15.44}$$

doesn't work, since it does not separate the states ρ.

To see why this is the case consider our usual example for $\mathcal{H} = \mathbb{C}^4$ and consider two density matrices

$$\rho_1 = \begin{pmatrix} 3/4 & 0 & 0 & 0 \\ 0 & 1/4 & 0 & 0 \\ 0 & 0 & 0 & 0 \\ 0 & 0 & 0 & 0 \end{pmatrix}, \qquad \rho = \begin{pmatrix} 3/5 & 0 & 0 & 0 \\ 0 & 2/5 & 0 & 0 \\ 0 & 0 & 0 & 0 \\ 0 & 0 & 0 & 0 \end{pmatrix}.$$

Then the condition for $\mathrm{tr}(\rho\hat{P}) = 1$ is that $\hat{P} \geq \hat{P}_1 + \hat{P}_2$ where $\hat{P}_1 = \mathrm{diag}(1, 0, 0, 0)$ and $\hat{P}_2 = \mathrm{diag}(0, 1, 0, 0)$. Similarly the condition for $\mathrm{tr}(\rho_1\hat{P}) = 1$ is that $\hat{P} \geq \hat{P}_1 + \hat{P}_2$. It follows that ρ and ρ_1 have the same support, namely $\hat{P}_1 + \hat{P}_2$. However,

[3] Recall that given the Alexandrov topology on $\mathcal{V}(\mathcal{H})$ then $Sh(\mathcal{V}(\mathcal{H})) \simeq Sets^{\mathcal{V}(\mathcal{H})}$. So, in the following we will alternate freely between sheaves and presheaves.

[4] Note that here we have added a suffix *org* to indicate original, since we will now change the formulation of truth objects.

from the definition of truth values, the element $v(A \in \Delta, \rho)(V)$ depends only on the support of ρ, and ρ_1, thus it is not possible to distinguish the two.

On the other hand, truth values that do separate the density matrices ρ are the one parameter family of truth values

$$v(A \in \Delta, \rho)^r(V) := \{V' \subseteq V | \mathrm{tr}(\rho \hat{E}[A \in \Delta])_{V'} \geq r\} \in \underline{\Omega}_V^{Sh(\mathcal{V}(\mathcal{H}))} \tag{15.45}$$

for $r \in (0, 1]$. As we can see from the above formula, such truth values introduce truth probabilities different from one. In particular, what (15.45) represents is the truth value for the proposition $A \in \Delta$ to be true with probability at least r.

However, the problem with such one parameter family is that it represents a collection of objects, one for each r. So we need somehow to group these objects together and show that such a family can be considered as a single object. This can be done by enlarging the topos $Sh(\mathcal{V}(\mathcal{H}))$.

Summarising, the main steps needed, in order to correctly express probabilities as truth values of quantum propositions in an appropriate topos are:

(1) Define the truth values which separate the density matrices so as to be an object in the topos. This, as it was hinted to above, can be done by enlarging the topos we are working with.
(2) Define a correct probability measure on the state space $\underline{\Sigma}$.
(3) Find a way to relate (1) and (2).

15.4 Measure on the Topos State Space

We are now interested in constructing a measure μ on the state space $\underline{\Sigma}$. As such, it should somehow define a size or weight for each sub-object of $\underline{\Sigma}$. However we will not consider all sub-objects of $\underline{\Sigma}$ but only a collection of them which we define as "measurable". These collection of "measurable" sub-objects of $\underline{\Sigma}$ will be the collection of all clopen sub-objects of $\underline{\Sigma}$ [24]. The reason being that we are again taking ideas from classical physics. There we know that a proposition is defined as a measurable subset of the state space. Hence in the topos formulation of quantum theory we ascribe the status of "measurable" to all clopen sub-objects of $\underline{\Sigma}$, i.e., all propositions. It could be the case that one can consider a larger collection of sub-objects, but surely $Sub_{cl}(\underline{\Sigma})$ is the minimal such collection. In any case we will define a probability measure on $Sub_{cl}(\underline{\Sigma})$, such that $\underline{\Sigma}$ will have measure 1 and $\underline{0}$ will have measure zero.

As we will see, the measure defined on $\underline{\Sigma}$ will be such that there exists a bijective correspondence between such a measure and states of the quantum system, i.e. $\mu \mapsto \rho_\mu$ is a bijection[5] (for ρ a density matrix).

---

[5] It should be noted at this point that the correspondence between measures and state is present also in the context of classical physics. In fact in that case, as we have seen in previous chapters, a pure state (i.e. a point) is identified with the Dirac measure, while a general state is simply a probability measure on the state space. Such a probability measure assigns a number in the interval $[0, 1]$ called the weight and that in a sense tells you the 'size' of the measurable set.

A suitable measure on $\underline{\Sigma}$ is defined as follows:
for each density matrix ρ we have

$$\begin{aligned} \mu_\rho : Sub_{cl}(\underline{\Sigma}) &\to \Gamma\underline{[0,1]^{\succeq}} \\ \underline{S} = (\underline{S}_V)_{V\in\mathcal{V}(\mathcal{H})} &\mapsto \mu_\rho(\underline{S}) := \big(\rho(\hat{P}_{\underline{S}_V})\big)_{V\in\mathcal{V}(\mathcal{H})} \end{aligned} \tag{15.46}$$

where $\underline{[0,1]^{\succeq}}$ is the sheaf of order-reversing functions from $\mathcal{V}(\mathcal{H})$ to $[0,1]$. Since we are considering von-Neumann algebras of bounded operators, it follows that in this case $\rho(\hat{P}_{\underline{S}_V})_{V\in\mathcal{V}(\mathcal{H})} = (\mathrm{tr}(\rho\hat{P}_{\underline{S}_V}))_{V\in\mathcal{V}(\mathcal{H})}$.

Recall that $\hat{P}_{S_V} = \mathfrak{S}^{-1}(\underline{S}_V)$ where

$$\mathfrak{S} : P(V) \to Sub_{cl}(\underline{\Sigma})_V. \tag{15.47}$$

The detailed definition of the sheaf $\underline{[0,1]^{\succeq}}$ is as follows:

Definition 15.3 The presheaf $\underline{[0,1]^{\succeq}} : \mathcal{V}(\mathcal{H}) \to Sets$ is defined on:

1. Objects: for each context V we obtain $\underline{[0,1]^{\succeq}_V} := \{f : \downarrow V \to [0,1] | f$ is order reversing$\}$ which is a set of order-reversing functions.
2. Morphisms: given $i : V' \subseteq V$ then the corresponding presheaf map is

$$\begin{aligned} \underline{[0,1]^{\succeq}}(i_{V'V}) : \underline{[0,1]^{\succeq}_V} &\to \underline{[0,1]^{\succeq}_{V'}} \\ f &\mapsto f_{|\downarrow V'}. \end{aligned} \tag{15.48}$$

Thus the measure μ_ρ defined in (15.46) takes a clopen sub-object of $\underline{\Sigma}$ and defines a global element of $\underline{[0,1]^{\succeq}}$, i.e. $\mu_\rho(\underline{S}) \in \Gamma\underline{[0,1]^{\succeq}}$. Thus, for each $V \in \mathcal{V}(\mathcal{H})$

$$\begin{aligned} \mu_\rho(\underline{S})(V) : \underline{1}_V &\to \underline{[0,1]^{\succeq}} \\ \{*\} &\mapsto \mu_\rho(\underline{S})(V)\big(\{*\}\big) := \mu_\rho(\underline{S})(V) \end{aligned} \tag{15.49}$$

where $\mu_\rho(\underline{S})(V) : \downarrow V \to [0,1]$ is an order reversing function. By applying $\mu_\rho(\underline{S})$ to all contexts we end up with an order reversing function

$$\mu_\rho(\underline{S}) : \mathcal{V}(\mathcal{H}) \to [0,1] \tag{15.50}$$

such that for each $V \in \mathcal{V}(\mathcal{H})$, $\mu_\rho(\underline{S})(V)$ defines the expectation value, with respect to ρ, of the projection operators $\hat{P}_{\underline{S}_V}$ to which the sub-object $\underline{S}$ corresponds to in V. Therefore, given two contexts $V' \subseteq V$, since $\hat{P}_{S_{V'}} \geq \hat{P}_{S_V}$, then $\mathrm{tr}(\rho\hat{P}_{S_{V'}}) \geq \mathrm{tr}(\rho\hat{P}_{S_V})$.

It is worth mentioning some properties of μ_ρ:

1. For each $V \in \mathcal{V}(\mathcal{H})$ then $\mu_\rho(\underline{0})(V) = \rho(\hat{0}) = 0$, therefore globally

$$\mu_\rho(\underline{0}) = (0)_{V\in\mathcal{V}(\mathcal{H})}. \tag{15.51}$$

2. For each $V \in \mathcal{V}(\mathcal{H})$ then $\mu_\rho \underline{\Sigma}(V) = \rho(\hat{\mathbb{1}}) = 1$, therefore globally

$$\mu_\rho(\underline{\Sigma}) = (1)_{V \in \mathcal{V}(\mathcal{H})}. \tag{15.52}$$

3. Given two disjoint clopen sub-objects $\underline{S}$ and $\underline{T}$ of $\underline{\Sigma}$ we then have for each context $V \in \mathcal{V}(\mathcal{H})$ that

$$\begin{aligned}\mu_\rho(\underline{S} \vee \underline{T})(V) &= \mu_\rho\big((\underline{S} \vee \underline{T})_V\big)\\ &= \mu_\rho(\underline{S}_V \cup \underline{T}_V)\\ &= \rho(\hat{P}_{\underline{S}_V} \vee \hat{P}_{\underline{T}_V})\\ &= \rho(\hat{P}_{\underline{S}_V} + \hat{P}_{\underline{T}_V})\\ &= \rho(\hat{P}_{\underline{S}_V}) + \rho\hat{P}_{\underline{T}_V})\\ &= \mu_\rho(\underline{S})(V) + \mu_\rho(\underline{T})(V).\end{aligned} \tag{15.53}$$

It follows that globally

$$\mu_\rho(\underline{S} \vee \underline{T}) = \mu_\rho(\underline{S})(V) + \mu_\rho(\underline{T}). \tag{15.54}$$

This is the property of finite additivity.

4. A generalisation of the above property, i.e. a property of which (15.53) is a special case is the following: given two arbitrary clopen sub-objects $\underline{S}$ and $\underline{T}$ of $\underline{\Sigma}$, for all $V \in \mathcal{V}(\mathcal{H})$ we obtain

$$\begin{aligned}\mu_\rho(\underline{S} \vee \underline{T})(V) + \mu_\rho(\underline{S} \wedge \underline{T})(V) &= t\rho(\hat{P}_{\underline{S}_V} \vee \hat{P}_{\underline{T}_V}) + \rho(\hat{P}_{\underline{S}_V} \wedge \hat{P}_{\underline{T}_V})\\ &= \rho(\hat{P}_{\underline{S}_V} \vee \hat{P}_{\underline{T}_V} + \hat{P}_{\underline{S}_V} \wedge \hat{P}_{\underline{T}_V})\\ &= \rho(\hat{P}_{\underline{S}_V} + \hat{P}_{\underline{T}_V})\\ &= \rho(\hat{P}_{\underline{S}_V}) + \rho(\hat{P}_{\underline{T}_V})\\ &= \mu_\rho(\underline{S})(V) + \mu_\rho(\underline{T})(V)\end{aligned} \tag{15.55}$$

which gives globally

$$\mu_\rho(\underline{S} \vee \underline{T}) + \mu_\rho(\underline{S} \wedge \underline{T}) = \mu_\rho(\underline{S}) + \mu_\rho(\underline{T}). \tag{15.56}$$

5. Because the collection of clopen sub-objects of $\underline{\Sigma}$ forms a Heyting algebra and not a Boolean algebra it follows that, since

$$\underline{S} \vee \neg\underline{S} < \underline{\Sigma} \tag{15.57}$$

then

$$\mu_\rho(\underline{S} \vee \neg\underline{S}) < (1)_{V \in \mathcal{V}(\mathcal{H})}. \tag{15.58}$$

6. The closest one can get to σ-additivity is *local σ-additivity*: given a countable infinite family $(\underline{S}_i)_{i\in I}$ for clopen sub-objects such that, for each context $V \in \mathcal{V}(\mathcal{H})$ the clopen subsets $(\underline{S}_i)_V \subseteq \underline{\Sigma}_V$ (for all $i \in I$) are pairwise disjoint, then we have

$$\left(\mu_\rho\left(\bigvee_{i\in I}\underline{S}_i\right)\right)(V) = \rho\left(\bigvee_{i\in I}\hat{P}_{\underline{S}_i}\right) = \rho\left(\sum_{i\in I}\hat{P}_{\underline{S}_i}\right) = \sum_{i\in I}\mu_\rho(\underline{S}_i)(V). \quad (15.59)$$

When such a condition is satisfied the associated state ρ is a normal state.

15.5 Deriving a State from a Measure

So far we have defined a particular measure given a quantum state ρ. We are now interested in doing the opposite, since in the end we want to show that there is a bijection between the two. First of all we would like to give an abstract characterisation of a measure with no reference to a state and, then, show how such a measure can uniquely determine a state ρ.

The abstract characterisation of a measure is as follows:

Definition 15.4 A measure μ on the state space $\underline{\Sigma}$ is a map

$$\begin{aligned} \mu : Sub_{cl}(\underline{\Sigma}) &\to \Gamma\underline{[0,1]^{\geq}} \\ \underline{S} = (\underline{S}_V)_{V\in\mathcal{V}(\mathcal{H})} &\mapsto \left(\mu(\underline{S}_V)\right)_{V\in\mathcal{V}(\mathcal{H})} \end{aligned} \quad (15.60)$$

such that, the following conditions holds:

1. $\mu(\underline{\Sigma}) = 1_{\mathcal{V}(\mathcal{H})}$.
2. For all $\underline{S}$ and $\underline{T}$ in $Sub_{cl}(\underline{\Sigma})$ then $\mu(\underline{S} \vee \underline{T}) + \mu(\underline{S} \wedge \underline{T}) = \mu(\underline{S}) + \mu(\underline{T})$.

Given such an abstract definition of measure all properties defined in the previous section follow. We now want to show that each such measure μ, uniquely determines a state ρ_μ. Since above we have shown that each state ρ has determined a measure, then combining the two results we end up with a bijective correspondence between the space of states ρ and the space of measures μ on $\underline{\Sigma}$.

Theorem 15.2 *Given a measure μ, as defined above, then there exist a unique state ρ "associated" to that measure.*

Proof Let us define

$$\begin{aligned} m : P(\mathcal{H}) &\to [0,1] \\ \hat{P} &\mapsto m(\hat{P}) \end{aligned} \quad (15.61)$$

such that (i) $m(\hat{\mathbb{1}})=1$; (ii) if $\hat{P}\hat{Q}=0$ (are orthogonal) then $m(\hat{P}\vee\hat{Q})=m(\hat{P}+\hat{Q})=m(\hat{P})+m(\hat{Q})$. Such a map is called, in the literature, a *finitely additive probability measure on the projections of* V. We now want to define a unique such finite additive measure given the probability measure μ. To this end we define

$$m(\hat{P}):=\mu(\underline{S}_{\hat{P}})(V)=\mu(\underline{S}_{\hat{P}})_V \tag{15.62}$$

where $m(\hat{\mathbb{1}}):=1$. In the above formula $\underline{S}$ is some clopen sub-object $\underline{S}\subseteq\underline{\Sigma}$ such that for V we have $\mathfrak{S}^{-1}(\underline{S}_V)=\hat{P}$.

In order for this definition to be well defined it has to be independent on which sub-object of $\underline{\Sigma}$ is chosen to represent the projection operator. In fact it could be the case that, for a contexts V and V', we have that $\underline{S}_V$ and $\underline{T}_{V'}$ both correspond to the same projection operator $\hat{P}=\mathfrak{S}^{-1}(\underline{S}_V)=\mathfrak{S}^{-1}(\underline{T}_{V'})$. If this is the case then we need to prove that $\mu(\underline{S}_{\hat{P}})_V=\mu(\underline{T}_{\hat{P}})_{V'}$. To show this we first take the case for which there exist two sub-objects of $\underline{\Sigma}$, such that they correspond to the same projection operators at the same context[6] V, i.e. $\underline{T}_V=\underline{S}_V$. We then obtain that

$$\begin{aligned}
\mu(\underline{S})(V)+\mu(\underline{T})(V) &\overset{(15.4)}{=} \mu(\underline{S}\vee\underline{T})(V)+\mu(\underline{S}\wedge\underline{T})(V)\\
&= \mu(\underline{S}\vee\underline{T})_V+\mu(\underline{S}\wedge\underline{T})_V\\
&= \mu(\underline{S}_V\cup\underline{T}_V)+\mu(\underline{S}_V\cap\underline{T}_V)\\
&= \mu(\underline{S}_V)+\mu(\underline{S}_V)\\
&= \mu(\underline{S})(V)+\mu(\underline{S})(V)
\end{aligned} \tag{15.63}$$

therefore

$$\mu(\underline{S})(V)=\mu(\underline{T})(V). \tag{15.64}$$

So, for the same context, we are able to prove the result. What about different contexts? Let us consider the case in which $\underline{S}_V=\underline{T}_{V'}$ and both correspond to the projection operator $\hat{P}$. This means that $\hat{P}\in V$ and $\hat{P}\in V'$ therefore $\hat{P}\in V\cap V'$ and $\delta^o(\hat{P})_V=\underline{S}_V=\underline{T}_V$. We should note that, although $\delta^o(\hat{P})_{V'\cap V}=\hat{P}$, it is not the case that $\mathfrak{S}^{-1}(\underline{S}_{V'\cap V})=\hat{P}$ or $\mathfrak{S}^{-1}(\underline{T}_{V'\cap V})=\hat{P}$. Given such situation we obtain

$$\begin{aligned}
\mu(\underline{S})(V) &\overset{(15.64)}{=} \mu\big(\delta^o(\hat{P})\big)(V)\\
&= \mu\big(\delta^o(\hat{P})\big)\big(V'\cap V\big)\\
&= \mu\big(\delta^o(\hat{P})\big)\big(V'\big)\\
&\overset{(15.64)}{=} \mu(\underline{T})\big(V'\big).
\end{aligned} \tag{15.65}$$

Hence, the definition in (15.62) is well defined. We next have to show that m is actually a finitely additive probability measure on the projections in $\mathcal{H}$. To this end

[6] In our usual sample of $\mathbb{C}^4$ this is the case for $\delta^o(\hat{P}_1)_{V_{\hat{P}_2}}$ and $\delta^o(\hat{P}_3)_{V_{\hat{P}_2}}$.

let us consider two orthogonal projection operators $\hat{P}$ and $\hat{Q}$, such that $\mathfrak{S}^{-1}(\underline{T}_V) = \hat{Q}$ and $\mathfrak{S}^{-1}(\underline{S}_V) = \hat{P}$. Then $(\underline{S} \vee \underline{T})_V$ corresponds to the projection operator $\hat{P} \vee \hat{Q}$. We then obtain

$$\begin{aligned} m(\hat{P} \vee \hat{Q}) &:= \mu(\underline{S} \vee \underline{T})(V) \\ &= \mu(\underline{S})(V) + \mu(\underline{T})(V) + \mu(\underline{S} \wedge \underline{T})(V) \\ &= \mu(\underline{S})(V) + \mu(\underline{T})(V) \\ &=: m(\hat{P}) + m(\hat{Q}). \end{aligned} \tag{15.66}$$

Finally it is easy to show that $m(\hat{0}) = 0$. These results together prove that, given a measure μ on $\underline{\Sigma}$, we can uniquely defined a finitely additive probability measure on the projections in $\mathcal{H}$. Through the generalised version [73] of Gleason's theorem[7] it is possible to show that such a probability measure corresponds to a state ρ_μ. Thus we have the following chain: $\mu \mapsto m \mapsto \rho_\mu$. □

We have so managed to show that there exists a one to one correspondence between states and measures as defined in Definition 15.4. We are now ready to tackle issue number (3): how to relate such a measure with a generalised version of the truth values in (15.45).

One thing to notice is that the measure in definition (15.46) is defined on all clopen sub-objects of the state space $\underline{\Sigma}$, not only those deriving from the process of daseinisation, while the truth objects in (15.42), (15.44) and thus the truth values in (15.41), (15.43), are only defined on those particular sub-objects which are derived from daseinisation. Thus another issues to solve is the following:

(4) Enlarge the truth objects in (15.42), (15.44) so as to include all clopen sub-objects of the state space $\underline{\Sigma}$.

We will tackle this issue in the following section.

---

[7] Gleason Theorem tells us that the only possible probability measures on Hilbert spaces of dimension at least 3 are measures of the form $\mu(P) = \mathrm{tr}(\rho P)$, where ρ is a positive semidefinite self-adjoint operator of unit trace. This theorem was extended to a von-Neumann algebra $\mathcal{N}$ in [73] where the author shows that, provided $\mathcal{N}$ contains no direct summand of type I_2, then every finitely additive probability measure on $\mathcal{P}(\mathcal{N})$ can be uniquely extended to a state on $\mathcal{N}$.

Given a Hilbert space $\mathcal{H}$, the general form of Gleason theorem is as follows:

Theorem 15.3 Assume that $\dim((\mathcal{H}) \geq 3)$ and let μ be a σ-additive probability measure on $P(\mathcal{B}(\mathcal{H}))$ then the following three statements hold:

1. μ is completely additive.
2. μ has support.
3. There exists a positive operator $x \in \mathcal{B}(\mathcal{H})$ of trace class, such that $\mathrm{tr}(x) = 1$ and $\mu(e) = \mathrm{tr}(xe)$ for $e \in P(\mathcal{B}(\mathcal{H}))$.

15.6 New Truth Object

We now would like to define a truth object which takes into account all clopen sub-objects of the state space $\underline{\Sigma}$, not only those coming from daseinisation. This is because the measure μ described in previous sections is defined in all clopen sub-objects of $\underline{\Sigma}$ which in this context correspond to measurable sub-objects.

Both the truth objects ${}^{org}\underline{\mathbb{T}}^{|\psi\rangle}$ and ${}^{org}\underline{\mathbb{T}}^{\rho}$ are defined in such a way such that the global sections $\Gamma({}^{org}\underline{\mathbb{T}}^{|\psi\rangle})$ give all the clopen sub-objects of $\underline{\Sigma}$ coming from daseinisation.

We now would like to enlarge $\mathbb{T}^{|\psi\rangle}$ such that the global sections of the enlarged truth object give us all the sub-objects of the state space, i.e. what we are looking for is an object $\underline{\mathbb{T}}$, such that

1. $\Gamma\underline{\mathbb{T}} = Sub_{cl}\underline{\Sigma}$.
2. ${}^{org}\underline{\mathbb{T}}^{|\psi\rangle} \in \Gamma\underline{\mathbb{T}}$.

We will define these new truth objects for both the pure state case and the density matrix case.

15.6.1 Pure State Truth Object

As mentioned above, we would like to be able to define truth values for all clopen sub-objects of $\underline{\Sigma}$, not just those coming from daseinisation. Thus the truth object $\mathbb{T}^{|\psi\rangle}$, now, has to be defined as a general sub-object of $P_{cl}(\underline{\Sigma})$, such that $\Gamma\mathbb{T}^{|\psi\rangle} = Sub_{cl}(\underline{\Sigma})$.

A possible choice for such a truth object, as referred to a state $|\psi\rangle$, is the sheaf $\underline{\mathbb{T}}^{|\psi\rangle}$, such that for a given context V we have[8]

$$\underline{\mathbb{T}}_V^{|\psi\rangle} := \{\underline{S} \in Sub_{cl}(\underline{\Sigma}_{|\downarrow V}) | \forall V' \subseteq V, |\psi\rangle\langle\psi| \leq \hat{P}_{S_{V'}}\}. \tag{15.67}$$

In words, this truth object represents the condition in terms of which any sub-object $\underline{S}$ of the state space $\underline{\Sigma}$ can be said to be true. In particular, for a sub-object $\underline{S} \subseteq \underline{\Sigma}$ to be totally true in a given state $|\psi\rangle$ it has to be such that for all $V \in \mathcal{V}(\mathcal{H})$ (i.e. locally) it corresponds to a projection operator $\hat{P}_{S_V} \in P(V)$ which corresponds to an available proposition in V. This proposition is then required to be true with respect to $|\psi\rangle$, i.e. $\langle\psi|\hat{P}_{S_V}|\psi\rangle = 1$ (equivalently $|\psi\rangle\langle\psi| \leq \hat{P}_{S_V}$).

The corresponding global elements are

$$\Gamma(\underline{\mathbb{T}}^{|\psi\rangle}) = \{\underline{S} \in Sub_{cl}(\underline{\Sigma}) | \forall V' \subseteq \mathcal{V}(\mathcal{H}), |\psi\rangle\langle\psi| \leq \hat{P}_{S_V}\}. \tag{15.68}$$

From the above definitions it is easy to see that indeed conditions 1. and 2. above are satisfied, namely

$$\Gamma(\underline{\mathbb{T}}^{|\psi\rangle}) = Sub_{cl}\underline{\Sigma}$$

---

[8]Note that $\underline{\Sigma}_{|\downarrow V}$ indicates the sheaf $\underline{\Sigma}$ defined on the lower set $\downarrow V$.

and

$$^{org}\underline{\mathbb{T}}^{|\psi\rangle} \in \Gamma\underline{\mathbb{T}}^{|\psi\rangle}.$$

So we have managed to solve problem (4) and define a state dependent truth object for all sub-objects of the state space, not only those derived from daseinisation. This implies that now the expression $[[\underline{S} \in \underline{\mathbb{T}}^{|\psi\rangle}]]$ makes sense and can be evaluated also for $\underline{S} \neq \underline{\delta P}$.

The truth value in (15.41) now becomes

$$v(A \in \Delta, |\psi\rangle) = [[\underline{\delta(\hat{E}[A \in \Delta])} \in \underline{\mathbb{T}}^{|\psi\rangle}]]. \tag{15.69}$$

15.6.2 Density Matrix Truth Object

The enlarged truth object, as referred to a density matrix ρ, is the sheaf $\underline{\mathbb{T}}^{\rho,r}$ such that, for each context $V \in \mathcal{V}(\mathcal{H})$ we have

$$\underline{\mathbb{T}}^{\rho,r}_V := \{\underline{S} \in Sub_{cl}(\underline{\Sigma}_{|\downarrow V})|\forall V' \subseteq V, \operatorname{tr}(\rho \hat{P}_{S_{V'}}) \geq r\}. \tag{15.70}$$

The corresponding global elements are

$$\Gamma(\underline{\mathbb{T}}^{\rho,r}) := \{\underline{S} \in Sub_{cl}(\underline{\Sigma})|\forall V \subseteq \mathcal{V}(\mathcal{H}), \operatorname{tr}(\rho \hat{P}_{S_V}) \geq r\}. \tag{15.71}$$

These global elements indicate which (general) propositions, represented by sub-objects of the state space, are true with probability at least r.

Moreover, if we have two distinct numbers $r_1 \leq r_2 \leq 1$, then

$$\underline{\mathbb{T}}^{\rho,r_2} \subseteq \underline{\mathbb{T}}^{\rho,r_1}.$$

Proof From the definition in (15.70), for all $V \in \mathcal{V}(\mathcal{H})$, $\underline{\mathbb{T}}^{\rho,r_2}_V$ is a family of clopen sub-objects $\underline{S} \subseteq \underline{\Sigma}_{\downarrow V}$ such that, for all $V' \in \downarrow V$, we have $\operatorname{tr}(\rho \hat{P}_{S_{V'}}) \geq r_2$. However $r_1 \leq r_2$ therefore $\operatorname{tr}(\rho \hat{P}_{S_{V'}}) \geq r_1$ for all $V' \in \downarrow V$. It follows that $\underline{\mathbb{T}}^{\rho,r_2} \subseteq \underline{\mathbb{T}}^{\rho,r_1}$. □

This means that the collection of sub-objects (propositions) of the state space, which are true with probability at least r_1, are more than the collection of sub-objects (propositions) of the state space which are true with a bigger probability (at least r_2).

Similarly, as above, conditions 1 and 2 are trivially satisfied by the newly defined object (15.70).

Therefore, also for a density matrix truth object we have managed to solve problem (4), i.e. we have managed to define a density matrix dependent truth object for all sub-objects of the state space, not only those derived from daseinisation.

Thus the expression $[[\underline{S} \in \underline{\mathbb{T}}^{\rho,r}]]$ makes sense and can be evaluated for all $\underline{S} \in Sub(\underline{\Sigma})$.

Equation (15.45) now becomes

$$v(A \in \Delta, \rho)^r(V) = \big[\big[\delta\big(\hat{E}[A \in \Delta]\big) \in \underline{\mathbb{T}}^{\rho,r}\big]\big]_V$$
$$:= \big\{V' \subseteq V \big| \mathrm{tr}\big(\rho \hat{E}[A \in \Delta]\big)_{V'} \geq r\big\} \in \underline{\Omega}_V^{Sh(\mathcal{V}(\mathcal{H}))}. \quad (15.72)$$

15.7 Generalised Truth Values

Let us try tackling issue number (1) at the end of Sect. 15.3, i.e. how to put together the one parameter family of truth values so as to be itself an object in a topos. To this end one needs to extend the topos from $Sh(\mathcal{V}(\mathcal{H}))$ to $Sh(\mathcal{V}(\mathcal{H}) \times (0,1)_L)$, so that now the stages/contexts are pairs $\langle V, r\rangle$. Such a category $\mathcal{V}(\mathcal{H}) \times (0,1)_L$ can be given the structure of a poset as follow:

$$\big\langle V', r'\big\rangle \leq \langle V, r\rangle \quad \text{iff} \quad V' \leq V, \quad \text{and} \quad r' \leq r. \quad (15.73)$$

Being a poset $\mathcal{V}(\mathcal{H}) \times (0,1)_L$ is equipped with the (lower) Alexander topology, where the basic opens are $(\downarrow V \times (0,r))$.

This enlargement makes sense intuitively, since we are trying to combine the topos definition of probabilities outlined in Sect. 15.1, which makes use of the topos $Sh((0,1)_L)$ and the concept of truth values which makes use of the topos $Sh(\mathcal{V}(\mathcal{H}))$. In particular, what we are trying to combine is the following:

- The one parameter family of truth values

$$v(A \in \Delta, \rho)^r(V) = \big\{V' \subseteq V \big| \mathrm{tr}\big(\rho\delta\big(\hat{E}[A \in \Delta]\big)_{V'}\big) \geq r\big\} \in \underline{\Omega}_V^{Sh(\mathcal{V}(\mathcal{H}))} \quad (15.74)$$

  which gives us sieves for each stage V in $\underline{\Omega}^{Sh(\mathcal{V}(\mathcal{H}))}$.
- The topos definition of probabilities

$$\big[\big[\underline{S} \in \underline{\mathbb{T}}^{\mu}\big]\big](0,r) := \big\{\big(0,r'\big) \leq (0,r) | \mu(S) \geq \big(0,r'\big)\big\} \in \underline{\Omega}^{Sh((0,1)_L)} \quad (15.75)$$

  which gives us, for each context $(0,r)$, a sieve in $\underline{\Omega}^{Sh((0,1)_L)}$.

To combine, meaningfully, the two objects above we first need to go back to the way in which topos probabilities were implemented for classical physics. In this context we ended up with the commuting diagram

$$\begin{array}{ccc}
Sub(X) & \xrightarrow{\quad \mu \quad} & [0,1] \\
\Big\downarrow{\scriptstyle \Delta} & & \Big\downarrow{\scriptstyle l} \\
Sub_{Sh((0,1)_L)}(\underline{X}) & \xrightarrow{\quad \epsilon^{\mu} \quad} & \Gamma\underline{\Omega}^{Sh(\mathcal{V}(\mathcal{H})\times(0,1)_L)}
\end{array}$$

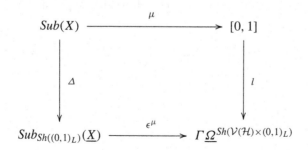

We are now looking for a quantum analogue of this. We already have the definition of a measure on the state space. The next step is to define the map of Heyting algebras

$$l : \Gamma\underline{[0,1]}^{\succeq} \to \Gamma\underline{\Omega}^{Sh(\mathcal{V}(\mathcal{H})\times(0,1)_L)}. \tag{15.76}$$

This cane be defined as:

$$\begin{aligned} l : \Gamma\underline{[0,1]}^{\succeq} &\to \Gamma\underline{\Omega}^{Sh(\mathcal{V}(\mathcal{H})\times(0,1)_L)} \\ \gamma &\mapsto l(\gamma) \end{aligned} \tag{15.77}$$

where $l(\gamma)(\langle V,r\rangle) := \{\langle V',r'\rangle \leq \langle V,r\rangle | \gamma(V') \geq r'\}$.

Since γ is a nowhere increasing function, $\{\langle V',r'\rangle \leq \langle V,r\rangle | \gamma(V') \geq r'\}$ is a sieve on $\langle V,r\rangle \in \mathcal{O}(\mathcal{V}(\mathcal{H}) \times (0,1)_L)$.

Moreover, to each density matrix ρ and to each projection operator $\hat{P}$ there corresponds a global element $\gamma_{\hat{P},\rho} \in \Gamma\underline{[0,1]}^{\succeq}$ defined by

$$\gamma_{\hat{P},\rho}(V) := \mathrm{tr}\big(\rho\delta^o(\hat{P})_V\big). \tag{15.78}$$

When applying the map l to such a section, for each context $\langle V,r\rangle$ we obtain:

$$\begin{aligned} l(\gamma_{\hat{P},\rho})(\langle V,r\rangle) &= \{\langle V',r'\rangle \leq \langle V,r\rangle | \gamma_{\hat{P},\rho}(V') \geq r'\} \\ &= \{\langle V',r'\rangle \leq \langle V,r\rangle | \mathrm{tr}\big(\rho\delta^o(\hat{P})_{V'}\big) \geq r'\}. \end{aligned} \tag{15.79}$$

If $\hat{P} = \delta([\hat{E}[A \in \Delta]])$ then we get

$$\begin{aligned} l(\gamma_{\delta(\hat{E}[A\in\Delta]),\rho})(\langle V,r\rangle) &= \{\langle V',r'\rangle \leq \langle V,r\rangle | \gamma_{\delta(\hat{E}[A\in\Delta]),\rho}(V') \geq r'\} \\ &= \{\langle V',r'\rangle \leq \langle V,r\rangle | \mathrm{tr}\big(\rho\delta\big(\hat{E}[A \in \Delta]\big)_{V'}\big) \geq r'\}. \end{aligned} \tag{15.80}$$

An important result is the following

Theorem 15.4 *The map l, as defined in* (15.77), *separates the elements in $\Gamma\underline{[0,1]}^{\succeq}$, i.e. it is injective* [24].

Proof Assume that $\gamma_1 \neq \gamma_2$. Then there will exist a context V_i such that $\gamma_1(V_i) \neq \gamma_2(V_i)$. Let us assume that $\gamma_1(V_i) > \gamma_2(V_i)$. If we then apply the l map we obtain, respectively,

$$l(\gamma_2)\big(\langle V_i,\gamma_2(V_i)\rangle\big) := \{\langle V',r'\rangle \leq \langle V_i,\gamma_2(V_i)\rangle | \gamma_2(V') \geq r'\} = \downarrow\!\langle V_i,\gamma_2(V_i)\rangle\} \tag{15.81}$$

and

$$l(\gamma_1)\big(\langle V_i,\gamma_2(V_i)\rangle\big) := \{\langle V',r'\rangle \leq \langle V_i,\gamma_2(V_i)\rangle | \gamma_1(V') \geq r'\} \subset \downarrow\!\langle V_i,\gamma_2(V_i)\rangle \tag{15.82}$$

therefore $l(\gamma_1) \neq l(\gamma_2)$. □

Since the map l represents the quantum analogue of the classical map $l : [0,1] \to \Gamma(\underline{\Omega}^{(0,1)})$ used to define probabilities we would like to check if the quantum l has the same properties, in particular if it preserves joins. To this end consider a family $(\gamma_i)_{i\in I}$ of global sections of $\underline{[0,1]}^{\geq}$. For each context $(\langle V,r\rangle)$ we then obtain:

$$\begin{aligned} l\Big(\bigvee_i \gamma_i\Big)(\langle V,r\rangle) &= \Big\{\langle V',r'\rangle \leq \langle V,r\rangle \Big| \Big(\bigvee_i \gamma_i\Big)(V') \geq r'\Big\} \\ &= \Big\{\langle V',r'\rangle \leq \langle V,r\rangle \,|\, \sup_i \gamma_i(V') \geq r'\Big\} \\ &= \bigcup_i \{\langle V',r'\rangle \leq \langle V,r\rangle | \gamma_i(V') \geq r'\} \\ &= \bigvee_i l(\gamma_i)(\langle V,r\rangle). \end{aligned} \tag{15.83}$$

It follows that

$$l\Big(\bigvee_i \gamma_i\Big) = \bigvee_i l(\gamma_i). \tag{15.84}$$

Now that we have both maps μ and l we can combine the two sieves defined in (15.74) and (15.75) in a unique sieve in $\underline{\Omega}^{Sh(\mathcal{V}(\mathcal{H})\times(0,1)_L)}$ by composing the maps μ^ρ and l as follows

$$l \circ \mu^\rho\big(\underline{\delta(\hat{P})}\big) \in \Gamma\big(\underline{\Omega}^{Sh(\mathcal{V}(\mathcal{H})\times(0,1)_L)}\big). \tag{15.85}$$

Therefore, for each context $\langle V,r\rangle$ we obtain

$$\begin{aligned} l \circ \mu^\rho\big(\underline{\delta(\hat{P})}\big)(\langle V,r\rangle) &\overset{(15.78)}{=} l(\gamma_{\delta(\hat{P}),\rho})(\langle V,r\rangle) \\ &= \{\langle V',r'\rangle \leq \langle V,r\rangle | \gamma_{\delta(\hat{P}),\rho}(V') \geq r'\} \\ &= \{\langle V',r'\rangle \leq \langle V,r\rangle | \mathrm{tr}(\rho\delta(\hat{P})_{V'}) \geq r'\}. \end{aligned} \tag{15.86}$$

In other words, for each context $\langle V,r\rangle$ we obtain the following truth value:

$$\begin{aligned} v(A\in\Delta,\rho)(\langle V,r\rangle) &:= \{\langle V',r'\rangle \leq \langle V,r\rangle | \mathrm{tr}(\rho\delta(\hat{E}[A\in\Delta])_{V'}) \geq r\} \\ &:= \{\langle V',r'\rangle \leq \langle V,r\rangle | \mu^\rho(\delta(\hat{E}[A\in\Delta])_{V'}) \geq r\}. \end{aligned} \tag{15.87}$$

Mathematically, this is a well defined element (sieve) of $\underline{\Omega}^{Sh(\mathcal{V}(\mathcal{H})\times(0,1)_L)}$. Moreover, it was shown that such a truth value separates the density matrices and, therefore, separates the measures [24].

However, in order to give physical meaning to the above expression, as representing the truth value of a quantum proposition regarding a physical system, we need to translate all the other objects which were defined in the topos $Sh(\mathcal{V}(\mathcal{H}))$ to objects in the topos $Sh(\mathcal{V}(\mathcal{H}) \times (0,1)_L)$. This will be done by defining the quantum analogues of the maps Δ and ϵ which we are still missing. In this way we can effectively obtain the quantum analogue of the commuting classical diagram above. In particular, the quantum map Δ will be identified through the geometric morphisms $p_1^* : Sh(\mathcal{H}(\mathcal{V})) \to Sh(\mathcal{V}(\mathcal{H}) \times (0,1)_L)$ induced by the projection map $p_1 : \mathcal{V}(\mathcal{H}) \times (0,1)_L \to \mathcal{V}(\mathcal{H})$. We then obtain the following:

- *State space*

$$\underline{\Sigma}^{Sh(\mathcal{V}(\mathcal{H})\times(0,1)_L)}_{\langle V,r\rangle} := \left(p_1^* \underline{\Sigma}^{Sh(\mathcal{V}(\mathcal{H}))}\right)_{\langle V,r\rangle} = \underline{\Sigma}^{Sh(\mathcal{V}(\mathcal{H}))}_{p_1(\langle V,r\rangle)} = \underline{\Sigma}^{Sh(\mathcal{V}(\mathcal{H}))}_{V}.$$

- *Propositions*

$$\underline{\delta(\hat{P})}^{Sh(\mathcal{V}(\mathcal{H})\times(0,1)_L)}_{\langle V,r\rangle} := \left(p_1^* \underline{\delta(\hat{P})}^{Sh(\mathcal{V}(\mathcal{H}))}\right)_{\langle V,r\rangle} = \underline{\delta(\hat{P})}^{Sh(\mathcal{V}(\mathcal{H}))}_{p_1(\langle V,r\rangle)} = \underline{\delta(\hat{P})}^{Sh(\mathcal{V}(\mathcal{H}))}_{V}.$$

- *Truth Object*

$$\underline{\mathbb{T}}^{\rho}_{Sh(\mathcal{V}(\mathcal{H})\times(0,1)_L)}\big(\langle V,r\rangle\big) := \underline{\mathbb{T}}^{\rho,r}V = \big\{\underline{S} \in Sub_{cl}(\underline{\Sigma}_{|\downarrow V}) \big| \forall V' \subseteq V,\ \mathrm{tr}(\rho \hat{P}_{\underline{S}_{V'}}) \geq r\big\}.$$

Equation (15.87) now becomes

$$\begin{aligned} v(A \in \Delta, \rho)\big(\langle V,r\rangle\big) &= [[p_1^* \underline{\delta\big(\hat{E}[A \in \Delta]\big)} \in \underline{\mathbb{T}}^{\rho}]]\big(\langle V,r\rangle\big) \\ &:= \big\{\langle V',r'\rangle \leq \langle V,r\rangle \big| \mathrm{tr}\big(\rho\delta\big(\hat{E}[A \in \Delta]\big)_{V'}\big) \geq r\big\} \\ &:= \big\{\langle V',r'\rangle \leq \langle V,r\rangle \big| \mu^{\rho}\big(\delta\big(\hat{E}[A \in \Delta]\big)_{V'}\big) \geq r\big\}. \end{aligned} \tag{15.88}$$

Having done this we can complete our commuting diagram by defining

$$\begin{aligned} \epsilon^{\rho} : Sub_{cl}\big(\underline{\Sigma}^{Sh(\mathcal{V}(\mathcal{H})\times(0,1)_L)}\big) &\to \Gamma \underline{\Omega}^{Sh(\mathcal{V}(\mathcal{H})\times(0,1)_L)} \\ \underline{S} &\mapsto [[\underline{S} \in \underline{\mathbb{T}}^{\rho}]]. \end{aligned} \tag{15.89}$$

Thus the quantum version of the above diagram is

$$\begin{array}{ccc} Sub_{cl}(\underline{\Sigma}^{Sh(\mathcal{V}(\mathcal{H}))}) & \xrightarrow{\ \mu^{\rho}\ } & \Gamma\underline{[0,1]}^{\geq} \\ \Big\downarrow{\scriptstyle p_1^*} & & \Big\downarrow{\scriptstyle l} \\ Sub_{cl}(\underline{\Sigma}^{Sh(\mathcal{V}(\mathcal{H})\times(0,1)_L)}) & \xrightarrow{\ \epsilon^{\rho}\ } & \Gamma\underline{\Omega}^{Sh(\mathcal{V}(\mathcal{H})\times(0,1)_L)} \end{array}$$

We then have that

$$\epsilon^{\rho}\big(p_1^* \underline{\delta(\hat{P})}\big) = [[p_1^* \underline{\delta(\hat{P})} \in \underline{\mathbb{T}}^{\rho}]] = l \circ \mu^{\rho}\big(\underline{\delta(\hat{P})}\big). \tag{15.90}$$

This equation gives the precise statement of the relation between truth values and probability measures in a topos. Moreover, we know that, given a normal state ρ then the measure μ^ρ has the local σ-additivity property such that, for a countable increasing family of clopen sub-objects $(\underline{S}_i)_{i\in N}$ then

$$\mu^\rho\Big(\bigvee_i \underline{S}_i\Big)=\bigvee_i \mu^\rho(\underline{S}_i). \tag{15.91}$$

If we then combine this result with the fact that l preserves joins we then get that

$$(l\circ\mu^\rho)\Big(\bigvee_i \underline{S}_i\Big)=\bigvee_i (l\circ\mu^\rho)(\underline{S}_i). \tag{15.92}$$

However from the commutativity of the above diagram it follows that

$$(\epsilon^\rho\circ p_1^*)\Big(\bigvee_i \underline{S}_i\Big)=(l\circ\mu^\rho)\Big(\bigvee_i \underline{S}_i\Big)=\bigvee_i (l\circ\mu^\rho)(\underline{S}_i). \tag{15.93}$$

This represents the logical reformulation of local σ-additivity.

So we have seen that even in the quantum case there is a clear relation between truth values and probability measures.

Thus, in both classical and quantum theory, probabilities can be faithfully expressed in terms of truth values in sheaf topoi with an intuitionistic logic.

Chapter 16
Group Action in Topos Quantum Theory

When analysing the origin of the twisted presheaves, it is clear that the reason we do get a twist is because the group moves the abelian algebras around, i.e. the group action is defined on the base category itself. Thus a possible way of avoiding the occurrence of twists is by imposing that the group does not act on the base category. The category of abelian von-Neumann sub-algebras with no group acting on it will be denoted $\mathcal{V}_f(\mathcal{H})$, where we have added the subscript f (for fixed) to distinguish this situation from the case in which the group does act. Obviously, if one just defined sheaves over $\mathcal{V}_f(\mathcal{H})$, then there would be no group action at all, therefore something extra is needed. As we will see this 'extra' will be the introduction of an intermediate category which will be used as an intermediate base category. On such an intermediate category the group is allowed to act, thus the sheaves defined over it will admit a group action. Once this is done, everything is "pushed down" to the fixed category $\mathcal{V}_f(\mathcal{H})$. As we will see, the sheaves defined in this way will admit a group action which now takes place at an intermediate stage, but will not produce any twists since the final base category stays fixed.

16.1 The Sheaf of Faithful Representations

In our new approach we still use the poset $\mathcal{V}(\mathcal{H})$ as the base category but now we 'forget' the group action. We will consider the collection, $Hom_{faithful}(\mathcal{V}_f(\mathcal{H}), \mathcal{V}(\mathcal{H}))$, of all faithful poset representations of $\mathcal{V}_f(\mathcal{H})$ in $\mathcal{V}(\mathcal{H})$ that come from the action of some transformation group G. Thus we have the collection of all homomorphisms $\phi_g : \mathcal{V}_f(\mathcal{H}) \to \mathcal{V}(\mathcal{H})$, $g \in G$, such that

$$\phi_g(V) := \hat{U}_g V \hat{U}_{g^{-1}}.$$

For some unitary representation $\hat{U}_g$ of g.

We can 'localise' $Hom_{faithful}(\mathcal{V}_f(\mathcal{H}), \mathcal{V}(\mathcal{H}))$ by considering for each V the set $Hom_{faithful}(\downarrow V, \mathcal{V}(\mathcal{H}))$. It is easy to see that this, actually, defines a *presheaf* over $\mathcal{V}(\mathcal{H})$, which we will denote $Hom_{faithful}(\mathcal{V}_f(\mathcal{H}), \mathcal{V}(\mathcal{H}))$.

C. Flori, *A First Course in Topos Quantum Theory*, Lecture Notes in Physics 868,
DOI 10.1007/978-3-642-35713-8_16,

Now, for each algebra V there exists the fixed point group

$$G_{FV} := \{g \in G|\forall v \in V\ \hat{U}_g v \hat{U}_{g^{-1}} = v\}.$$

This implies that the collection of all faithful representations for each V is actually the quotient space G/G_{FV} since the group homomorphisms $\phi : G \to GL(V)$ has to be injective. This will not be the case if we also consider the elements of G_{FV}, since each such element would give the same homomorphism.

Thus for each V we have that

$$\underline{Hom}_{faithful}\big(\underline{\mathcal{V}_f(\mathcal{H}), \mathcal{V}(\mathcal{H})}\big)_V := Hom_{faithful}\big(\downarrow V, \underline{\mathcal{V}(\mathcal{H})}\big) \cong G/G_{FV}.$$

As we will shortly see, there is a presheaf, $\underline{G/G_F}$, such that, as presheaves,

$$\underline{Hom}_{faithful}\big(\mathcal{V}_f(\mathcal{H}), \mathcal{V}(\mathcal{H})\big) \cong \underline{G/G_F}$$

whose local components are defined above. In the rest of this book, unless otherwise specified, $\underline{Hom}(\mathcal{V}_f(\mathcal{H}), \mathcal{V}(\mathcal{H}))$ will mean $\underline{Hom}_{faithful}(\mathcal{V}_f(\mathcal{H}), \mathcal{V}(\mathcal{H}))$.

Lemma 16.1 *G_{FV} is a normal subgroup of G_V.*

Proof Consider an element $g \in G_{FV}$, then given any other element $g_i \in G_V$ we consider the element $g_i g g_i^{-1}$. Such an element acts on each $v \in V$ as follows:

$$\begin{aligned}
\hat{U}_{g_i g g_i^{-1}} v \hat{U}_{(g_i g g_i^{-1})^{-1}} &= \hat{U}_{g_i g g_i^{-1}} v \hat{U}_{g_i g^{-1} g_i^{-1}} \\
&= \hat{U}_{g_i} \hat{U}_g \hat{U}_{g_i^{-1}} v \hat{U}_{g_i} \hat{U}_{g^{-1}} \hat{U}_{g_i^{-1}} \\
&= \hat{U}_{g_i} \hat{U}_g v' \hat{U}_{g^{-1}} \hat{U}_{g_i^{-1}} \\
&= \hat{U}_{g_i} v' \hat{U}_{g_i^{-1}} \\
&= \hat{U}_{g_i} \hat{U}_{g_i^{-1}} v \hat{U}_{g_i} \hat{U}_{g_i^{-1}} \\
&= v
\end{aligned} \tag{16.1}$$

where $v' \in V$ because $g_i \in G_V$. □

We note *en passant* that G is a principal fibre bundle over G/G_{FV} with fiber G_{FV}.

For us, the interesting aspect of the collection G_{FV}, $V \in \mathcal{V}(\mathcal{H})$ is that, unlike the collection of stability groups G_V, $V \in \mathcal{V}(\mathcal{H})$ it forms a presheaf over $\mathcal{V}_f(\mathcal{H})$ (or $\mathcal{V}(\mathcal{H})$). This is because, given $V' \subseteq V$ then $G_{FV} \subseteq G_{FV'}$.

Definition 16.1 The presheaf $\underline{G_F}$ over $\mathcal{V}_f(\mathcal{H})$ has on

- Objects: for each $V \in \mathcal{V}_f(\mathcal{H})$ we define set $\underline{G}_{FV} := G_{FV} = \{g \in G|\forall v \in V\ \hat{U}_g v \hat{U}_{g^{-1}} = v\}$.

– Morphisms: given a map $i : V' \to V$ in $\mathcal{V}_f(\mathcal{H})$ $(V' \subseteq V)$ we obtain the morphism $\underline{G}_F(i) : \underline{G}_{FV} \to \underline{G}_{FV'}$ as subgroup inclusion.

The morphisms $\underline{G}_F(i) : \underline{G}_{FV} \to \underline{G}_{FV'}$ are well defined since if $V' \subseteq V$, then, clearly $G_{FV} \subseteq G_{FV'}$. Associativity is obvious.

We now define the presheaf $\underline{G/G_F}$ as follows:

Definition 16.2 The presheaf $\underline{G/G_F}$ is defined as the presheaf with

– Objects: for each $V \in \mathcal{V}_f(\mathcal{H})$ we assign $(\underline{G/G_F})_V := G/G_{FV} \cong Hom(\downarrow V, \mathcal{V}(\mathcal{H}))$. An element of G/G_{FV} is an orbit $w_V^g := \{g \cdot G_{FV}\}$ which corresponds to the unique homeomorphism ϕ^g.
– Morphisms: Given a morphisms $i_{V'V} : V' \to V$ $(V' \subseteq V)$ in $\mathcal{V}_f(\mathcal{H})$ we define

$$\begin{aligned} \underline{G/G_F}(i_{V'V}) : G/G_{FV} &\to G/G_{FV'} \\ w_V^g &\mapsto \underline{G/G_F}(i_{V'V})\big(w_V^g\big) \end{aligned} \tag{16.2}$$

as the projection maps $\pi_{V'V}$ of the fibre bundles

$$G_{FV'}/G_{FV} \to G/G_{FV} \to G/G_{FV'} \tag{16.3}$$

with fibre isomorphic to $G_{FV'}/G_{FV}$.

What this means is that to each $w_{V'}^g = g \cdot G_{FV'} \in G/G_{FV'}$ one obtains in G/G_{FV} the fibre

$$\begin{aligned} \sigma_V^g := \pi_{V'V}^{-1}(g \cdot G_{FV'}) &= \big\{g_i(g \cdot G_{FV}) | \forall g_i \in G_{FV'}\big\} \\ &= \big\{l_{g_i} \cdot w_V^g \big| \forall g_i \in G_{FV'}\big\} \\ &= \big\{w_V^{g_i g} \big| g_i \in G_{FV'}\big\}. \end{aligned} \tag{16.4}$$

In the above expression we have used the usual action of the group G on an orbit:

$$l_{g_i} \cdot w_V^g = g_i \cdot (g \cdot G_{FV}) = g_i \cdot g \cdot G_{FV} =: w_V^{g_i g}. \tag{16.5}$$

The fibre σ_V^g is obviously isomorphic to $G_{FV'}/G_{FV}$. Thus the projection map $\pi_{V'V}$ projects

$$\pi_{V'V}\big(\sigma_V^g\big) = g \cdot G_{FV'} = w_{V'}^g \tag{16.6}$$

such that for individual elements we have

$$\underline{G/G_F}(i_{V'V})\big(w_V^g\big) := \pi_{V'V}\big(\sigma_V^g\big) = w_{V'}^g. \tag{16.7}$$

Note that when $g_i \in G_{FV'}$, but $g_i \notin G_{FV}$, then $w_V^g = g \cdot G_{FV}$ and $w_{V'}^g = g \cdot G_{FV'} = g_i g G_{FV'} = w_{V'}^{g_i g}$. Therefore $\underline{G/G_F}(i_{V'V}) w_V^g = w_{V'}^g = w_{V'}^{g_i g}$.

It should be noted that the morphisms in the presheaf $\overline{G/G_F}$ can also be defined in terms of the homeomorphisms $Hom(\mathcal{V}_f(\mathcal{H}), \mathcal{V}(\mathcal{H}))$. Namely, given an equivalence class w_V^g we obtain the associated homeomorphisms ϕ_g, such that

$$\underline{G/G_F}(i_{V'V})\phi_g := \phi_{g|V'}. \tag{16.8}$$

We will now define another presheaf which we will then show to be isomorphic to $\underline{G/G_F}$. To this end we first of all have to introduce the constant presheaf $\underline{G}$ which represents the group presheaf. This is defined as follows:

Definition 16.3 The presheaf $\underline{G}$ over $\mathcal{V}_f(\mathcal{H})$ is defined on

- Objects: for each context V, $\underline{G}_V$ is simply the entire group, i.e. $\underline{G}_V = G$.
- Morphisms: given a morphisms $i : V' \subseteq V$ in $\mathcal{V}(\mathcal{H})$, the corresponding morphisms $\underline{G}_V \to \underline{G}_{V'}$ is simply the identity map.

We are now ready to define the new presheaf.

Definition 16.4 The presheaf $\underline{G}/\underline{G_F}$ over $\mathcal{V}_f(\mathcal{H})$ is defined on:

- Objects. For each $V \in \mathcal{V}_f(\mathcal{H})$ we obtain $(\underline{G}/\underline{G_F})_V := G/G_{FV}$ since equivalence relations are computed context-wise.
- Morphisms. For each map $i : V' \subseteq V$ we obtain the morphisms

$$\begin{aligned} (\underline{G}/\underline{G_F})_V &\to (\underline{G}/\underline{G_F})_{V'} \\ G/G_{FV} &\to G/G_{FV'}. \end{aligned} \tag{16.9}$$

These are defined to be the projection maps $\pi_{V'V}$ of the fibre bundles

$$G_{FV'}/G_{FV} \to G/G_{FV} \to G/G_{FV'} \tag{16.10}$$

with fibre isomorphic to $G_{FV'}/G_{FV}$.

From the above definition it is trivial to show the following theorem:

Theorem 16.1

$$\underline{G/G_F} \simeq \underline{G}/\underline{G_F}. \tag{16.11}$$

Proof We construct the map $k : \underline{G/G_F} \to \underline{G}/\underline{G_F}$ such that, for each context V we have

$$\begin{aligned} k_V : \underline{G/G_F}_V &\to \underline{G}/\underline{G_F}_V \\ G/G_{FV} &\mapsto G/G_{FV}. \end{aligned} \tag{16.12}$$

□

16.2 Changing Base Category

We know from Chap. 14 that given a sheaf over a poset there corresponds an étale bundle. In our case the sheaf in question is $\underline{G/G_F}$ with corresponding étale bundle $p : \Lambda\underline{G/G_F} \to \mathcal{V}_f(\mathcal{H})$ where $\Lambda\underline{G/G_F}$ is the étale space. We will now equip the étale space $\Lambda(\underline{G/G_F}) = \coprod_{V\in\mathcal{V}_F(\mathcal{H})}(\underline{G/G_F})_V$ with a poset structure.

The most obvious poset structure to use would be the partial order given by restriction, i.e., $w_V \leq w_{V'}$ iff $V' \subseteq V$ and $w^g_V = w^g_{V'}|_V$ or equivalently $g \cdot G_{FV} = g \cdot (G_{FV} \cap G_{FV'})$. We could write this last condition as an inclusion of sets as follows: $w_V \subseteq w'_V$ $(g \cdot G_{FV} \subseteq g \cdot G_{FV'})$. However this poset structure would not give a presheaf if we were to use it as the base category, rather it would give a covariant functor. To solve this problem we adopt the *order dual* of the partially ordered set, this is the same set but equipped with the inverse order which is itself a partial order. We thus define the ordering on $\Lambda(\underline{G/G_F})$ as follows:

Lemma 16.2 *Given two orbits $w^g_V \in G/G_{FV}$ and $w^g_{V'} \in G/G_{FV'}$ we define the partial ordering $\leq$, by defining*[1]

$$w^g_{V'} \leq w^g_V$$

iff

$$\begin{aligned} V' &\subseteq V \\ w^g_V &\subseteq w^g_{V'}. \end{aligned} \tag{16.13}$$

Note that the last condition is equivalent to $w^g_V = w^g_{V'}|_V$ $(g \cdot G_{FV} = g \cdot (G_{FV} \cap G_{FV'}))$.

It should be noted, though, that if $w^g_V = w^g_{V'}|_V$ then $\underline{G/G_F}(i_{V'V})(w^g_V) = \underline{G/G_F}(i_{V'V})(w^g_{V'}|_V) = w^g_{V'}$. In other words it is also possible to define the partial ordering in terms of the presheaf maps defined above, i.e.,

$$w^g_V \geq w^g_{V'} \quad \text{iff} \quad w^g_{V'} = \underline{G/G_F}(i_{V'V})w^g_V. \tag{16.14}$$

We now show that the ordering defined on $\Lambda(\underline{G/G_F})$ is indeed a partial order.

Proof 1. *Reflexivity*: trivially $w^g_V \leq w^g_V$ for all $w^g_V \in \Lambda\underline{G/G_F}$.

2. *Transitivity*: if $w^g_{V_1} \leq w^g_{V_2}$ and $w^g_{V_2} \leq w^g_{V_3}$ then $V_1 \subseteq V_2$ and $V_2 \subseteq V_3$. From the partial ordering on $\mathcal{V}(\mathcal{H})$ it follows that $V_1 \subseteq V_3$. Moreover from the definition of ordering on $\Lambda(\underline{G/G_F})$ we have that $w^g_{V_2} = w^g_{V_1}|_{V_2}$ and $w^g_{V_3} = w^g_{V_2}|_{V_3}$ which implies that $w^g_{V_3} = w^g_{V_1}|_{V_3}$. It follows that $w_{V_1} \leq w_{V_3}$.

[1] Note that given $w^{g_1}_V$ and $w^{g_2}_V$ such that $g_1 \neq g_2$ then it is not possible to define an ordering between $w^{g_1}_V$ and $w^{g_2}_V$.

3. *Antisymmetry*: if $w_{V_1} \leq w_{V_2}$ and $w_{V_2} \leq w_{V_1}$, it implies that $V_1 \leq V_2$ and $V_2 \leq V_1$ which, by the partial ordering on $\mathcal{V}(\mathcal{H})$ implies that $V_1 = V_2$. Moreover the above conditions imply that $w_{V_1} = w_{V_2}|_{V_1}$ and $w_{V_2} = w_{V_1}|_{V_2}$, which implies that $w_{V_1} = w_{V_2}$. □

Given the previously defined isomorphism $Hom(\downarrow V, \mathcal{V}(\mathcal{H})) \cong (\underline{G/G_F})_V$ for each V, then to each equivalence class w_V^g there is associated a particular homeomorphism $\phi_g : \downarrow V \to \mathcal{V}(\mathcal{H})$. Even though w_V^g is an equivalence class, each element in it will give the same ϕ_g, i.e. it will pick out the same $V_i \in \mathcal{V}(\mathcal{H})$. This is because the equivalence relation is defined in terms of the fixed point group for V.

Therefore it is also possible to define the ordering relation on $\Lambda(\underline{G/G_F})$ in terms of the homeomorphisms ϕ_i^g. First of all we introduce the bundle space $\Lambda J \simeq \Lambda(\underline{G/G_F})$ which is essentially the same as $\Lambda(\underline{G/G_F})$, but whose elements are now the maps ϕ_i^g, i.e., $\Lambda J = \Lambda(\underline{Hom}(\mathcal{V}_f(\mathcal{H}), \mathcal{V}(\mathcal{H})))$. The associated bundle map is $p_J : \Lambda J \to \mathcal{V}_f(\mathcal{H})$. We then define the ordering on ΛJ as $\phi_i^g \leq \phi_j^g$ iff

$$p_J(\phi_i^g) \subseteq p_J(\phi_j^g) \tag{16.15}$$

and

$$\phi_i^g = \phi_j^g|_{p_J(\phi_i^g)}. \tag{16.16}$$

We now need to show that this does indeed define a partial order on ΛJ.

Proof 1. *Reflexivity*: trivially $\phi_i^g \leq \phi_i^g$ since $p_J(\phi_i^g) \subseteq p_J(\phi_i^g)$ and $\phi_i^g = \phi_i^g$.

2. *Transitivity*: if $\phi_i^g \leq \phi_j^g$ and $\phi_j^g \leq \phi_k^g$ then $p_J(\phi_i^g) \subseteq p_J(\phi_j^g)$ and $p_J(\phi_j^g) \subseteq p_J(\phi_k^g)$, therefore $p_J(\phi_i^g) \subseteq p_J(\phi_k^g)$. Moreover we have that $\phi_i^g = \phi_j^g|_{p_J(\phi_i^g)}$ and $\phi_j^g = \phi_k^g|_{p_J(\phi_j^g)}$, therefore $\phi_i^g = \phi_k^g|_{p_J(\phi_i^g)}$.

3. *Antisymmetry*: if $\phi_i^g \leq \phi_j^g$ and $\phi_j^g \leq \phi_i^g$ it implies that $p_J(\phi_i^g) \subseteq p_J(\phi_j^g)$ and $p_J(\phi_j^g) \subseteq p_J(\phi_i^g)$, thus $p_J(\phi_i^g) = p_J(\phi_j^g)$. Moreover we have that $\phi_i^g = \phi_j^g|_{p_J(\phi_i^g)}$ and $\phi_j^g = \phi_i^g|_{p_J(\phi_j^g)}$, therefore $\phi_i^g = \phi_j^g$. □

Given this ordering we can now define the corresponding ordering on $\Lambda(\underline{G/G_F})$ as $w_{V_i}^g \leq w_{V_j}^g$ iff $\phi_i^g \leq \phi_j^g$. We have again used the fact that to each w_V^g there is associated a unique homeomorphism $\phi_V^g : \downarrow V \to \mathcal{V}(\mathcal{H})$.

16.3 From Sheaves on the Old Base Category to Sheaves on the New Base Category

In what follows we will move freely between the language of presheaves and that of sheaves which we will both denote as $\underline{A}$. Which of the two is being used should be clear from the context. The reason we are able to do this is because our base categories are posets (see Chap. 14).

We are now interested in 'transforming' all the physically relevant sheaves on $\mathcal{V}(\mathcal{H})$ to sheaves over $\Lambda(\underline{G/G_F})$. Therefore we are interested in finding a functor $I : Sh(\mathcal{V}(\mathcal{H})) \to Sh(\Lambda(\underline{G/G_F}))$. As a first attempt we define:

$$\begin{aligned} I : Sh\big(\mathcal{V}(\mathcal{H})\big) &\to Sh\big(\Lambda(\underline{G/G_F})\big) \\ \underline{A} &\mapsto I(\underline{A}) \end{aligned} \tag{16.17}$$

such that for each context $w_V^g \simeq \phi^g$ we define

$$\big(I(\underline{A})\big)_{w_V^g} := \underline{A}_{\phi^g(V)} = \big((\phi^g)^*(\underline{A})\big)(V) \tag{16.18}$$

where $\phi^g : \downarrow V \to \mathcal{V}(\mathcal{H})$ is the unique homeomorphism associated with the equivalence class $w_V^g = g \cdot G_{FV}$ and $(\phi^g)^* : Sh(\mathcal{V}(\mathcal{H})) \to Sh(\downarrow V)$ is the inverse image part of the geometric morphisms induced by ϕ^g.

We then need to define the morphisms. Thus, given $i_{w_{V'}^g, w_V^g} : w_{V'}^g \to w_V^g$ ($w_{V'}^g \leq w_V^g$) with corresponding homeomorphisms $\phi_1^g \leq \phi_2^g$ ($\phi_1^g \in Hom(\downarrow V', \mathcal{V}(\mathcal{H}))$ and $\phi_2^g \in Hom(\downarrow V, \mathcal{V}(\mathcal{H}))$) we have the associated morphisms $I\underline{A}(i_{w_{V'}^g, w_V^g}) : (I(\underline{A}))_{w_V^g} \to (I(\underline{A}))_{w_{V'}^g}$ defined as

$$\big(I\underline{A}(i_{w_{V'}^g, w_V^g})\big)(a) = \big(I\underline{A}(i_{\phi_2^g, \phi_1^g})\big)(a) := \underline{A}_{\phi_2^g(V), \phi_1^g(V')}(a) \tag{16.19}$$

for all $a \in \underline{A}_{\phi_2^g(V)}$. In the above equation $V = p_J(\phi_2^g)$ and $V' = p_J(\phi_2^g)$.[2] Moreover, since $\phi_1^g \leq \phi_2^g$ is equivalent to the condition $w_{V'}^g \leq w_V^g$, then $\phi_1^g(V') \subseteq \phi_2^g(V)$ and $\phi_1^g = \phi_2^g|_{V'}$. Therefore, effectively, each $(I\underline{A}(i_{w_{V'}^g, w_V^g})) \in Sh(\Lambda(\underline{G/G_F}))$ is defined in terms of the maps $\underline{A}_{\phi_2^g(V), \phi_1^g(V')} \in Sh(\mathcal{V}(\mathcal{H}))$.

Theorem 16.2 *The map $I : Sh(\mathcal{V}(\mathcal{H})) \to Sh(\Lambda(\underline{G/G_F}))$ is a functor defined as follows*:

(i) *Objects*: $(I(\underline{A}))_{w_V^g} := \underline{A}_{\phi^g(V)} = ((\phi^g)^*(\underline{A}))(V)$. *If $w_{V'}^g \leq w_V^g$ with associated homeomorphisms $\phi_1^g \leq \phi_2^g$ ($\phi_1^g \in Hom(\downarrow V', \mathcal{V}(\mathcal{H}))$ and $\phi_2^g \in Hom(\downarrow V, \mathcal{V}(\mathcal{H}))$), then*

$$\big(I\underline{A}(i_{w_{V'}^g, w_V^g})\big) = I\underline{A}(i_{\phi_1^g, \phi_2^g}) := \underline{A}_{\phi_2^g(V), \phi_1^g(V')} : \underline{A}_{\phi_2^g(V)} \to \underline{A}_{\phi_1^g(V')}$$

where $V = p_J(\phi_2^g)$ and $V' = p_J(\phi_1^g)$.

---

[2]Recall that $p_J : (\Lambda J = \Lambda(\underline{Hom}(\mathcal{V}_f(\mathcal{H}), \mathcal{V}(\mathcal{H})))) \to \mathcal{V}_f(\mathcal{H})$.

(ii) *Morphisms*: *if we have a morphisms* $f : \underline{A} \to \underline{B}$ *in* $Sh(\mathcal{V}(\mathcal{H}))$ *we then define the corresponding morphisms in* $Sh(\Lambda(\underline{G/G_F}))$ *as*

$$
\begin{aligned}
I(f)_{w_V^g} &: I(\underline{A})_{w_V^g} \to I(\underline{B})_{w_V^g} \\
f_{p_J(\phi_1^g)} &: \underline{A}_{\phi_1^g(p_J(\phi_1^g))} \to \underline{B}_{\phi_1^g(p_J(\phi_1^g))}.
\end{aligned}
\tag{16.20}
$$

Proof Consider an arrow $f : \underline{A} \to \underline{B}$ in $Sh(\mathcal{V}(\mathcal{H}))$ so that, for each $V \in \mathcal{V}(\mathcal{H})$, the local component is $f_V : \underline{A}_V \to \underline{B}_V$ with commutative diagram

$$
\begin{array}{ccc}
\underline{A}_{V_1} & \xrightarrow{f_{V_1}} & \underline{B}_{V_1} \\
{\scriptstyle \underline{A}_{V_1V_2}}\downarrow & & \downarrow{\scriptstyle \underline{B}_{V_1V_2}} \\
\underline{A}_{V_2} & \xrightarrow[f_{V_2}]{} & \underline{B}_{V_2}
\end{array}
$$

for all pairs V_1, V_2 with $V_2 \subseteq V_1$. Now, suppose that $w_{V_2}^g \leq w_{V_1}^g$ with associated homeomorphisms $\phi_2^g \leq \phi_1^g$, such that (i) $p_J(\phi_2^g) \subseteq p_J(\phi_1^g)$; and (ii) $\phi_2^g = \phi_1^g|_{p_J(\phi_2^g)}$. We want to show that the action of the I functor gives the commutative diagram

$$
\begin{array}{ccc}
I(\underline{A})_{w_{V_1}^g} & \xrightarrow{I(f)_{w_{V_1}^g}} & I(\underline{B})_{w_{V_1}^g} \\
{\scriptstyle I(\underline{A})(i_{w_{V_1}^g, w_{V_2}^g})}\downarrow & & \downarrow{\scriptstyle I(\underline{B})(i_{w_{V_1}^g, w_{V_2}^g})} \\
I(\underline{A})_{w_{V_2}^g} & \xrightarrow[I(f)_{w_{V_2}^g}]{} & I(\underline{B})_{w_{V_2}^g}
\end{array}
$$

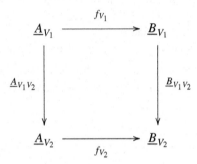

for all $V_2 \subseteq V_1$. By applying the definitions we get

$$\begin{array}{ccc}
\underline{A}_{\phi_1^g(p_J(\phi_1^g))} & \xrightarrow{f_{\phi_1^g(p_J(\phi_1^g))}} & \underline{B}_{\phi_1^g(p_J(\phi_1^g))} \\
\downarrow {\scriptstyle \underline{A}_{\phi_1^g(p_J(\phi_1^g)),\phi_2^g(p_J(\phi_2^g))}} & & \downarrow {\scriptstyle \underline{B}_{\phi_1^g(p_J(\phi_1^g)),\phi_2^g(p_J(\phi_2^g))}} \\
\underline{A}_{\phi_2^g(p_J(\phi_2^g))} & \xrightarrow[f_{\phi_2^g(p_J(\phi_2^g))}]{} & \underline{B}_{\phi_2^g(p_J(\phi_2^g))}
\end{array}$$

which is commutative. Therefore $I(f)$ is a well defined arrow in $Sh(\Lambda\underline{G/G_F})$ from $I(\underline{A})$ to $I(\underline{B})$.

Given two arrows f, g in $Sh(\mathcal{V}(\mathcal{H}))$ then it follows that:

$$I(f \circ g) = I(f) \circ I(g). \tag{16.21}$$

This proves that I is a functor from $Sh(\mathcal{V}(\mathcal{H}))$ to $Sh(\Lambda(\underline{G/G_F}))$. □

From the above definition of the functor I we immediately have the following corollary:

Corollary 16.1 *The functor I preserves monic arrows.*

Proof Given a monic arrow $f : \underline{A} \to \underline{B}$ in $Sh(\mathcal{V}(\mathcal{H}))$ then by definition, for any w_V^g

$$I(f)_{w_V^g} : I(\underline{A})_{w_V^g} \to I(\underline{B})_{w_V^g} \tag{16.22}$$

$$f_{\phi^g(p_J(\phi^g))} : \underline{A}_{\phi^g(p_J(\phi^g))} \to \underline{B}_{\phi^g(p_J(\phi^g))}. \tag{16.23}$$

Thus $I(f)_{w_V^g}$ is monic since $f_{\phi^g(p_J(\phi^g))}$ is. □

Similarly, we can show the following:

Corollary 16.2 *The functor I preserves epic arrows.*

Proof Given an epic arrow $f : \underline{A} \to \underline{B}$ in $Sh(\mathcal{V}(\mathcal{H}))$ then by definition for any w_V^g

$$\begin{aligned} I(f)_{w_V^g} &: I(\underline{A})_{w_V^g} \to I(\underline{B})_{w_V^g} \\ f_{\phi^g(p_J(\phi^g))} &: \underline{A}_{\phi^g(p_J(\phi^g))} \to \underline{B}_{\phi^g(p_J(\phi^g))}. \end{aligned} \tag{16.24}$$

Thus $I(f)_{w_V^g}$ is epic since $f_{\phi^g(p_J(\phi^g))}$ is. □

We would now like to know how such a functor behaves with respect to the terminal object. To this end we define the following corollary:

Corollary 16.3 *The functor I preserves the terminal object.*

Proof The terminal object in $Sh(\mathcal{V}(\mathcal{H}))$ is the objects $\underline{1}_{Sh(\mathcal{V}(\mathcal{H}))}$ such that to each element $V \in \mathcal{V}(\mathcal{H})$ it associates the singleton set $\{*\}$ while the maps are simply the identity maps. We now apply the I functor to such an object obtaining

$$I(\underline{1}_{Sh(\mathcal{V}(\mathcal{H}))})_{w_V^g} := (\underline{1}_{Sh(\mathcal{V}(\mathcal{H}))})_{\phi^g(p_J(\phi^g))} = \{*\} \tag{16.25}$$

where ϕ^g is the unique homeomorphism associated to the coset w_V^g.

Hence it follows that $I(\underline{1}_{Sh(\mathcal{V}(\mathcal{H}))}) = \underline{1}_{Sh(\Lambda(\underline{G/G_F}))}$. □

We now check whether I preserves the initial object. We recall that the initial object in $Sh(\mathcal{V}(\mathcal{H}))$ is simply the sheaf $\underline{O}_{Sh(\mathcal{V}(\mathcal{H}))}$ which assigns to each element V the empty set $\{\emptyset\}$. We then have

$$I(\underline{O}_{Sh(\mathcal{V}(\mathcal{H}))})_{w_V^g} := (\underline{O}_{Sh(\mathcal{V}(\mathcal{H}))})_{\phi^g(p_J(\phi^g))} = \{\emptyset\} \tag{16.26}$$

where $\phi^g \in Hom(\downarrow V, \mathcal{V}(\mathcal{H}))$ is the unique homeomorphism associated with the coset w_V^g.

It follows that:

$$I(\underline{O}_{Sh(\mathcal{V}(\mathcal{H}))}) = \underline{O}_{Sh(\Lambda(\underline{G/G_F}))}. \tag{16.27}$$

From the above proof it transpires that the reason the functor I preserves monic, epic, terminal object, and initial object is manly due to the fact that the action of I is defined component-wise as $(I(\underline{A}))_\phi := \underline{A}_{\phi(V)}$ for $\phi \in Hom(\downarrow V, \mathcal{V}(\mathcal{H}))$. In particular, it can be shown that I preserves all limits and colimits.

Theorem 16.3 *The functor I preserves limits.*

In order to prove the above theorem we first of all have to recall some general results and definitions. To this end consider two categories $\mathcal{C}$ and $\mathcal{D}$, such that there exists a functor between them $F : \mathcal{C} \to \mathcal{D}$. For a small index category J we consider diagrams of type J in both $\mathcal{C}$ and $\mathcal{D}$, i.e. elements in the functor categories $\mathcal{C}^J$ and $\mathcal{D}^J$, respectively.[3] The functor F then induces a functor between these diagrams as follows:

$$\begin{aligned} F^J : \mathcal{C}^J &\to \mathcal{D}^J \\ A &\mapsto F^J(A) \end{aligned} \tag{16.28}$$

[3]Note that here an element $\mathcal{D}^J$ is called a diagram of type J since, intuitively, it 'representes' the category J onto the category $\mathcal{D}$.

such that $(F^J(A))(j) := F(A(j))$. In this setting a cone (with vertex $C \in \mathcal{C}$) of a diagram $A \in \mathcal{C}^J$ consists of an object $C \in \mathcal{C}$ and a natural transformation $f : C \to A$ such that, the following diagram commutes for all $j, k \in J$

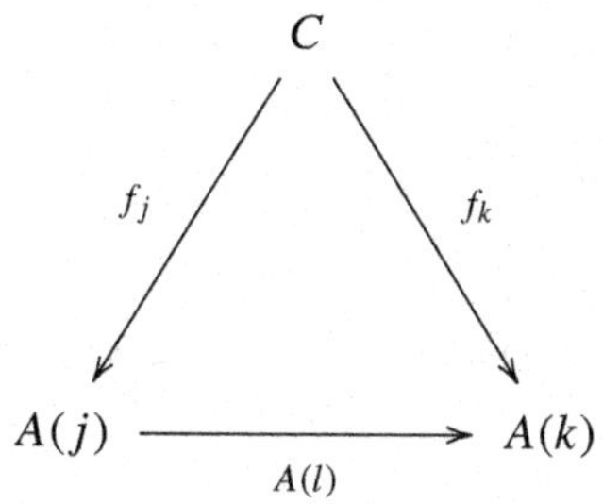

where $l : j \to k$.

We then call a cone $p : B \to A$ *limit* to A when for any cone $f : C \to A$ there exists a unique map $g : C \to B$, such that, for all $j \in J$ the following diagram commutes:

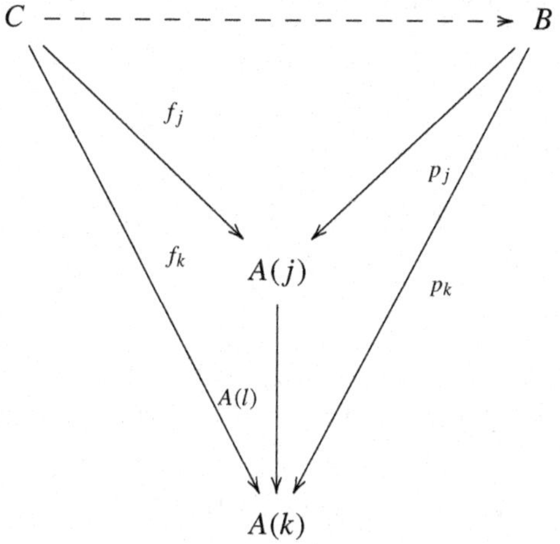

We will often denote the limit $p : B \to A$ by $B = \lim_{\leftarrow J} A$. We now introduce the functor

$$\begin{aligned} \lim_{\leftarrow J} : \mathcal{C}^J &\to \mathcal{C} \\ A &\mapsto \lim_{\leftarrow J} A \end{aligned} \tag{16.29}$$

which assigns to each diagram A of type J in $\mathcal{C}$, its limit $\lim_{\leftarrow J}(A) \in \mathcal{C}$. This functor is well defined iff all diagrams in $\mathcal{C}^J$ have limits.

Assuming that indeed limits of type J exist in $\mathcal{C}$ and $\mathcal{D}$ we obtain the following diagram:

$$\begin{array}{ccc} \mathcal{C}^J & \xrightarrow{\lim_{\leftarrow J}} & \mathcal{C} \\ {\scriptstyle F^J}\downarrow & & \downarrow{\scriptstyle F} \\ \mathcal{D}^J & \xrightarrow[\lim_{\leftarrow J}]{} & \mathcal{D} \end{array}$$

By the universal properties of limits we obtain the *natural transformation*

$$\alpha_J : F \circ \lim_{\leftarrow J} \rightarrow \lim_{\leftarrow J} \circ F^J. \tag{16.30}$$

One then says that F preserves limits if α_J is a *natural isomorphisms.*

Let us now go back to our main task of proving Theorem 16.3. To this end we need to show that the functor I preserves limits. Following the discussion above this is equivalent to showing that the map

$$\alpha_J : I \circ \lim_{\leftarrow J} \rightarrow \lim_{\leftarrow J} \circ I^J \tag{16.31}$$

is a natural isomorphisms. Here I^J is

$$\begin{aligned} I^J : \big(Sh\big(\mathcal{V}(\mathcal{H})\big)\big)^J &\rightarrow \big(Sh\big(\Lambda\big(\underline{G/G_F}\big)\big)\big)^J \\ A &\mapsto I^J(A) \end{aligned} \tag{16.32}$$

where $(I^J(A)(j))_\phi := I(A(j))_\phi$.

The proof of α_J being a natural isomorphisms will utilise a result derived in [18] where it is shown that for any diagram $A : J \rightarrow \mathcal{C}^{\mathcal{D}}$ of type J in $\mathcal{C}^{\mathcal{D}}$ the following isomorphism holds

$$\Big(\lim_{\leftarrow J} A\Big) D \simeq \lim_{\leftarrow J} A_D \quad \forall D \in \mathcal{D} \tag{16.33}$$

where $A_D : J \rightarrow \mathcal{C}$ is a diagram in $\mathcal{D}$. With these results in mind we are now ready to prove Theorem 16.3.

Proof Let us consider a diagram $A : J \rightarrow \mathbf{Sets}^{\mathcal{V}(\mathcal{H})}$ of type J in $\mathbf{Sets}^{\mathcal{V}(\mathcal{H})}$:

$$\begin{aligned} A : J &\rightarrow \mathbf{Sets}^{\mathcal{V}(\mathcal{H})} \\ j &\mapsto A(j) \end{aligned} \tag{16.34}$$

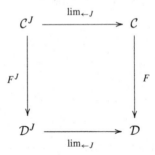

where $A(j)(V) := A_V(j)$ for $A_V : j \rightarrow \mathbf{Sets}$ a diagram in *Sets*. Assume that L is a limit of type J for A, i.e. $L : \mathcal{V}(\mathcal{H}) \rightarrow \mathbf{Sets}$ such that $\lim_{\leftarrow J} A = J$. We then construct the diagram

$$\begin{array}{ccc}
(\mathbf{Sets}^{\mathcal{V}(\mathcal{H})})^J & \xrightarrow{\lim_{\leftarrow J}} & \mathbf{Sets}^{\mathcal{V}(\mathcal{H})} \\
\Big\downarrow {\scriptstyle I^J} & & \Big\downarrow {\scriptstyle I} \\
(\mathbf{Sets}^{\Lambda(\underline{G/G_F})})^J & \xrightarrow[\lim_{\leftarrow J}]{} & \mathbf{Sets}^{\Lambda(\underline{G/G_F})}
\end{array}$$

with associated natural transformation

$$\alpha_J : I \circ \lim_{\leftarrow J} \to \lim_{\leftarrow J} \circ I^J. \tag{16.35}$$

For each diagram $A : J \to \mathbf{Sets}^{\mathcal{V}(\mathcal{H})}$ and element $\phi \in \Lambda(\underline{G/G_F})$ we obtain

$$\Big(I \circ \lim_{\leftarrow J}(A)\Big)_\phi = \Big(I\Big(\lim_{\leftarrow J} A\Big)\Big)_\phi := \Big(\lim_{\leftarrow J} A\Big)_{\phi(V)} \simeq \lim_{\leftarrow J} A_{\phi(V)} \tag{16.36}$$

where $A_{\phi(V)} : J \to \mathbf{Sets}$ is such that $A_{\phi(V)}(j) = A(j)(\phi(V))$.[4]

On the other hand

$$\Big(\Big(\lim_{\leftarrow J} \circ I^J\Big) A\Big)_\phi = \Big(\lim_{\leftarrow J}\big(I^J(A)\big)\Big)_\phi \simeq \lim_{\leftarrow J}\big(I^J(A)\big)_\phi = \lim_{\leftarrow J} A_{\phi(V)} \tag{16.37}$$

where

$$\begin{aligned} I^J(A) : J &\to \mathbf{Sets}^{\Lambda(\underline{G/G_F})} \\ j &\mapsto I^J(A)(j) \end{aligned} \tag{16.38}$$

such that for all $\phi \in \Lambda(\underline{G/G_F})$ we have $(I^J(A(j)))_\phi = (I(A(j)))_\phi = A(j)_{\phi(V)}$.

It follows that

$$I \circ \lim_{\leftarrow J} \simeq \lim_{\leftarrow J} \circ I^J. \tag{16.39}$$

□

Similarly one can show that

Theorem 16.4 *The functor I preserves all colimits.*

---

[4]Recall that $A : J \to \mathbf{Sets}^{\mathcal{V}(\mathcal{H})}$ is such that $A_V(j) = A(j)(V)$, therefore $(I(A(j)))_\phi := A(j)_{\phi(V)} = A_{\phi(V)}(j)$.

Since colimits are simply duals to the limits, the proof of this theorem is similar to the proof given above. However, for completeness sake we will, nonetheless, report it here.

Proof We first of all construct the analogue of the diagram above:

$$\begin{array}{ccc}
(\mathbf{Sets}^{\mathcal{V}(\mathcal{H})})^J & \xrightarrow{\lim_{\to J}} & \mathbf{Sets}^{\mathcal{V}(\mathcal{H})} \\
\Big\downarrow{\scriptstyle I^J} & & \Big\downarrow{\scriptstyle I} \\
(\mathbf{Sets}^{\Lambda(\underline{G/G_F})})^J & \xrightarrow[\lim_{\to J}]{} & \mathbf{Sets}^{\Lambda(\underline{G/G_F})}
\end{array}$$

where $\lim_{\to J} : (\mathbf{Sets}^{\mathcal{V}(\mathcal{H})})^I \to \mathbf{Sets}^{\mathcal{V}(\mathcal{H})}$ represents the map which assigns colimits to all diagrams in $(\mathbf{Sets}^{\mathcal{V}(\mathcal{H})})^I$.

We now need to show that the associated natural transformation

$$\beta_J : I \circ \lim_{\to J} \to \lim_{\to J} \circ I^J \tag{16.40}$$

is a natural isomorphisms.

For any diagram $A \in (Sets^{\mathcal{V}(\mathcal{H})})^I$ and $\phi \in \Lambda(\underline{G/G_F})$ we compute

$$\Big(I \circ \lim_{\to J}(A)\Big)_\phi = \Big(I\Big(\lim_{\to J} A\Big)\Big)_\phi = \Big(\lim_{\to J} A\Big)_{\phi(V)} \simeq \lim_{\to J} A_{\phi(V)} \tag{16.41}$$

where $(\lim_{\to J} A)_{\phi(V)} \simeq \lim_{\to J} A_{\phi(V)}$ is the dual of (16.33). On the other hand

$$\Big(\Big(\lim_{\to J} \circ I^J\Big)(A)\Big)_\phi = \Big(\lim_{\to J}\big(I^J(A)\big)\Big)_\phi \simeq \lim_{\to J}\big(I^J(A)\big)_\phi = \lim_{\to J} A_{\phi(V)}. \tag{16.42}$$

It follows that indeed β_J is a natural isomorphisms. □

Now we would like to check whether I is a left adjoint. The Adjoint Functor Theorem as applied to Grothendieck toposes [94] states that any colimit preserving functor between Grothendieck topoi has a right adjoint and any limit preserving functor has a left adjoint, so I has both adjoints. The construction of this functor is left as an exercise.

16.4 The Adjoint Pair

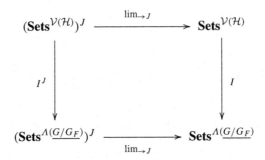

As discussed in Chap. 14, given a map $f : X \to Y$ between topological spaces X and Y we obtain a geometric morphism, whose inverse and direct image are, respectively,

$$\begin{aligned} f^* &: Sh(Y) \to Sh(X) \\ f_* &: Sh(X) \to Sh(Y). \end{aligned} \tag{16.43}$$

We also know that $f^* \dashv f_*$, i.e., f^* is the left-adjoint of f_*. If f is an étale map, however, there also exists the left adjoint $f!$ to f^*, namely

$$f! : Sh(X) \to Sh(Y) \tag{16.44}$$

with $f! \dashv f^* \dashv f_*$.

In Theorem 16.5 below we will show that

$$f!(p_A : A \to X) = f \circ p_A : A \to Y \tag{16.45}$$

so that we combine the étale bundle $p_A : A \to X$ with the étale map $f : X \to Y$, to give the étale bundle $f \circ p_A : A \to Y$. Here we have used the fact that sheaves can be defined in terms of étale bundles. In fact in Chap. 14 it was shown that there exists an equivalence of categories $Sh(X) \simeq Etale(X)$ for any topological space X.

Theorem 16.5 *Given the étale map $f : X \to Y$ the left adjoint functor $f! : Sh(X) \to Sh(Y)$ is defined as follows:*

$$f!(p_A : A \to X) = f \circ p_A : A \to Y \tag{16.46}$$

for $p_A : A \to Y$ an étale bundle.

Proof In the proof we will first define the functor $f!$ for general presheaves, then restrict our attention to the case of sheaves ($Sh(X) \subseteq \mathbf{Sets}^{X^{op}}$) and f étale.

Consider the map $f : X \to Y$, this gives rise to the functor $f! : \mathbf{Sets}^{X^{op}} \to \mathbf{Sets}^{Y^{op}}$. The standard definition of $f!$ is as follows:

$$f! := - \otimes_X \left({}_f Y^\bullet\right) \tag{16.47}$$

such that, for any object $A \in \mathbf{Sets}^{X^{op}}$ we have

$$f!(A) = A \otimes_X \left({}_f Y^\bullet\right). \tag{16.48}$$

This is a presheaf in $\mathbf{Sets}^{Y^{op}}$, thus for each element $y \in Y$ we obtain the set

$$\left(A \otimes_X \left({}_f Y^\bullet\right)\right) y := A \otimes_X \left({}_f Y^\bullet\right)(-, y) \tag{16.49}$$

where $({}_f Y^\bullet)$ is the presheaf

$$\left({}_f Y^\bullet\right) : X \times Y^{op} \to \mathbf{Sets}. \tag{16.50}$$

This presheaf derives from the composition of $f \times id_{Y^{op}} : X \times Y^{op} \to Y \times Y^{op}$ ($(f \times id_{Y^{op}})^* : \mathbf{Sets}^{Y \times Y^{op}} \to \mathbf{Sets}^{X \times Y^{op}}$) with the bi-functor ${}^{\bullet}Y^{\bullet} : Y \times Y^{op} \to \mathbf{Sets}$; $(y, y') \mapsto Hom_Y(y', y)$, i.e.,

$$\left({}_fY^{\bullet}\right) := (f \times id_{Y^{op}})^*\left({}^{\bullet}Y^{\bullet}\right) = {}^{\bullet}Y^{\bullet} \circ (f \times id_{Y^{op}}). \tag{16.51}$$

Now coming back to our situation we then have the restricted functor

$$\begin{aligned} \left({}_fY^{\bullet}\right)(-, y) : (X, y) &\to \mathbf{Sets} \\ (x, y) &\mapsto \left({}_fY^{\bullet}\right)(x, y) \end{aligned} \tag{16.52}$$

which, from the definition given above is

$$\begin{aligned} \left({}_fY^{\bullet}\right)(x, y) &= {}^{\bullet}Y^{\bullet} \circ (f \times id_{Y^{op}})(x, y) \\ &= {}^{\bullet}Y^{\bullet}\big(f(x), y\big) = Hom_Y\big(y, f(x)\big). \end{aligned} \tag{16.53}$$

Therefore, putting all the results together we have that for each $y \in Y$ we obtain $A \otimes_X ({}_fY^{\bullet})(-, y)$, defined for each $x \in X$ as

$$A(-) \otimes_X \left({}_fY^{\bullet}\right)(x, y) := A(x) \otimes_X Hom_Y\big(y, f(x)\big). \tag{16.54}$$

This represents the presheaf A defined over the element x, plus a collection of maps in Y mapping the original y to the image of x via f.

In particular $A(-) \otimes_X ({}_fY^{\bullet}) = A(-) \otimes_X Hom_Y(y, f(-))$ represents the following equaliser:

$$\begin{array}{ccc} \coprod_{x,x'} A(x) \times Hom_X(x', x) \times Hom_Y(y, f(x')) & \overset{\tau}{\underset{\theta}{\rightrightarrows}} & \coprod_x A(x) \times Hom_Y(y, f(x)) \\ & & \Big\downarrow \sigma \\ & & A(-) \otimes_X Hom_Y(y, f(-)) \end{array}$$

such that, given a triplet $(a, g, h) \in A(x) \times Hom_X(x', x) \times Hom_Y(y, f(x'))$, we then obtain that[5]

$$\tau(a, g, h) = (ag, h) = \theta(a, g, h) = (a, gh). \tag{16.55}$$

Therefore, form the above equivalence conditions, $A(-) \otimes_X Hom_Y(y, f(-))$ is the quotient space of $\coprod_x A(x) \times Hom_Y(y, f(x))$.

---

[5]Note that $g : x' \to x$ and $A(g) : A(x) \to A(x')$, therefore $ag := A(g)(a)$ for $a \in A(x)$, while $gh := f(g) \circ h$ for $h : y \to f(x')$.

We now consider the situation in which A is a sheaf on X, in particular it is an étale bundle $p_A : A \to X$ and f is an étale map which means that it is a local homeomorphism, i.e. for each $x \in X$ there is an open set V, such that $x \in V$ and $f_{|V} : V \to f(V)$ is a homeomorphism. It follows that for each $x_i \in V$ there is a unique element y_i such that $f_{|V}(x_i) = y_i$. In particular for each $V \subset X$ then $f_{|V}(V) = U$ for some $U \subset Y$.

Note that, since the condition of being a homeomorphism is only local, it can be the case that $f_{|V_i}(V_i) = f_{|V_j}(V_j)$ even if $V_i \neq V_j$. However in these cases the restricted étale maps have to agree on the intersections, i.e. $f_{|V_i}(V_i \cap V_j) = f_{|V_j}(V_i \cap V_j)$.

Let us now consider an open set V with local homeomorphism $f_{|V}$. In this setting each element $y_i \in f_{|V}(V)$ will be of the form $f(x_i)$ for a unique x_i. Moreover, if we consider two open sets $V_1, V_2 \subseteq V$, then to each map $V_1 \to V_2$ in X, with associated bundle map $A(V_2) \to A(V_1)$, there corresponds a map $f_{|V}(V_1) \to f_{|V}(V_2)$ in Y. Therefore, evaluating $A(-) \otimes_X Hom_Y(-, f(-))$ at the open set $f_{|V}(V) \subset Y$ we get, for each $V_i \subseteq V$, the equivalence classes $[A(V_i) \times_X Hom_Y(f_{|V}(V), f_{|V}(V_i))]$ where $A(V_j) \times_X Hom_Y(f_{|V}(V), f_{|V}(V_j)) \simeq A(V_k) \times_X Hom_Y(f_{|V}(V), f_{|V}(V_k))$ iff there exists a map $f_{|V}(V_j) \to f_{|V}(V_k)$ which combines with $f_{|V}(V) \to f_{|V}(V_j)$, giving $f_{|V}(V) \to f_{|V}(V_k)$ and corresponding bundle map $A(V_k) \to A(V_j) \to A(V)$, given by the map $V \to V_j \to V_k$ in X. A moment of thought reveals that such an equivalence class is nothing but $p_A^{-1}(V)$ (the fibre of p_A at V) with associated fibre maps induced by the base maps.

We will now denote such an equivalence class by $[A(V) \times_X Hom_Y(f_{|V}(V), f_{|V}(V))]$ since, obviously, in each equivalence class there will be the element $A(V) \times_X Hom_Y(f_{|V}(V), f_{|V}(V))$.

Applying the same procedure for each open set $V_i \subset X$, we can obtain two cases:

(i) $f_{|V_i}(V_i) = U \neq f_V(V)$. In this case we simply get an independent equivalence class for U.
(ii) If $f_{|V_i}(V_i) = U = f_V(V)$ and there is no map $i : V \to V_i$ in X then, in this case, for U, we obtain two distinct equivalence classes $[A(V_i) \times_X Hom_Y(f_{|V_i}(V_i), f_{|V_i}(V_i))]$ and $[A(V) \times_X Hom_Y(f_{|V}(V), f_{|V}(V))]$.

Thus the sheaf $A(-) \otimes_X ({}_f Y^\bullet)$ is defined for each open set $f_V(V) \subset Y$ as the set $[A(V) \times_X Hom_Y(f_{|V}(V), f_{|V}(V))] \simeq A(V)$ and, for each map $f_{V'}(V') \to f_V(V)$ in Y (with associated map $V' \to V$ in X), the corresponding maps $[A(V) \times_X Hom_Y(f_{|V}(V), f_{|V}(V))] \simeq A(V) \to [A(V') \times_X Hom_Y(f_{|V'}(V)', f_{|V'}(V'))] \simeq A(V')$.

This is precisely what the étale bundle $f \circ p_A : A \to Y$ is. □

Now that we understand the action of $f!$ on sheaves we will try and understand its action of functions. To this end, let us go back to étale bundles. Given a map $\alpha : A \to B$ of étale bundles over X, we obtain the map $f!(\alpha) : f!(A) \to f!(B)$ which is defined as follows. We start with the collection of fibre maps $\alpha_x : A_x \to B_x$, $x \in X$, where $A_x := p_A^{-1} A(\{x\})$. Then, for each $y \in Y$ we want to define the maps $f!(\alpha)_y$:

$f!(A)_y \to f!(B)_y$, i.e., $f!(\alpha)_y : p_A^{-1}(A(f^{-1}\{y\})) \to p_B^{-1}(B(f^{-1}\{y\}))$. This are defined as

$$f!(\alpha)_y(a) := \alpha_{p_A(a)}(a) \tag{16.56}$$

for all $a \in f!(A)_y = p_A^{-1}(A(f^{-1}\{a\}))$.

For the case of interest we obtain the left adjoint functor $p_J! : Sh(\Lambda(\underline{G/G_F})) \to Sh(\mathcal{V}_f(\mathcal{H}))$ of $p_J^* : Sh(\mathcal{V}_f(\mathcal{H})) \to Sh(\Lambda(\underline{G/G_F}))$. The existence of such a functor enables us to define the composite functor

$$F := p_J! \circ I : Sh\big(\mathcal{V}(\mathcal{H})\big) \to Sh\big(\mathcal{V}_f(\mathcal{H})\big). \tag{16.57}$$

Such a functor sends all the original sheaves we had defined over $\mathcal{V}(\mathcal{H})$ to new sheaves over $\mathcal{V}_f(\mathcal{H})$. Denoting the sheaves over $\mathcal{V}_f(\mathcal{H})$ as $\underline{\breve{A}}$ we have

$$\underline{\breve{\Sigma}} := F(\underline{\Sigma}) = p_J! \circ I(\underline{\Sigma}). \tag{16.58}$$

What happens to the terminal object? Given $\underline{1}_{\mathcal{V}(\mathcal{H})}$ we obtain

$$F(\underline{1}_{\mathcal{V}(\mathcal{H})}) = p_J! \circ I(\underline{1}_{\mathcal{V}(\mathcal{H})}) = p_J!(\underline{1}_{\Lambda(\underline{G/G_F})}). \tag{16.59}$$

Now the étale bundle associated to the sheaf $\underline{1}_{\Lambda(\underline{G/G_F})}$ is $p_1 : \Lambda(\{*\}) \to \Lambda(\underline{G/G_F})$ where $\Lambda(\{*\})$ represents the collection of singletons, one for each $w_V^g \in \Lambda(\underline{G/G_F})$. Obviously the étale bundle $p_1 : \Lambda(\{*\}) \to \Lambda(\underline{G/G_F})$ is nothing but $\Lambda(\underline{G/G_F})$. Thus by applying the definition of $p_J!$ we then get

$$p_J!(\underline{1}_{\Lambda(\underline{G/G_F})}) = \underline{G/G_F}. \tag{16.60}$$

It follows that the functor F does not preserve the terminal object therefore it can not be a right adjoint. In fact we would like F to be left adjoint.

We have seen above that the functor I preserves *colimits* (initial object) and *limits* which implies that I is both left and right adjoint. Since $F = p_J! \circ I$ and $p_J!$ is left adjoint, thus preserves *colimits*, it follows that F is left adjoint and thus preserve *colimits*.

Of particular importance to us is the following: each object $\underline{A} \in Sh(\mathcal{V}(\mathcal{H}))$ has associated to it the unique arrow $!\underline{A} : \underline{A} \to \underline{1}_{\mathcal{V}(\mathcal{H})}$. This arrow is epic, thus $F(!\underline{A}) : F(\underline{A}) \to F(\underline{1}_{\mathcal{V}(\mathcal{H})})$ is also epic. In particular we obtain

$$\begin{aligned} F(!\underline{A}) : F(\underline{A}) &\to F(\underline{1}_{\mathcal{V}(\mathcal{H})}) \\ \underline{\breve{A}} &\to \underline{G/G_F} \end{aligned} \tag{16.61}$$

such that for each $V \in \mathcal{V}(\mathcal{H})$ we get

$$\begin{aligned} \underline{\breve{A}}_V &\to (\underline{G/G_F})_V \\ \coprod_{w_V^g \in (\underline{G/G_F})_V} \underline{A}_{w_V^g} &\to G/G_{FV}. \end{aligned} \tag{16.62}$$

However, since we are considering sub-objects of the state object presheaf $\breve{\underline{\Sigma}}$ we would like the F functor to preserve also monic arrows. And indeed it does.

Lemma 16.3 *The functor* $F : Sh(\mathcal{V}(\mathcal{H})) \to Sh(\mathcal{V}_f(\mathcal{H}))$ *preserves monics.*

Proof Let $i : \underline{A} \to \underline{B}$ be a monic arrow in $Sh(\mathcal{V}(\mathcal{H}))$, then we have that

$$F(i) = p_J!\big(I(i)\big). \tag{16.63}$$

However, the I functor preserves monics, as a consequence $I(i)$ is monic in $Sh(\Lambda(\underline{G/G_F}))$.

Moreover, from the defining equation (16.56), it follows that, if $f : X \to Y$ is étale and $p_A : A \to X$ is étale then, since $i : A \to B$ is monic, so is $f!(i) : f!(A) \to f!(B)$. Therefore, applying this reasoning to our case it follows that $F(i) = p_J!(I(i))$ is monic. □

16.5 From Sheaves over $\mathcal{V}(\mathcal{H})$ to Sheaves over $\mathcal{V}(\mathcal{H}_f)$

Now that we have defined the functor F we will map all the sheaves in our original formalism $(Sh(\mathcal{V}(\mathcal{H})))$ to sheaves over $\mathcal{V}_f(\mathcal{H})$. We will then analyse how the truth values behave under such mappings.

16.5.1 Spectral Sheaf

Given the spectral sheaf $\underline{\Sigma} \in Sh(\mathcal{V}(\mathcal{H}))$ we define the following:

$$\breve{\underline{\Sigma}} := F(\underline{\Sigma}) = p_I! \circ I(\underline{\Sigma}). \tag{16.64}$$

This will be our new spectral sheaf whose definition, given below, is in terms of the corresponding presheaf (which we will still denote $\breve{\underline{\Sigma}}$). Once again we have used the correspondence between sheaves and presheaves, derived by the fact that the base category is a poset (see Chap. 14)

Definition 16.5 The spectral presheaf $\breve{\underline{\Sigma}}$ is defined on

– Objects: for each $V \in \mathcal{V}_f(\mathcal{H})$ we have

$$\breve{\underline{\Sigma}}_V := \coprod_{w_V^{g_i} \in \Lambda(\underline{G/G_F})_V} \underline{\Sigma}_{w_V^{g_i}} \simeq \coprod_{\phi_i \in Hom(\downarrow V, \mathcal{V}(\mathcal{H}))} \underline{\Sigma}_{\phi(V)} \tag{16.65}$$

which represents the disjoint union of the Gel'fand spectrum of all algebras related to V via a group transformation.

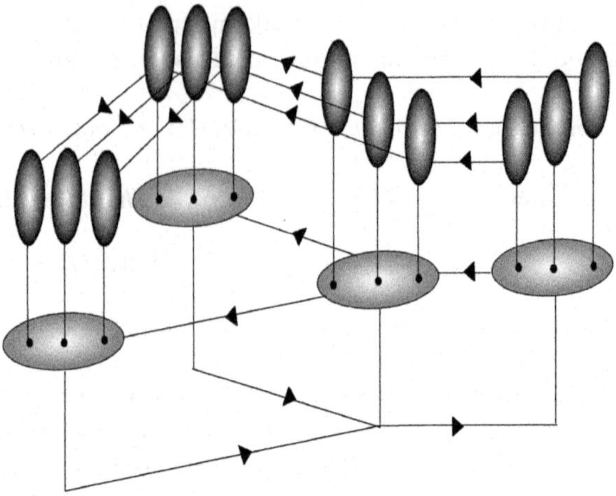

Fig. 16.1 Pictorial representation of the "two level" spectral presheaf

– Morphisms: given a morphism $i : V' \to V, (V' \subseteq V)$ in $\mathcal{V}_f(\mathcal{H})$ the corresponding spectral presheaf morphism is

$$\begin{aligned} \breve{\underline{\Sigma}}(i_{V'V}) : \breve{\underline{\Sigma}}_V &\to \breve{\underline{\Sigma}}_{V'} \\ \coprod_{\phi_i \in Hom(\downarrow V, \mathcal{V}(\mathcal{H}))} \underline{\Sigma}_{\phi_i(V)} &\to \coprod_{\phi_j \in Hom(\downarrow V', \mathcal{V}(\mathcal{H}))} \underline{\Sigma}_{\phi(V')} \end{aligned} \tag{16.66}$$

such that, given $\lambda \in \underline{\Sigma}_{\phi_i(V)}$, we obtain $\breve{\underline{\Sigma}}(i_{V'V})(\lambda) := \underline{\Sigma}_{\phi_i(V),\phi_j(V')}\lambda = \lambda_{|\phi_j(V')}$.

Thus, in effect $\breve{\underline{\Sigma}}(i_{V'V})$, is actually a coproduct of morphisms $\underline{\Sigma}_{\phi_i(V),\phi_j(V')}$, one for each $\phi_i \in Hom(\downarrow V, \mathcal{V}(\mathcal{H}))$.

From the above definition it is clear that the new spectral sheaf contains the information of all possible representations of a given abelian von-Neumann algebra at the same time. It is such an idea that might reveal itself fruitful, when considering how quantisation is defined in a topos. A pictorial representation of the new state space is given in Fig. 16.1

16.5.1.1 Topology on the State Space

We would now like to analyse what kind of topology the sheaf $\breve{\underline{\Sigma}} := F(\underline{\Sigma})$ has. We know that for each $V \in \mathcal{V}_f(\mathcal{H})$ we obtain the collection $\coprod_{w_V^{g_i} \in G/G_{FV}} \underline{\Sigma}_{w_V^{g_i}}$, where each $\underline{\Sigma}_{w_V^{g_i}} := \underline{\Sigma}_{\phi^{g_i}(V)}$ is equipped with the spectral topology. Thus we could equip $\breve{\underline{\Sigma}}$ with the disjoint union topology or with the spectral topology. In order to understand the spectral topology we should recall that the functor $F : Sh(\mathcal{V}(\mathcal{H})) \to Sh(\mathcal{V}_f(\mathcal{H}))$ preserves monics, thus if $\underline{S} \subseteq \underline{\Sigma}$, then $\breve{\underline{S}} := F(\underline{S}) \subseteq \breve{\underline{\Sigma}} := F(\underline{\Sigma})$. We can then define the spectral topology on $\breve{\underline{\Sigma}}$ as follows:

Definition 16.6 The spectral topology on $\underline{\breve{\Sigma}}$ has, as basis, the collection of clopen sub-objects $\underline{\breve{S}} \subseteq \underline{\breve{\Sigma}}$ which are defined for each $V \in \mathcal{V}_f(\mathcal{H})$ as

$$\underline{\breve{S}}_V := \coprod_{w_V^{g_i} \in G/G_{FV}} \underline{S}_{w_V^{g_i}} = \coprod_{\phi_i \in Hom(\Downarrow V, \mathcal{V}(\mathcal{H}))} \underline{S}_{\phi_i(V)}. \tag{16.67}$$

That the above is indeed a basis is easy to prove and it is left as an exercise for the reader.

From the definition it follows that on each element $\underline{\Sigma}_{w_V^{g_i}}$ of the stalks we retrieve the standard spectral topology. Moreover the map $p : \coprod_{w_V^{g_i} \in \Lambda(\underline{G/G_F})} \underline{\Sigma}_{w_V^{g_i}} \to \mathcal{V}_f(\mathcal{H})$ is continuous since $p^{-1}(\downarrow V) := \coprod_{w_{V'}^{g_i} \in \downarrow w_V^{g_i} | \forall w_V^{g_i} \in G/G_{FV}} \underline{\Sigma}_{w_{V'}^{g_i}}$ is the clopen sub-object which has value $\coprod_{w_{V'}^{g_i} \in G/G_{FV'}} \underline{\Sigma}_{w_{V'}^{g_i}}$, at each context $V' \in \downarrow V$ and $\emptyset$ everywhere else.

Similarly, as it was the case for the topology on $\underline{\Sigma} \in Sh(\mathcal{V}(\mathcal{H}))$, the spectral topology defined above is weaker than the product topology and it has the advantage that it takes into account both the 'vertical' topology on the fibres (derived from the fact that they are isomorphic to G/G_{V_i}) and the 'horizontal' topology on the base space $\mathcal{V}_f(\mathcal{H})$ (which is the Alexandroff topology). See Sect. 9.1.3.

A moment of thought will reveal that also with respect to the disjoint union topology the map p is continuous, however because of the above argument, from now on we will use the spectral topology on the spectral presheaf.

16.5.2 Quantity Value Object

We are now interested in mapping the quantity value object $\underline{\mathbb{R}}^{\leftrightarrow} \in Sh(\mathcal{V}(\mathcal{H}))$ to an object in $Sh(\mathcal{V}_f(\mathcal{H}))$ via the F functor. We thus define:

Definition 16.7 The quantity value object $\underline{\breve{R}}^{\leftrightarrow} := F(\underline{\mathbb{R}}^{\leftrightarrow}) = p_I! \circ I(\underline{\mathbb{R}}^{\leftrightarrow})$ is an $\mathbb{R}$-valued presheaf of order-preserving and order-reversing functions on $\mathcal{V}_f(\mathcal{H})$ defined as follows:

– On objects $V \in \mathcal{V}_f(\mathcal{H})$ we have

$$\left(F(\underline{\mathbb{R}}^{\leftrightarrow})\right)_V := \coprod_{\phi_i \in Hom(\Downarrow V, \mathcal{V}(\mathcal{H}))} \underline{\mathbb{R}}^{\leftrightarrow}_{\phi_i(V)} \tag{16.68}$$

where each[6]

$$\underline{\mathbb{R}}^{\leftrightarrow}_{\phi_i(V)} := \{(\mu,\nu) | \mu \in OP(\downarrow\phi_i(V),\mathbb{R}),\ \mu \in OR(\downarrow\phi_i(V),\mathbb{R}),\ \mu \leq \nu\}. \quad (16.70)$$

The downward set $\downarrow\phi_i(V)$ comprises all the sub-algebras $V' \subseteq \phi_i(V)$. The condition $\mu \leq \nu$ implies that for all $V' \in \downarrow\phi_i(V)$, $\mu(V') \leq \nu(V')$.

- On morphisms $i_{V'V} : V' \to V$ $(V' \subseteq V)$ we get:

$$\begin{aligned} \underline{\breve{R}}^{\leftrightarrow}(i_{V'V}) : \underline{\breve{R}}^{\leftrightarrow}_V &\to \underline{\breve{R}}^{\leftrightarrow}_{V'} \\ \coprod_{\phi_i \in Hom(\downarrow V, \mathcal{V}(\mathcal{H}))} \underline{\mathbb{R}}^{\leftrightarrow}_{\phi_i(V)} &\to \coprod_{\phi_j \in Hom(\downarrow V', \mathcal{V}(\mathcal{H}))} \underline{\mathbb{R}}^{\leftrightarrow}_{\phi_j(V')} \end{aligned} \quad (16.71)$$

where for each element $(\mu,\nu) \in \underline{\mathbb{R}}^{\leftrightarrow}_{\phi_i(V)}$ we obtain

$$\begin{aligned} \underline{\breve{R}}^{\leftrightarrow}(i_{V'V})(\mu,\nu) &:= \underline{R}^{\leftrightarrow}(i_{\phi_i(V),\phi_j(V')})(\mu,\nu) \\ &= (\mu_{|\phi_j(V')}, \nu_{|\phi_j(V')}) \end{aligned} \quad (16.72)$$

where $\mu_{|\phi_j(V')}$ denotes the restriction of μ to $\downarrow\phi_j(V') \subseteq \downarrow\phi_i(V)$ and, analogously, for $\nu_{|\phi_j(V')}$.

---

[6]Note that it is possible to define a topology on $\underline{\breve{\mathbb{R}}}$. To this end let us consider the set

$$\mathcal{R} = \coprod_{V \in \mathcal{V}_f(\mathcal{H})} \underline{\breve{\mathbb{R}}}^{\leftrightarrow}_V = \bigcup_{V \in \mathcal{V}_f(\mathcal{H})} \{V\} \times \underline{\breve{\mathbb{R}}}^{\leftrightarrow}_V \quad (16.69)$$

where each $\underline{\breve{\mathbb{R}}}^{\leftrightarrow}_V := \coprod_{\phi_i \in Hom(\downarrow V, \mathcal{V}(\mathcal{H}))} \underline{\mathbb{R}}^{\leftrightarrow}_{\phi_i(V)}$.

This set represents a bundle over $\mathcal{V}_f(\mathcal{H})$ with bundle map $p_{\mathcal{R}} : \mathcal{R} \to \mathcal{V}_f(\mathcal{H})$; $p_{\mathcal{R}}(\mu,\nu) = V = p_J(\phi_i)$, where V is the context such that $(\mu,\nu) \in \underline{\mathbb{R}}^{\leftrightarrow}_{\phi_i(V)}$. In this setting $p^{-1}_{\mathcal{R}}(V) = \underline{\breve{\mathbb{R}}}^{\leftrightarrow}_V$ are the fibres of the map $p_{\mathcal{R}}$. We would like to define a topology on $\mathcal{R}$ with the minimal requirement that the map $p_{\mathcal{R}}$ is continuous. We know that the category $\mathcal{V}_f(\mathcal{H})$ has the Alexandroff topology whose basis open sets are of the form $\downarrow V$ for some $V \in \mathcal{V}_f(\mathcal{H})$. Thus we are looking for a topology such that the pullback $p^{-1}_{\underline{\mathbb{R}}}(\downarrow V) := \coprod_{V' \in \downarrow V} \underline{\breve{\mathbb{R}}}_{V'}$ is open in $\mathcal{R}$.

Given the correspondence between sheaves and étale bundles, we know that each $\underline{\mathbb{R}}^{\leftrightarrow}$ is equipped with the discrete topology in which all sub-objects are open (in particular each $\underline{\mathbb{R}}^{\leftrightarrow}_V$ has the discrete topology). Since the F functor preserves monics, if $\underline{Q} \subseteq \underline{\mathbb{R}}^{\leftrightarrow}$ is open, then $F(\underline{Q}) \subseteq F(\underline{\mathbb{R}}^{\leftrightarrow})$ is open, where $F(\underline{Q}) := \coprod_{\phi_i \in Hom(\downarrow V, \mathcal{V}(\mathcal{H}))} \underline{Q}_{\phi_i(V)}$.

Therefore we define a sub-sheaf, $\underline{\breve{Q}}$, of $\underline{\breve{\mathbb{R}}}^{\leftrightarrow}$ to be *open* if for each $V \in \mathcal{V}_f(\mathcal{H})$ the set $\underline{\breve{Q}}_V \subseteq \underline{\breve{\mathbb{R}}}_V$ is open, i.e., each $\underline{Q}_{\phi_i(V)} \subseteq \underline{\mathbb{R}}^{\leftrightarrow}_{\phi_i(V)}$ is open in the discrete topology on $\underline{\mathbb{R}}^{\leftrightarrow}_{\phi_i(V)}$. It follows that the sheaf $\underline{\breve{\mathbb{R}}}^{\leftrightarrow}$ gets induced the discrete topology in which all sub-objects are open. In this setting the 'horizontal' topology on the base category $\mathcal{V}_f(\mathcal{H})$ would be accounted for by the sheave maps.

For each $\downarrow V$ we then obtain the open set $p^{-1}_{\underline{\mathbb{R}}}(\downarrow V)$ which has value $\underline{\breve{\mathbb{R}}}_{V'}$ at contexts $V' \in \downarrow V$ and $\emptyset$ everywhere else.

16.5.3 Truth Values

We now want to see what happens to the truth values when they are mapped via the functor F. In particular, given the sub-object classifier $\underline{\Omega}^{\mathcal{V}(\mathcal{H})} \in Sh(\mathcal{V}(\mathcal{H}))$ we want to know what $F(\underline{\Omega}^{\mathcal{V}(\mathcal{H})})$ is. Since

$$F\big(\underline{\Omega}^{\mathcal{V}(\mathcal{H})}\big) = p_J! \circ I\big(\underline{\Omega}^{\mathcal{V}(\mathcal{H})}\big) \tag{16.73}$$

we first of all need to analyse what $I(\underline{\Omega}^{\mathcal{V}(\mathcal{H})})$ is. Applying the definition for each $w_V^{g_i} \in \Lambda(\underline{G/G_F})$ we obtain

$$\big(I\big(\underline{\Omega}^{\mathcal{V}(\mathcal{H})}\big)\big)_{w_V^{g_i}} := \underline{\Omega}^{\mathcal{V}(\mathcal{H})}_{\phi_i(V)} \tag{16.74}$$

where $\phi_i \in Hom(\downarrow V, \mathcal{V}(\mathcal{H}))$ is the unique homeomorphism associated to the equivalence class $w_V^{g_i} \in G/G_{FV}$. If we then consider another element $w_V^{g_j} \in G/G_{FV}$, we then have

$$\big(I\big(\underline{\Omega}^{\mathcal{V}(\mathcal{H})}\big)\big)_{w_V^{g_j}} := \underline{\Omega}^{\mathcal{V}(\mathcal{H})}_{\phi_j(V)} \tag{16.75}$$

where now $\phi_i(V) \neq \phi_j(V)$. What this implies is that once we apply the functor $p_J!$, to push everything down to $\mathcal{V}_f(\mathcal{H})$, the distinct elements $\underline{\Omega}^{\mathcal{V}(\mathcal{H})}_{\phi_i(V)}$ and $\underline{\Omega}^{\mathcal{V}(\mathcal{H})}_{\phi_j(V)}$ will be pushed down to the same V, since both $\phi_i, \phi_j \in Hom(\downarrow V, \mathcal{V}(\mathcal{H})) = p_J^{-1}(V)$. It follows that, for every $V \in \mathcal{V}_f(\mathcal{H})$, $F(\underline{\Omega}^{\mathcal{V}(\mathcal{H})})$ is defined as

$$\begin{aligned} F\big(\underline{\Omega}^{\mathcal{V}(\mathcal{H})}\big)_V &:= \coprod_{w_V^{g_i} \in G/G_{FV}} \underline{\Omega}^{\mathcal{V}(\mathcal{H})}_{w_V^{g_i}} \simeq \bigcup_{w_V^{g_i} \in G/G_{FV}} \{w_V^{g_i}\} \times \underline{\Omega}^{\mathcal{V}(\mathcal{H})}_{w_V^{g_i}} \\ &\simeq \bigcup_{\phi_i \in Hom(\downarrow V, \mathcal{V}(\mathcal{H}))} \{\phi_i\} \times \underline{\Omega}^{\mathcal{V}(\mathcal{H})}_{\phi_i(V)}. \end{aligned} \tag{16.76}$$

Thus it seems that for each $V \in \mathcal{V}_f(\mathcal{H})$, $F(\underline{\Omega}^{\mathcal{V}(\mathcal{H})})_V$ assigns the disjoint union of the collection of sieves for each algebra $V_i \in \mathcal{V}(\mathcal{H})$ such that $V_i = \phi_i(V)$, where ϕ_i is the unique homeomorphisms associated to each $w_V^{g_i} \in G/G_{FV}$. This leads to the following theorem:

Theorem 16.6 $F(\underline{\Omega}^{\mathcal{V}(\mathcal{H})}) \simeq \underline{G/G_F} \times \underline{\Omega}^{\mathcal{V}(\mathcal{H})}$.

Before proving the above theorem it should be noted that $\mathcal{V}_f(\mathcal{H}) \simeq \mathcal{V}(\mathcal{H})$ since $\mathcal{V}_f(\mathcal{H})$ and $\mathcal{V}(\mathcal{H})$ are, in fact, the same categories only that, in the former, there is no group action on it. Thus it also follows trivially that $\underline{\Omega}^{\mathcal{V}_f(\mathcal{H})} \simeq \underline{\Omega}^{\mathcal{V}(\mathcal{H})}$. Having said that we can now proceed to prove the above theorem.

Proof For each $V \in \mathcal{V}_f(\mathcal{H})$ we define the map

$$\begin{aligned} i_V : F(\underline{\Omega}^{\mathcal{V}(\mathcal{H})})_V &\to G/G_{FV} \times \underline{\Omega}_V^{\mathcal{V}(\mathcal{H})} \\ S &\mapsto (w_V^{g_i}, l_{g_i^{-1}} S) \end{aligned} \tag{16.77}$$

where $S \in \underline{\Omega}_{w_V^{g_i}}^{\mathcal{V}(\mathcal{H})} = \underline{\Omega}_{\phi_i(V)}^{\mathcal{V}(\mathcal{H})}$ for $\phi_i \in Hom(\downarrow V, \mathcal{V}(\mathcal{H}))$ and $\phi_i(V) := l_{g_i} V$ while $l_{g_i^{-1}} S \in \underline{\Omega}_V^{\mathcal{V}(\mathcal{H})}$.

Such a map is one to one since if $(w_V^{g_i}, l_{g_i^{-1}} S_1) = (w_V^{g_i}, l_{g_i^{-1}} S_2)$ then $l_{g_i^{-1}} S_1 = l_{g_i^{-1}} S_2$ and $S_1 = S_2$. The fact that it is 'onto' follows from the definition.

Now, for each $V \in \mathcal{V}(\mathcal{H})$, we construct the map

$$\begin{aligned} j : G/G_{FV} \times \underline{\Omega}_V^{\mathcal{V}(\mathcal{H})} &\to F(\underline{\Omega}^{\mathcal{V}(\mathcal{H})})_V \\ (w_V^{g_i}, S) &\mapsto l_{g_i}(S) \end{aligned} \tag{16.78}$$

where $S \in \underline{\Omega}_V^{\mathcal{V}(\mathcal{H})}$ and $l_{g_i} S \in \underline{\Omega}_{l_{g_i} V}^{\mathcal{V}(\mathcal{H})}$ for $l_{g_i} V = \phi_i(V)$, thus $l_{g_i} S \in \underline{\Omega}_{w_V^{g_i}}^{\mathcal{V}(\mathcal{H})}$.

A moment of thought reveals that $j = i^{-1}$. □

A straightforward consequence of Theorem 16.6 is

Theorem 16.7 $\underline{\Omega}^{\mathcal{V}_f(\mathcal{H})} \simeq F(\underline{\Omega}^{\mathcal{V}(\mathcal{H})})/\underline{G}$.

Before proving the above theorem we, first of all, need to understand, precisely, what $F(\underline{\Omega}^{\mathcal{V}(\mathcal{H})})/\underline{G}$ is.

We already know that for presheaves over $\mathcal{V}_f(\mathcal{H})$ the group action is at the level of the base category $\Lambda(\underline{G/G_F})$. In particular for each $g \in G$ we have

$$\big(l_g^*(\underline{\Omega}^{\mathcal{V}(\mathcal{H})})\big)_{\phi(V)} := \underline{\Omega}_{l_g(\phi(V))}^{\mathcal{V}(\mathcal{H})} \tag{16.79}$$

where $\phi \in Hom(\downarrow V, \mathcal{V}(\mathcal{H}))$. Therefore by defining for each $V \in \mathcal{V}(\mathcal{H})$ the equivalence relation on $\coprod_{\phi_i \in Hom(\Downarrow V, \mathcal{V}(\mathcal{H}))} \underline{\Omega}_{\phi_i(V)}^{\mathcal{V}(\mathcal{H})} =: (F(\underline{\Omega}^{\mathcal{V}(\mathcal{H})}))_V$ by the action of G, the elements in $(F(\underline{\Omega}^{\mathcal{V}(\mathcal{H})}))_V/G_V = (\coprod_{\phi_i \in Hom(\Downarrow V, \mathcal{V}(\mathcal{H}))} \underline{\Omega}_{\phi_i(V)}^{\mathcal{V}(\mathcal{H})})/G$ will be equivalence classes of sieves, i.e.,

$$[S_i] := \{l_g(S_i) | g \in G\} \tag{16.80}$$

for each $S_i \in \underline{\Omega}_{\phi_i(V)}^{\mathcal{V}(\mathcal{H})} \in \coprod_{\phi_i \in Hom(\Downarrow V, \mathcal{V}(\mathcal{H}))} \underline{\Omega}_{\phi_i(V)}^{\mathcal{V}(\mathcal{H})}$. In the above we have used the action of the group G on sieves which is defined as $l_g S := \{l_g V' | V' \in S\}$. We are now ready to define the presheaf $F(\underline{\Omega}^{\mathcal{V}(\mathcal{H})})/\underline{G}$.

Definition 16.8 The Presheaf $F(\underline{\Omega}^{\mathcal{V}(\mathcal{H})})/\underline{G}$ is defined:

- On objects: for each context $V \in \mathcal{V}_f(\mathcal{H})$ we have the object

$$\left(F\left(\underline{\Omega}^{\mathcal{V}(\mathcal{H})}\right)\right)_V / G_V := \left(\coprod_{\phi_i \in Hom(\downarrow V, \mathcal{V}(\mathcal{H}))} \underline{\Omega}^{\mathcal{V}(\mathcal{H})}_{\phi_i(V)} \right) \Big/ (G) \tag{16.81}$$

whose elements are equivalence classes of sieves $[S_i]$, i.e., $S_1, S_2 \in [S_i]$ iff $S_1 = l_g S_2$ for some $g \in G$, $S_2 \in \underline{\Omega}^{\mathcal{V}(\mathcal{H})}_{\phi_i(V)}$ and $S_1 = \underline{\Omega}^{\mathcal{V}(\mathcal{H})}_{l_g\phi_i(V)}$. Therefore, each equivalence class will contain only one sieve for each algebra. This definition of equivalence condition follows from the fact that the group action of G moves each set $\underline{\Omega}^{\mathcal{V}(\mathcal{H})}_{\phi_i(V)}$ to another set $\underline{\Omega}^{\mathcal{V}(\mathcal{H})}_{l_g\phi_i(V)}$ in the same stork $F(\underline{\Omega}^{\mathcal{V}(\mathcal{H})})_V$, i.e. the group action is at the level of the base category $\Lambda(\underline{G/G_F})$ and acts vertically along the fibres of $p_J : \Lambda(\underline{G/G_F}) \to \mathcal{V}_f(\mathcal{H})$.
- On morphisms: for each $V' \subseteq V$ we then have the corresponding morphisms

$$\begin{aligned} \alpha_{VV'} : \left(\coprod_{\phi_i \in Hom(\downarrow V, \mathcal{V}(\mathcal{H}))} \underline{\Omega}^{\mathcal{V}(\mathcal{H})}_{\phi_i(V)} \right) \Big/ (G) &\to \left(\coprod_{\phi_j \in Hom(\downarrow V', \mathcal{V}(\mathcal{H}))} \underline{\Omega}^{\mathcal{V}(\mathcal{H})}_{\phi_j(V')} \right) \Big/ (G) \\ [S] &\mapsto \alpha_{VV'}\big([S]\big) := \big[S \cap \downarrow V'\big] \end{aligned} \tag{16.82}$$

where $[S \cap \downarrow V'] := \{l_g(S \cap V') | g \in G\}$, and we choose, as the representative for the equivalence class, $S \in \underline{\Omega}^{\mathcal{V}(\mathcal{H})}_V$ for $V = \phi_i(V)$, where $\phi_i \in Hom(\downarrow V, \mathcal{V}(\mathcal{H}))$ is associated to some $g \in G_V$.

We can now prove Theorem 16.7.

Proof We want to show that the functor

$$\beta : \underline{\Omega}^{\mathcal{V}_f(\mathcal{H})} \to F\left(\underline{\Omega}^{\mathcal{V}(\mathcal{H})}\right) / (\underline{G}) \tag{16.83}$$

is an isomorphism.

In particular for each context $V \in \mathcal{V}(\mathcal{H})$ we define

$$\begin{aligned} \beta_V : \underline{\Omega}^{\mathcal{V}_f(\mathcal{H})}_V &\to F\left(\underline{\Omega}^{\mathcal{V}(\mathcal{H})}\right)_V / (\underline{G})_V \\ S &\mapsto [S] \end{aligned} \tag{16.84}$$

where $[S]$ denotes the equivalence class to which the sieve S belongs to, i.e., $[S] := \{l_g S | g \in G\}$.

First we need to show that β is indeed a functor, i.e., we need to show that, for all pairs $V' \subseteq V$, the following diagram commutes

$$\begin{array}{ccc}
\underline{\Omega}_V^{\mathcal{V}_f(\mathcal{H})} & \xrightarrow{\beta_V} & F(\underline{\Omega}^{\mathcal{V}(\mathcal{H})})_V/G \\
\Big\downarrow {\scriptstyle \underline{\Omega}^{\mathcal{V}_f(\mathcal{H})}(i_{V'V})} & & \Big\downarrow {\scriptstyle \alpha_{VV'}} \\
\underline{\Omega}_{V'}^{\mathcal{V}_f(\mathcal{H})} & \xrightarrow{\beta_{V'}} & F(\underline{\Omega}^{\mathcal{V}(\mathcal{H})})_V/G
\end{array}$$

Thus for each S we obtain, for one direction,

$$\big(\beta_{V'} \circ \underline{\Omega}^{\mathcal{V}_f(\mathcal{H})}(i_{V'V})\big)(S) = \beta_{V'}\big(S \cap \downarrow V'\big) = \big[S \cap \downarrow V'\big] \tag{16.85}$$

where the first equality follows from the definition of the sub-object classifier $\underline{\Omega}^{\mathcal{V}_f(\mathcal{H})}$ [32].

Going the opposite direction we get

$$(\alpha_{VV'} \circ \beta_V)S = \alpha_{VV'}[S] = \big[S \cap \downarrow V'\big]. \tag{16.86}$$

It follows that indeed the above diagram commutes. Now that we have showed that β is a functor we need to show that it is an isomorphisms. Consider each individual component β_V, $V \in \mathcal{V}_f(\mathcal{H})$.

1. *The map β_V is one-to-one.*

Given $S_1, S_2 \in \underline{\Omega}_V^{\mathcal{V}_f(\mathcal{H})}$, if $\beta_V(S_1) = \beta_V(S_2)$ then $[S_1] = [S_2]$, thus both S_1 and S_2 belong to the same equivalence class. Each equivalence class is of the form $[S] = \{l_g S | g \in G\}$, therefore $S_1 = l_g S_2$ for some $g \in G$. From the definition of the group action it follows that for each $V \in \mathcal{V}(\mathcal{H})$, $\underline{\Omega}_V^{\mathcal{V}_f(\mathcal{H})}$ will have at most one sieve in each equivalence class. Thus if $[S_1] = [S_2]$ and both $S_1, S_2 \in \underline{\Omega}_V^{\mathcal{V}_f(\mathcal{H})}$, then $S_1 = S_2$.

2. *The map β_V is onto.* This follows at once from the definition.

3. *The map β_V has an inverse.*

We now need to define an inverse. We choose

$$\gamma : F\big(\underline{\Omega}^{\mathcal{V}(\mathcal{H})}\big)/G \to \underline{\Omega}^{\mathcal{V}_f(\mathcal{H})} \tag{16.87}$$

such that for each context we get

$$\begin{aligned} \gamma_V : F\big(\underline{\Omega}^{\mathcal{V}(\mathcal{H})}\big)_V/G &\to \underline{\Omega}_V^{\mathcal{V}_f(\mathcal{H})} \\ [S] &\mapsto [S] \cap V \end{aligned} \tag{16.88}$$

where $[S] \cap V := \{l_g(S) \cap \downarrow V | g \in G\}$ represents the only sieve in the equivalence class which belongs to $\underline{\Omega}_V^{\mathcal{V}_f(\mathcal{H})}$. First of all we have to prove that this is indeed a functor, i.e. we need to show that, for each $V' \subseteq V$, the following diagram commutes:

$$\begin{array}{ccc}
F(\underline{\Omega}^{\mathcal{V}(\mathcal{H})})_V/G & \xrightarrow{\gamma_V} & \underline{\Omega}_V^{\mathcal{V}_f(\mathcal{H})} \\
\downarrow{\scriptstyle \alpha_{VV'}} & & \downarrow{\scriptstyle \underline{\Omega}^{\mathcal{V}_f(\mathcal{H})}(i_{V'V})} \\
F(\underline{\Omega}^{\mathcal{V}(\mathcal{H})})_V/G & \xrightarrow{\gamma_{V'}} & \underline{\Omega}^{\mathcal{V}_f(\mathcal{H})}V'
\end{array}$$

Chasing the diagram around for each $[S] \in F(\underline{\Omega}^{\mathcal{V}(\mathcal{H})})/G$ we obtain

$$\underline{\Omega}^{\mathcal{V}_f(\mathcal{H})}(i_{V'V}) \circ \gamma_V\big([S]\big) = \underline{\Omega}^{\mathcal{V}_f(\mathcal{H})}(i_{V'V})\big([S] \cap V\big) = \big([S] \cap V\big) \cap V' = [S] \cap V'. \tag{16.89}$$

On the other hand we have

$$\gamma_{V'} \circ \alpha_{VV'}[S] = \gamma_{V'}\big[S \cap \downarrow V'\big] = \big[S \cap \downarrow V'\big] \cap V' = [S] \cap V' \tag{16.90}$$

where the last equality follows since $[S \cap \downarrow V'] \cap V' := \{l_g(S \cap \downarrow V') | g \in G\} \cap V'$ and the only sieve in $[S]$ belonging to $\underline{\Omega}_{V'}^{\mathcal{V}_f(\mathcal{H})}$ is $S \cap \downarrow V'$. Therefore the map γ is a functor.

It now remains to show that, for each $V \in \mathcal{V}_f(\mathcal{H})$ and each $S \in \underline{\Omega}_V^{\mathcal{V}(\mathcal{H})}$, γ_V is the inverse of β_V. Thus

$$\gamma_V \circ \beta_V(S) = \gamma_V\big([S]\big) = [S] \cap V = S \tag{16.91}$$

where the last equality follows from the fact that in each equivalence class of sieves there is one and only one sieve referred to each context V. On the other hand we have

$$\beta_V \circ \gamma_V\big([S]\big) = \beta_V \circ \big([S] \cap V\big) = \beta_V(S) = [S]. \tag{16.92}$$

The functor β is indeed an isomorphism. □

16.6 Group Action on the New Sheaves

We would now like to analyse what the group action on the new sheaves is. In particular, we will show how the action of the group $\underline{G}$ on the sheaves, defined on $\mathcal{V}_f(\mathcal{H})$ via the F functor, will not induce twisted sheaves.

16.6.1 Spectral Sheaf

The action of the group $\underline{G}$ on the new spectral sheaf $\underline{\breve{\Sigma}} := F(\underline{\Sigma})$ is given by the following map:

$$\underline{G} \times \underline{\breve{\Sigma}} \to \underline{\breve{\Sigma}} \tag{16.93}$$

defined for each context $V \in \mathcal{V}_f(\mathcal{H})$ as

$$\underline{G}_V \times \underline{\breve{\Sigma}}_V \to \underline{\breve{\Sigma}}_V \tag{16.94}$$

$$(g, \lambda) \mapsto l_g \lambda \tag{16.95}$$

where $\underline{\breve{\Sigma}}_V := \coprod_{\phi_i \in Hom(\Downarrow V, \mathcal{V}(\mathcal{H}))} \underline{\Sigma}_{\phi_i(V)}$ such that if $\lambda \in \underline{\Sigma}_{\phi_i(V)}$ we define $l_g \lambda \in l_g \underline{\Sigma}_{\phi_i(V)} := \underline{\Sigma}_{l_g(\phi_i(V))}$ by

$$\big(l_g(\lambda)\big)\hat{A} := \big\langle \lambda, \hat{U}(g)^{-1}\hat{A}\hat{U}(g)\big\rangle \tag{16.96}$$

for all $g \in G$, $\hat{A} \in V_{sa}$(self-adjoint operators in V) and $V \in \mathcal{V}(\mathcal{H})$.

However from the definition of $\underline{\breve{\Sigma}}$, both $\underline{\Sigma}_{\phi_i(V)}$ and $\underline{\Sigma}_{l_g(\phi_i(V))}$ belong to the same stalk, i.e., belong to $\underline{\breve{\Sigma}}_V$.

We then obtain a well defined group action which does not induce twisted presheaves.

We would now like to check whether such a group action is continuous with respect to the spectral topology, i.e., if the map

$$\rho : \underline{G} \times \underline{\breve{\Sigma}} \to \underline{\breve{\Sigma}} \tag{16.97}$$

is continuous. In particular we want to check if, for each $V \in \mathcal{V}_f(\mathcal{H})$, the local component

$$\rho_V : \underline{G}_V \times \underline{\breve{\Sigma}}_V \to \underline{\breve{\Sigma}}_V \tag{16.98}$$

is continuous, i.e., if $\rho_V^{-1}\underline{\breve{S}}_V = \rho_V^{-1}(\coprod_{\phi_i \in Hom(\Downarrow V, \mathcal{V}(\mathcal{H}))} \underline{S}_{\phi_i(V)})$ is open for $\underline{\breve{S}}_V$ open.

$$\begin{aligned}\rho_V^{-1}\Bigg(\coprod_{\phi_i \in Hom(\Downarrow V, \mathcal{V}(\mathcal{H}))} \underline{S}_{\phi_i(V)}\Bigg) &= \big\{(g_j, \underline{S}_{\phi_i(V)}) | l_{g_j}(\underline{S}_{\phi_i(V)}) \in \underline{\breve{S}}_V\big\} \\ &= (G, \underline{\breve{S}}_V)\end{aligned} \tag{16.99}$$

where $l_{g_j}(\underline{S}_{\phi_i(V)}) := \underline{S}_{l_{g_j}\phi_i(V)} = \underline{S}_{l_{g_j}(\phi_i(V))}$. It follows that indeed the action is continuous.

Moreover it seems that the sub-objects $\underline{\breve{S}}$ actually remain invariant under the group action. In fact, for each $V \in \mathcal{V}_f(\mathcal{H})$, $\underline{\breve{S}}_V = \coprod_{\phi_i \in Hom(\downarrow V, \mathcal{V}(\mathcal{H}))} \underline{S}_{\phi_i(V)}$ and the set $Hom(\downarrow V, \mathcal{V}(\mathcal{H}))$ contains all G related homeomorphisms, i.e. all $l_{g_j}(\phi_i)$ $\forall g_j \in G$, such that $l_{g_j}(\phi)(V) := l_{g_j}(\phi(V))$.

It follows that the sub-objects $\underline{\breve{S}} \subseteq \underline{\breve{\Sigma}}$ are invariant under the group action.

This is an important result when considering propositions which are identified with clopen sub-objects coming from daseinisation. In this context the group action is defined, for each $V \in \mathcal{V}_f(\mathcal{H})$, as:

$$\underline{G}_V \times \underline{\delta \breve{P}}_V \to \underline{\delta \breve{P}}_V$$
$$\underline{G}_V \times \coprod_{\phi_i \in Hom(\Downarrow V, \mathcal{V}(\mathcal{H}))} \delta^o(\hat{P})_{\phi_i(V)} \to \coprod_{\phi_i \in Hom(\Downarrow V, \mathcal{V}(\mathcal{H}))} \delta^o(\hat{P})_{\phi_i(V)} \tag{16.100}$$
$$\big(g, \delta^o(\hat{P})_{\phi_i(V)}\big) \mapsto \delta^o\big(\hat{U}_g \hat{P} \hat{U}_g^{-1}\big)_{l_g(\phi_i(V))}.$$

Thus for each $g \in G$ we get a collection of transformations, each similar to those obtained in the original formalism. However, since the effect of such a transformation is to move the objects around within a stalk, when considering the action of the entire G, the stalk, as an entire set, remains invariant, i.e., the collection of local component of the propositions stays the same.

Moreover the fact that individual sub-objects $\underline{\breve{S}} \subseteq \underline{\breve{\Sigma}}$ are invariant under the group action, implies that the action $\underline{G} \times \underline{\breve{\Sigma}} \to \underline{\breve{\Sigma}}$ is not transitive. In fact the transitivity of the action of the group action on a sheaf is defined as follows:

Definition 16.9 Given a group $\underline{G}$, we say that the action of $\underline{G}$ on any other sheaf $\underline{A}$ is transitive iff there are no invariant sub-objects of A.

Thus, although the group actions moves the elements around in each stalk, it never moves elements in between different stalks, thus each sub-object is left invariant.

16.6.2 Sub-object Classifier

We now are interested in defining the group action on the sub-object classifier $\underline{\Omega}^{\mathcal{V}_f(\mathcal{H})}$. However, by definition, there is no action on such object. The only action which could be defined would be the action on $\underline{\breve{\Omega}} := F(\underline{\Omega}^{\mathcal{V}(\mathcal{H})})$. In this case, for each $V \in \mathcal{V}_f(\mathcal{H})$, we have

$$\alpha_V : \underline{G}_V \times \underline{\breve{\Omega}}_V \to \underline{\breve{\Omega}}_V$$
$$\underline{G}_V \times \coprod_{w_V^{g_i} \in G/G_{FV}} \underline{\Omega}_{w_V^{g_i}} \to \coprod_{w_V^{g_i} \in G/G_{FV}} \underline{\Omega}_{w_V^{g_i}} \tag{16.101}$$
$$\underline{G}_V \times \coprod_{\phi_i \in Hom(\Downarrow V, \mathcal{V}(\mathcal{H}))} \underline{\Omega}_{\phi_i(V)} \to \coprod_{\phi_i \in Hom(\Downarrow V, \mathcal{V}(\mathcal{H}))} \underline{\Omega}_{\phi_i(V)} \tag{16.102}$$
$$(g, S) \mapsto l_g(S)$$

where $l_g(S) := \{l_g V | V \in S\}$.

If $S \in \underline{\Omega}_{\phi_i(V)} \in \coprod_{\phi_i \in Hom(\Downarrow V, \mathcal{V}(\mathcal{H}))} \underline{\Omega}_{\phi_i(V)}$, then $l_g(S)$ is a sieve on $l_g\phi_i(V)$, i.e., $l_g(S) \in \underline{\Omega}_{l_g\phi_i(V)} \in \coprod_{\phi_i \in Hom(\Downarrow V, \mathcal{V}(\mathcal{H}))} \underline{\Omega}_{\phi_i(V)}$.

It follows that the action of the group $\underline{G}$ is to move sieves around in each stalk but never to move sieves to different stalks.

The next question is to define a topology on $\underline{\breve{\Omega}}$ and check whether the action is continuous or not.

A possible topology would be the topology whose basis are the collection of open sub-sheaves of $\underline{\breve{\Omega}}$. If we assume that each $\underline{\Omega}$ has the discrete topology coming from the fact that it can be seen as an étale bundle, then the topology on $\underline{\breve{\Omega}}$ will be the topology in which each sub-sheaf is open, i.e., the discrete topology.

Given such a topology we would like to check if the group action is continuous. To this end we need to show that $\alpha_{\overline{V}}^{-1}(\underline{\breve{S}}_V)$ is open for $\breve{S}_V$ being an open sub-object. We recall that $\underline{\breve{S}}_V = \coprod_{\phi_i \in Hom(\Downarrow V, \mathcal{V}(\mathcal{H}))} \underline{S}_{\phi_i(V)}$. We then obtain

$$\begin{aligned} \alpha_{\overline{V}}^{-1}(\underline{\breve{S}}_V) &= \{(g, S) | l_g(S) \in \underline{\breve{S}}_V\} \\ &= (\underline{G}_V, \underline{\breve{S}}_V) \end{aligned} \tag{16.103}$$

which is open.

16.6.3 Quantity Value Object

We would now like to analyse how the group acts on the new quantity value object $\underline{\breve{\mathbb{R}}}^{\leftrightarrow}$. This is defined via the map

$$\underline{G} \times \underline{\breve{\mathbb{R}}}^{\leftrightarrow} \to \underline{\breve{\mathbb{R}}}^{\leftrightarrow} \tag{16.104}$$

which, for each $V \in \mathcal{V}_f(\mathcal{H})$, has local components

$$\begin{aligned} \underline{G}_V \times \underline{\breve{\mathbb{R}}}^{\leftrightarrow}_V &\to \underline{\breve{\mathbb{R}}}^{\leftrightarrow}_V \\ \underline{G}_V \times \coprod_{\phi_i \in Hom(\Downarrow V, \mathcal{V}(\mathcal{H}))} \underline{\mathbb{R}}^{\leftrightarrow}_{\phi_i(V)} &\to \coprod_{\phi_i \in Hom(\Downarrow V, \mathcal{V}(\mathcal{H}))} \underline{\mathbb{R}}^{\leftrightarrow}_{\phi_i(V)} \\ (g, (\mu, \nu)) &\mapsto (l_g\mu, l_g\nu) \end{aligned} \tag{16.105}$$

where $(\mu, \nu) \in \underline{\mathbb{R}}^{\leftrightarrow}_{\phi_i(V)}$, while $(l_g\mu, l_g\nu) \in \underline{\mathbb{R}}^{\leftrightarrow}_{l_g(\phi_i(V))}$. Therefore $l_g\mu : \downarrow l_g(\phi_i(V)) \to \mathbb{R}$ and $l_g\nu : \downarrow l_g(\phi_i(\nu)) \to \mathbb{R}$.

As it can be easily deduced, even in this case the action of the $\underline{G}$ group is to map elements around in the same stalk but never to map elements between different stalks. Thus, even in this case, we do not obtain twisted sheaves.

We would now like to check whether the group action is continuous with respect to the discrete topology on $\underline{\breve{\mathbb{R}}}$. Therefore we have to check whether for $V \in \mathcal{V}_f(\mathcal{H})$

the following map is continuous:

$$\begin{aligned}\Phi_V : \underline{G}_V \times \underline{\breve{\mathbb{R}}}_V &\to \underline{\breve{\mathbb{R}}}_V \\ \big(g, (\mu, \nu)\big) &\to (l_g\mu, l_g\nu).\end{aligned} \tag{16.106}$$

A typical open set in $\underline{\breve{\mathbb{R}}}_V$ is of the form $\underline{\breve{Q}}_V := \coprod_{\phi_i \in Hom(\downarrow V, \mathcal{V}(\mathcal{H}))} \underline{Q}_{\phi_i(V)}$ where each $\underline{Q}_{\phi_i(V)} \subseteq \underline{\mathbb{R}}^{\leftrightarrow}_{\phi_i(V)}$ is open. Therefore

$$\begin{aligned}\Phi_V^{-1}(\underline{\breve{Q}}_V) &= \big\{g_i, (\mu, \nu) | (l_{g_i}\mu, l_{g_i}\nu) \in \underline{\breve{Q}}_V\big\} \\ &= (G, \underline{\breve{Q}}_V).\end{aligned} \tag{16.107}$$

Thus the group action with respect to the discrete topology is continuous.

16.6.4 Truth Object

The new truth value object for pure states obtained through the action of the F functor is

$$\underline{\breve{\mathbb{T}}}^{|\psi\rangle} := F\big(\underline{\mathbb{T}}^{|\psi\rangle}\big) \tag{16.108}$$

which is defined as follows:

Definition 16.10 The truth object $F(\underline{\mathbb{T}}^{|\psi\rangle})$ is the presheaf defined on

- Objects: for each $V \in \mathcal{V}_f(\mathcal{H})$ we get

$$F\big(\underline{\mathbb{T}}^{|\psi\rangle}\big) := \coprod_{\phi_i \in Hom(\downarrow V, \mathcal{V}(\mathcal{H}))} \underline{\mathbb{T}}^{|\psi\rangle}_{\phi_i(V)} \tag{16.109}$$

  where $\underline{\mathbb{T}}^{|\psi\rangle}_{\phi_i(V)} := \{\hat{\alpha} \in P(\phi_i(V)) | \langle\psi|\hat{\alpha}|\psi\rangle = 1\}$ and $P(\phi_i(V))$ denotes the collection of all projection operators in $\phi_i(V)$.
- Morphisms: given $V' \subseteq V$ the corresponding map is

$$\underline{\breve{\mathbb{T}}}^{|\psi\rangle}(i_{V'V}) : \coprod_{\phi_i \in Hom(\downarrow V, \mathcal{V}(\mathcal{H}))} \underline{\mathbb{T}}^{|\psi\rangle}_{\phi_i(V)} \to \coprod_{\phi_j \in Hom(\downarrow V', \mathcal{V}(\mathcal{H}))} \underline{\mathbb{T}}^{|\psi\rangle}_{\phi_j(V')} \tag{16.110}$$

  such that, given $\underline{S} \in \underline{\mathbb{T}}^{|\psi\rangle}_{\phi_i(V)}$, then

$$\underline{\breve{\mathbb{T}}}^{|\psi\rangle}(i_{V'V})\underline{S} := \underline{\mathbb{T}}^{|\psi\rangle}(i_{\phi_i(V), \phi_j(V')})\underline{S} = \underline{S}_{|\phi_j(V')} \tag{16.111}$$

  where $\phi_j \leq \phi_i$ thus $\phi_j(V') \subseteq \phi_i(V)$ and $\phi_j(V') = \phi_{i\,|V'}(V')$.

In order to define the truth object for density matrices we need to change the topos as it was done in [24] and explained in details in Chap. 15, but this time replacing the category $\mathcal{V}(\mathcal{H})$ with $\mathcal{V}_f(\mathcal{H})$. In particular we need to go to the topos $Sh(\mathcal{V}_f(\mathcal{H}) \times (0,1)_L)$. This is done by first defining the map $pr_1 : \mathcal{V}_f(\mathcal{H}) \times (0,1)_L \to \mathcal{V}_f(\mathcal{H})$, which gives rise to the geometric morphisms, whose inverse image part is $pr_1^* : Sh(\mathcal{V}_f(\mathcal{H})) \to Sh(\mathcal{V}_f(\mathcal{H}) \times (0,1)_L)$.

We then compose the two functors

$$Sh(\mathcal{V}(\mathcal{H})) \xrightarrow{F} Sh(\mathcal{V}_f(\mathcal{H})) \xrightarrow{pr_1^*} Sh(\mathcal{V}_f(\mathcal{H}) \times (0,1)_L). \tag{16.112}$$

It is such a functor which is used to map our original truth object (for each density matrix ρ) $\underline{\mathbb{T}}^\rho \in Sh(\mathcal{V}(\mathcal{H}))$ to our new truth object $\breve{\underline{\mathbb{T}}}^\rho \in Sh(\mathcal{V}_f(\mathcal{H}) \times (0,1)_L)$:

$$\breve{\underline{\mathbb{T}}}^\rho := pr_1^* \circ F(\underline{\mathbb{T}}^\rho). \tag{16.113}$$

The definition is as follows:

Definition 16.11 The truth object presheaf $\breve{\underline{\mathbb{T}}}^\rho$ is defined on

- Objects: for each pair $(V, r) \in \mathcal{V}_f(\mathcal{H}) \times (0,1)_L$ we obtain

$$\breve{\underline{\mathbb{T}}}^\rho_{(V,r)} := \coprod_{\phi_i \in Hom(\Downarrow V, \mathcal{V}(\mathcal{H}))} \underline{\mathbb{T}}^{\rho_{\phi_i}}_{(\phi_i(V),r)} \tag{16.114}$$

  where $\rho_{\phi_i} : \phi_i(V) \to \mathbb{C}$ represents a state on the algebra $\phi_i(V)$ and each

$$\underline{\mathbb{T}}^{\rho_{\phi_i}}_{(\phi_i(V),r)} = \{\underline{S} \in Sub(\underline{\Sigma}_{\downarrow \phi_i(V)}) | \forall V_k \subseteq \phi_i(V), \rho_{\phi_i}(\hat{P}_{\underline{S}_{V_k}}) \geq r\}. \tag{16.115}$$

  In this case, since we are considering density matrices, these are associated to normal states of the von Neuamnn algebra, hence $\rho_{\phi_i}(\hat{P}_{\underline{S}_{V_k}}) = \mathrm{tr}(\rho_{\phi_i} \hat{P}_{\underline{S}_{V_k}})$.
- Morphisms: given a map $i : (V', r') \leq (V, r)$ (iff $V' \subseteq V$ and $r' \leq r$), then the corresponding map is

$$\breve{\underline{\mathbb{T}}}^\rho(i) : \coprod_{\phi_i \in Hom(\Downarrow V, \mathcal{V}(\mathcal{H}))} \underline{\mathbb{T}}^{\rho_{\phi_i}}_{(\phi_i(V),r)} \to \coprod_{\phi_j \in Hom(\Downarrow V', \mathcal{V}(\mathcal{H}))} \underline{\mathbb{T}}^{\rho_{\phi_j}}_{(\phi_j(V'),r')} \tag{16.116}$$

  such that given a sub-object $\underline{S} \in \underline{\mathbb{T}}^{\rho_{\phi_i}}_{(\phi_i(V),r)}$ we get

$$\breve{\underline{\mathbb{T}}}^\rho(i)\underline{S} := \underline{\mathbb{T}}^{\rho_{\phi_i}}(i_{\phi_i(V),\phi_j(V')})\underline{S} = \underline{S}_{|\phi_j(V')} \tag{16.117}$$

  where $\phi_j \leq \phi_i$ thus $\phi_j(V') \subseteq \phi_i(V)$ and $\phi_j(V') = (\phi_i)_{|\phi_j}(V')$. Obviously now the condition on the restricted sub-object is $\rho(\hat{P}_{S_{V''}}) = \mathrm{tr}(\rho \hat{P}_{S_{V''}}) \geq r'$ where $V'' \subseteq \phi_j(V')$. However such a condition is trivially satisfied since $r' \leq r$.

We would now like to define the group action on such an object. Thus, we define the following:

$$\underline{G} \times \underline{\breve{\mathbb{T}}}^{\rho} \to \underline{\breve{\mathbb{T}}}^{\rho} \tag{16.118}$$

such that for each context $V \in \mathcal{V}_f(\mathcal{H})$ we obtain

$$\begin{aligned} \underline{G}_V \times \underline{\breve{\mathbb{T}}}^{\rho}_V &\to \underline{\breve{\mathbb{T}}}^{\rho}_V \\ (g, \underline{S}) &\mapsto l_g(\underline{S}) \end{aligned} \tag{16.119}$$

where $\underline{S} \in \underline{\mathbb{T}}^{\rho\phi_i}_{\phi_i(V)} \in \underline{\breve{\mathbb{T}}}^{\rho}_V$, while $l_g(\underline{S}) \in \underline{\mathbb{T}}^{\rho l_g(\phi_i)}_{l_g(\phi_i(V))} \in \underline{\breve{\mathbb{T}}}^{\rho}_V$. Here $\underline{\mathbb{T}}^{\rho l_g(\phi_i)}_{l_g(\phi_i(V))} = l_g \underline{\mathbb{T}}^{\rho\phi_i}_{\phi_i(V)}$.

Therefore also for the truth object the action of the group is to map elements around in a given stalk, but never to map elements in between stalks.

16.7 New Representation of Physical Quantities

We are now interested in understanding the action of the F functor on physical quantities. We thus define the following:

$$F\big(\breve{\delta}(\hat{A})\big) : \underline{\breve{\Sigma}} \to \underline{\breve{\mathbb{R}}}^{\leftrightarrow} \tag{16.120}$$

which, at each context V, is defined as

$$F\big(\breve{\delta}(\hat{A})\big)_V : \coprod_{\phi_i \in Hom(\Downarrow V, \mathcal{V}(\mathcal{H}))} \underline{\Sigma}_{\phi_i(V)} \to \coprod_{\phi_i \in Hom(\Downarrow V, \mathcal{V}(\mathcal{H}))} \underline{\mathbb{R}}^{\leftrightarrow}_{\phi_i(V)} \tag{16.121}$$

such that for a given $\lambda \in \underline{\Sigma}_{\phi_i(V)}$ we obtain

$$\begin{aligned} F\big(\breve{\delta}(\hat{A})\big)_V(\lambda) &:= \breve{\delta}(\hat{A})_{\phi_i(V)}(\lambda) \\ &= \big(\breve{\delta}^i(\hat{A})_{\phi_i(V)}(\cdot), \breve{\delta}^o(\hat{A})_{\phi_i(V)}(\cdot)\big)(\lambda) = (\mu_\lambda, \nu_\lambda). \end{aligned} \tag{16.122}$$

Thus, in effect, the map $F(\breve{\delta}(\hat{A}))_V$ is a coproduct of maps of the form $\breve{\delta}(\hat{A})_{\phi_i(V)}$ for all $\phi_i \in Hom(\downarrow V, \mathcal{V}(\mathcal{H}))$.

From this definition it is straightforward to understand how the group acts on such physical quantities. In particular, for each context $V \in \mathcal{V}_f(\mathcal{H})$ we obtain a collection of maps

$$F\big(\breve{\delta}(\hat{A})\big)_V : \coprod_{\phi_i \in Hom(\Downarrow V, \mathcal{V}(\mathcal{H}))} \underline{\Sigma}_{\phi_i(V)} \to \coprod_{\phi_i \in Hom(\Downarrow V, \mathcal{V}(\mathcal{H}))} \underline{\mathbb{R}}^{\leftrightarrow}_{\phi_i(V)}. \tag{16.123}$$

The group action is to map each individual maps of such a collection into one another. Thus, for example, if we consider the component

$$\breve{\delta}(\hat{A})_{\phi_i(V)} : \underline{\Sigma}_{\phi_i(V)} \to \underline{\mathbb{R}}^{\leftrightarrow}_{\phi_i(V)} \tag{16.124}$$

by acting on it with an element g of the group we would obtain

$$\begin{aligned} l_g\big(\breve{\delta}(\hat{A})_{\phi_i(V)}\big) : l_g\underline{\Sigma}_{\phi_i(V)} &\to l_g\underline{\mathbb{R}}^{\leftrightarrow}_{\phi_i(V)} \\ \underline{\Sigma}_{l_g(\phi_i(V))} &\to \underline{\mathbb{R}}^{\leftrightarrow}_{l_g(\phi_i(V))}. \end{aligned} \tag{16.125}$$

Let us now analyse what $l_g(\breve{\delta}(\hat{A}))_{\phi_i(V)}$ is.

We know that it is comprised of two functions, namely

$$l_g\big(\breve{\delta}(\hat{A})_{\phi_i(V)}\big) = \big(l_g\big(\breve{\delta}^i(\hat{A})\big)_{\phi_i(V)}\big)(\cdot), \big(l_g\big(\breve{\delta}^o(\hat{A})\big)_{\phi_i(V)}\big)(\cdot)\big). \tag{16.126}$$

We will consider each of them separately. Given $\lambda \in \underline{\Sigma}_{l_g(\phi_i(V))}$ we obtain

$$\begin{aligned} l_g\big(\breve{\delta}^i(\hat{A})_{\phi_i(V)}\big)(\lambda) &= \lambda\big(l_g\big(\delta^i(\hat{A})_{\phi_i(V)}\big)\big) \\ &= \lambda\big(\hat{U}_g\big(\delta^i(\hat{A})_{\phi_i(V)}\big)\hat{U}_g^{-1}\big) \\ &= \lambda\big(\delta^i\big(\hat{U}_g\hat{A}\hat{U}_g^{-1}\big)_{l_g(\phi_i(V))}\big) \\ &= \breve{\delta}^i\big(\hat{U}_g\hat{A}\hat{U}_g^{-1}\big)_{l_g(\phi_i(V))}(\lambda). \end{aligned} \tag{16.127}$$

Similarly for the order reversing function we obtain

$$l_g\big(\breve{\delta}^o(\hat{A})_{\phi_i(V)}\big)(\lambda) = \breve{\delta}^o\big(\hat{U}_g\hat{A}\hat{U}_g^{-1}\big)_{l_g(\phi_i(V))}(\lambda). \tag{16.128}$$

Hence putting the two results together we have

$$l_g\big(\breve{\delta}(\hat{A})_{\phi_i(V)}\big) = \big(\breve{\delta}\big(\hat{U}_g\hat{A}\hat{U}_g^{-1}\big)_{l_g(\phi_i(V))}\big). \tag{16.129}$$

This is the topos analogue of the standard transformation of self-adjoint operators in the canonical formalism of quantum theory. In particular, given a self-adjoint operator $\breve{\delta}(\hat{A})$ its local component in the context V is $\breve{\delta}(\hat{A})_V$. This 'represents' the pair of self-adjoint operators $(\delta^i(\hat{A})_V, \delta^o(\hat{A})_V)$ which live in V. By acting with a unitary transformation we obtain the transformed quantity $l_g(\breve{\delta}(\hat{A}))$ with local component $(\breve{\delta}(\hat{U}_g\hat{A}\hat{U}_g^{-1})_{l_g V})$, $V \in \mathcal{V}_f(\mathcal{H})$. Such a quantity represents the pair $(\delta^i(\hat{U}_g\hat{A}\hat{U}_g^{-1})_{l_g(V)}, \delta^o(\hat{U}_g\hat{A}\hat{U}_g^{-1})_{l_g(V)})$ of self-adjoint operators living in the transformed context $l_g(V)$.

Chapter 17
Topos History Quantum Theory

Consistent-history quantum theory was developed as an attempt to deal with closed systems in quantum mechanics. Such innovation is needed because the standard Copenhagen interpretation is incapable of describing the universe as a whole, since the existence of an external observer is required.

Griffiths [57, 62], Omn'es [58–61] and Gell-Mann and Hartle [54–56] approached this problem by proposing a new way of looking at quantum mechanics and quantum field theory, in which the fundamental objects are 'consistent' sets of histories. Using this approach it is then possible to make sense of the Copenhagen concept of probabilities even though no external observer is present. A key facet of this approach is that it is possible to assign probabilities to history propositions rather than just to propositions at a single time.

The possibility of making such an assignment rests on the so-called *decoherence functional* (see Sect. 17.1) which is a complex-valued functional, $d : \mathcal{UP} \times \mathcal{UP} \to \mathbb{C}$, where $\mathcal{UP}$ is the space of history propositions. Roughly speaking, the decoherence functional selects those sets of histories whose elements do not 'interfere' with each other pairwise (i.e., pairs of histories α, β, such that $d(\alpha, \beta) = 0$ if $\alpha \neq \beta$). A set $C = \{\alpha, \beta, \ldots, \gamma\}$ of history propositions is said to be *consistent* if C is complete[1] and $d(\alpha, \beta) = 0$ for all pairs of non equal histories in C. Given a consistent set C, the value $d(\alpha, \alpha)$ for any $\alpha \in C$ is interpreted as the probability of the history α being realized. This set can be viewed 'classically' in so far as the logic of such a set is necessarily Boolean.

Although this approach overcomes many conceptual problems related to applying the Copenhagen interpretation of quantum mechanics to a closed system, there is still the problem of how to deal with the plethora of different consistent sets of histories. In fact a typical decoherence functional will give rise to many consistent sets, some of which are incompatible with each other, in the sense that they cannot be joined to form a larger set.

[1] A set $C = \{\alpha, \beta, \ldots, \gamma\}$ is said to be complete if all history are pairwise disjoint and their logical 'or' forms the unit history.

C. Flori, *A First Course in Topos Quantum Theory*, Lecture Notes in Physics 868,
DOI 10.1007/978-3-642-35713-8_17, © Springer-Verlag Berlin Heidelberg 2013

In the literature, two main ways have been suggested for dealing with this problem, the first of which is to try and select a particular set which is realized in the physical world because of some sort of physical criteria. An attempt along these lines was put forward by Gell-Mann and Hartle in [54] where they postulated the existence of a measure of the quasi-classicality of a consistent set, and which, they argued, is sharply peaked.

A different approach is to accept the plethora of consistent sets and interpret them some sort of many-worlds view. This was done by Isham in [52, 53]. The novelty of his approach [43] is that, by using a different mathematical structure, namely 'topos theory', he was able to give a rigorous mathematical definition of the concept of many worlds. In particular, he exploited the mathematical structure of the collection of *all* complete sets of history propositions, to construct a logic that can be used to interpret the probabilistic predictions of the theory when all consistent sets are taken into account simultaneously, i.e., a many-worlds viewpoint.

The logic so defined has the particular feature that:

1. It is manifestly 'contextual' in regard to complete sets of propositions (not necessarily consistent).
2. It is multi-valued (i.e., the set of truth values is larger than just {true, false}).
3. In sharp distinction from standard quantum logic, it is *distributive*.

Using this new, topos-based logic, Isham assigned generalised truth values to the probability of realizing a given history proposition. These type of propositions are called 'second level' and are of the form "the probability of a history α being true is p". In defining these truth values Isham makes use of the notion of 'd-consistent Boolean algebras W^d', which are the algebras associated with consistent sets. The philosophy of his approach, therefore, was to translate into the language of topos theory the existing formalism of consistent histories, but in such a way that all consistent sets are considered at once.

In this formalism the notion of probability is still involved because of the use of second-level propositions that refer to the probability of realizing a history. Therefore, the notion of a decoherence functional is still central in Isham's approach, since it is only in terms of this quantity that the probabilities of histories are determined

However the approach we will adopt here is different. We start with the topos formulation of quantum theory as discussed so far in the book and extend it by introducing a time component. This 'temporal' extension will give rise to a history version of quantum theory. As we shall see, this new formalism departs from consistent-history theory in the fact that it does *not* make use of the notion of consistent sets and, thus of a decoherence functional. This result is striking since the notion of a decoherence functional is an essential feature in all of the history formalisms that have been suggested so far.

In deriving this new topos version of history theory we had in mind that, in the consistent-history approach to quantum mechanics there is no explicit state-vector reduction process induced by measurements (see [36] for an analysis of state-vector reduction in terms of topos theory). This suggests postulating that: given a state $|\psi\rangle_{t_1}$ at time t_1, the truth value of a proposition $A_1 \in \Delta_1$ at time t_1 should not influence the truth value of a proposition $A_2 \in \Delta_2$ with respect to the state, $|\psi\rangle_{t_2} = \hat{U}(t_2, t_1)|\psi\rangle_{t_1}$, at some later time t_2.

Thus for a history proposition of the form "the quantity A_1 has a value in Δ_1 at time t_1, and then the quantity A_2 has a value in Δ_2 at time $t_1 = 2$, and then ..." it should be possible to determine its truth value in terms of the individual (generalised) truth values of the constituent single-time propositions, as defined in Chap. 12. Thus our goal is to use topos theory to define truth values of sequentially-connected propositions, i.e., a time-ordered sequence of proposition, each of which refers to a single time.

As we will see, the possibility of doing this depends on how entanglement is taken into consideration. In fact, it is possible to encode the concept of entanglement entirely in the elements ('contexts') of the base category with which we are working. In particular, when entanglement is not taken into account the context category is just a product category. In this situation it is straightforward to exhibit a direct dependence between the truth values of a history proposition, both *homogeneous* and *inhomogeneous*, and the truth values of its constituent single-time components.

Moreover, in this case it is possible to identify all history propositions with certain sub-objects which are the categorical products of the appropriate pullbacks of the sub-objects, that represent the single-time propositions. It follows that, when entanglement is not considered, a precise mathematical relation between history propositions and their individual components subsists, even for inhomogeneous propositions. This is an interesting feature of the topos formalism of history theories since it implies that, in order to correctly represent history propositions as sequentially connected proposition, it suffices to use a topos in which the notion of entanglement is absent. However, if we were to use the full topos in which entanglement is present then a third type of history propositions would arise, namely *entangled inhomogeneous propositions*. It is precisely such propositions that can not be defined in terms of sequentially connected single-time propositions. This is a consequence of the fact that projection operators onto entangled states cannot be viewed, in the context of history theory, as inhomogeneous propositions.

Our goal, in this chapter, is to construct a topos formulation of quantum history theory as defined in the HPO formalism.[2] In particular, HPO history propositions will be considered as entities to which the Döring-Isham topos procedure can be applied. Since the set of HPO history-propositions forms a temporal logic, the possibility arises of representing such histories as sub-objects in a certain topos, which contains a temporal logic formed by the Heyting algebras of certain sub-objects in the single-time topoi. In this chapter we will develop such a logic. Moreover, we

[2]The acronym 'HPO' stands for 'history projection operator' and was the name given by Isham to his own (non-topos based) approach to consistent-history quantum theory. This approach is distinguished by the fact that any history proposition is represented by a projection operator in a new Hilbert space, that is the tensor product of the Hilbert spaces at the constituent times. In the older approaches, a history proposition is represented by a sum of products of projection operators, and this is almost always *not* itself a projection operator. Thus the HPO formalism is a natural framework with which to realise 'temporal quantum logic'.

will also develop a temporal logic of truth values and discuss the extent to which the evaluation map, which assigns truth values to propositions, does or does not preserve all the temporal connectives. An interesting feature of the topos analogue of the HPO formalism of quantum history theory is that, although it is possible to represent such a formalism within a topos in which the notion of entanglement is present (full topos), in order to correctly define history propositions and their truth values, we have to resort to the intermediate topos. Specifically we need to pull back history propositions as expressed in the full topos to history propositions as expressed in the intermediate topos in which the notion of entanglement is not present. This is necessary since history propositions per se are defined as sequentially connected single-time propositions and such a definition makes sense only in a topos in which the context category is a product category (intermediate topos). It is precisely to such an intermediate topos that the correct temporal logic of Heyting algebras belongs.

The absence of the concept of probability is consistent with the philosophical motivation that underlines the idea, in the first place, of using topos theory to describe quantum mechanics. Namely, the need to find an alternative to the instrumentalism that lies at the heart of the Copenhagen interpretation of quantum mechanics. In this respect, to maintain the use of a decoherence functional would conflict with the basic philosophical premises of the topos approach to quantum theory. In fact, as it will be shown, the topos formulation of quantum history theory does not employ a decoherence functional, and the associated concept of 'consistency' is absent.

This is an advantage since it avoids the problem of the plethora of incompatible consistent history sets. In fact, the novelty of this approach rests precisely on the fact that, although all possible history propositions are taken into consideration, when defining the logical structure in terms of which truth values are assigned to history propositions, there is no need to introduce the notion of consistent sets.

17.1 A Brief Introduction to Consistent Histories

Consistent histories theory was born as an attempt to describe closed systems in quantum mechanics, partly for the desire to construct quantum theories of cosmology. In fact, as explained in Chap. 2 the Copenhagen interpretation of quantum mechanics cannot be applied to closed systems, since it rests on the notion of probabilities defined in terms of a sequence of repeated measurements by an external observer. Thus it enforces a cosmologically inappropriate division between system and observer. The consistent-history formulation avoids this division, since it assigns probabilities without making use of the measurements and the associated state vector reductions.

In the standard Copenhagen interpretation of quantum theory, probability assignments to sequences of measurements are computed using the von Neumann reduction postulate which, roughly speaking, determines a measurement-induced change in the density matrix that represents the state. Therefore, to give meaning to probabilities, the notion of measurement-induced, state vector reduction is essential.

The consistent history formalism was developed in order to make sense of probability assignments but without invoking the notion of measurement. This requires introducing the decoherence functional d, which is a map from the space of all histories to the complex numbers. Specifically, given two histories (sequences of projection operators) $\alpha = (\hat{\alpha}_{t_1}, \hat{\alpha}_{t_2}, \dots, \hat{\alpha}_{t_n})$ and $\beta = (\hat{\beta}_{t_1}, \hat{\beta}_{t_2}, \dots, \hat{\beta}_{t_n})$ the decoherence functional is defined as

$$d_{\rho,\hat{H}}(\alpha, \beta) = \mathrm{tr}\big(\tilde{C}_\alpha^\dagger \rho \tilde{C}_\beta\big) = \mathrm{tr}\big(\hat{C}_\alpha^\dagger \rho \hat{C}_\beta\big) \tag{17.1}$$

where ρ is the initial density matrix, $\hat{H}$ is the Hamiltonian and $\tilde{C}_\alpha$ represents the 'class operator' which is defined in terms of the Schrodinger-picture projection operator α_{t_i} as

$$\tilde{C}_\alpha := \hat{U}(t_0, t_1)\alpha_{t_1}\hat{U}(t_1, t_2)\alpha_{t_2} \cdots \hat{U}(t_{n-1}, t_n)\alpha_{t_n}\hat{U}(t_n, t_0). \tag{17.2}$$

Thus $\tilde{C}_\alpha$ represents the history proposition "α_{t_1} is true at time t_1, and then α_{t_2} is true at time $t_2, \dots,$ and then α_{t_n} is true at time t_n". It is worth noting that, the class operator can be written as the product of Heisenberg-picture projection operators in the form $\hat{C}_\alpha = \hat{\alpha}_{t_n}(t_n)\hat{\alpha}_{t_{n-1}}(t_{n-1}) \cdots \hat{\alpha}_{t_1}(t_1)$. Generally speaking this is not itself a projection operator.

The physical meaning associated to the quantity $d(\alpha, \alpha)$ is that it is the probability of the history α being realized. However, this interpretation can only be ascribed in a non-contradictory way if the history α belongs to a special set of histories, namely a *consistent set* which is a set $\{\alpha^1, \alpha^2, \dots, \alpha^n\}$ of histories that do not interfere with each other, i.e. $d(\alpha_i, \alpha_j) = 0$ for all $i, j = 1, \dots, n$. Only within a consistent set does the definition of consistent histories have any physical meaning. In fact, it is only within a given consistent set that the probability assignments are consistent. Each decoherence functional defines such a consistent set(s).

For an in-depth analysis of the axioms and definition of consistent-history theory the reader can refer to [51, 53, 64, 70, 71] and references therein. For the present analysis only the following definitions are needed.

1. A *homogeneous history* is any sequentially-ordered sequence of projection operators $\hat{\alpha}_1, \hat{\alpha}_2, \dots, \hat{\alpha}_n$.
2. The definition of the join $\vee$ is straightforward when the two histories have the same time support and differ in their values only at one point t_i. In this case $\alpha \vee \beta := (\alpha_{t_1}, \alpha_{t_2}, \dots, \alpha_{t_i} \vee \beta_{t_i}, \dots, \alpha_{t_n}) = (\beta_{t_1}, \beta_{t_2}, \dots, \beta_{t_i} \vee \alpha_{t_i}, \dots, \beta_{t_n})$ is a homogeneous history and satisfies the relation $\hat{C}_{\alpha\vee\beta} = \hat{C}_\alpha \vee \hat{C}_\beta := \hat{C}_\alpha + \hat{C}_\beta$. The problem arises when the time supports are different, in particular when the two histories α and β are disjoint. The join of such histories would take us outside the class of homogeneous histories. Similarly, the negation of a homogeneous history would not itself be a homogeneous history.
3. An *inhomogeneous history* arises when two disjoint homogeneous histories are joined using the logical connective "or" ($\vee$) or when taking the negation ($\neg$) of a history proposition. Specifically, given two disjoint homogeneous histories α and β we can meaningfully talk about the inhomogeneous histories $\alpha \vee \beta$ and $\neg\alpha$. Such histories are generally not just a sequence of projection operators, but

when computing the decoherence functional they are represented by the operator $\hat{C}_{\alpha\vee\beta} := \hat{C}_\alpha \vee \hat{C}_\beta$ and $\hat{C}_{\neg\alpha} := \hat{1} - \hat{C}_\alpha$.

Gell-Mann and Hartle, tried to solve the problem of representing inhomogeneous histories using path integrals on the configuration space, Q, of the system. In this formalism the histories α and β are seen as subsets of the paths of Q. Then a pair of histories is said to be disjoint if they are disjoint subsets of the path space Q. Seen as path integrals, the additivity property of the decoherence functional is easily satisfied, namely

$$d(\alpha \vee \beta, \gamma) = d(\alpha, \gamma) + d(\beta, \gamma) \tag{17.3}$$

where γ is any subset of the path space Q.

Similarly, the negation of a history proposition $\neg\alpha$ is represented by the complement of the subset α of Q, therefore

$$d(\neg\alpha, \gamma) = d(1, \gamma) - d(\alpha, \gamma) \tag{17.4}$$

where 1 is the unit history.[3]

The above properties in (17.3) and (17.4) are well defined in the context of path integrals. But what happens when defining the decoherence functional on a string of projection operators? Gell-Mann and Hartle solved this problem by postulating the following definitions for the class operators when computing decoherence functionals:

$$\begin{aligned} \tilde{C}_{\alpha\vee\beta} &:= \tilde{C}_\alpha + \tilde{C}_\beta \\ \tilde{C}_{\neg\alpha} &:= 1 - \tilde{C}_\alpha \end{aligned} \tag{17.5}$$

if α and β are disjoint histories. The right hand side of these equations are indeed operators that represent $\alpha \vee \beta$ and $\neg\alpha$ when computing the decoherence functional, but as objects in the consistent-history formalism, it is not really clear what $\alpha \vee \beta$ and $\neg\alpha$ are.

In fact, as defined above, a homogeneous history is a time ordered sequence of projection operators, but there is no analogue definition for $\alpha \vee \beta$ or $\neg\alpha$. One might try to define the inhomogeneous histories $\neg\alpha$ and $\alpha \vee \beta$ component-wise so that, for a simple two-time history $\alpha = (\hat{\alpha}_{t_1}, \hat{\alpha}_{t_2})$, we would have

$$\neg\alpha = \neg(\hat{\alpha}_{t_1}, \hat{\alpha}_{t_2}) := (\neg\hat{\alpha}_{t_1}, \neg\hat{\alpha}_{t_2}). \tag{17.6}$$

However, this definition of the negation operation is wrong. Since α is the temporal proposition "α_1 is true at time t_1, and then α_2 is true at time t_2", which we shall write as $\hat{\alpha}_{t_1} \sqcap \hat{\alpha}_{t_2}$. It is then intuitively clear that the negation of this proposition should be

$$\neg(\hat{\alpha}_{t_1} \sqcap \hat{\alpha}_{t_2}) = (\neg\hat{\alpha}_{t_1} \sqcap \hat{\alpha}_{t_2}) \vee (\hat{\alpha}_{t_1} \sqcap \neg\hat{\alpha}_{t_2}) \vee (\neg\hat{\alpha}_{t_1} \sqcap \neg\hat{\alpha}_{t_2}) \tag{17.7}$$

which is not in any obvious sense the same as (17.6).

[3]The *unit history* is the history which is always true.

A similar problem arises with the "or" ($\vee$) operation: given two homogenous histories (α_1, α_2) and (β_1, β_2), the "or" operation defined component-wise is

$$(\alpha_1, \alpha_2) \vee (\beta_1, \beta_2) := (\alpha_1 \vee \beta_1, \alpha_2 \vee \beta_2). \tag{17.8}$$

This history would be true (realized) if both $(\alpha_1 \vee \beta_1)$ and $(\alpha_2 \vee \beta_2)$ are true, which implies that either an element in each of the pairs (α_1, α_2) and (β_1, β_2) is true, or both elements in either of the pairs (α_1, α_2) and (β_1, β_2) are true. But this contradicts with the actual meaning of the proposition $(\alpha_1, \alpha_2) \vee (\beta_1, \beta_2)$, which states that either history (α_1, α_2) is realized or history (β_1, β_2) is realized. In fact the 'or' in the proposition $(\alpha_1, \alpha_2) \vee (\beta_1, \beta_2)$ should really be as follows:

$$(\alpha_1 \sqcap \alpha_2) \vee (\beta_1 \sqcap \beta_2) = \big(\neg(\alpha_1 \sqcap \alpha_2) \wedge (\beta_1 \sqcap \beta_2)\big) \vee \big((\alpha_1 \sqcap \alpha_2) \wedge \neg(\beta_1 \sqcap \beta_2)\big). \tag{17.9}$$

Thus, for the proposition $(\alpha_1 \sqcap \alpha_2) \vee (\beta_1 \sqcap \beta_2)$ to be true, both elements in either of the pairs $(\alpha_1 \sqcap \alpha_2)$ and $(\beta_1 \sqcap \beta_2)$ have to be true, but not all four elements at the same time.

If instead we had the history proposition $(\alpha_1 \vee \beta_1) \sqcap (\alpha_2 \vee \beta_2)$, this would be equivalent to

$$\begin{aligned}(\alpha_1 \vee \beta_1) \sqcap (\alpha_2 \vee \beta_2) &:= (\alpha_1 \sqcap \alpha_2) \vee (\alpha_1 \sqcap \beta_2) \vee (\beta_1 \sqcap \beta_2) \vee (\beta_1 \sqcap \alpha_2) \\ &\geq (\alpha_1 \sqcap \alpha_2) \vee (\beta_1 \sqcap \beta_2). \end{aligned} \tag{17.10}$$

This shows that it is not possible to define inhomogeneous histories component-wise. Moreover, the appeal to path integrals when defining $\tilde{C}_{\alpha\vee\beta}$ is realization-dependent and does not uncover what $\tilde{C}_{\alpha\vee\beta}$ actually is.

However, the right hand sides of (17.5) have a striking similarity to the single-time propositions in quantum logic. In fact, given two single-time propositions P and Q which are disjoint, the proposition $P \vee Q$ is simply represented by the projection operator $\hat{P} + \hat{Q}$; similarly, the negation $\neg P$ is represented by the operator $\hat{1} - \hat{P}$.

This similarity of the single-time propositions with the right hand side of (17.5) suggests that somehow it should be possible to identify history propositions with projection operators. Obviously, these projection operators cannot be the class operators since, generally, these are not projection operators.

The claim that a logic for consistent histories can be defined, such that each history proposition is represented by a projection operator on some Hilbert space, is also motivated by the fact that the statement that a certain history is "realized" is, itself, a proposition. Therefore, the set of all such histories could possess a lattice structure similar to the lattice of single-time propositions in standard quantum logic.

These considerations led Isham to construct the, so-called, HPO formalism. In this new formalism of consistent histories it is possible to identify the entire set $\mathcal{UP}$ with the projection lattice of some 'new' Hilbert space. In the following section we will describe this formalism in more detail.

17.2 The HPO Formulation of Consistent Histories

As shown in the previous section, the identification of a homogeneous history α, as a projection operator on the direct sum $\bigoplus_{t\in\{t_1,t_2,\ldots,t_n\}}\mathcal{H}_t$ of n copies of the Hilbert space $\mathcal{H}$ does not lead to a satisfactory definition of a quantum logic for histories.

A solution to this problem was put forward in [65] where an alternative formulation of consistent histories, namely the HPO (History Projection Operator) formulation was introduced. The key idea is to identify homogeneous histories with tensor products of projection operators: i.e., $\alpha = \hat{\alpha}_{t_1}\otimes\hat{\alpha}_{t_2}\otimes\cdots\otimes\hat{\alpha}_{t_n}$. This definition was motivated by the fact that, unlike a normal product, a *tensor* product of projection operators is itself a projection operators, since

$$(\hat{\alpha}_{t_1}\otimes\hat{\alpha}_{t_2})^2 = (\hat{\alpha}_{t_1}\otimes\hat{\alpha}_{t_2})(\hat{\alpha}_{t_1}\otimes\hat{\alpha}_{t_2}) := \hat{\alpha}_{t_1}\hat{\alpha}_{t_1}\otimes\hat{\alpha}_{t_2}\hat{\alpha}_{t_2}$$

$$= \hat{\alpha}^2_{t_1}\otimes\hat{\alpha}^2_{t_2} \tag{17.11}$$

$$= \hat{\alpha}_{t_1}\otimes\hat{\alpha}_{t_2} \tag{17.12}$$

and

$$(\hat{\alpha}_{t_1}\otimes\hat{\alpha}_{t_2})^\dagger := \hat{\alpha}^\dagger_{t_1}\otimes\hat{\alpha}^\dagger_{t_2} \tag{17.13}$$

$$= \hat{\alpha}_{t_1}\otimes\hat{\alpha}_{t_2}. \tag{17.14}$$

For this alternative definition of a homogeneous history, the negation operation coincides with (17.7):

$$\neg(\hat{\alpha}_{t_1}\otimes\hat{\alpha}_{t_2}) = \hat{1}\otimes\hat{1} - \hat{\alpha}_{t_1}\otimes\hat{\alpha}_{t_2}$$

$$= (\hat{1}-\hat{\alpha}_{t_1})\otimes\hat{\alpha}_{t_2} + \hat{\alpha}_{t_1}\otimes(\hat{1}-\hat{\alpha}_{t_2}) + (1-\hat{\alpha}_{t_1})\otimes(1-\hat{\alpha}_{t_2})$$

$$= \neg\hat{\alpha}_{t_1}\otimes\hat{\alpha}_{t_2} + \hat{\alpha}_{t_1}\otimes\neg\hat{\alpha}_{t_2} + \neg\hat{\alpha}_{t_1}\otimes\neg\hat{\alpha}_{t_2}. \tag{17.15}$$

Moreover, given two disjoint homogeneous histories $\alpha = (\hat{\alpha}_{t_1},\hat{\alpha}_{t_2})$ and $\beta = (\hat{\beta}_{t_1},\hat{\beta}_{t_2})$, then, since $\hat{\alpha}_{t_1}\hat{\beta}_{t_1} = 0$ and/or $\hat{\alpha}_{t_2}\hat{\beta}_{t_2} = 0$ it follows that the projection operators that represent the two propositions are themselves disjoint, i.e., $(\hat{\alpha}_{t_1}\otimes\hat{\alpha}_{t_2})(\hat{\beta}_{t_1}\otimes\hat{\beta}_{t_2}) = 0$. It is now possible to define $\alpha\vee\beta$ as

$$(\hat{\alpha}_{t_1}\otimes\hat{\alpha}_{t_2})\vee(\hat{\beta}_{t_1}\otimes\hat{\beta}_{t_2}) := (\hat{\alpha}_{t_1}\otimes\hat{\alpha}_{t_2}) + (\hat{\beta}_{t_1}\otimes\hat{\beta}_{t_2}). \tag{17.16}$$

In the HPO formalism, homogeneous histories are represented by 'homogeneous' projection operators in the lattice $P(\bigotimes_{t\in\{t_1,t_2,\ldots,t_n\}}\mathcal{H}_t)$, while inhomogeneous histories are represented by inhomogeneous operators. For example, $\hat{P}_1\otimes\hat{P}_2\vee\hat{R}_1\otimes\hat{R}_2 = \hat{P}_1\otimes\hat{P}_2 + \hat{R}_2\otimes\hat{R}_2$ would be the join of the two elements $\hat{P}_1\otimes\hat{P}_2$ and $\hat{R}_2\otimes\hat{R}_2$, as defined in the lattice $P(\bigotimes_{t\in\{t_1,t_2\}}\mathcal{H}_t)$.

Mathematically, the introduction of the tensor product is quite natural. In fact, in the general history formalism a homogenous history is an element of $\bigoplus_{t\in\{t_1,t_2,\ldots,t_n\}}P(\mathcal{H}_t)\subset\bigoplus_{t\in\{t_1,t_2,\ldots,t_n\}}B(\mathcal{H}_t)$ which is a vector space. The vector space structure of $\bigoplus_{t\in\{t_1,t_2,\dots,t_n\}} B(\mathcal{H}_t)$ is utilized when defining the decoherence functional, since the map $(\hat{\alpha}_{t_1}, \hat{\alpha}_{t_2}, \dots, \hat{\alpha}_{t_n}) \to \mathrm{tr}(\hat{\alpha}_{t_1}(t_1)\hat{\alpha}_{t_2}(t_2)\cdots\hat{\alpha}_{t_n}(t_n))$ is multilinear.

However, tensor products are defined through the universal factorization property, namely: given a finite collection of vector spaces $V_1, V_2, \dots, V_n$, any multilinear map $\mu : V_1 \times V_2 \times \cdots \times V_n \to W$ uniquely factorizes through a tensor product, i.e., the diagram

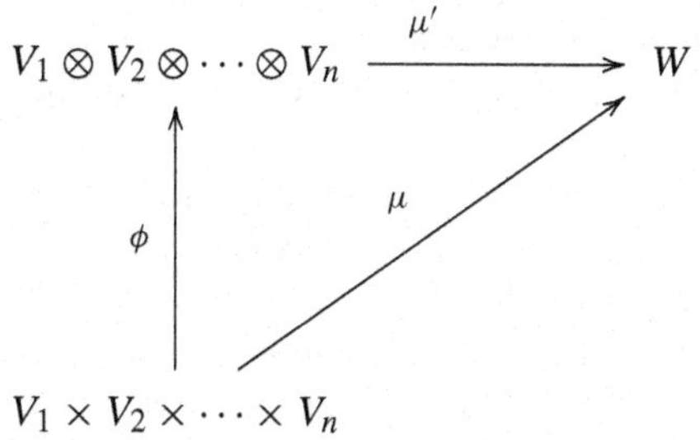

commutes. Thus the map $\phi : (\hat{\alpha}_{t_1}, \hat{\alpha}_{t_2}, \dots, \hat{\alpha}_{t_n}) \mapsto \hat{\alpha}_{t_1} \otimes \hat{\alpha}_{t_2} \otimes \cdots \otimes \hat{\alpha}_{t_n}$ arises naturally.

At the level of algebras, the map ϕ is defined is the obvious way as

$$\phi : \bigoplus_{t\in\{t_1,t_2,\dots,t_n\}} B(\mathcal{H}_t) \to \bigotimes_{t\in\{t_1,t_2,\dots,t_n\}} B(\mathcal{H}_t). \tag{17.17}$$

This map is many-to-one since $(\lambda A) \otimes (\lambda^{-1} B) = A \otimes B$. However, if we restrict only to $\bigoplus_{t\in\{t_1,t_2\cdots t_n\}} P(\mathcal{H}_t) \subseteq \bigoplus_{t\in\{t_1,t_2,\dots,t_n\}} B(\mathcal{H}_t)$ then the map becomes one-to-one, since for all projection operators $\hat{P} \in \bigoplus_{t\in\{t_1,t_2,\dots,t_n\}} P(\mathcal{H}_t)$, $\lambda\hat{P}$ $(\lambda \neq 0,\ \hat{P} \neq \hat{0})$ is a projection operator if and only if $\lambda = 1$.

In this scheme, the decoherence functional is computed using the map

$$D : \bigotimes_{t\in\{t_1,t_2,\dots,t_n\}} B(\mathcal{H}) \to B(\mathcal{H}) \tag{17.18}$$

$$(\hat{A}_1 \otimes \hat{A}_2 \otimes \cdots \otimes \hat{A}_n) \mapsto \big(\hat{A}_n(t_n)\hat{A}_{n-1}(t_{n-1})\cdots\hat{A}_1(t_1)\big). \tag{17.19}$$

Since this map is linear, it can be extended to include inhomogeneous histories. Furthermore, the class operators $\hat{C}$ can be defined as a map from the projectors on the Hilbert space $\bigotimes_{t\in\{t_1,t_2,\dots,t_n\}} \mathcal{H}$, seen as a subset of all linear operators on $\bigotimes_{t\in\{t_1,t_2,\dots,t_n\}} \mathcal{H}$ to the operators on $\mathcal{H}$

$$\hat{C}_\alpha := D\big(\phi(\alpha)\big) \tag{17.20}$$

and again extended to inhomogeneous histories by linearity.

This map satisfies the relations $\tilde{C}_{\alpha\vee\beta} = \tilde{C}_\alpha \vee \tilde{C}_\beta$ and $\tilde{C}_{\neg\alpha} = 1 - \tilde{C}_\alpha$ and, hence, their justification by path integrals is no longer necessary.

The HPO formalism can be extended to non-finite temporal supports by using an infinite (continuous if necessary) tensor product of copies of $B(\mathcal{H})$. The interested reader can refer to [53].

17.3 The Temporal Logic of Heyting Algebras of Sub-objects

In this section we begin to consider sequences of propositions at different times, these are commonly called 'homogeneous histories'. The goal is to assign truth value to such propositions using a temporal extension of the topos formalism discussed in Chaps. 11–12.

As previously stated, in the consistent-history program, a central goal is to get rid of the idea of state vector reductions induced by measurements. The absence of the state-vector reduction process implies that, given a state $\psi(t_0)$ at time t_0, the truth value (if there is one) of a proposition "$A_0 \in \Delta_0$" with respect to $\psi(t_0)$, should not influence the truth value of a proposition "$A_1 \in \Delta_1$" with respect to $\psi(t_1) = \hat{U}(t_1, t_0)\psi(t_0)$ which is the evolved state at time t_1. This suggests that, if it existed, the truth value of a homogeneous history should be computable from the truth values of the constituent single-time propositions.

Of course, such truth values do not exist in standard quantum theory, however, they *do* in the topos approach to quantum theory. Furthermore, since there is no explicit state reduction in that scheme, it seems reasonable to try to assign a generalised truth value to a homogeneous history, by employing the topos truth values that can be assigned to the constituent single-time propositions at each of the time points in the temporal support of the proposition.

With this in mind let us consider the (homogeneous) history proposition $\hat{\alpha} =$ "the quantity A_1 has a value in Δ_1 at time t_1 and then the quantity A_2 has value in Δ_2 at time t_2 and then ... and then the quantity A_n has value in Δ_n at time t_n", which is a time-ordered sequence of different propositions at different given times (we are assuming that $t_1 < t_2 < \cdots < t_n$). Thus α represents a homogeneous history. Symbolically, we can write α as

$$\alpha = (A_1 \in \Delta_1)_{t_1} \sqcap (A_2 \in \Delta_2)_{t_2} \sqcap \cdots \sqcap (A_n \in \Delta_n)_{t_n} \tag{17.21}$$

where the symbol '$\sqcap$' is the temporal connective 'and then'.

The first thing we need to understand is how to ascribe some sort of 'temporal structure' to the Heyting algebras of sub-objects of the spectral presheaves at the relevant times. What we are working towards here is the notion of the 'tensor product' of Heyting algebras. As a first step towards motivating the definition, let us reconsider the history theory of classical physics in this light.

For classical history theory, the topos under consideration is **Sets**. In this case the state spaces Σ_i, for each time t_i, are topological spaces and we can focus on their Heyting algebras of open sets. For simplicity we will concentrate on two-time histories, but the arguments generalise, at once, to any histories whose temporal support is a finite set.

Thus, consider propositions α_1, β_1 at time t_1 and α_2, β_2 at time t_2, and let[4] $S_1, S_1' \in Sub_{op}(\Sigma_1)$ and $S_2, S_2' \in Sub_{op}(\Sigma_2)$ be the open subsets[5] that represent them. Now consider the homogeneous history propositions $\alpha_1 \sqcap \alpha_2$ and $\beta_1 \sqcap \beta_2$, and the inhomogeneous proposition $\alpha_1 \sqcap \alpha_2 \vee \beta_1 \sqcap \beta_2$. Heuristically, this proposition is true (or the history is *realized*) if either history $\alpha_1 \sqcap \alpha_2$ is realized, or history $\beta_1 \sqcap \beta_2$ is realised. In the classical history theory, $\alpha_1 \sqcap \alpha_2$ and $\beta_1 \sqcap \beta_2$ are represented by the subsets (of $\Sigma_1 \times \Sigma_2$) $S_1 \times S_2$ and $S_1' \times S_2'$, respectively. However, it is clearly not possible to represent the inhomogeneous proposition $(\alpha_1 \sqcap \alpha_2) \vee (\beta_1 \sqcap \beta_2)$ by any subset of $\Sigma_1 \times \Sigma_2$, which is itself of the product form $O_1 \times O_2$.

What if, instead, we considered the proposition $(\alpha_1 \vee \beta_1) \sqcap (\alpha_2 \vee \beta_2)$, which is represented by the sub-object $S_1 \cup S_1' \times S_2 \cup S_2'$: symbolically, we would write

$$(\alpha_1 \vee \beta_1) \sqcap (\alpha_2 \vee \beta_2) \mapsto S_1 \cup S_1' \times S_2 \cup S_2'. \tag{17.22}$$

This history has a different meaning from $(\alpha_1 \sqcap \alpha_2) \vee (\beta_1 \sqcap \beta_2)$, since it indicates that at time t_1 either proposition α_1 or β_1 is realized and, subsequently, at time t_2 either α_2 or β_2 is realized. It is clear intuitively that we then have the equation

$$(\alpha_1 \vee \beta_1) \sqcap (\alpha_2 \vee \beta_2) := (\alpha_1 \sqcap \beta_2) \vee (\alpha_1 \sqcap \alpha_2) \vee (\beta_1 \sqcap \alpha_2) \vee (\beta_1 \sqcap \beta_2). \tag{17.23}$$

The question that arises now is how to represent these inhomogeneous histories in such a way that (17.23) is somehow satisfied when using the representation of $(\alpha_1 \vee \beta_1) \sqcap (\alpha_2 \vee \beta_2)$ in (17.22).

The point is that if we take just the product $Sub_{op}(\Sigma_1) \times Sub_{op}(\Sigma_2)$ then we cannot represent inhomogeneous histories and, therefore, cannot find a realisation of the right hand side of (17.23). However, in the case at hand the answer is obvious, since we know that $Sub_{op}(\Sigma_1) \times Sub_{op}(\Sigma_2)$ does not exhaust the open sets in the topological space $\Sigma_1 \times \Sigma_2$. By itself, $Sub_{op}(\Sigma_1) \times Sub_{op}(\Sigma_2)$ is the collection of open sets in the *disjoint union* of Σ_1 and Σ_2, not the Cartesian product.

In fact, as we know, the subsets of $\Sigma_1 \times \Sigma_2$ in $Sub_{op}(\Sigma_1) \times Sub_{op}(\Sigma_2)$ actually form a *basis* for the topology on $\Sigma_1 \times \Sigma_2$: i.e., an arbitrary open set can be written as a *union* of elements of $Sub_{op}(\Sigma_1) \times Sub_{op}(\Sigma_2)$. It is then clear that the representation of the inhomogeneous history $(\alpha_1 \sqcap \alpha_2) \vee (\beta_1 \sqcap \beta_2)$ is

$$(\alpha_1 \sqcap \alpha_2) \vee (\beta_1 \sqcap \beta_2) \mapsto S_1 \times S_1' \cup S_2 \times S_2'. \tag{17.24}$$

It is easy to check that (17.23) is satisfied in this representation.

It is not being too fanciful to imagine that we have here made the transition from the product Heyting algebra $Sub_{op}(\Sigma_1) \times Sub_{op}(\Sigma_2)$ to a *tensor* product; i.e., we can tentatively postulate the relation

$$Sub_{op}(\Sigma_1) \otimes Sub_{op}(\Sigma_2) \simeq Sub_{op}(\Sigma_1 \times \Sigma_2). \tag{17.25}$$

---

[4] We will denote the set of open subsets of a topological space, X, by $Sub_{op}(X)$.

[5] Arguably, it is more appropriate to represent propositions in classical physics with Borel subsets, not just open ones. However, we will not go into this subtlety here.

The task now is to see if some meaning can be given, in general, to the tensor product of Heyting algebras and, if so, if it is compatible with (17.25). Fortunately this is indeed possible although it is easier to do this in the language of *frames*, rather than Heyting algebras. Frames are easier to handle in so far as the negation operation is not directly present. However, each frame gives rise to a unique Heyting algebra and vice versa (see below). So, nothing is lost in this way.

All this is described in detail in the book by Vickers [72]. In particular, we have the following definition.

Definition 17.1 A frame A is a poset such that the following are satisfied

1. Every subset has a join.
2. Every finite subset has a meet.
3. *Frame distributivity*: $x \wedge \bigvee Y = \bigvee \{x \wedge y : y \in Y\}$
   i.e., binary meets distribute over joins. Here $\bigvee Y$ represents the join of the subset $Y \subseteq A$.

We now come to something that is of fundamental importance in our discussion of topos temporal logic, namely, the definition of the tensor product of two frames:

Definition 17.2 [72] Given two frames A and B, the tensor product $A \otimes B$ is defined to be the frame represented by the following presentation

$$\mathcal{T}\langle a \otimes b, a \in A \text{ and } b \in B|$$

$$\bigwedge_i (a_i \otimes b_i) = \left(\bigwedge_i a_i\right) \otimes \left(\bigwedge_i b_i\right) \tag{17.26}$$

$$\bigvee_i (a_i \otimes b) = \left(\bigvee_i a_i\right) \otimes b \tag{17.27}$$

$$\bigvee_i (a \otimes b_i) = a \otimes \left(\bigvee_i b_i\right). \tag{17.28}$$

In other words, we form the formal products, $a \otimes b$, of elements $a \in A$, $b \in B$ and subject them to the relations in (17.26)–(17.28). Our intention is to use the tensor product as the temporal connective, $\sqcap$, meaning 'and then'. It is straightforward to show that (17.26)–(17.28) are indeed satisfied with this interpretation when '$\vee$' and '$\wedge$' are interpreted as 'or' and 'and', respectively.

We note that there are injective maps

$$\begin{aligned} i : A &\to A \otimes B \\ a &\mapsto a \otimes \text{true} \end{aligned} \tag{17.29}$$

and

$$\begin{aligned} j : B &\to A \otimes B \\ b &\mapsto \text{true} \otimes b. \end{aligned} \tag{17.30}$$

These frame constructions are easily translated into the setting of Heyting algebras with the aid of the following theorem [72].

Theorem 17.1 *Every frame A defines a complete Heyting algebra (cHa) in such a way that the operations $\wedge$ and $\vee$ are preserved, and the implication relation $\to$ is defined as follows:*

$$a \to b = \bigvee \{c : c \wedge a \leq b\}. \tag{17.31}$$

Frame distributivity implies that $(a \to b) \wedge a \leq b$, from which it follows

$$c \leq a \to b \quad \text{iff} \quad c \wedge a \leq b. \tag{17.32}$$

This is the definition of the relative pseudo-complement in the Heyting algebra. If we then substitute b with 0 we obtain the definition of the pseudo-complement: $\neg a := (a \to 0)$ (see Sect. 7.7).

Now that we have the definition of the tensor product of frames and, hence, the definition of the tensor product of Heyting algebras, we are ready to analyze quantum history propositions in terms of topos theory.

Within a topos framework, propositions are identified with sub-objects of the spectral presheaf. For example, given two systems S_1 and S_2, whose Hilbert spaces are $\mathcal{H}_1$ and $\mathcal{H}_2$ respectively, the propositions concerning each system are identified with elements of $Sub(\underline{\Sigma}^{\mathcal{H}_1})$ and $Sub(\underline{\Sigma}^{\mathcal{H}_2})$ respectively, via the process of 'daseinisation'. We will return, later, to the daseinisation of history propositions but, for the time being, we will often, with a slight abuse of language, talk about elements of $Sub(\underline{\Sigma})$ as 'being' propositions rather than as 'representing propositions via the process of daseinisation'.

With this in mind, since both $Sub(\underline{\Sigma}^{\mathcal{H}_1})$ and $Sub(\underline{\Sigma}^{\mathcal{H}_2})$ are Heyting algebras, it is possible to use definition (17.2) to define the tensor product $Sub(\underline{\Sigma}^{\mathcal{H}_1}) \otimes Sub(\underline{\Sigma}^{\mathcal{H}_2})$ which is itself a Heyting algebra. We propose to use such tensor products to represent the temporal logic of history propositions.

Because of the existence of a one-to-one correspondence between Heyting algebras and frames, in the following we will first develop a temporal logic for frames in quantum theory and, then, generalize to a temporal logic for Heyting algebras by utilizing Theorem 17.1. Thus we will consider $Sub(\underline{\Sigma}^{\mathcal{H}_1})$, $Sub(\underline{\Sigma}^{\mathcal{H}_2})$ and $Sub(\underline{\Sigma}^{\mathcal{H}_1}) \otimes Sub(\underline{\Sigma}^{\mathcal{H}_2})$ as frames rather than Heyting algebras, thereby not taking into account the logical connectives of implication and negation. These will then be reintroduced by applying Theorem 17.1.

Definition 17.3 $Sub(\underline{\Sigma}^{\mathcal{H}_1}) \otimes Sub(\underline{\Sigma}^{\mathcal{H}_2})$ is the frame whose generators are of the form $\underline{S_1} \otimes \underline{S_2}$ for $\underline{S_1} \in Sub(\underline{\Sigma}^{\mathcal{H}_1})$ and $\underline{S_2} \in Sub(\underline{\Sigma}^{\mathcal{H}_2})$, and such that the following relations are satisfied:

$$\bigwedge_{i\in I}(\underline{S_1^i}\otimes \underline{S_2^i}) = \Big(\bigwedge_{i\in I}\underline{S_1^i}\Big)\otimes\Big(\bigwedge_{j\in I}\underline{S_2^j}\Big) \tag{17.33}$$

$$\bigvee_{i\in I}(\underline{S_1^i}\otimes \underline{S_2}) = \Big(\bigvee_{i\in I}\underline{S_1^i}\Big)\otimes\underline{S_2} \tag{17.34}$$

$$\bigvee_{i\in I}(\underline{S_1}\otimes \underline{S_2^i}) = \underline{S_1}\otimes\Big(\bigvee_{i\in I}\underline{S_2^i}\Big) \tag{17.35}$$

for an arbitrary index set I.

From the above definition it follows that a general element of $Sub(\underline{\Sigma}^{\mathcal{H}_1})\otimes Sub(\underline{\Sigma}^{\mathcal{H}_2})$ will be of the form $\bigvee_{i\in I}(\underline{S_1^i}\otimes\underline{S_2^i})$.

17.4 Realising the Tensor Product in a Topos

We propose to use, via daseinisation, the Heyting algebra $Sub(\underline{\Sigma}^{\mathcal{H}_1})\otimes Sub(\underline{\Sigma}^{\mathcal{H}_2})$ to represent the temporal logical structure with which to handle (two-times) history propositions in the setting of topos theory. A homogeneous history $\alpha_1\sqcap\alpha_2$ will be represented by the daseinised quantity $\underline{\delta(\hat{\alpha}_1)}\otimes\underline{\delta(\hat{\alpha}_2)}$ and the inhomogeneous history $(\alpha_1\sqcap\alpha_2)\vee(\beta_1\sqcap\beta_2)$ by $\underline{\delta(\hat{\alpha}_1)}\otimes\underline{\delta(\hat{\alpha}_2)}\vee\underline{\delta(\hat{\beta}_1)}\otimes\underline{\delta(\hat{\beta}_2)}$, i.e. we denote

$$(\alpha_1\sqcap\alpha_2)\vee(\beta_1\sqcap\beta_2)\mapsto\underline{\delta(\hat{\alpha}_1)}\otimes\underline{\delta(\hat{\alpha}_2)}\vee\underline{\delta(\hat{\beta}_1)}\otimes\underline{\delta(\hat{\beta}_2)}. \tag{17.36}$$

Here, the '$\vee$' refers to the 'or' operation in the Heyting algebra $Sub(\underline{\Sigma}^{\mathcal{H}_1})\otimes Sub(\underline{\Sigma}^{\mathcal{H}_2})$.

Our task now is to relate this, purely-algebraic representation, with one that involves sub-objects of an object in some topos. We suspect that there should be some connection with $Sub(\underline{\Sigma}^{\mathcal{H}_1\otimes\mathcal{H}_2})$ but, at this stage, it is not clear what this can be. What we need is a topos in which there is some object whose Heyting algebra of sub-objects is isomorphic to $Sub(\underline{\Sigma}^{\mathcal{H}_1})\otimes Sub(\underline{\Sigma}^{\mathcal{H}_2})$. The connection with $Sub(\underline{\Sigma}^{\mathcal{H}_1\otimes\mathcal{H}_2})$ will then hopefully become clear.

Of course, in classical physics the analogue of $\underline{\Sigma}^{\mathcal{H}_1\otimes\mathcal{H}_2}$ is just the Cartesian product $\Sigma_1\times\Sigma_2$ and, then, as we have indicated above, we have the relation $Sub_{op}(\Sigma_1)\otimes Sub_{op}(\Sigma_2)\simeq Sub_{op}(\Sigma_1\times\Sigma_2)$. This suggests that, in the quantum case, we should start by looking at the 'product' $\underline{\Sigma}^{\mathcal{H}_1}\times\underline{\Sigma}^{\mathcal{H}_2}$. However, here we immediately encounter the problem that $\underline{\Sigma}^{\mathcal{H}_1}$ and $\underline{\Sigma}^{\mathcal{H}_2}$ are objects in *different* topoi,[6] and so we cannot just take their 'product' in the normal categorial way.

To get around this let us consider, heuristically, what defining something like '$\underline{\Sigma}^{\mathcal{H}_1} \times \underline{\Sigma}^{\mathcal{H}_2}$' entails. The fact that $\mathbf{Sets}^{\mathcal{H}_1}$ and $\mathbf{Sets}^{\mathcal{H}_2}$ are independent topoi strongly suggests that we will need something in which the contexts are *pairs* $\langle V_1, V_2\rangle$, where $V_1 \in Ob(\mathcal{V}(\mathcal{H}_1))$ and $V_2 \in Ob(\mathcal{V}(\mathcal{H}_2))$. In other words, the base category for our new presheaf topos will be the product category $\mathcal{V}(\mathcal{H}_1) \times \mathcal{V}(\mathcal{H}_2)$, defined as follows:

Definition 17.4 The category $\mathcal{V}(\mathcal{H}_1) \times \mathcal{V}(\mathcal{H}_2)$ is such that

- Objects: the objects are pairs of abelian von Neumann sub-algebras $\langle V_1, V_2\rangle$ with $V_1 \in \mathcal{V}(\mathcal{H}_1)$ and $\mathcal{V}(\mathcal{H}_2)$
- Morphisms: given two such pairs, $\langle V_1, V_2\rangle$ and $\langle V_1', V_2'\rangle$, there exist an arrow $l : \langle V_1', V_2'\rangle \to \langle V_1, V_2\rangle$ if and only if $V_1' \subseteq V_1$ and $V_2' \subseteq V_2$; i.e., if and only if there exists a morphism $i_1 : V_1' \to V_1$ in $\mathcal{V}(\mathcal{H}_1)$ and a morphism $i_2 : V_2' \to V_2$ in $\mathcal{V}(\mathcal{H}_2)$.

This product category $\mathcal{V}(\mathcal{H}_1) \times \mathcal{V}(\mathcal{H}_2)$ is related to the constituent categories, $\mathcal{V}(\mathcal{H}_1)$ and $\mathcal{V}(\mathcal{H}_2)$ by the existence of the functors

$$p_1 : \mathcal{V}(\mathcal{H}_1) \times \mathcal{V}(\mathcal{H}_2) \to \mathcal{V}(\mathcal{H}_1) \tag{17.37}$$

$$p_2 : \mathcal{V}(\mathcal{H}_1) \times \mathcal{V}(\mathcal{H}_2) \to \mathcal{V}(\mathcal{H}_2) \tag{17.38}$$

which are defined in the obvious way. These gives rise to the geometric morphisms between the topoi[7] $\mathbf{Sets}^{\mathcal{V}(\mathcal{H}_1)^{op}}$, $\mathbf{Sets}^{\mathcal{V}(\mathcal{H}_2)^{op}}$ and $\mathbf{Sets}^{(\mathcal{V}(\mathcal{H}_1)\times\mathcal{V}(\mathcal{H}_2))^{op}}$, namely,

$$p_1 : \mathbf{Sets}^{(\mathcal{V}(\mathcal{H}_1)\times\mathcal{V}(\mathcal{H}_2))^{op}} \to \mathbf{Sets}^{\mathcal{V}(\mathcal{H}_1)^{op}} \tag{17.39}$$

$$p_2 : \mathbf{Sets}^{(\mathcal{V}(\mathcal{H}_1)\times\mathcal{V}(\mathcal{H}_2))^{op}} \to \mathbf{Sets}^{\mathcal{V}(\mathcal{H}_2)^{op}} \tag{17.40}$$

with associated left-exact functors

$$p_1^* : \mathbf{Sets}^{\mathcal{V}(\mathcal{H}_1)^{op}} \to \mathbf{Sets}^{(\mathcal{V}(\mathcal{H}_1)\times\mathcal{V}(\mathcal{H}_2))^{op}} \tag{17.41}$$

$$p_2^* : \mathbf{Sets}^{\mathcal{V}(\mathcal{H}_2)^{op}} \to \mathbf{Sets}^{(\mathcal{V}(\mathcal{H}_1)\times\mathcal{V}(\mathcal{H}_2))^{op}}. \tag{17.42}$$

The above enables us to give a meaningful definition of the 'product' of $\underline{\Sigma}^{\mathcal{H}_1}$ and $\underline{\Sigma}^{\mathcal{H}_2}$ as

$$\underline{\Sigma}^{\mathcal{H}_1} \times \underline{\Sigma}^{\mathcal{H}_2} := p_1^*(\underline{\Sigma}^{\mathcal{H}_1}) \times p_2^*(\underline{\Sigma}^{\mathcal{H}_2}) \tag{17.43}$$

[6] Of course, in the case of temporal logic, the Hilbert spaces $\mathcal{H}_1$ and $\mathcal{H}_2$ are isomorphic and, hence, so are the associated topoi. However, their structural roles in the temporal logic are clearly different. In fact, in the closely related situation of composite systems it will generally be the case that $\mathcal{H}_1$ and $\mathcal{H}_2$ are *not* isomorphic. Therefore, in the following, we will not exploit this particular isomorphism.

[7] We are here exploiting the trivial fact that, for any pair of categories $\mathcal{C}_1, \mathcal{C}_2$, we have $(\mathcal{C}_1 \times \mathcal{C}_2)^{op} \simeq \mathcal{C}_1{}^{op} \times \mathcal{C}_2{}^{op}$.

where the '×' on the right hand side of (17.43) is the standard categorial product in the topos $\mathbf{Sets}^{(\mathcal{V}(\mathcal{H}_1)\times\mathcal{V}(\mathcal{H}_2))^{op}}$.

We will frequently write the product, $p_1^*(\underline{\Sigma}^{\mathcal{H}_1}) \times p_2^*(\underline{\Sigma}^{\mathcal{H}_2})$ in the simpler-looking form '$\underline{\Sigma}^{\mathcal{H}_1} \times \underline{\Sigma}^{\mathcal{H}_2}$' but it must always be born in mind that what it is really meant is the more complex form on the right hand side of (17.43). The topos $\mathbf{Sets}^{(\mathcal{V}(\mathcal{H}_1)\times\mathcal{V}(\mathcal{H}_2))^{op}}$ will play an important role in what follows. We will call it the 'intermediate topos' for reasons that will shortly appear.

We have argued that (two-times) history propositions, both homogeneous and inhomogeneous, should be represented in the Heyting algebra $Sub(\underline{\Sigma}^{\mathcal{H}_1}) \otimes Sub(\underline{\Sigma}^{\mathcal{H}_2})$ and we now want to assert that the topos that underlies such a possibility is precisely the intermediate topos $\mathbf{Sets}^{(\mathcal{V}(\mathcal{H}_1)\times\mathcal{V}(\mathcal{H}_2))^{op}}$.

The first thing to notice is that the constituent single-time propositions can be represented in the pullbacks $p_1^*(\underline{\Sigma}^{\mathcal{H}_1})$ and $p_2^*(\underline{\Sigma}^{\mathcal{H}_2})$ to the topos $\mathbf{Sets}^{(\mathcal{V}(\mathcal{H}_1)\times\mathcal{V}(\mathcal{H}_2))^{op}}$, since we have, for example, for the functor p_1,

$$p_1^*(\underline{\Sigma}^{\mathcal{H}_1})_{\langle V_1, V_2\rangle} := \underline{\Sigma}^{\mathcal{H}_1}_{V_1} \tag{17.44}$$

for all stages $\langle V_1, V_2\rangle$. Further more

$$p_1^*(\underline{\Sigma}^{\mathcal{H}_1}) \times p_2^*(\underline{\Sigma}^{\mathcal{H}_2})_{\langle V_1, V_2\rangle} := \underline{\Sigma}^{\mathcal{H}_1}_{V_1} \times \underline{\Sigma}^{\mathcal{H}_2}_{V_2} \tag{17.45}$$

so that it is clear that we can represent two-time homogeneous histories in this intermediate topos.

However, at this point everything looks similar to the corresponding classical case. In particular we have

$$Sub\big(p_1^*(\underline{\Sigma}^{\mathcal{H}_1})\big) \times Sub\big(p_2^*(\underline{\Sigma}^{\mathcal{H}_2})\big) \subset Sub\big(p_1^*(\underline{\Sigma}^{\mathcal{H}_1}) \times p_2^*(\underline{\Sigma}^{\mathcal{H}_2})\big) \tag{17.46}$$

which is a proper subset relation because, as it is clear from (17.45) the general sub-object of $\underline{\Sigma}^{\mathcal{H}_1} \times \underline{\Sigma}^{\mathcal{H}_2} := p_1^*(\underline{\Sigma}^{\mathcal{H}_1}) \times p_2^*(\underline{\Sigma}^{\mathcal{H}_2})$ will be a '$\vee$' of product sub-objects in the Heyting algebra $Sub(\underline{\Sigma}^{\mathcal{H}_1}) \times Sub(\underline{\Sigma}^{\mathcal{H}_2})$. In fact, we have the following theorem:

Theorem 17.2 *There is an isomorphism of Heyting algebras*

$$Sub(\underline{\Sigma}^{\mathcal{H}_1}) \otimes Sub(\underline{\Sigma}^{\mathcal{H}_2}) \simeq Sub(\underline{\Sigma}^{\mathcal{H}_1} \times \underline{\Sigma}^{\mathcal{H}_2}). \tag{17.47}$$

In order to show there is an isomorphism between the algebras we will first construct an isomorphism between the associated frames, the application of Theorem 17.1 will then lead to the desired isomorphisms between Heyting algebras. Because of the fact that the tensor product is given in terms of relations on product elements, it suffices to define h on products $\underline{S_1} \otimes \underline{S_2}$ and show that the function, thus defined, preserves these relations

The actual definition of h is the obvious one:

$$\begin{aligned} h : Sub(\underline{\Sigma}^{\mathcal{V}(\mathcal{H}_1)}) \otimes Sub(\underline{\Sigma}^{\mathcal{V}(\mathcal{H}_2)}) &\to Sub(\underline{\Sigma}^{\mathcal{H}_1} \times \underline{\Sigma}^{\mathcal{H}_2}) \\ \underline{S}_1 \otimes \underline{S}_2 &\mapsto \underline{S}_1 \times \underline{S}_2 (:= p_1^*\underline{S_1} \times p_2^*\underline{S_2}) \end{aligned} \tag{17.48}$$

and the main thing is to show that (17.33) are preserved by h. To this end let us consider the following:

$$\begin{aligned}
h\Big(\bigvee_i \underline{S}_1^i \otimes \underline{S}_2\Big)_{\langle V_1\ V_2\rangle} &\overset{(17.34)}{=} h\Big(\bigvee_i (\underline{S}_1^i) \otimes \underline{S}_2\Big) \\
&\overset{\text{Def. of } h \text{ on generators}}{=} \bigvee_i (\underline{S}_1^i) \times \underline{S}_2 \\
&\overset{\text{Frame law in } Sub(\underline{\Sigma}^{\mathcal{H}_1} \times \underline{\Sigma}^{\mathcal{H}_2})}{=} \bigvee_i (\underline{S}_1^i \times \underline{S}_2) \\
&\overset{\text{Def. of } h \text{ on generators}}{=} \bigvee_i \big(h(\underline{S}_1^i \otimes \underline{S}_2)\big)
\end{aligned} \tag{17.49}$$

where on the third line we have made use of the fact that $Sub(\underline{\Sigma}^{\mathcal{H}_1} \times \underline{\Sigma}^{\mathcal{H}_2})$ is a frame, thus the analogous of (17.34) holds.

It follows that:

$$h\Big(\bigvee_i \underline{S}_1^i \otimes \underline{S}_2\Big) = \bigvee_i h(\underline{S}_1^i \otimes \underline{S}_2). \tag{17.50}$$

There is a very similar proof of

$$h\Big(\bigvee_i \underline{S}_1 \otimes \underline{S}_2^i\Big) = \bigvee_i h(\underline{S}_1 \otimes \underline{S}_2^i). \tag{17.51}$$

Moreover

$$\begin{aligned}
h\Big(\bigwedge_{i\in I} (\underline{S}_1^i \otimes \underline{S}_2^i)\Big) &\overset{\text{Eq. (17.33)}}{=} h\Big(\bigwedge_{i\in I} \underline{S}_1^i \otimes \bigwedge_{j\in I} \underline{S}_2^j\Big) \\
&\overset{\text{Def. of } h \text{ on generators}}{=} \bigwedge_{i\in I} \underline{S}_1^i \times \bigwedge_{j\in I} \underline{S}_2^j \\
&\overset{\text{Frame law in } Sub(\underline{\Sigma}^{\mathcal{H}_1} \times \underline{\Sigma}^{\mathcal{H}_2})}{=} \bigwedge_{i\in I} (\underline{S}_1^i \times \underline{S}_2^i) \\
&\overset{\text{Def. of } h \text{ on generators}}{=} \bigwedge_{i\in I} \big(h(\underline{S}_1^i \otimes \underline{S}_2^i)\big)
\end{aligned} \tag{17.52}$$

from which it follows that

$$h\Big(\bigwedge_{i\in I} \underline{S}_1^i \otimes \underline{S}_2^i\Big) = \bigwedge_{i\in I} h(\underline{S^i}_1 \otimes \underline{S^i}_2) \tag{17.53}$$

as required.

The injectivity of h is obvious. The surjectivity follows from the fact that any element $\underline{R}$, of $Sub(\underline{\Sigma}^{\mathcal{H}_1} \times \underline{\Sigma}^{\mathcal{H}_2})$ can be written as $\underline{R} = \bigvee_{i \in I}(\underline{S_1^i} \times \underline{S_2^i}) = \bigvee_{i \in I} h(\underline{S_1^i} \otimes \underline{S_2^i}) = h(\bigvee_{i \in I} \underline{S_1^i} \otimes \underline{S_2^i})$ (because h is a homomorphism of frames).

Thus the frames $Sub(\underline{\Sigma}^{\mathcal{H}_1}) \otimes Sub(\underline{\Sigma}^{\mathcal{H}_2})$ and $Sub(\underline{\Sigma}^{\mathcal{H}_1} \times \underline{\Sigma}^{\mathcal{H}_2})$ are isomorphic. The isomorphisms of the associated Heyting algebras then follows from Theorem 17.1.

17.5 Entangled Stages

The discussion above reinforces the idea that homogeneous history propositions can be represented by sub-objects of products of pullbacks of single-time spectral presheaves.

However, in this setting there can be no notion of entanglement of contexts since the contexts are just pairs $\langle V_1, V_2 \rangle$; i.e., objects in the product category $\mathcal{V}(\mathcal{H}_1) \times \mathcal{V}(\mathcal{H}_2)$. To recover 'context entanglement' one needs to use the context category $\mathcal{V}(\mathcal{H}_1 \otimes \mathcal{H}_2)$, some of whose objects are simple tensor products $V_1 \otimes V_2$ (which, presumably, relate in some way to the pair $\langle V_1, V_2 \rangle$) but, others, are 'entangled' algebras.[8] Evidently, the discussion above does not apply to contexts of this more general type.

To explore this further consider the following functor:

$$\theta : \mathcal{V}(\mathcal{H}_1) \times \mathcal{V}(\mathcal{H}_2) \to \mathcal{V}(\mathcal{H}_1 \otimes \mathcal{H}_2) \tag{17.54}$$

$$\langle V_1, V_2 \rangle \mapsto V_1 \otimes V_2 \tag{17.55}$$

where (17.55) refers to the action on the objects in the category $\mathcal{V}(\mathcal{H}_1) \times \mathcal{V}(\mathcal{H}_2)$, the action on the arrows is obvious.

This gives rise to a geometric morphism, θ, between topoi, and an associated left-exact functor, θ^*:

$$\theta : \mathbf{Sets}^{\mathcal{V}(\mathcal{H}_1)} \times \mathbf{Sets}^{\mathcal{V}(\mathcal{H}_2)} \to \mathbf{Sets}^{\mathcal{V}(\mathcal{H}_1 \otimes \mathcal{H}_2)} \tag{17.56}$$

$$\theta^* : \mathbf{Sets}^{\mathcal{V}(\mathcal{H}_1 \otimes \mathcal{H}_2)} \to \mathbf{Sets}^{\mathcal{V}(\mathcal{H}_1)} \times \mathbf{Sets}^{\mathcal{V}(\mathcal{H}_2)}. \tag{17.57}$$

In particular, we can consider the pullback $\theta^*(\underline{\Sigma}^{\mathcal{H}_1 \otimes \mathcal{H}_2})$ which, on pairs of contexts, is:

$$\left(\theta^* \underline{\Sigma}^{\mathcal{H}_1 \otimes \mathcal{H}_2}\right)_{\langle V_1, V_2 \rangle} := \left(\underline{\Sigma}^{\mathcal{H}_1 \otimes \mathcal{H}_2}\right)_{\theta \langle V_1, V_2 \rangle} = \underline{\Sigma}^{\mathcal{H}_1 \otimes \mathcal{H}_2}_{V_1 \otimes . V_2}. \tag{17.58}$$

Thus the pullback, $\theta^*(\underline{\Sigma}^{\mathcal{H}_1 \otimes \mathcal{H}_2})$ of the spectral presheaf of $\mathcal{H}_1 \otimes \mathcal{H}_2$ to the intermediate topos $\mathbf{Sets}^{(\mathcal{V}(\mathcal{H}_1) \times \mathcal{V}(\mathcal{H}_2))^{op}}$ completely reproduces $\underline{\Sigma}^{\mathcal{H}_1 \otimes \mathcal{H}_2}$ at contexts of the tensor-product form $V_1 \otimes V_2$.

---

[8] It should be noted that, in this context, by entangled algebra W we mean any algebra which can not be written in the form of a pure tensor product, i.e. $W \neq V_1 \otimes V_2$.

However, it is clear that, for all contexts V_1, V_2 we have

$$\underline{\Sigma}^{\mathcal{H}_1\otimes\mathcal{H}_2}_{V_1\otimes V_2} \cong \underline{\Sigma}^{\mathcal{H}_1}_{V_1} \times \underline{\Sigma}^{\mathcal{H}_2}_{V_2} \tag{17.59}$$

since we can define an isomorphic function

$$\mu : \underline{\Sigma}^{\mathcal{H}_1}_{V_1} \times \underline{\Sigma}^{\mathcal{H}_2}_{V_2} \to \underline{\Sigma}^{\mathcal{H}_1\otimes\mathcal{H}_2}_{V_1\otimes V_2} \tag{17.60}$$

where, for all $\hat{A} \otimes \hat{B} \in V_1 \otimes V_2$, we have

$$\mu\big(\langle\lambda_1, \lambda_2\rangle\big)(\hat{A} \otimes \hat{B}) := \lambda_1(\hat{A})\lambda_2(\hat{B}). \tag{17.61}$$

The fact that, for all contexts of the form $V_1 \otimes V_2$, we have $\underline{\Sigma}^{\mathcal{H}_1\otimes\mathcal{H}_2}_{V_1\otimes V_2} \cong \underline{\Sigma}^{\mathcal{H}_1}_{V_1} \times \underline{\Sigma}^{\mathcal{H}_2}_{V_2}$, it means that,

$$\theta^*\big(\underline{\Sigma}^{\mathcal{H}_1\otimes\mathcal{H}_2}\big) \simeq \underline{\Sigma}^{\mathcal{H}_1} \times \underline{\Sigma}^{\mathcal{H}_2} \tag{17.62}$$

in the intermediate topos $\mathbf{Sets}^{(\mathcal{V}(\mathcal{H}_1)\times\mathcal{V}(\mathcal{H}_2))^{op}}$. Thus, in the topos $\mathbf{Sets}^{(\mathcal{V}(\mathcal{H}_1)\times\mathcal{V}(\mathcal{H}_2))^{op}}$, the product $\underline{\Sigma}^{\mathcal{H}_1} \times \underline{\Sigma}^{\mathcal{H}_2}$ is essentially the spectral presheaf $\underline{\Sigma}^{\mathcal{H}_1\otimes\mathcal{H}_2}$ but restricted to contexts of the form $V_1 \otimes V_2$. Hence $\mathbf{Sets}^{(\mathcal{V}(\mathcal{H}_1)\times\mathcal{V}(\mathcal{H}_2))^{op}}$ is an 'intermediate' stage in the progression from the pair of topoi $\mathbf{Sets}^{\mathcal{V}(\mathcal{H}_1)^{op}}$, $\mathbf{Sets}^{\mathcal{V}(\mathcal{H}_2)^{op}}$ to the topos $\mathbf{Sets}^{\mathcal{V}(\mathcal{H}_1\otimes\mathcal{H}_2)^{op}}$, associated with the full tensor-product Hilbert space $\mathcal{H}_1 \otimes \mathcal{H}_2$. This explains why we have called $\mathbf{Sets}^{(\mathcal{V}(\mathcal{H}_1)\times\mathcal{V}(\mathcal{H}_2))^{op}}$ the 'intermediate' topos.

The choice of $\mathbf{Sets}^{(\mathcal{V}(\mathcal{H}_1)\times\mathcal{V}(\mathcal{H}_2))^{op}}$, as the appropriate topos to use in the setting of quantum temporal logic reflects the fact that, although the full topos for quantum history theory is $\mathbf{Sets}^{\mathcal{V}(\mathcal{H}_1\otimes\mathcal{H}_2)^{op}}$, nevertheless, to account for both homogeneous and inhomogeneous history propositions it suffices to use the intermediate topos. However, if we do use the full topos $\mathbf{Sets}^{\mathcal{V}(\mathcal{H}_1\otimes\mathcal{H}_2)^{op}}$, a third type of history proposition arises. These 'entangled, inhomogeneous propositions' cannot be reached/defined by single-time propositions connected through temporal logic.

The existence of such propositions is a consequence of the fact that in the topos $\mathbf{Sets}^{\mathcal{V}(\mathcal{H}_1\otimes\mathcal{H}_2)^{op}}$, the *context* category $\mathcal{V}(\mathcal{H}_1 \otimes \mathcal{H}_2)$ contains 'entangled' abelian Von Neumann sub-algebras W, which cannot be reduced to a pure tensor product $W_1 \otimes W_2$. For such contexts it is not possible to define a clear relation between a history proposition and individual single-time propositions.

To clarify what is going on let us return for a moment to the HPO formalism of consistent history theory. There, a time-ordered sequence of individual time propositions (i.e., a homogeneous history) is identified with the tensor product of projection operators $\hat{P}_1 \otimes \hat{P}_2 \otimes \cdots \otimes \hat{P}_n$. We get a form of 'entanglement' when we consider inhomogeneous propositions $\hat{P}_1 \otimes \hat{P}_2 \vee \hat{P}_3 \otimes \hat{P}_4$ that cannot be written as $\hat{Q}_1 \otimes \hat{Q}_2$. However, this type of entanglement, which comes from logic, is not exactly the same as the usual entanglement of quantum mechanics (although there are close connections).

To understand this further consider a simple example in ordinary quantum theory of an entangled pair of spin-up, spin-down particles. A typical entangled state is

$$|\uparrow\rangle|\downarrow\rangle - |\downarrow\rangle|\uparrow\rangle \tag{17.63}$$

and the projector operator associated with this state is

$$\hat{P}_{\text{entangled}} = \big(|\uparrow\rangle|\downarrow\rangle - |\downarrow\rangle|\uparrow\rangle\big)\big(\langle\uparrow|\langle\downarrow| - \langle\downarrow|\langle\uparrow|\big). \tag{17.64}$$

However, the projection operator $\hat{P}_{\text{entangled}}$ is not the same as the projection operator $\hat{P}_{ud} \vee \hat{P}_{du}$ where $\hat{P}_{ud} := (|\uparrow\rangle|\downarrow\rangle)(\langle\downarrow|\langle\uparrow|)$ and $\hat{P}_{du} := (|\downarrow\rangle|\uparrow\rangle)(\langle\uparrow|\langle\downarrow|)$. This implies that $\hat{P}_{\text{entangled}} \neq \hat{P}_{ud} \vee \hat{P}_{du}$.

When translated to the history situation, this implies that a projection operator onto an entangled state in $\mathcal{H}_1 \otimes \mathcal{H}_2$, cannot be viewed as being an inhomogeneous history proposition: it is something different. The precise temporal-logic meaning, if any, of these entangled projectors remains to be seen.

17.6 Direct Product of Truth Values

We are now interested in defining truth values for history propositions. In single-time topos quantum theory, truth values are assigned through the evaluation map, which is a state-dependent map from the algebra of history propositions to the Heyting algebra of truth values. In the history case, for this map to be well-defined it has to map the temporal structure of the Heyting algebras of sub-objects to some temporal structure of the algebras of truth values. In the following section we will analyse how this mapping takes place.

Let us consider a homogeneous history proposition $\hat{\alpha} =$ "the quantity A_1 has a value in Δ_1 at time t_1, and then the quantity A_2 has a value in Δ_2 at time $t_1 = 2$, and then ... and then the quantity A_n has a value in Δ_n at time t_n". Symbolically, we can write α as

$$\alpha = (A_1 \in \Delta_1)_{t_1} \sqcap (A_2 \in \Delta_2)_{t_2} \sqcap \cdots \sqcap (A_n \in \Delta_n)_{t_n} \tag{17.65}$$

where the symbol '$\sqcap$' is the temporal connective 'and then'.

In the HPO formalism, α is represented by a tensor product of the spectral projection operators $\hat{E}[A_k \in \Delta_k]$, associated with each single-time proposition "$A_k \in \Delta_k$", $k = 1, 2, \ldots, n$:

$$\hat{\alpha} = \hat{E}[A_1 \in \Delta_1]_{t_1} \otimes \hat{E}[A_2 \in \Delta_2]_{t_2} \otimes \cdots \otimes \hat{E}[A_n \in \Delta_n]_{t_n}. \tag{17.66}$$

In order to ascribe a topos truth value to the homogeneous history α, we will first consider the truth values of the individual, single-time propositions "$(A_1 \in \Delta_1)_{t_1}$", "$(A_2 \in \Delta_2)_{t_2}$", ..., "$(A_n \in \Delta_n)_{t_n}$". These truth values are elements of $\Gamma \underline{\Omega}^{\mathcal{H}_{t_k}}$, $k = 1, 2, \ldots, n$: i.e. global sections of the sub-object classifier in the appropriate topos,

$\mathbf{Sets}^{\mathcal{V}(\mathcal{H}_{t_k})^{op}}$. We will analyse how these truth values can be combined to obtain a truth value for the entire history proposition α. For the sake of simplicity we will restrict ourselves to two-time propositions, but the extension to n-time slots is trivial.

Since there is no state-vector reduction, one can hope to define the truth value of the entire history $\alpha := (A_1 \in \Delta_1)_{t_1} \sqcap (A_2 \in \Delta_2)_{t_2}$ in terms of the truth values of the individual propositions at times t_1 and t_2. In particular, since we are conjecturing that the truth values at the two times are independent of each other, we expect an equation of the form[9]

$$v\big((A_1 \in \Delta_1)_{t_1} \sqcap (A_2 \in \Delta_2)_{t_2}; |\psi\rangle_{t_1}\big) = v\big(A_1 \in \Delta_1; |\psi\rangle_{t_1}\big) \sqcap v\big(A_2 \in \Delta_2; |\psi\rangle_{t_2}\big) \tag{17.67}$$

where $|\psi\rangle_{t_2}$ is the unitary evolution of $|\psi\rangle_{t_2}$. The '$\sqcap$' on the right hand side remains to be defined as some sort of temporal connective on the Heyting algebras $\mathbf{Sets}^{\mathcal{V}(\mathcal{H}_{t_1})^{op}}$ and $\mathbf{Sets}^{\mathcal{V}(\mathcal{H}_{t_1})^{op}}$.

However, at this point we hit the problem that $v(A_1 \in \Delta_1; |\psi\rangle_{t_1})$ and $v(A_2 \in \Delta_2; |\psi\rangle_{t_2})$ are global elements of the sub-object classifiers $\underline{\Omega}^{\mathcal{H}_{t_1}}$ and $\underline{\Omega}^{\mathcal{H}_{t_2}}$ in the topoi $\mathbf{Sets}^{\mathcal{V}(\mathcal{H}_{t_1})^{op}}$ and $\mathbf{Sets}^{\mathcal{V}(\mathcal{H}_{t_2})^{op}}$, respectively. Since these topoi are different from each other, it is not obvious how the '$\sqcap$' operation on the right hand side of (17.67) is to be defined.

On the other hand, since $\Gamma\underline{\Omega}^{\mathcal{H}_{t_1}}$ and $\Gamma\underline{\Omega}^{\mathcal{H}_{t_2}}$ are Heyting algebras, we can take their tensor product $\Gamma\underline{\Omega}^{\mathcal{H}_{t_1}} \otimes \Gamma\underline{\Omega}^{\mathcal{H}_{t_2}}$. By analogy with what we did earlier with the Heyting algebras of sub-objects of the spectral presheaves, it is natural to interpret the '$\sqcap$' on the right hand side of (17.67) as this tensor product, so that we end up with the plausible looking equation

$$v\big((A_1 \in \Delta_1)_{t_1} \sqcap (A_2 \in \Delta_2)_{t_2}; |\psi\rangle_{t_1}\big) = v\big(A_1 \in \Delta_1; |\psi\rangle_{t_1}\big) \otimes v\big(A_2 \in \Delta_2; |\psi\rangle_{t_2}\big). \tag{17.68}$$

[9] Since there is no state-vector reduction, the existence of an operation $\sqcap$ between truth values that satisfies (17.67) is plausible. In fact, unlike the normal logical connective '$\wedge$', the meaning of the temporal connective '$\sqcap$' implies that the propositions it connects do not 'interfere' with each other, since they are asserted at different times: it is thus a sensible first guess to assume that their truth values are independent.

The distinction between the temporal connective '$\sqcap$' and the logical connective '$\wedge$' has been discussed in details in various papers by Stachow and Mittelstaedt [66–69]. In these papers they analyse quantum logic using the ideas of game theory. In particular, they define logical connectives in terms of sequences of subsequent moves of possible attacks and defences. They also introduce the concept of 'commensurability property', which essentially defines the possibility of quantities being measured at the same time or not. The definition of *logical connectives* involves both possible attacks and defences, as well as the satisfaction of the commensurability property since logical connectives relate propositions which refer to the same time. On the other hand, the definition of *sequential connectives* does not need the introduction of the commensurability properties, since sequential connectives refer to propositions defined at different times and, thus, can always be evaluated together. The commensurability property introduced by Stachow and Mittelstaedt can be seen as the game theory analogue of the commutation relation between operators in quantum theory. We note that the same type of analysis can be applied as a justification of Isham's choice of the tensor product as temporal connective in the HPO theory.

The problem now is to find a topos for which the Heyting algebra $\Gamma\underline{\Omega}^{\mathcal{H}_{t_1}} \otimes \Gamma\underline{\Omega}^{\mathcal{H}_{t_2}}$ is well defined. This is reminiscent of the problem we encountered earlier when trying to represent inhomogeneous histories in a topos, and the answer is the same: pull everything back to the intermediate topos $\mathbf{Sets}^{(\mathcal{V}(\mathcal{H}_{t_1})\times\mathcal{V}(\mathcal{H}_{t_2}))^{op}}$. Specifically, let us define

$$\underline{\Omega}^{\mathcal{H}_{t_1}} \times \underline{\Omega}^{\mathcal{H}_{t_2}} := p_1^*\big(\underline{\Omega}^{\mathcal{H}_{t_1}}\big) \times p_2^*\big(\underline{\Omega}^{\mathcal{H}_{t_2}}\big) \tag{17.69}$$

which is an object in $\mathbf{Sets}^{(\mathcal{V}(\mathcal{H}_{t_1})\times\mathcal{V}(\mathcal{H}_{t_2}))^{op}}$. In fact, it is easy to check that it is the *sub-object classifier* in the intermediate topos, and is defined at stage $\langle V_1, V_2\rangle \in Ob(\mathcal{V}(\mathcal{H}_{t_1}) \times \mathcal{V}(\mathcal{H}_{t_2}))$ by

$$\big(\underline{\Omega}^{\mathcal{H}_{t_1}} \times \underline{\Omega}^{\mathcal{H}_{t_2}}\big)_{\langle V_1, V_2\rangle} := \underline{\Omega}_{V_1}^{\mathcal{H}_{t_1}} \times \underline{\Omega}_{V_2}^{\mathcal{H}_{t_2}} \tag{17.70}$$

and we have the important result that there is an isomorphism

$$\begin{aligned} j : \Gamma\underline{\Omega}^{\mathcal{H}_{t_1}} \otimes \Gamma\underline{\Omega}^{\mathcal{H}_{t_2}} \to \Gamma\big(\underline{\Omega}^{\mathcal{H}_{t_1}} \times \underline{\Omega}^{\mathcal{H}_{t_2}}\big) &:= \Gamma\big(p_1^*\big(\underline{\Omega}^{\mathcal{H}_{t_1}}\big) \times p_2^*\big(\underline{\Omega}^{\mathcal{H}_{t_2}}\big)\big) \\ &\simeq \Gamma\big(p_1^*\big(\underline{\Omega}^{\mathcal{H}_{t_1}}\big)\big) \times \Gamma\big(p_2^*\big(\underline{\Omega}^{\mathcal{H}_{t_2}}\big)\big) \end{aligned} \tag{17.71}$$

given by

$$j(\omega_1 \otimes \omega_2)\big(\langle V_1, V_2\rangle\big) := \big\langle \omega_1(V_1), \omega_2(V_2)\big\rangle. \tag{17.72}$$

The proof of this result is similar to that of Theorem 17.2 and will not be written out here.

For us, the significant implication of this result is that the truth value, $v((A_1 \in \Delta_1)_{t_1} \sqcap (A_2 \in \Delta_2)_{t_2}; |\psi\rangle_{t_1})$, of the history proposition $(A_1 \in \Delta_1)_{t_1} \sqcap (A_2 \in \Delta_2)_{t_2}$ can be regarded as an element of the Heyting algebra $\Gamma(\underline{\Omega}^{\mathcal{H}_{t_1}} \times \underline{\Omega}^{\mathcal{H}_{t_2}})$, whose 'home' is the intermediate topos $\mathbf{Sets}^{(\mathcal{V}(\mathcal{H}_{t_1})\times\mathcal{V}(\mathcal{H}_{t_2}))^{op}}$. Therefore a more accurate way of writing (17.68) is

$$v\big((A_1 \in \Delta_1)_{t_1} \sqcap (A_2 \in \Delta_2)_{t_2}; |\psi\rangle_{t_1}\big) = j\big(v\big(A_1 \in \Delta_1; |\psi\rangle_{t_1}\big) \otimes v\big(A_2 \in \Delta_2; |\psi\rangle_{t_2}\big)\big). \tag{17.73}$$

17.7 The Representation of HPO Histories

In this section we will pull together what it has been said above, in order to obtain a topos analogue of the HPO formalism of quantum history theory.

First we recall that in the HPO formalism a history proposition $\alpha = \alpha_1 \sqcap \alpha_2$, is identified with the tensor product of the projection operators $\hat{\alpha}_1$ and $\hat{\alpha}_2$ representing the single-time propositions α_1 and α_2, respectively, i.e., $\hat{\alpha} = \hat{\alpha}_1 \otimes \hat{\alpha}_2$. One main motivation for introducing the tensor product was a desire to make sense of the negation operation of homogeneous history propositions, as given intuitively by (17.7).

In fact, in the original approaches to consistent-histories theory the temporal connective 'and then' was simply associated to the operator product: thus the proposition $\alpha = \alpha_1 \sqcap \alpha_2$ was represented by $\hat{\alpha} = \hat{\alpha}_1 \hat{\alpha}_2$. But this identification loses any logical meaning since, given projection operators $\hat{P}$ and $\hat{Q}$, the product $\hat{P}\hat{Q}$ is generally not itself a projection operator.

However, if one defines the sequential connective $\sqcap$ in terms of the tensor product, such that $\alpha = \alpha_1 \sqcap \alpha_2$ is represented by $\hat{\alpha} = \hat{\alpha}_1 \otimes \hat{\alpha}_2$, then $\hat{\alpha}$ *is* a projection operator. Furthermore, one obtains the right definition for the negation operation specifically

$$\neg(\hat{\alpha}_1 \otimes \hat{\alpha}_2) = (\neg\hat{\alpha}_1 \otimes \hat{\alpha}_2) + (\hat{\alpha}_1 \otimes \neg\hat{\alpha}_2) + (\neg\hat{\alpha}_1 \otimes \neg\hat{\alpha}_2) \tag{17.74}$$

where we identify $+$ with $\vee$.[10]

We will now proceed by considering history propositions as defined by the HPO formalism as individual entities and, then, apply the topos machinery defined throughout this book to derive a topos version of the history formalism. Thus (i) the 'and then', $\sqcap$, on the right hand side of (17.67) is represented by the tensor products of the Heyting algebras $\Gamma\underline{\Omega}^{\mathcal{H}_{t_1}}$ and $\Gamma\underline{\Omega}^{\mathcal{H}_{t_2}}$ (as in (17.68)); (ii) the 'and then' on the left hand side of (17.67) will be initially represented by the tensor product of the associated spectral projectors (i.e., using the HPO formalism) and, then, it will be 'daseinised' to become the tensor product between the Heyting algebras $Sub(\underline{\Sigma}^{\mathcal{H}_{t_1}})$ and $Sub(\underline{\Sigma}^{\mathcal{H}_{t_2}})$.

In the previous sections we have argued that (two-time) inhomogeneous history propositions can be represented as sub-objects of the spectral presheaf in the intermediate topos $\mathbf{Sets}^{(\mathcal{V}(\mathcal{H}_1)\times\mathcal{V}(\mathcal{H}_2))^{op}}$. In particular, the homogeneous history $\alpha_1 \sqcap \alpha_2$ is represented by the presheaf $\underline{\delta(\hat{\alpha}_1)} \otimes \underline{\delta(\hat{\alpha}_2)} \subseteq \underline{\Sigma}^{\mathcal{H}_{t_1}} \times \underline{\Sigma}^{\mathcal{H}_{t_2}} \simeq \theta^*(\underline{\Sigma}^{\mathcal{H}_{t_1}} \otimes \underline{\Sigma}^{\mathcal{H}_{t_2}})$. On the other hand, the HPO-representative, $\hat{\alpha}_1 \otimes \hat{\alpha}_2$, belongs to $\mathcal{H}_{t_1} \otimes \mathcal{H}_{t_2}$ and, hence, its daseinisation, $\underline{\delta(\hat{\alpha}_1 \otimes \hat{\alpha}_2)}$, is a sub-object of the spectral presheaf $\underline{\Sigma}^{\mathcal{H}_{t_1}\otimes\mathcal{H}_{t_2}}$, which is an object in the topos $\mathbf{Sets}^{\mathcal{V}(\mathcal{H}_{t_1}\otimes\mathcal{H}_{t_2})^{op}}$. As such, $\underline{\delta(\hat{\alpha}_1 \otimes \hat{\alpha}_2)}$ is defined, at every stage, in $\mathcal{V}(\mathcal{H}_{t_1} \otimes \mathcal{H}_{t_2})$, including entangled ones. However, since by its very nature the tensor product $\underline{\delta(\hat{\alpha}_1)} \otimes \underline{\delta(\hat{\alpha}_2)}$ is defined only in the intermediate topos $\mathbf{Sets}^{(\mathcal{V}(\mathcal{H}_{t_1})\times\mathcal{V}(\mathcal{H}_{t_2}))^{op}}$, in order to compare it with $\underline{\delta(\hat{\alpha}_1 \otimes \hat{\alpha}_2)}$ it is necessary to first pullback the latter to the intermediate topos, using the geometric morphism θ^* defined in (17.56). However, having done this, it is easy to prove that

$$\theta^*\big(\underline{\delta(\hat{\alpha}_1 \otimes \hat{\alpha}_2)}\big)_{\langle V_1, V_2\rangle} = \underline{\delta(\hat{\alpha}_1)}_{V_1} \otimes \underline{\delta(\hat{\alpha}_2)}_{V_2} \tag{17.75}$$

for all $\langle V_1, V_2\rangle \in \mathcal{V}(\mathcal{H}_{t_1}) \times \mathcal{V}(\mathcal{H}_{t_2})$. In fact, from the definition of θ^* we have that $\theta^*(\underline{\delta(\hat{\alpha}_1 \otimes \hat{\alpha}_2)})_{\langle V_1, V_2\rangle} = (\underline{\delta(\hat{\alpha}_1 \otimes \hat{\alpha}_2)})_{\theta(\langle V_1, V_2\rangle)} = (\underline{\delta(\hat{\alpha}_1 \otimes \hat{\alpha}_2)})_{V_1 \otimes V_2}$. By then

[10] This is correct since the projectors which appear on the right hand side of the equation are pairwise orthogonal, thus the 'or', $\vee$, can be replaced by the summation operation $+$ of projector operators.

applying the definition of outer daseinisation, (17.75) follows. A marginally less accurate way of writing this equation is

$$\underline{\delta(\hat{\alpha}_1 \otimes \hat{\alpha}_2)_{V_1 \otimes V_2}} = \underline{\delta(\hat{\alpha}_1)}_{V_1} \otimes \underline{\delta(\hat{\alpha}_2)}_{V_2}. \tag{17.76}$$

We need to be able to daseinise inhomogeneous histories as well as homogeneous ones but, fortunately, here we can exploit one of the important features of daseinisation, namely: that it preserves the '$\vee$'-operation, i.e., at any stage V we have $\delta(\hat{Q}_1 \vee \hat{Q}_2)_V = \delta(\hat{Q}_1)_V \vee \delta(\hat{Q}_2)_V$. Thus, for an inhomogeneous history of the form $\alpha := (\alpha_1 \sqcap \alpha_2) \vee (\beta_1 \sqcap \beta_2)$ we have the topos representation

$$\begin{aligned}\underline{\delta(\hat{\alpha})} &= \underline{\delta(\hat{\alpha}_1 \otimes \hat{\alpha}_2 \vee \hat{\beta}_1 \otimes \hat{\beta}_2)} \\ &= \underline{\delta(\hat{\alpha}_1 \otimes \hat{\alpha}_2)} \vee \underline{\delta(\hat{\beta}_1 \otimes \hat{\beta}_2)}\end{aligned} \tag{17.77}$$

which, using (17.76), can be rewritten as

$$\underline{\delta(\hat{\alpha})}_{V_1 \otimes V_2} = \underline{\delta(\hat{\alpha}_1)}_{V_1} \otimes \underline{\delta(\hat{\alpha}_2)}_{V_2} \cup \underline{\delta(\hat{\beta}_1)}_{V_1} \otimes \underline{\delta(\hat{\beta}_2)}_{V_2}. \tag{17.78}$$

Let us now consider a specific two-time history $\alpha := (A_1 \in \Delta_1)_{t_1} \sqcap (A_2 \in \Delta_2)_{t_2}$ and try to determine its truth value in terms of the truth values of the single-time propositions by which it is composed. Let the initial state be $|\psi\rangle_{t_1} \in \mathcal{H}_{t_1}$ and let us first construct the truth value of the proposition "$(A_1 \in \Delta_1)_{t_1}$" (with associated spectral projector $\hat{E}[A_1 \in \Delta_1]$) in the state $|\psi\rangle_{t_1}$. To do this we must construct the pseudo-state associated with $|\psi\rangle_{t_1}$. This is defined at each context $V \in Ob(\mathcal{V}(\mathcal{H}_{t_1}))$ as

$$\underline{\mathfrak{w}}^{|\psi\rangle_{t_1}}_V := \underline{\delta\big(|\psi\rangle_{t_1}\,{}_{t_1}\langle\psi|\big)}_V$$

which form the components of the presheaf $\underline{\mathfrak{w}}^{|\psi\rangle_{t_1}} \subseteq \underline{\Sigma}^{\mathcal{H}_1}$. The truth value of the proposition "$(A_1 \in \Delta_1)_{t_1}$" at stage V_1, given the pseudo-state $\underline{\mathfrak{w}}^{|\psi\rangle_{t_1}}$, is then the global element of $\underline{\Omega}^{\mathcal{H}_{t_1}}$ given by

$$\begin{aligned} v\big(A_1 \in \Delta_1; |\psi\rangle_{t_1}\big)(V_1) &= \big\{V' \subseteq V_1 \,\big|\, \underline{\mathfrak{w}}^{|\psi\rangle_{t_1}}_{V'} \subseteq \underline{\delta\big(\hat{E}[A_1 \in \Delta_1]\big)}_{V'}\big\} & (17.79)\\ &= \big\{V' \subseteq V_1 \,\big|\, {}_{t_1}\langle\psi|\delta\big(\hat{E}[A_1 \in \Delta_1]\big)_{V'}|\psi\rangle_{t_1} = 1\big\} & (17.80)\end{aligned}$$

for all $V_1 \in Ob(\mathcal{V}(\mathcal{H}_{t_1}))$.

As there is no state-vector reduction in the topos quantum theory, the next step is to evolve the state $|\psi\rangle_{t_1}$ to time t_2 using the usual, unitary time-evolution operator $\hat{U}(t_1, t_2)$, thus $|\psi\rangle_{t_2} = \hat{U}(t_1, t_2)|\psi\rangle_{t_1}$. Of course this vector still lies in $\mathcal{H}_{t_1}$. However, in the spirit of the HPO formalism, we will take its isomorphic copy (but still denoted $|\psi\rangle_{t_2}$) in the Hilbert space $\mathcal{H}_{t_2} \simeq \mathcal{H}_{t_1}$.

We now consider the truth value of the proposition "$(A_2 \in \Delta_2)_{t_2}$" in this evolved state $|\psi\rangle_{t_2}$. To do so we employ the pseudo-state

$$\begin{aligned}\underline{\mathfrak{w}}^{|\psi\rangle_{t_2}}_{V_2} = \underline{\mathfrak{w}}^{\hat{U}(t_2,t_1)|\psi\rangle_{t_1}}_{V_2} &= \underline{\delta\big(|\psi\rangle_{t_2}\,{}_{t_2}\langle\psi|\big)}_{V_2} \\ &= \underline{\delta\big(\hat{U}(t_2,t_1)|\psi\rangle_{t_1}\,{}_{t_1}\langle\psi|\,\hat{U}(t_2,t_1)^{-1}\big)}_{V_2}\end{aligned} \tag{17.81}$$

at all stages $V_2 \in Ob(\mathcal{V}(\mathcal{H}_2))$. Then the truth value of the proposition "$(A_2 \in \Delta_2)_{t_2}$" (with associated spectral projector $\hat{E}[A_2 \in \Delta_2]$) at stage $V_2 \in Ob(\mathcal{V}(\mathcal{H}_2))$ is

$$\begin{aligned} v\big(A_2 \in \Delta_2; |\psi\rangle_{t_2}\big)(V_2) &= \big\{V' \subseteq V_2 \,|\, \underline{\mathfrak{w}}^{|\psi\rangle_{t_2}}_{V'} \subseteq \underline{\delta\big(\hat{E}[A_2 \in \Delta_2]\big)}_{V'}\big\} \\ &= \big\{V' \subseteq V_2 \,|\, {}_{t_2}\langle\psi|\,\delta\big(\hat{E}[A_2 \in \Delta_2]\big)_{V'}|\psi\rangle_{t_2} = 1\big\}. \end{aligned} \tag{17.82}$$

We would now like to define truth values of daseinised history propositions of the form $\underline{\delta(\hat{\alpha}_1 \otimes \hat{\alpha}_2)}$. To do so we need to construct the appropriate pseudo-states. A state in the tensor product Hilbert space $\mathcal{H}_{t_1} \otimes \mathcal{H}_{t_2}$ is represented by $|\psi\rangle_{t_1} \otimes |\psi\rangle_{t_2}$ where, for reasons explained above, $|\psi\rangle_{t_2} = \hat{U}(t_2, t_1)|\psi\rangle_{t_1}$. To each such tensor product of states we can associate the tensor product pseudo-state:

$$\underline{\mathfrak{w}}^{|\psi\rangle_{t_1} \otimes |\psi\rangle_{t_2}} := \underline{\delta\big(|\psi_{t_1} \otimes \psi_{t_2}\rangle\langle\psi_{t_2} \otimes \psi_{t_1}|\big)} = \underline{\delta\big(|\psi\rangle_{t_1}\, {}_{t_1}\langle\psi| \otimes |\psi\rangle_{t_2}\, {}_{t_2}\langle\psi|\big)}. \tag{17.83}$$

However, for contexts $V_1 \otimes V_2 \in Ob(\mathcal{V}(\mathcal{H}_1 \otimes \mathcal{H}_2))$ we have

$$\underline{\mathfrak{w}}^{|\psi\rangle_{t_1}}_{V_1} \otimes \underline{\mathfrak{w}}^{|\psi\rangle_{t_2}}_{V_2} = \underline{\delta\big(|\psi\rangle_{t_1}\, {}_{t_1}\langle\psi|\big)}_{V_1} \otimes \underline{\delta\big(|\psi\rangle_{t_2}\, {}_{t_2}\langle\psi|\big)}_{V_2} \tag{17.84}$$

$$= \underline{\delta\big(|\psi\rangle_{t_1}\, {}_{t_1}\langle\psi| \otimes |\psi\rangle_{t_2}\, {}_{t_2}\langle\psi|\big)}_{V_1 \otimes V_2} \tag{17.85}$$

so that

$$\underline{\mathfrak{w}}^{|\psi\rangle_{t_1}}_{V_1} \otimes \underline{\mathfrak{w}}^{|\psi\rangle_{t_2}}_{V_2} = \underline{\mathfrak{w}}^{|\psi\rangle_{t_1} \otimes |\psi\rangle_{t_2}}_{V_1 \otimes V_2} \tag{17.86}$$

or, slightly more precisely

$$\underline{\mathfrak{w}}^{|\psi\rangle_{t_1}}_{V_1} \otimes \underline{\mathfrak{w}}^{|\psi\rangle_{t_2}}_{V_2} = \theta^*\big(\underline{\mathfrak{w}}^{|\psi\rangle_{t_1} \otimes |\psi\rangle_{t_2}}\big)_{\langle V_1, V_2\rangle}. \tag{17.87}$$

Given the pseudo-state $\underline{\mathfrak{w}}^{|\psi\rangle_{t_1}} \otimes \underline{\mathfrak{w}}^{|\psi\rangle_{t_2}} \in Sub_{\mathrm{cl}}(\underline{\Sigma}^{\mathcal{H}_{t_1}}) \otimes Sub_{\mathrm{cl}}(\underline{\Sigma}^{\mathcal{H}_{t_2}})$ we want to consider the truth value of the sub-objects of the form $\underline{S}_1 \otimes \underline{S}_2$ (more precisely, of the homogeneous history proposition represented by this sub-object) as a global element of $\underline{\Omega}^{\mathcal{H}_{t_1}} \times \underline{\Omega}^{\mathcal{H}_{t_2}}$. This is given by

$$\begin{aligned} &v\big(\underline{\mathfrak{w}}^{|\psi\rangle_{t_1}} \otimes \underline{\mathfrak{w}}^{|\psi\rangle_{t_2}} \subseteq \underline{S}_1 \otimes \underline{S}_2\big)\big(\langle V_1, V_2\rangle\big) \\ &\quad := \big\{\langle V_1', V_2'\rangle \subseteq \langle V_1, V_2\rangle \,\big|\, \big(p_1^*\big(\underline{\mathfrak{w}}^{|\psi\rangle_{t_1}}\big) \times p_2^*\big(\underline{\mathfrak{w}}^{|\psi\rangle_{t_2}}\big)\big)_{\langle V_1', V_2'\rangle} \subseteq (\underline{S}_1 \times \underline{S}_2)_{\langle V_1', V_2'\rangle}\big\} \\ &\quad \simeq \big\{V_1' \subseteq V_1 \,|\, \underline{\mathfrak{w}}^{|\psi\rangle_{t_1}}_{V_1'} \subseteq (\underline{S}_1)_{V_1'}\big\} \times \big\{V_2' \subseteq V_2 \,|\, \underline{\mathfrak{w}}^{|\psi\rangle_{t_1}}_{V_2'} \subseteq (\underline{S}_1)_{V_2'}\big\} \qquad (17.88) \\ &\quad = \big\langle v\big(\underline{\mathfrak{w}}^{|\psi\rangle_{t_1}} \subseteq \underline{S}_1\big)(V_1), v\big(\underline{\mathfrak{w}}^{|\psi\rangle_{t_2}} \subseteq \underline{S}_2\big)(V_2)\big\rangle \\ &\quad = j\big(v\big(\underline{\mathfrak{w}}^{|\psi\rangle_{t_1}} \subseteq \underline{S}_1\big) \otimes v\big(\underline{\mathfrak{w}}^{|\psi\rangle_{t_2}} \subseteq \underline{S}_2\big)\big)\big(\langle V_1, V_2\rangle\big) \qquad (17.89) \end{aligned}$$

where $j : \Gamma\underline{\Omega}^{\mathcal{H}_{t_1}} \otimes \Gamma\underline{\Omega}^{\mathcal{H}_{t_2}} \to \Gamma(\underline{\Omega}^{\mathcal{H}_{t_1}} \times \underline{\Omega}^{\mathcal{H}_{t_2}})$ is discussed in (17.72). Thus we have

$$v\big(\underline{\mathfrak{w}}^{|\psi\rangle_{t_1}} \otimes \underline{\mathfrak{w}}^{|\psi\rangle_{t_2}} \subseteq \underline{S}_1 \otimes \underline{S}_2\big) = j\big(v\big(\underline{\mathfrak{w}}^{|\psi\rangle_{t_1}} \subseteq \underline{S}_1\big) \otimes v\big(\underline{\mathfrak{w}}^{|\psi\rangle_{t_2}} \subseteq \underline{S}_2\big)\big) \tag{17.90}$$

where the link with (17.67) is clear. In particular, for the homogenous history $\alpha := (A_1 \in \Delta_1)_{t_1} \sqcap (A_2 \in \Delta_2)_{t_2}$ we have the generalised truth value

$$\begin{aligned} &v\big((A_1 \in \Delta_1)_{t_1} \sqcap (A_2 \in \Delta_2)_{t_2}; |\psi\rangle_{t_1}\big) \\ &= v\big(\underline{\mathfrak{w}}^{|\psi\rangle_{t_1}} \otimes \underline{\mathfrak{w}}^{|\psi\rangle_{t_2}} \subseteq \underline{\delta\big(\hat{E}[A_1 \in \Delta_1]\big)} \otimes \underline{\delta\big(\hat{E}[A_2 \in \Delta_2]\big)}\big) \\ &= j\Big(v\big(\underline{\mathfrak{w}}^{|\psi\rangle_{t_1}} \subseteq \underline{\delta\big(\hat{E}[A_1 \in \Delta_1]\big)}\big) \otimes v\big(\underline{\mathfrak{w}}^{|\psi\rangle_{t_2}} \subseteq \underline{\delta\big(\hat{E}[A_2 \in \Delta_2]\big)}\big)\Big). \end{aligned} \tag{17.91}$$

This can be extended to inhomogeneous histories with the aid of (17.78).

The discussion above shows that the topos scheme for quantum theory can be extended to include propositions about the history of the system in time. A rather striking feature of the scheme is the way that the tensor product of projectors, used in the HPO history formalism is, 'reflected' in the existence of a tensor product between the Heyting algebras of sub-objects of the relevant presheaves. Or, to put it another way, a type of 'temporal logic' of Heyting algebras can be constructed using the definition of the Heyting-algebra tensor product.

As we have seen, the topos to use for all this is the 'intermediate topos' $\mathbf{Sets}^{(\mathcal{V}(\mathcal{H}_{t_1})\times\mathcal{V}(\mathcal{H}_{t_2}))^{op}}$ of presheaves over the category $\mathcal{V}(\mathcal{H}_{t_1}) \times \mathcal{V}(\mathcal{H}_{t_2})$. The all-important spectral presheaves in this topos is essentially the presheaf $\underline{\Sigma}^{\mathcal{H}_{t_1}\otimes\mathcal{H}_{t_2}}$ in the topos $\mathbf{Sets}^{\mathcal{V}(\mathcal{H}_{t_1}\otimes\mathcal{H}_{t_2})^{op}}$, but restricted to 'product' stages $V_1 \otimes V_2$ for $V_1 \in Ob(\mathcal{V}(\mathcal{H}_{t_1}))$ and $V_2 \in Ob(\mathcal{V}(\mathcal{H}_{t_2}))$. This restricted presheaf can be understood as a 'product' $\underline{\Sigma}^{\mathcal{H}_{t_1}} \times \underline{\Sigma}^{\mathcal{H}_{t_2}}$. In this context a key result is our proof in Theorem 17.2 of the existence of a Heyting algebra isomorphism $h : Sub(\underline{\Sigma}^{\mathcal{H}_{t_1}}) \otimes Sub(\underline{\Sigma}^{\mathcal{H}_{t_2}}) \rightarrow Sub(\underline{\Sigma}^{\mathcal{H}_{t_1}} \times \underline{\Sigma}^{\mathcal{H}_{t_2}})$.

Moreover, as we have shown, the evaluation map of history propositions maps the temporal structure of history propositions to the temporal structure of truth values in such a way that, the temporal-logic properties are preserved.

A fundamental feature of the topos analogue of the HPO formalism developed in this Chapter is that the notion of consistent sets, and thus of the decoherence functional, plays no role. In fact, as shown above, truth values can be ascribed to *any* history proposition independently of whether it belongs to a consistent set or not. Ultimately, this is because the topos formulation of quantum theory makes no fundamental use of the notion of probabilities, which are such a central notion in the (instrumentalist) Copenhagen interpretation of quantum theory. Instead, the topos approach deals with 'generalised' truth values in the Heyting algebra of global elements of the sub-object classifier. For this reason the theory is 'neo-realist'.

Reiterating, the standard consistent histories approach makes use of the Copenhagen concept of probabilities which must satisfy the classical summation rules and, thus, can only be applied to "classical" sets of histories i.e., consistent sets of histories defined using the decoherence functional. The topos formulation of the HPO formalism abandons the concept of probabilities and replaces them with truth values defined at particular stages (abelian Von Neumann sub-algebras). These stages are interpreted as the classical snapshots of the theory. In this framework there is no need for the notion of consistent set and, consequently, of decoherence functional. Thus the topos formulation of consistent histories avoids the issue of having many incompatible, consistent sets of proposition and can assign truth values to any history proposition.

It is interesting to note that, also in the consistent history formulation of *classical* physics we do not have the notion of decoherence functional since, in this case, no history interferes with any other. Considering that one of the aims of re-expressing quantum theory in terms of topos theory was to make it "look like" classical physics, it would seem that, at least as far as the notion of decoherence functional is involved, the resemblance has been successfully demonstrated.

Chapter 18
Normal Operators

From Chap. 16 we know that self-adjoint operators are represented as arrows, from the state space $\underline{\Sigma}$ to the quantity value object $\underline{\mathbb{R}}^{\leftrightarrow}$. The natural question to ask is whether such a representation can be extended to all normal operators. To this end one needs to, first of all, define the topos analogue of the complex numbers. Of course there exists the trivial object $\underline{\mathbb{C}}$ but this, as we will see, can not be identified with the complex number object since (a) it does not reduce to $\underline{\mathbb{R}}^{\leftrightarrow}$, and (b) the presheaf maps in $\underline{\mathbb{C}}$ are the identity maps which do not respect the ordering induced by the daseinisation of normal operators, defined in the next section. To choose a suitable complex number object, an ordering on the complex numbers $\mathbb{C}$ has to be defined. This ordering has to respect the ordering of the spectra of the normal operators induced by the ordering of the self-adjoint operators comprising them. Such an ordering will then allow us to define the complex number object $\underline{\mathbb{C}}^{\leftrightarrow}$ as a sheaf of pairs of order-preserving and order reversing functions. Consequently, normal operators are represented as maps from the state space to the newly defined object $\underline{\mathbb{C}}^{\leftrightarrow}$. The complete analysis on how this is done can be found in [49], here we will only report the essential steps.

18.1 Spectral Ordering of Normal Operators

The first step in creating a possible complex number object in topos quantum theory is to define an ordering of the complex numbers, which agrees with the ordering of the spectra of the normal operators. In particular, let us consider a normal operator $\hat{C} = \hat{A} + i\hat{B}$, where $\hat{A} = \int_{\mathbb{R}} \gamma \hat{E}^{\hat{A}}_{\gamma}$ and $\hat{B} = \int_{\mathbb{R}} \sigma \hat{E}^{\hat{B}}_{\sigma}$ are the self-adjoint operators comprising it. We can then write $\hat{C}$ as:

$$\hat{C} = \int_{\mathbb{R}} \gamma d\hat{E}^{\hat{A}}_{\gamma} + \int_{\mathbb{R}} i\sigma d\hat{E}^{\hat{B}}_{\sigma} = \int_{\mathbb{R}} (\gamma + i\sigma) d\hat{E}^{\hat{A}}_{\gamma} d\hat{E}^{\hat{B}}_{\sigma}. \tag{18.1}$$

However, normal operators can also be written as:

$$\hat{C} = \int_{\mathbb{C}} \lambda d\hat{E}^{\hat{C}}_{\lambda}. \tag{18.2}$$

C. Flori, *A First Course in Topos Quantum Theory*, Lecture Notes in Physics 868,
DOI 10.1007/978-3-642-35713-8_18, © Springer-Verlag Berlin Heidelberg 2013

To define a relation between those two expressions we utilise the following theorem:

Theorem 18.1 *Given a bounded operator $\hat{A} = \hat{C} + i\hat{B}$ on $\mathcal{H}$, there exists a family of projection operators $\{\hat{P}(\varepsilon, \eta) := \hat{P}_1(\varepsilon)\hat{P}_2(\eta) | (\varepsilon, \eta) \in \mathbb{R}^2\}$ which commute with $\hat{A}$, where $\{\hat{P}_1(\varepsilon) | \varepsilon \in \mathbb{R}\}$ is the spectral family of $\hat{C}$ and $\{\hat{P}_2(\eta) | \eta \in \mathbb{R}\}$ is the spectral family of $\hat{B}$. We then say that $\{P(\varepsilon, \eta) := \hat{P}_1(\varepsilon)\hat{P}_2(\eta) | (\varepsilon, \eta) \in \mathbb{R}^2\}$ is the spectral family of $\hat{A}$. Such a family has the following properties*:

(a) $\hat{P}(\varepsilon, \eta)\hat{P}(\varepsilon', \eta') = \hat{P}(\min\{\varepsilon, \varepsilon'\}, \min\{\eta, \eta'\})$ *for all* $(\varepsilon, \eta) \in \mathbb{R}^2$ *and* $(\varepsilon', \eta') \in \mathbb{R}^2$;
(b) $\hat{P}(\varepsilon, \eta) = 0$ *for all* $\varepsilon < -\|A\|$ *or* $\eta < -\|A\|$ *where* $\|A\|$ *is the Frobenius norm*;
(c) $\hat{P}(\varepsilon, \eta) = I$ *for all* $\varepsilon \geq \|A\|$ *and* $\eta \geq \|A\|$;
(d) $\hat{P}(\varepsilon + 0, \eta + 0) = \hat{P}(\varepsilon, \eta)$ *for all* $(\varepsilon, \eta) \in \mathbb{R}^2$;
(e)

$$\hat{A} = \int_{-\infty}^{\infty}\int_{-\infty}^{\infty} (\varepsilon + i\eta) d\big(\hat{P}_1(\varepsilon)\hat{P}_2(\eta)\big) = \int_{-\infty}^{\infty}\int_{-\infty}^{\infty} (\varepsilon + i\eta) d\big(\hat{P}(\varepsilon, \eta)\big). \quad (18.3)$$

The proof of this theorem can be found in [74].

Given this definition of spectral decomposition for normal operators, it is now possible to define a spectral ordering.

Definition 18.1 Given two normal operators $\hat{E} = \hat{D} + i\hat{F} = \int_{-\infty}^{\infty}\int_{-\infty}^{\infty}(\alpha + i\beta) d(\hat{Q}_1(\alpha)\hat{Q}_2(\beta))$ and $\hat{A} = \int_{-\infty}^{\infty}\int_{-\infty}^{\infty}(\varepsilon + i\eta) d(\hat{P}_1(\varepsilon)\hat{P}_2(\eta))$, we then define the spectral order as follows:

$$\hat{A} \geq_s \hat{E} \quad \text{iff} \quad \hat{P}_1(\varepsilon)\hat{P}_2(\eta) \leq \hat{Q}_1(\varepsilon)\hat{Q}_2(\eta) \quad \text{for all } (\varepsilon, \eta) \in \mathbb{R}^2. \quad (18.4)$$

Equivalently, it is possible to define the ordering in terms of subspaces of the Hilbert space.

Definition 18.2 Given two normal operators $\hat{E} = \hat{D} + i\hat{F} = \int_{-\infty}^{\infty}\int_{-\infty}^{\infty}(\alpha + i\beta) d(\hat{Q}_1(\alpha)\hat{Q}_2(\beta))$ and $\hat{A} = \int_{-\infty}^{\infty}\int_{-\infty}^{\infty}(\varepsilon + i\eta) d(\hat{P}_1(\varepsilon)\hat{P}_2(\eta))$, denoting by $\mathcal{M}_{\hat{P}_i}$ the subspace of the Hilbert space on which $\hat{P}_i$ projects, we then define the spectral order as follows:

$$\hat{A} \geq_s \hat{E} \quad \text{iff} \quad \mathcal{M}_{\hat{P}_1(\varepsilon)} \cap \mathcal{M}_{\hat{P}_2(\eta)} \subseteq \mathcal{M}_{\hat{Q}_1(\alpha)} \cap \mathcal{M}_{\hat{Q}_2(\beta)}. \quad (18.5)$$

However, property (a) of Theorem 18.1 implies that for any two points $(\varepsilon, \eta) \in \mathbb{R}^2$ and $(\varepsilon', \eta') \in \mathbb{R}^2$ then

$$\hat{P}(\varepsilon, \eta)\hat{P}\big(\varepsilon', \eta'\big) = \hat{P}(\varepsilon, \eta) \quad \text{for } \varepsilon \leq \varepsilon' \text{ and } \eta \leq \eta' \quad (18.6)$$

therefore, we could define

$$\hat{P}(\varepsilon,\eta) \leq \hat{P}(\varepsilon',\eta') \quad \text{for } \varepsilon \leq \varepsilon' \text{ and } \eta \leq \eta' \quad \text{iff} \quad \hat{P}(\varepsilon,\eta)\hat{P}(\varepsilon',\eta') = \hat{P}(\varepsilon,\eta). \tag{18.7}$$

The above reasoning shows that, the spectral ordering of normal operators, is intimately connected to the ordering of its self-adjoint components.

We are now in a position to define a suitable ordering for the complex numbers.

Definition 18.3 Given two complex numbers $\lambda_1 = \epsilon_1 + i\eta_1$ and $\lambda = \epsilon + i\eta$, then we say that

$$\lambda_1 \geq \lambda \quad \text{if } (\epsilon_1 + \eta_1) \geq (\epsilon + \eta) \tag{18.8}$$

where $(\epsilon_1 + \eta_1) \geq (\epsilon + \eta)$ obeys the usual ordering of the reals.

This implies the following definition:

Definition 18.4 Given two normal operators $\hat{A} = \hat{C} + i\hat{B}$ and $\hat{E} = \hat{D} + i\hat{F}$, if $\hat{A} \geq_s \hat{E}$, with respect to the spectral order of normal operators defined in equations (18.4) and (18.5), then $\lambda(\hat{C} + i\hat{B}) \geq \lambda(\hat{D} + i\hat{F})$.

18.1.1 Example

To illustrate what has been said so far we will give a simple example. Let us consider a two-state system with non-self-adjoint operator:

$$\hat{O}_z = \begin{pmatrix} 1 & 0 \\ 0 & -i \end{pmatrix} \tag{18.9}$$

and its norm squared

$$\hat{O}_z^2 = \begin{pmatrix} 1 & 0 \\ 0 & 1 \end{pmatrix}. \tag{18.10}$$

We can decompose $\hat{O}_z$ into two matrices $\hat{A}$ and $\hat{B}$ with eigen-projectors $\hat{P}_1 = \text{diag}(0, 1)$, $\hat{P}_2 = \text{diag}(1, 0)$ and $\hat{Q}_1 = \text{diag}(-1, 0)$ and $\hat{Q}_2 = \text{diag}(0, -1)$ as follows:

$$\hat{O}_z = \hat{A} + i\hat{B} = \begin{pmatrix} 1 & 0 \\ 0 & 0 \end{pmatrix} + i\begin{pmatrix} 0 & 0 \\ 0 & -1 \end{pmatrix} \tag{18.11}$$

where

$$\hat{A} = \int_{-\infty}^{\infty} \epsilon\, d\big(\hat{P}_i(\epsilon)\big) \quad \text{and} \quad \hat{B} = \int_{-\infty}^{\infty} \eta d\big(\hat{Q}_i(\eta)\big). \tag{18.12}$$

The spectral family for $\lambda = \epsilon + i\eta$ is then as follows:

$$\hat{E}_\lambda^{\hat{O}_z} = \begin{cases} \hat{0}\hat{0} & \text{if } \epsilon < 0, \eta < -1 \\ \hat{0}\hat{Q}_2 & \text{if } \epsilon < 0, -1 \leq \eta < 0 \\ \hat{0}(\hat{Q}_1 + \hat{Q}_2) & \text{if } \epsilon < 0, 0 \leq \eta \\ \hat{P}_2\hat{0} & \text{if } 0 \leq \epsilon < 1, \eta < -1 \\ \hat{P}_2\hat{Q}_2 & \text{if } 0 \leq \epsilon < 1, -1 \leq \eta < 0 \\ \hat{P}_2(\hat{Q}_1 + \hat{Q}_2) & \text{if } 0 \leq \epsilon < 1, 0 \leq \eta \\ (\hat{P}_1 + \hat{P}_2)\hat{0} & \text{if } 1 \leq \epsilon, \eta < -1 \\ (\hat{P}_1 + \hat{P}_2)\hat{Q}_2 & \text{if } 1 \leq \epsilon, -1 \leq \eta < 0 \\ (\hat{P}_1 + \hat{P}_2)(\hat{Q}_1 + \hat{Q}_2) & \text{if } 1 \leq \epsilon, 0 \leq \eta \end{cases} \tag{18.13}$$

and

$$\hat{E}_\lambda^{\hat{O}_z^2} = \begin{cases} \hat{0}\hat{0} & \text{if } \epsilon < 1, \eta < 0 \\ \hat{0}\hat{Q} & \text{if } \epsilon < 1, 0 \leq \eta \\ (\hat{P}_1 + \hat{P}_2)\hat{0} & \text{if } \epsilon \geq 1, \eta < 0 \\ (\hat{P}_1 + \hat{P}_2)\hat{Q} & \text{if } \epsilon \geq 1, 0 \leq \eta \end{cases}$$

where $\hat{Q} \equiv \hat{Q}_1 + \hat{Q}_2$. By comparing the spectral families in a piecewise manner, one can see that for any breakdown of ϵ, η, we have that $\hat{E}_\lambda^{\hat{O}_z} \geq \hat{E}_\lambda^{\hat{O}_z^2}$ and so $\hat{O}_z \leq_S \hat{O}_z^2$.

We now compare this ordering of the operators $\hat{O}_z$ and $\hat{O}_z^2$ to that obtained using definition (18.3). To this end we consider the spectral decomposition of the associated self-adjoint operators. In particular for $\hat{O}_z$ the associated self-adjoint operator is

$$\hat{A} + \hat{B} = \begin{pmatrix} 1 & 0 \\ 0 & -1 \end{pmatrix} \tag{18.14}$$

while the operator $\hat{O}_z^2$ stays the same. The spectral decomposition of these operators are, respectively,

$$\hat{E}_\lambda^{\hat{A}+\hat{B}} = \begin{cases} \hat{0} & \text{if } \lambda < -1 \\ \hat{Q}_2 & \text{if } -1 \leq \lambda < 1 \\ (\hat{P}_2 + \hat{Q}_2) & \text{if } 1 \leq \lambda \end{cases}$$

and

$$\hat{E}_\lambda^{|\hat{O}_z^2|} = \begin{cases} \hat{0} & \text{if } \lambda < 1 \\ (\hat{P}_2 + \hat{Q}_2) & \text{if } \lambda \geq 1. \end{cases}$$

We can see that the natural spectral ordering implies

$$\begin{cases} \hat{E}_\lambda^{|\hat{O}_z^2|} = \hat{E}_\lambda^{\hat{A}+\hat{B}} & \text{if } \lambda < -1 \\ \hat{E}_\lambda^{|\hat{O}_z^2|} \leq \hat{E}_\lambda^{\hat{A}+\hat{B}} & \text{if } -1 \leq \lambda < 1 \\ \hat{E}_\lambda^{|\hat{O}_z^2|} = \hat{E}_\lambda^{\hat{A}+\hat{B}} & \text{if } 1 \leq \lambda \end{cases}$$

so $\hat{A}+\hat{B} \leq_s |\hat{O}_z^2|$ and, by Definition 18.3, $\hat{O}_z \leq_s \hat{O}_z^2$. Therefore, both treatments are equivalent.

18.2 Normal Operators in a Topos

Our aim is to define normal operators as maps, from the state space $\underline{\Sigma}$ to the topos analogue of the complex numbers, in analogy with how self-adjoint operators are defined in a topos. However, since the spectrum of self-adjoint operators is real, they are defined as maps from the state space $\underline{\Sigma}$ to the (real) quantity value object $\underline{\mathbb{R}}^{\leftrightarrow}$, see Sect. 13.5. The definition of such maps relied on the concept of daseinisation of self-adjoint operators. We will now try to extend this notion to include also normal operators. To achieve this we will utilise the Gel'fand representation theorem for normal operators [75], which states the following:

for a closed *-sub-algebra $\mathcal{A}$ generated by a normal operator $\hat{T}$, $\hat{T}^*$ and the identity element, there exists a mapping onto the space of $\Sigma(\mathcal{A})$ of all continuous functional on $\mathcal{A}$ given by

$$\mathcal{A} \to C\big(\Sigma(\mathcal{A})\big)$$
$$\hat{T} \to \bar{T} \equiv \lambda_C(\hat{T})$$

where $C(\Sigma(\mathcal{A}))$ is the space of complex continuous functions on $\Sigma(\mathcal{A})$.

For the case at hand $\mathcal{A} = V$ while $\Sigma(\mathcal{A}) = \underline{\Sigma}_V$ is the spectrum of the abelian von Neumann algebra V. We can therefore define the inner and outer daseinisation of normal operators in the same manner as for self-adjoint operators, namely in terms of the Gel'fand transforms:

$$\overline{\delta^o(\hat{C})_V} : \underline{\Sigma}_V \to \mathbb{C} \tag{18.15}$$

$$\overline{\delta^i(\hat{C})_V} : \underline{\Sigma}_V \to \mathbb{C}. \tag{18.16}$$

However, in the case of self-adjoint operator we know that for a sub-context $V' \subseteq V$, then $\delta^o(\hat{A})_{V'} \geq \delta^o(\hat{A})_V$ and $\delta^i(\hat{A})_{V'} \leq \delta^i(\hat{A})_V$. These relations imply that the respective Gel'fand transforms satisfies $\overline{\delta^o(\hat{A})_{V'}}(\lambda_{|V'}) \geq \overline{\delta^o(\hat{A})_V}(\lambda)$ and $\overline{\delta^i(\hat{A})_{V'}}(\lambda_{|V'}) \leq \overline{\delta^i(\hat{A})_V}(\lambda)$. We want similar relations to hold for normal operators. This is indeed the case. In fact, given any normal operator $\hat{A}$ in $\mathcal{B}(\mathcal{H})$, then there exists a unique spectral family $\{\hat{P}(\varepsilon,\eta) := \hat{P}_1(\varepsilon)\hat{P}_2(\eta)|(\varepsilon,\eta) \in \mathbb{R}^2\}$ of projection operators which comprise its spectral decomposition.

By applying the exact same proof as used in [76], it can be shown that

$$\lambda = \varepsilon + i\eta \to \delta^i\big(\hat{P}(\varepsilon,\eta)\big)_V \tag{18.17}$$

$$\lambda = \varepsilon + i\eta \to \bigwedge_{\mu>\lambda} \delta^o\big(\hat{P}(\varepsilon',\eta')\big)_V \tag{18.18}$$

represent spectral families. Here $\mu = \varepsilon' + i\eta'$ and the ordering $\mu > \lambda$ is given in Definition 18.3.

Using this result it is possible to define the daseinisation of normal operators as follows:

Definition 18.5 Let $\hat{A} = \hat{C} + i\hat{D}$ be an arbitrary normal operator, then the outer and inner daseinisations of $\hat{A}$ are defined in each sub-context V as:

$$\delta^o(\hat{A})_V := \int_{\mathbb{C}} \lambda d\big(\delta^i(\hat{E}_\lambda)_V\big) = \int_{\mathbb{C}} (\varepsilon + i\eta) d\big(\delta^i\big(\hat{P}(\varepsilon,\eta)_V\big)\big) \tag{18.19}$$

$$\delta^i(\hat{A})_V := \int_{\mathbb{C}} \lambda d\Big(\bigwedge_{\mu>\lambda} \delta^o(\hat{E}_\mu)_V\Big) = \int_{\mathbb{C}} (\varepsilon + i\eta) d\Big(\bigwedge_{\mu>\lambda} \delta^o\big(\hat{P}(\varepsilon',\eta)_V\big)\Big) \tag{18.20}$$

respectively.

Since for all $V \in \mathcal{V}(\mathcal{H})$

$$\delta^i(\hat{P} \wedge \hat{Q})_V \le \delta^i(\hat{P})_V \wedge \delta^i(\hat{Q})_V \tag{18.21}$$

$$\delta^o(\hat{P} \wedge \hat{Q})_V \ge \delta^o(\hat{P})_V \wedge \delta^o(\hat{Q})_V \tag{18.22}$$

it follows that

$$\delta^o(\hat{A})_V \ge \delta^o(\hat{C})_V + i\delta^o(\hat{D})_V \tag{18.23}$$

$$\delta^i(\hat{A})_V \le \delta^i(\hat{C})_V + i\delta^i(\hat{D})_V. \tag{18.24}$$

Moreover, from the definition of inner and outer daseinisation of projection operators, for all $V \in \mathcal{V}(\mathcal{H})$ and all $\lambda \in \mathbb{C}$ we have

$$\bigwedge_{\mu\ge\lambda} \delta^o\big(\hat{E}(\mu)\big)_V \ge \delta^i\big(\hat{E}(\lambda)\big)_V. \tag{18.25}$$

From the definition of spectral order it then follows that:

$$\delta^i(\hat{A})_V \le_s \delta^o(\hat{A})_V. \tag{18.26}$$

We would now like to analyse the spectrum of these operators. As a consequence of the spectral theorem and the fact that both $\delta^i(\hat{A})_V$ and $\delta^o(\hat{A})_V$ are in V, it is possible to represent them through the Gel'fand transform:

$$\overline{\delta^o(\hat{A})_V} : \underline{\Sigma}_V \to sp\big(\delta^o(\hat{A})_V\big) \tag{18.27}$$

$$\overline{\delta^i(\hat{A})_V} : \underline{\Sigma}_V \to sp\big(\delta^i(\hat{A})_V\big). \tag{18.28}$$

Since $sp(\delta^o(\hat{A})_V) \subseteq \mathbb{C}$ and $sp(\delta^i(\hat{A})_V) \subseteq \mathbb{C}$ we can generalise the above maps to

$$\overline{\delta^o(\hat{A})_V} : \underline{\Sigma}_V \to \mathbb{C} \tag{18.29}$$

$$\overline{\delta^i(\hat{A})_V} : \underline{\Sigma}_V \to \mathbb{C}. \tag{18.30}$$

However, the relation $\delta^i(\hat{A})_V \leq_s \delta^o(\hat{A})_V$ implies that for all $V \in \mathcal{V}(\mathcal{H})$, $\overline{\delta^i(\hat{A})_V}(\lambda) \leq \overline{\delta^o(\hat{A})_V}(\lambda)$ (where, again, the ordering is the one defined in Definition 18.3).

Moreover, as we go to smaller sub-algebras $V' \subseteq V$, since $\delta^o(\hat{E}(\lambda))_{V'} \geq \delta^o(\hat{E}(\lambda))_V$ and $\delta^i(\hat{E}(\lambda))_{V'} \leq \delta^i(\hat{E}(\lambda))_V$, then $\delta^i(\hat{A})_V \geq_s \delta^i(\hat{A})_{V'}$ while $\delta^o(\hat{A})_V \leq_s \delta^o(\hat{A})_{V'}$. Thus, inner daseinisation preserves the order, while outer daseinisation reverses the order. As a consequence we obtain that

$$\overline{\delta^o(\hat{A})_V}(\lambda) \leq \overline{\delta^o(\hat{A})_{V'}}(\lambda_{|V'}) \tag{18.31}$$

$$\overline{\delta^i(\hat{A})_V}(\lambda) \geq \overline{\delta^i(\hat{A})_{V'}}(\lambda_{|V'}) \tag{18.32}$$

where $\overline{\delta^o(\hat{A})_V}(\lambda) := \lambda(\delta^o(\hat{A})_V)$ while $\overline{\delta^o(\hat{A})_{V'}}\lambda_{|V'} := \lambda_{|V'}(\delta^o(\hat{A})_{V'})$.

18.2.1 Example

Building on to our previous example we can explore the daseinisation of the operator

$$\hat{O}_z = \begin{pmatrix} 1 & 0 \\ 0 & -i \end{pmatrix} = \begin{pmatrix} 1 & 0 \\ 0 & 0 \end{pmatrix} + i \begin{pmatrix} 0 & 0 \\ 0 & -1 \end{pmatrix}. \tag{18.33}$$

As before, we have four projector operators $\hat{P}_1, \hat{P}_2, \hat{Q}_1, \hat{Q}_2$ along with the projectors $\hat{0}$ and $\hat{1}$.

Given Definition 18.5, in order to compute $\delta^i(\hat{O}_z)$ and $\delta^o(\hat{O}_z)$, we simply need to compute the daseinisation of the projection operators in the spectral decomposition of $\hat{O}_z$. Thus our problem reduces to computing the daseinisations $\delta^{o,i}(\hat{P}_1)_V$, $\delta^{o,i}(\hat{P}_2)_V$, $\delta^{o,i}(\hat{Q}_1)_V$ and $\delta^{o,i}(\hat{Q}_2)_V$.

Let us choose a sub-context V spanned by the projection operators $\hat{P}_1 + \hat{P}_2, \hat{Q}_1, \hat{Q}_2$. Then, the only non-trivial outer daseinisations are $\delta^o(\hat{P}_1)_V = \delta^o(\hat{P}_2)_V = \hat{P}_1 + \hat{P}_2$, while for inner daseinisation we get $\delta^o(\hat{P}_1)_V = \delta^o(\hat{P}_2)_V = \hat{0}$.

Recalling that the spectral family of the operator $\hat{E}_\lambda^{\hat{O}_z}$ is

$$\hat{E}_\lambda^{\hat{O}_z} = \begin{cases} \hat{0}\hat{0} & \text{if } \epsilon < 0, \eta < -1 \\ \hat{0}\hat{Q}_2 & \text{if } \epsilon < 0, -1 \leq \eta < 0 \\ \hat{0}(\hat{Q}_1 + \hat{Q}_2) & \text{if } \epsilon < 0, 0 \leq \eta \\ \hat{P}_2\hat{0} & \text{if } 0 \leq \epsilon < 1, \eta < -1 \\ \hat{P}_2\hat{Q}_2 & \text{if } 0 \leq \epsilon < 1, -1 \leq \eta < 0 \\ \hat{P}_2(\hat{Q}_1 + \hat{Q}_2) & \text{if } 0 \leq \epsilon < 1, 0 \leq \eta \\ (\hat{P}_1 + \hat{P}_2)\hat{0} & \text{if } 1 \leq \epsilon, \eta < -1 \\ (\hat{P}_1 + \hat{P}_2)\hat{Q}_2 & \text{if } 1 \leq \epsilon, -1 \leq \eta < 0 \\ (\hat{P}_1 + \hat{P}_2)(\hat{Q}_1 + \hat{Q}_2) & \text{if } 1 \leq \epsilon, 0 \leq \eta \end{cases}$$

we then obtain:

$$\delta^o\big(\hat{E}_\lambda^{\hat{O}_z}\big)_V = \begin{cases} \hat{0}\hat{0} & \text{if } \epsilon < 0, \eta < -1 \\ \hat{0}\hat{Q}_2 & \text{if } \epsilon < 0, -1 \leq \eta < 0 \\ \hat{0}(\hat{Q}_1 + \hat{Q}_2) & \text{if } \epsilon < 0, 0 \leq \eta \\ (\hat{P}_1 + \hat{P}_2)\hat{0} & \text{if } 0 \leq \epsilon, \eta < -1 \\ (\hat{P}_1 + \hat{P}_2)\hat{Q}_2 & \text{if } 0 \leq \epsilon, -1 \leq \eta < 0 \\ (\hat{P}_1 + \hat{P}_2)(\hat{Q}_1 + \hat{Q}_2) & \text{if } 0 \leq \epsilon, 0 \leq \eta \end{cases} \tag{18.34}$$

and

$$\delta^i\big(\hat{E}_\lambda^{\hat{O}_z}\big)_V = \begin{cases} \hat{0}\hat{0} & \text{if } \epsilon < 1, \eta < -1 \\ \hat{0}\hat{Q}_2 & \text{if } \epsilon < 1, -1 \leq \eta < 0 \\ \hat{0}(\hat{Q}_1 + \hat{Q}_2) & \text{if } \epsilon < 1, 0 \leq \eta \\ (\hat{P}_1 + \hat{P}_2)\hat{0} & \text{if } 0 \leq \epsilon, \eta < -1 \\ (\hat{P}_1 + \hat{P}_2)\hat{Q}_2 & \text{if } 0 \leq \epsilon, -1 \leq \eta < 0 \\ (\hat{P}_1 + \hat{P}_2)(\hat{Q}_1 + \hat{Q}_2) & \text{if } 0 \leq \epsilon, 0 \leq \eta. \end{cases} \tag{18.35}$$

We can see that for any ϵ, η, $\hat{E}_\lambda^{\hat{O}_z} \leq \delta^o(\hat{E}_\lambda^{\hat{O}_z})_V$ and, therefore, $\delta^o(\hat{O}_z)_V \leq_s \hat{O}_z$. Similarly, $\hat{E}_\lambda^{\hat{O}_z} \geq \delta^i(\hat{E}_\lambda^{\hat{O}_z})_V$ and $\delta^i(\hat{O}_z)_V \leq_s \hat{O}_z$.

18.3 Complex Number Object in a Topos

Complex numbers in a topos have been previously defined in various papers [18, 89, 90], however the definition of these objects did not take into account the spectra of normal operators. In the present situation, since our ultimate aim is to define normal operators as maps from the state space to the (complex) quantity value object, we have to resort to a different characterisation of complex numbers in a topos. This will be very similar to how the real quantity value object is defined.

Definition 18.6 The complex quantity value object is the presheaf $\underline{\mathbb{C}}^{\leftrightarrow}$ which has, as:

- Objects: for all contexts $V \in \mathcal{V}(\mathcal{H})$,

$$\underline{\mathbb{C}}^{\leftrightarrow}_V := \big\{(\mu, \nu)|\mu \in OP(\downarrow V, \mathbb{C}), \nu \in OR(\downarrow V, \mathbb{C}); \mu \leq \nu\big\} \tag{18.36}$$

  where OP denotes the set of order preserving functions, while OR the set of order reversing functions.
- Morphisms: given a map between contexts $i_{V'V} : V' \subseteq V$ the corresponding morphisms are

$$\underline{\mathbb{C}}^{\leftrightarrow}(i_{V'V}) : \underline{\mathbb{C}}^{\leftrightarrow}_V \to \underline{\mathbb{C}}^{\leftrightarrow}_{V'} \tag{18.37}$$

$$(\mu, \nu) \mapsto (\mu_{|V'}, \nu_{|V'}) \tag{18.38}$$

where $\mu_{V'} : \downarrow V' \to \mathbb{C}$ is simply the restriction of μ to the sub-context V'.

This definition suits our purpose, since it preserves the fact that under outer daseinisation we obtain the inequality $\overline{\delta^o(\hat{A})_V}(\lambda) \leq \overline{\delta^o(\hat{A})_{V'}}\lambda_{|V'}$ and, under inner daseinisation, we have $\overline{\delta^i(\hat{A})_V}(\lambda) \geq \overline{\delta^i(\hat{A})_{V'}}\lambda_{|V'}$.

Furthermore, we can use the isometry

$$\mathbb{R} \times \mathbb{R} \simeq \mathbb{C} \tag{18.39}$$

as a guideline to define, rigorously, the transformation between the quantity value object and the complex quantity value object. Recall that the quantity value object is a monoid (a semigroup with unit) and, as such, it is equipped with the summation operation

$$+ : \underline{\mathbb{R}}^{\leftrightarrow} \times \underline{\mathbb{R}}^{\leftrightarrow} \to \underline{\mathbb{R}}^{\leftrightarrow} \tag{18.40}$$

which is defined for each $V \in \mathcal{V}(\mathcal{H})$ as

$$+_V : \underline{\mathbb{R}}^{\leftrightarrow}_V \times \underline{\mathbb{R}}^{\leftrightarrow}_V \to \underline{\mathbb{R}}^{\leftrightarrow}_V \tag{18.41}$$

$$\big((\mu_1, \nu_1), (\mu_2, \nu_2)\big) \mapsto (\mu_1 + \mu_2, \nu_1 + \nu_2) = (\mu, \nu). \tag{18.42}$$

Here $(\mu_1 + \mu_2, \nu_1 + \nu_2)$ is defined, for each $V' \subseteq V$, as $(\mu_1(V) + \mu_2(V), \nu_1(V) + \nu_2(V))$.

We can make use of this to define the map

$$f_V : \underline{\mathbb{R}}^{\leftrightarrow}_V \times \underline{\mathbb{R}}^{\leftrightarrow}_V \to \underline{\mathbb{C}}^{\leftrightarrow}_V \tag{18.43}$$

$$\big((\mu_1, \nu_1), (\mu_2, \nu_2)\big) \mapsto (\mu_1 + i\mu_2, \nu_1 + i\nu_2). \tag{18.44}$$

Even in this case, for each $V' \subseteq V$, the above complex sum should be intended as:

$$\mu(V') := \mu_1(V') + i\mu_2(V') \tag{18.45}$$

$$\nu(V') := \nu_1(V') + i\nu_2(V') \tag{18.46}$$

for each context $V \in \mathcal{V}(\mathcal{H})$.

Thus the map f_V takes the pair $((\mu_1, \nu_1), (\mu_2, \nu_2)) \in \underline{R}^{\leftrightarrow} \times \underline{\mathbb{R}}^{\leftrightarrow}$ and maps it to the element $(\mu, \nu) \in \underline{\mathbb{C}}^{\leftrightarrow}$ consisting of a pair of order reversing and order preserving maps, such that for each $V \in \mathcal{V}(\mathcal{H})$, $\mu(V) := \mu_1(V) + i\mu_2(V)$ and $\nu(V') := \nu_1(V) + i\nu_2(V)$.

This induces a natural transformation

$$f : \underline{\mathbb{R}}^{\leftrightarrow} \times \underline{\mathbb{R}}^{\leftrightarrow} \to \underline{\mathbb{C}}^{\leftrightarrow} \tag{18.47}$$

between the quantity value object and the complex quantity value object.

First of all we will show that f is indeed a natural transformation. To this end we need to show that for all pairs $V' \subseteq V$ in $\mathcal{V}(\mathcal{H})$, the following diagram commutes:

$$\begin{array}{ccc}
\underline{\mathbb{R}}^{\leftrightarrow}_V \times \underline{\mathbb{R}}^{\leftrightarrow}_V & \xrightarrow{\;f_V\;} & \underline{\mathbb{C}}^{\leftrightarrow}_V \\
\Big\downarrow g & & \Big\downarrow h \\
\underline{\mathbb{R}}^{\leftrightarrow}_{V'} \times \underline{\mathbb{R}}^{\leftrightarrow}_{V'} & \xrightarrow[\;f_{V'}\;]{} & \underline{\mathbb{C}}^{\leftrightarrow}_{V'}
\end{array}$$

where h and g are the presheaf maps, i.e. we want to show that $h \circ f_V = f_{V'} \circ g$. Let us consider an element $((\mu_1, \nu_1), (\mu_2, \nu_2))$, chasing the diagram around in one direction we have

$$\begin{aligned}
h \circ f_V\big((\mu_1, \nu_1), (\mu_2, \nu_2)\big) &= h(\mu_1 + i\mu_2, \nu_1 + i\nu_2) \\
&= \big((\mu_1 + i\mu_2)_{|V'}, (\nu_1 + i\nu_2)_{|V'}\big)
\end{aligned} \tag{18.48}$$

where $(\mu_1 + i\mu_2)_{|V'} = (\mu_1)_{|V'} + i(\mu_2)_{|V'}$. On the other hand

$$\begin{aligned}
f_{V'} \circ g\big((\mu_1, \nu_1), (\mu_2, \nu_2)\big) &= f_{V'}\big((\mu_1, \nu_1)_{|V'}, (\mu_2, \nu_2)_{|V'}\big) \\
&= \big((\mu_1)_{|V'} + i(\mu_2)_{|V'}, (\nu_1)_{|V'} + i(\nu_2)_{|V'}\big).
\end{aligned} \tag{18.49}$$

Thus indeed f is a natural transformation.

Theorem 18.2 *f is* 1:1.

Proof If $f_V((\mu_1,\nu_1),(\mu_2,\nu_2)) = f_V((\mu_1',\nu_1'),(\mu_2',\nu_2'))$ then $(\mu_1 + i\mu_2, \nu_1 + i\nu_2) = (\mu_1' + i\mu_2', \nu_1' + i\nu_2')$. Therefore $(\mu_1 + i\mu_2) = (\mu_1' + i\mu_2')$ and $(\nu_1 + i\nu_2) = (\nu_1' + i\nu_2')$. By evaluating such maps at each $V \in \mathcal{V}(\mathcal{H})$ it follows that $((\mu_1,\nu_1),(\mu_2,\nu_2)) = ((\mu_1',\nu_1'),(\mu_2',\nu_2'))$. □

However, the map f is not onto. This is because the ordering that we defined on $\underline{\mathbb{C}}^{\leftrightarrow}$ is more general than the ordering coming from pairs of order reversing and order preserving maps. In fact, for $\mu_1 + i\mu_2 \leq \mu_3 + i\mu_4$ we only require that $\mu_1 + \mu_2 \leq \mu_3 + \mu_4$, not that $\mu_1 \leq \mu_3$ and $\mu_2 \leq \mu_4$. Obviously, the latter relation implies the former, but the converse is not true. It follows, then, that $\underline{\mathbb{R}}^{\leftrightarrow} \times \underline{\mathbb{R}}^{\leftrightarrow}$ will be isomorphic to a sub-object of $\underline{\mathbb{C}}^{\leftrightarrow}$.

Algebraic Properties of $\underline{\mathbb{C}}^{\leftrightarrow}$ Given the object $\underline{\mathbb{C}}^{\leftrightarrow}$ defined above, we are interested in analysing what types of properties it has. In particular, we know that $\mathbb{C}$ is a group and it is also a vector space over the reals. Can the same be said for $\underline{\mathbb{C}}^{\leftrightarrow}$? To this end we will first analyse whether the usual operations present in $\mathbb{C}$ are also present in $\underline{\mathbb{C}}^{\leftrightarrow}$.

1. *Conjugation.* The most obvious way of defining conjugation would be the following:
   for each $V \in \mathcal{V}(\mathcal{H})$ we have

$$\overset{*}{V} : \underline{\mathbb{C}}^{\leftrightarrow}_V \to \underline{\mathbb{C}}^{\leftrightarrow}_V \tag{18.50}$$

$$(\mu, \nu) \mapsto (\mu^*, \nu^*) \tag{18.51}$$

   where $\mu^*(V) := (\mu(V))^*$ However, if μ is order preserving it is not necessarily the case that μ^* is. This is related to the same problem which prevents us from defining subtraction in $\underline{\mathbb{R}}^{\leftrightarrow}$.
2. *Sum.*

Definition 18.7 The sum operation is defined to be a map $+ : \underline{\mathbb{C}}^{\leftrightarrow} \times \underline{\mathbb{C}}^{\leftrightarrow} \to \underline{\mathbb{C}}^{\leftrightarrow}$, such that for each $V \in \mathcal{V}(\mathcal{H})$ we have

$$+_V : \underline{\mathbb{C}}^{\leftrightarrow}_V \times \underline{\mathbb{C}}^{\leftrightarrow}_V \to \underline{\mathbb{C}}^{\leftrightarrow}_V \tag{18.52}$$

$$\big((\mu,\nu),(\mu',\nu')\big) \mapsto (\mu + \mu', \nu + \nu'). \tag{18.53}$$

Where $(\mu + \mu', \nu + \nu')(V') := ((\mu + \mu')(V'), (\nu + \nu')(V')) = (\mu(V') + \mu'(V'), \nu(V') + \nu'(V'))$ for all $V \in \mathcal{V}(\mathcal{H})$.

It is straightforward to see that, as defined above, the maps $(\mu + \mu', \nu + \nu')$ are a pair of order preserving and order reversing maps.

3. *Multiplication.* Similar to the case for $\underline{\mathbb{R}}^{\leftrightarrow}$ we can not define multiplication.
4. *Subtraction.* Similar to the case for $\underline{\mathbb{R}}^{\leftrightarrow}$ we can not define subtraction.

Because of the above properties $\underline{\mathbb{C}}^{\leftrightarrow}$ is only a monoid (semigroup with a unit). However, as shown in [49], it is possible to transform such a semigroup into a group through the process of Grothendieck k-extension, obtaining the object $k(\underline{\mathbb{C}}^{\leftrightarrow})$ which can be seen as a vector space over $\underline{\mathbb{R}}$.[1]

It would be of particular interest to understand what and if $\underline{\mathbb{C}}^{\leftrightarrow}$ has any topological properties. To this end we would have to define $\mathbb{C}^{\leftrightarrow}$ as an internal locale.[2] Work in this direction has been partially done in [47]. Here the authors introduce an alternative *internal* formulation of quantum topos theory. In particular, given a C^*-algebra $\mathcal{A}$, they define the internal[3] C^* algebra $\tilde{\mathcal{A}} \in Sets^{\mathcal{C}(\mathcal{A})}$ where $\mathcal{C}(\mathcal{A})$ is the category of abelian sub-algebras of $\mathcal{A}$ ordered by inclusion. Given such an internal algebra they construct its spectrum $\underline{\tilde{\Sigma}}$ and show that it is an internal locale. This enables them to define the (internal) topos of sheaves $Sh(\underline{\tilde{\Sigma}})$. They then construct the locale $\underline{\mathbb{R}}_{Sh(\underline{\tilde{\Sigma}})}$ which has as associated sheaf the sheaf $pt(\underline{\mathbb{R}}_{Sh(\underline{\tilde{\Sigma}})})$ of Dedekind Reals in $Sh(\underline{\tilde{\Sigma}})$. The detailed way in which $pt(\underline{\mathbb{R}}_{Sh(\underline{\tilde{\Sigma}})})$ is defined can be found in [18]. Given the internal local $\underline{\mathbb{R}}_{Sh(\underline{\tilde{\Sigma}})}$ it is now possible to construct the internal locale $\underline{\mathbb{C}}_{Sh(\underline{\tilde{\Sigma}})}$ whose associated sheaf would be the complex number object in $Sh(\underline{\tilde{\Sigma}})$ defined by $pt(\underline{\mathbb{C}}_{Sh(\underline{\tilde{\Sigma}})}) \simeq pt(\underline{\mathbb{R}}_{Sh(\underline{\tilde{\Sigma}})}) \times pt(\underline{\mathbb{R}}_{Sh(\underline{\tilde{\Sigma}})})$. A very in depth analysis of the complex number object can be found in [89, 90] and [91]. What is interesting for us is that the object $pt(\underline{\mathbb{R}}_{Sh(\underline{\tilde{\Sigma}})})$ was shown [47] to be related to $\underline{\mathbb{R}}^{\leftrightarrow}$ in the case of quantum theory. This would suggest that the object $\underline{C}^{\leftrightarrow}$ is related to $\underline{\mathbb{C}}_{Sh(\underline{\tilde{\Sigma}})}$. If this were the case, it would help unveil what, if any, topological properties $\underline{C}^{\leftrightarrow}$ has. Such an analysis, however, is not yet been done.

18.3.1 Domain-Theoretic Structure

In the recent paper [92, 93] the authors show how the quantity value object $\underline{\mathbb{R}}^{\leftrightarrow}$ can be given a domain theoretic structure. This then results in the fibres $\underline{\mathbb{R}}^{\leftrightarrow}_V$ being almost-bounded directed-complete posets (see later on for the appropriate definitions). We will now give a brief description on how the results given in [93] can be generalised for the complex quantity value object $\underline{\mathbb{C}}^{\leftrightarrow}$ defined above. As a first step we give the definition of a closed rectangle in the complex plane $\mathbb{C}$.

Definition 18.8 The set

$$T_{\alpha,\beta} = \{z \in \mathbb{C} | a \leq Re(z) \leq c, b \leq Im(z) \leq d; a, b, c, d \in \mathbb{R}\} \tag{18.54}$$

---

[1]The presheaf $\underline{\mathbb{R}}$ is defined on each $V \in \mathcal{V}(\mathcal{H})$ as $\underline{\mathbb{R}}_V := \mathbb{R}$ and given the maps $i : V' \subseteq V$ then the respective presheaf morphisms are simply the identity maps. See Definition 20.9.

[2]Recall that a local is essentially a frame (see Definition 17.1) but a morphisms between two locales is a morphisms between the frames which goes in the opposite direction.

[3]Note that in this internal approach one is working with co-presheaves instead of presheaves.

defines a closed rectangle in $\mathbb{C}$. Denoting $\alpha = a + ib$ and $\beta = c + id$, then the above closed rectangle is defined by the two points (α, β).

It is clear from the definition that $\alpha \leq \beta$ for the ordering in $\mathbb{C}$ defined in Definition 18.3. However this definition of closed rectangles does not account for the general case in which $\alpha \leq \beta$ if $a + b \leq c + d$ but not necessarily $a \leq c$ and $b \leq d$. To remedy this we slightly change the above definition of a closed rectangle, obtaining

Definition 18.9 Given any two points $\alpha, \beta \in \mathbb{C}$ such that $\alpha \leq \beta$ according to Definition 18.3, then the general closed rectangle "spanned" by them is

$$[\alpha, \beta] := \{z \in \mathbb{C} | Re(z) \in [Re(\alpha), Re(\beta)] \wedge Im(z) \in [Im(\alpha), Im(\beta)]\} \tag{18.55}$$

where $[Re(\alpha), Re(\beta)]$ is the closed line interval spanned by $Re(\alpha)$ and $Re(\beta)$ and similarly for $[Im(\alpha), Im(\beta)]$. Clearly, for a given α, β, $T_{\alpha,\beta} \equiv [\alpha, \beta]$.

Following the discussion of Sect. 18.3, it is clear that in the topos approach normal operators (as well as self-adjoint operators) are assigned an interval of values which are called "unsharp values". The set of such "unsharp complex values" is defined as follows

$$\mathbf{I}\mathbb{C} = \{[\alpha, \beta] | \alpha, \beta \in \mathbb{C}\ \alpha \leq \beta\}. \tag{18.56}$$

Clearly $\mathbb{C} \subset \mathbf{I}\mathbb{C}$ and it consists of all those intervals for which $\alpha = \beta$.

The claim is that, similarly as done in [93] for $\mathbf{I}\mathbb{R}$, the set $\mathbf{I}\mathbb{C}$ is a *domain* whose definition is given as follows:

Definition 18.10 A *domain* $\langle D, \sqsubseteq \rangle$ is a poset such that i) any directed set[4] has a supremum, i.e. it is a directed-complete poset (dcpo); ii) it is continuous: for any $d \in D$ one has $\bigsqcup^{\uparrow} \twoheaduparrow y = y$ where $\twoheaduparrow y := \{x \in D | x \ll y\}$.[5]

For the case of $\mathbf{I}\mathbb{C}$ we obtain the definition

Definition 18.11 The *complex interval domain* is the poset of closed rectangles in $\mathbb{C}$ partially ordered by reverse inclusion

$$\mathbf{I}\mathbb{C} := \langle \{[\alpha, \beta] | \alpha, \beta \in \mathbb{C}\}, \sqsubseteq := \supseteq \rangle \tag{18.57}$$

where $[\alpha, \beta] \sqsubseteq [\alpha_1, \beta_2]$ if $[\alpha_1, \beta_2] \subseteq [\alpha, \beta]$ in the sense that any complex number z lying in the rectangle $[\alpha_1, \beta_1]$ will also lie in the rectangle $[\alpha, \beta]$.

---

[4] A set P is directed if for any $x, y \in P$ there exists a $z \in P$ such that $x, y \sqsubseteq z$.

[5] Here $\bigsqcup^{\uparrow}$ indicates the supremum of a directed set and the relation $x \ll y$ indicates that x *approximates* y. In particular $x \ll y$ if, for any directed set S with a supremum, then $y \sqsubseteq \bigsqcup^{\uparrow} S \Rightarrow \exists s \in S : x \sqsubseteq s$.

Following the discussion given in [93], we can denote a rectangle as $x = [x_-, x_+]$ where x_- represents the left "end point" while x_+ represents the right "end point". In this way, for each function $f : X \to \mathbf{I}\mathbb{C}$ there corresponds a pair of functions $f_-, f_+ : X \to \mathbb{C}$ defined as $f_\pm(x) := (f(x))_\pm$ such that $f_- \leq f_+$ (pointwise order) and $f(x) = [f_-(x), f_+(x)]$. Conversely, for each pair of functions $g \leq h : X \to \mathbb{C}$ there corresponds a function $f : X \to \mathbf{I}\mathbb{C}$ such that $f_- = g$ and $f_+ = h$. The decomposition of each map $f : X \to \mathbf{I}\mathbb{C}$ into two maps can be explicitly stated by the following diagram

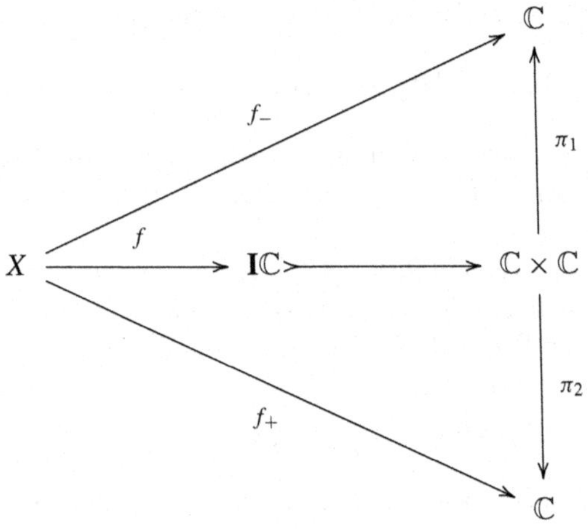

In order for $\mathbf{I}\mathbb{C}$ to be a well defined domain we need to show that it satisfies the definition of a domain. To achieve this we will need the generalised nested rectangle theorem

Theorem 18.3 *Given a sequence* $\{[\alpha, \beta]_n\}$ *of nested generalised closed rectangles (as defined above) such that* $\lim_{n\to\infty} l([\alpha, \beta]_n) = 0$[6] *then the following conditions hold*:

1. $[\alpha, \beta] = \bigcap_{i\in N}[\alpha, \beta]_i = z_0$ *for some* $z_0 \in \mathbb{C}$.
2. *Given* $\epsilon > 0$ *there is an* $m \in \mathbb{N}$ *such that*

$$[\alpha, \beta]_n \subset \{z : |z - z_0| < \epsilon\} \quad \forall n > m. \tag{18.58}$$

Proof To prove the first condition we resort to the theorem of nested intervals in $\mathbb{R}$. In particular the rectangle $[\alpha, \beta]_n$ has as boundary lines $Rel(z) = a_n$, $Rel(z) = c_n$, $Im(z) = b_n$, $Im(z) = d_n$ where $\alpha_n = a_n + ib_n$ and $\beta_n = c_n + id_n$. Therefore we have two sequences of nested intervals $R_n = [a_n, c_n]$ and $I_n = [b_n, d_n]$ such that

[6] Here $l([\alpha, \beta]_n)$ represents the length of the largest side of the rectangle and thus is defined as $l([\alpha, \beta]_n) = Max\{|a_n - c_n|, |b_n - d_n|\}$ where $\alpha = a + ib$ and $\beta = c + id$.

$[\alpha, \beta]_n = R_n \times I_n$. From the theorem of nested sequences of intervals it follows that $\bigcap_{i \in N} R_n = a$ and $\bigcap_{i \in N} I_n = d$ thus $\bigcap_{i \in N}[\alpha, \beta]_n = a + id = z_0$.

To prove condition 2) we choose an arbitrary element $z \in [\alpha, \beta]_n$. Then, given the existence of $[\alpha, \beta] = \bigcap_{i \in N}[\alpha, \beta]_i = z_0$ it follows that

$$|z - z_0|^2 \leq |\alpha_n - \beta_n|^2 = |a_n - c_n|^2 + |b_n - d_n|^2 = 2\big(l\big([\alpha, \beta]_n\big)\big)^2. \tag{18.59}$$

Since, by assumption, $\lim_{n\to\infty} l([\alpha, \beta]_n) = 0$, it follows that there exists an m such that given $\epsilon > 0$

$$\sqrt{2}l\big([\alpha, \beta]_n\big) < \epsilon, \quad \forall n > m. \tag{18.60}$$

Therefore

$$|z - z_0| \leq \sqrt{2}l\big([\alpha, \beta]_n\big) < \epsilon. \tag{18.61}$$

The above holds for any $z \in [\alpha, \beta]_n$ therefore $[\alpha, \beta]_n \subset \{z : |z - z_0| < \epsilon\}$. □

The reason we went through the trouble of stating the nested rectangle theorem is because we will use it when defining the supremum of directed subsets $S \in \mathbf{I}\mathbb{C}$. In particular, given such a directed set we then have

$$\bigsqcup^{\uparrow}[\alpha, \beta] = \bigcap[\alpha, \beta] = \big[\sup\{x_- | x \in [\alpha, \beta]\}, \inf\{x_+ | x \in [\alpha, \beta]\}\big]. \tag{18.62}$$

Thus for a sequence $[\alpha, \beta]_n$ of nested rectangles we simply get a point z_0 as the supremum.

Given the above, we can define the relation $\ll$ in $\mathbf{I}\mathbb{C}$ as follows

Definition 18.12 Given any two rectangles x, y then

$$x \ll y \quad \text{iff} \quad (x_- < y_-) \wedge (y_+ < x_+). \tag{18.63}$$

For $\mathbf{I}\mathbb{C}$ to be a well defined domain we need to show that $\bigcup^{\uparrow} \twoheaduparrow y = y$. Since $\twoheaduparrow y := \{x \in D | x \ll y\}$ then clearly

$$\bigcup^{\uparrow} \twoheaduparrow y = \bigcap\{x \in D | x \ll y\} = \bigcap\big\{x \in \mathbf{I}\mathbb{C} | (x_- < y_-) \wedge (y_+ < x_+)\big\} = y. \tag{18.64}$$

Since $\mathbf{I}\mathbb{C}$ is a domain it is a continuous poset and as such it comes with a Scott topology[7] whose basis opens are

$$\twoheaddownarrow[\alpha, \beta] := \big\{[\gamma, \delta] | \alpha < \gamma \leq \beta < \delta\big\} = \big\{\sigma \in \mathbf{I}\mathbb{C} | \sigma \subseteq (\alpha, \beta)\big\} \tag{18.65}$$

[7] We recall that, given a poset $\langle P, \leq \rangle$ a subset G is said to be Scott-open if (i) $x \in G \wedge x \leq y \Rightarrow y \in G$; (ii) for any directed set S with supremum then $\bigsqcup^{\uparrow} S \in G \Rightarrow \exists s \in S | s \in G$. In other words all supremums in G have a non-empty intersection with G.

where $(\alpha,\beta) := \{z \in \mathbb{C} | \alpha < z < \beta\}$ represents the general open rectangle "spanned" by α, β.[8] Recalling that any map $f : X \to \mathbf{I}\mathbb{C}$ can be decomposed into a left and right part f_- and f_+, respectively, such that $f_- \leq f_+$, we note that f is order preserving iff f_- is order preserving and f_+ is order reversing. This, similar to the case for the real quantity value object, suggests we re-write the complex valued object as follows

Definition 18.13 The complex value object $\underline{\mathbb{C}}^{\leftrightarrow}$ is defined on

1. Objects: for each $V \in \mathcal{V}(\mathcal{H})$ we obtain the set

$$\underline{\mathbb{C}}^{\leftrightarrow}_V = \{f : \downarrow V \to \mathbf{I}\mathbb{C} | f \text{ order-preserving}\}. \tag{18.66}$$

2. Morphisms: given $i_{V'V} : V' \subseteq V$ the corresponding morphism is

$$\underline{\mathbb{C}}^{\leftrightarrow}(i_{V'V}) : \underline{\mathbb{C}}^{\leftrightarrow}_V \to \underline{\mathbb{C}}^{\leftrightarrow}_{V'} \tag{18.67}$$

$$f \mapsto f|_{\downarrow V'}. \tag{18.68}$$

We now utilise the analogue of Proposition 4.2 in [93] as applied for the complex number object. The proof is identical to the case of the real valued object so we will omit it.

Proposition 18.1 *The global elements of $\underline{\mathbb{C}}^{\leftrightarrow}$ are in bijective correspondence with order-preserving functions from $\mathcal{V}(\mathcal{H})$ to $\mathbf{I}\mathbb{C}$.*

The new reformulation of the complex valued object together with the above proposition imply that that, for each $V \in \mathcal{V}(\mathcal{H})$

$$\underline{\mathbb{C}}^{\leftrightarrow}_V = OP(\downarrow V, \mathbf{I}\mathbb{C}) \tag{18.69}$$

$$\Gamma\underline{\mathbb{C}}^{\leftrightarrow} = OP\big(\mathcal{V}(\mathcal{H}), \mathbf{I}\mathbb{C}\big) \tag{18.70}$$

where $OP(\downarrow V, \mathbf{I}\mathbb{C})$ indicates the set of order preserving functions from the poset $\downarrow V$ to $\mathbf{I}\mathbb{C}$.

By equipping both $\downarrow V$ and $\mathcal{V}(\mathcal{H})$ with the Alexandroff topology and utilising Propositions 4.3 and 4.4 in [93] we arrive at the following results: for each $V \in \mathcal{V}(\mathcal{H})$

$$\underline{\mathbb{C}}^{\leftrightarrow}_V = C(\downarrow V, \mathbf{I}\mathbb{C}) \quad \text{and} \quad \underline{\mathbb{C}}^{\leftrightarrow}_V \text{ is an almost complete dcpo} \tag{18.71}$$

$$\Gamma\underline{\mathbb{C}}^{\leftrightarrow} = C(\mathcal{V}(\mathcal{H}), \mathbf{I}\mathbb{C}) \quad \text{and} \quad \Gamma\underline{\mathbb{C}}^{\leftrightarrow}_V \text{ is an almost complete dcpo.} \tag{18.72}$$

Thus, with the use of domain theory, we can understand the precise structure of the complex valued quantity object.

---

[8] Given this topology then it is clear that, as topological spaces $\mathbb{C} \simeq max\mathbf{I}\mathbb{C}$ where $\mathbb{C}$ is equipped with the (general) open rectangles topology and $max\mathbf{I}\mathbb{C}$ has the topology inherited by $\mathbf{I}\mathbb{C}$. The homeomorphisms can be see by the fact that $\Downarrow[\alpha,\beta] \cap max\mathbf{I}\mathbb{C} = \{\sigma \in \mathbf{I}\mathbb{C} | \sigma \subseteq (\alpha,\beta)\} \cap max\mathbf{I}\mathbb{C} = \{\{\beta\} | \beta \in (\alpha,\beta)\} = (\alpha,\beta)$.

Chapter 19
KMS States

The KMS state occurs in statistical analysis as a state describing the thermal equilibrium of many body quantum systems [77, 78]. For oscillator systems and under particular conditions, the KMS state turns out to coincide with Gibbs state [79–81]. Its significance has also been highlighted through a mathematical framework called the Tomita-Takesaki modular theory on von Neumann algebras [82–88]. The Tomita-Takesaki theory, on its own, was the focus of intense research activities in the mid 70's and after.

In [28] the topos analogues of the KMS state and KMS condition for an equilibrium state was defined. In particular, it was shown that, given an automorphism on a von-Neumann algebra $\mathcal{N}$, to which there corresponds a geometric morphism in the topos formulation, it is possible to define two KMS conditions: (i) the *external* condition, defined through a geometric morphism which does not respect the truth value object (topos state) fibration, but maps the topos KMS state onto a twisted KMS presheaf. (ii) The *internal*[1] condition, defined via an internal group object and which respects the presheaf structure of the topos state.

The interesting feature of the topos KMS state is that when considering the von-Neumann algebra $\mathcal{N} = B(\mathcal{H})$, utilised in the topos formulation of quantum theory, then the topos KMS state has a unique solution as a Gibbs state. This shows that the philosophy behind topos quantum theory, namely that quantum theory "looks like" classical physics in the appropriate topos, is indeed confirmed in the case of KMS states.

Before going into the details of how the derivation of the topos analogue of the KMS state was carried out, it is worth, briefly, to summarise what a KMS state is.

[1] It should be also noted that the notion of *external* and *internal* topos properties used in the present chapter, should not be confused with the notion of external and internal description, as used in [46, 47] and [48].

C. Flori, *A First Course in Topos Quantum Theory*, Lecture Notes in Physics 868, DOI 10.1007/978-3-642-35713-8_19,

19.1 Brief Review of the KMS State

The general notion of KMS state in a von-Neumann algebra $\mathcal{N}$ can be stated as follows:

Definition 19.1 Given a von Neumann algebra $\mathcal{N}$ equipped with a one parameter automorphism group $\{\alpha(t) | t \in \mathbb{R}\}$, then a state $\rho : \mathcal{N} \to \mathbb{C}$ is a KMS state if it is normal and satisfies the following properties:

1. Invariance under the automorphism group: $[\alpha(t) \cdot \rho](A) = \rho(\alpha(t)A) = \rho(A)$;
2. Given any two elements $A, B \in \mathcal{N}$, the function

$$F_{A,B}(t) = \rho\bigl(A\alpha(t)B\bigr) \tag{19.1}$$

for all $t \in \mathbb{R}$, has an extension to the strip $\{z = t + iy | t \in \mathbb{R}, y \in [0, \beta]\} \subset \mathbb{C}$, such that $F_{A,B}(z)$ is analytic in the open strip $(0, \beta)$ and continuous on its boundary. Moreover, it satisfies the boundary condition

$$F_{A,B}(t + i\beta) = \rho\bigl(\alpha(t)BA\bigr) \tag{19.2}$$

for $t \in \mathbb{R}$.

As shown in Proposition 5.3.7 of [28] the above definition is equivalent to the following

Definition 19.2 Given a von-Neumann algebra $\mathcal{N}$ with an automorphism group $\{\alpha(t) | t \in \mathbb{R}\}$, a state ρ on $\mathcal{N}$ is a KMS state if

$$\rho\bigl(A\alpha_{i\beta}(B)\bigr) = \rho(BA) \tag{19.3}$$

for all A, B in a σ-weakly dense α-invariant $*$-subalgebra of $\mathcal{N}$, which we denote $\mathcal{N}_\alpha$.

From the equivalence of these two definitions it follows that for all $A, B \in \mathcal{N}_\alpha$ then $F_{A,B}(t + i\beta) = \rho(A\alpha_{t+i\beta}B) = \rho(\alpha(t)BA)$. This is precisely condition (19.3) in fact $\rho(A\alpha_{t+i\beta}B) = \rho(A\alpha_{i\beta}(\alpha_t(B)))$ renaming $B' = \alpha_t(B)$ we obtain $\rho(A\alpha_{i\beta}B') = \rho(B'A)$.[2] In order to better understand the above definitions we recall certain constructs related to von-Neumann algebras (see, for example, [81–83]).

[2] Note that there are analogous definitions for KMS states for general C^*-algebras, not only von-Neumann algebras. In particular, the analogues of Definitions 19.1 and 19.2 are, respectively

Definition 19.3 Given a C^*-algebra $\mathcal{C}$ equipped with a one parameter automorphism group $\{\alpha(t) | t \in \mathbb{R}\}$, then a state $\rho : \mathcal{C} \to \mathbb{C}$ is a KMS state if it satisfies the following properties:

1. Invariance under the automorphism group: $[\alpha(t) \cdot \rho](A) = \rho(\alpha(t)A) = \rho(A)$;
2. Given any two elements $A, B \in \mathcal{C}$, the function

$$F_{A,B}(t) = \rho\bigl(A\alpha(t)B\bigr) \tag{19.4}$$

States on von Neumann Algebras Von Neumann algebras were already defined in Sect. 9.1.1 here, however, we will recall some definitions concerning states on these algebras.

Given a bounded linear functional $\varphi : \mathcal{A} \longrightarrow \mathbb{C}$ on $\mathcal{A}$ whose action, for $A \in \mathcal{A}$, is given by $\langle \varphi; A \rangle$, then φ is called a *state* on this algebra if (a) $\langle \varphi; A^* A \rangle \geq 0, \forall A \in \mathcal{A}$ and (b) $\langle \varphi; I_{\mathcal{H}} \rangle = 1$.

It is possible to classify states as follows:

- A state φ is said to be *faithful* if $\langle \varphi; A^* A \rangle > 0$ for all $A \neq 0$.
- A state is said to be *normal* if and only if there is a density matrix ϱ, such that $\langle \varphi \mid A \rangle = \mathrm{Tr}[\varrho A], \forall A \in \mathcal{A}$.
- A state is called a *vector state* if there exists a vector $\phi \in \mathcal{H}$, such that $\langle \varphi;\ A \rangle = \langle \phi \mid A\phi \rangle, \forall A \in \mathcal{A}$. Note that such a state is also normal.
- A vector $\psi \in \mathcal{H}$ is called *cyclic* for the von Neumann algebra if the set $\{A\psi \mid A \in \mathcal{A}\}$ is dense in $\mathcal{H}$.
- A vector $\psi \in \mathcal{H}$ is said to be *separating* for $\mathcal{A}$ if $A\psi = B\psi, \forall A, B \in \mathcal{A}$, if and only if $A = B$.

19.2 External KMS State

From Sect. 15.4 we recall that, for each state $\rho : \mathcal{N} \to \mathbb{C}$ of a von Neumann algebra $\mathcal{N}$, there corresponds a measure μ^ρ on the state space $\underline{\Sigma}$. If we consider a given KMS state, we would have an associated measure and it is precisely the properties of such a measure that we would like to understand. To achieve this goal we first

---

for all $t \in \mathbb{R}$, has an extension to the strip $\{z = t + iy \mid t \in \mathbb{R},\ y \in [0, \beta]\} \subset \mathbb{C}$, such that $F_{A,B}(z)$ is analytic in the open strip $(0, \beta)$ and continuous on its boundary. Moreover, it satisfies the boundary condition

$$F_{A,B}(t + i\beta) = \rho\big(\alpha(t) B A\big) \tag{19.5}$$

for $t \in \mathbb{R}$.

Thus here the only difference is that the state is not required to be normal. On the other hand we have

Definition 19.4 Given a C^*-algebra $\mathcal{C}$ with an automorphism group $\{\alpha(t) \mid t \in \mathbb{R}\}$, a state ρ on $\mathcal{C}$ is a KMS state if

$$\rho\big(A\alpha_{i\beta}(B)\big) = \rho(BA) \tag{19.6}$$

for all A, B in a norm dense α-invariant $*$-subalgebra of $\mathcal{C}$, which we denote $\mathcal{C}_\alpha$.

Differently form the case of a von Neumann algebra, for a general C^*-algebra the sub-algebra $\mathcal{C}_\alpha$ is required to be norm dense invariant rather than σ-weakly dense α-invariant. This is because while a C^*-algebra is closed in the norm topology a von-Neumann algebra is closed in the weak operator topology.

of all analyse the automorphisms group $\text{Aut} = \{\alpha(t) | t \in \mathbb{R}\}$ acting on $\mathcal{N}$ and defined in the previous section. Each element $\alpha(t) \in \text{Aut}$ can be extended to a functor[3] $\alpha(t) : \mathcal{V}(\mathcal{N}) \to \mathcal{V}(\mathcal{N})$; $V \mapsto \alpha(t)V$, which induces a geometric morphism whose inverse image part is

$$\alpha(t)^* : \mathbf{Sets}^{\mathcal{V}(\mathcal{N})^{op}} \to \mathbf{Sets}^{\mathcal{V}(\mathcal{N})^{op}} \tag{19.7}$$

$$\underline{A} \mapsto \alpha(t)^* \underline{A} \tag{19.8}$$

where, for each $V \in \mathcal{V}(\mathcal{N})$, $\alpha(t)^* \underline{A}_V := \underline{A}_{\alpha(t)(V)}$. Note that the automorphism group Aut is not itself an object in the topos. It is in this sense that we will call the ensuing results as being *external*.

Given the geometric morphisms (19.7), we derive the following proposition regarding a measure associated to a KMS state:

Proposition 19.1 (KMS measure) *Given a KMS state ρ, for all elements of the one parameter group* $\text{Aut} = \{\alpha(t) | t \in \mathbb{R}\}$, *the associated measure μ^ρ satisfies the following conditions*:

(C1) $\mu^\rho(\underline{S}) = \mu^\rho(\alpha(t)^* \underline{S})$, $\forall \underline{S} \in Sub(\underline{\Sigma})$
(C2) $\forall \underline{S}, \underline{T} \in Sub(\underline{\Sigma})$ *and* $V \in \mathcal{V}(\mathcal{H})$ *the function*

$$F_{\underline{T},\underline{S}}(t) := \left[\mu^\rho\left(\underline{T} \wedge \alpha(z)^* \underline{S}\right)\right](V) \tag{19.9}$$

for all $t \in \mathbb{R}$ has an extension to the strip $\{z = t + iy | \; t \in \mathbb{R}, \; y \in [0, \beta]\} \subset \mathbb{C}$ such that $F_{\underline{T},\underline{S}}(z)$ is analytic in the open strip $(0, \beta)$ and continuous on its boundary where it satisfies the following boundary condition

$$F_{\underline{T},\underline{S}}(t + i\beta) = \mu^\rho\left(\alpha^*(t)\underline{S} \wedge \underline{T}\right)(V). \tag{19.10}$$

From the definition of KMS states given in Definition 19.2, it follows that it is possible to give a precise characterisation of the function $F_{\underline{T},\underline{S}}(t + i\beta)$ for an appropriate domain. In particular, given any algebra V we consider the σ-weakly dense α invariant *-sub-algebra $V_\alpha \subseteq V$. For these contexts $F_{\underline{T},\underline{S}}(t + i\beta)$ can be defined as

$$F_{\underline{T},\underline{S}}(t + i\beta) := \mu^\rho\left(\underline{T} \wedge \alpha(t + i\beta)^* \underline{S}\right)(V_\alpha) \tag{19.11}$$

such that the boundary condition becomes

$$\mu^\rho\left(\underline{T} \wedge \alpha(t + i\beta)^* \underline{S}\right)(V_\alpha) = \mu^\rho\left(\alpha^*(t)\underline{S} \wedge \underline{T}\right)(V_\alpha). \tag{19.12}$$

Let us now analyse conditions C1 and C2, separately. In order to do this we first of all need to define how $\alpha(t)$ acts on a state ρ. For simplicity we will sometimes write $\alpha(t) = \alpha_t$. If ρ is a normal state, then we know that it can be defined as $\text{tr}(\varrho -)$ where ϱ is the associated density matrix. This implies that

$$\alpha_t \cdot \rho = \alpha_t \cdot \text{tr}(\varrho -) := \text{tr}(\alpha_t \varrho -) = \text{tr}\left(\varrho \alpha_t^{-1} -\right) = \rho \circ \alpha_t^{-1}. \tag{19.13}$$

---

[3]Recall that $\mathcal{V}(\mathcal{H})$ is the category of abelian sub-algebras of $\mathcal{N}$ ordered by inclusion.

Thus, for a general state we can define

$$\alpha_t \cdot \rho := \rho \circ \alpha_t^{-1}. \tag{19.14}$$

Having said this we can turn to the question of the significance of condition C1. In particular C1 means that, for all $V \in \mathcal{V}(\mathcal{N})$, one has

$$\begin{aligned}
\big[\mu^\rho\big(\alpha(t)^*\underline{S}\big)\big](V) &= \mu^\rho_{|V}\big(\underline{S} \circ \alpha(t)_{|_V}\big) := \rho_{|V} \circ \mathfrak{S}^{-1}_{|V}\big(\alpha(t)^*\underline{S}\big)(V) \\
&= \rho_{|V}\big(\hat{P}_{(\alpha(t)^*\underline{S})(V)}\big) = \rho_{|V} \circ \alpha(t)\hat{P}_{\underline{S}_V} \\
&= \alpha(t)^{-1}\rho_{|V} \circ \mathfrak{S}^{-1}_{|V}\underline{S}_V \\
&= \big[\mu^\rho(\underline{S})\big](V) = \rho_{|V} \circ \mathfrak{S}^{-1}_{|V}\underline{S}_V.
\end{aligned} \tag{19.15}$$

Our claim is that condition C1 on the measure represents the first condition of the KMS state ρ (see 1 in Definition 19.1), namely that ρ is invariant under the automorphism group. To verify this we, first of all, need to prove that the following diagram commutes:

$$\begin{array}{ccc}
\mathcal{P}(V) & \xrightarrow{\ \mathfrak{S}_V\ } & Sub_{cl}(\underline{\Sigma}_V) \\
\Big\downarrow{\scriptstyle \alpha(t)} & & \Big\downarrow{\scriptstyle \alpha(t)} \\
\mathcal{P}(\alpha(t)V) & \xrightarrow{\ \mathfrak{S}_{\alpha(t)V}\ } & Sub_{cl}((\alpha^*(t)\underline{\Sigma})_V)
\end{array}$$

The diagram in one direction yields, for a given $\hat{P} \in \mathcal{P}(V)$,

$$\begin{aligned}
[\alpha_t \circ \mathfrak{S}_V](\hat{P}) = \alpha_t(\underline{S}_{\hat{P}_V}) &= \big\{\alpha_t(\lambda) \in \underline{\Sigma}_{\alpha_t V} \,\big|\, [\alpha_t\lambda](\hat{P}) = 1\big\} \\
&= \big\{\lambda \in \underline{\Sigma}_{\alpha_t V} \,\big|\, \lambda\big(\alpha_t(\hat{P})\big) = 1\big\}.
\end{aligned} \tag{19.16}$$

In the other direction, we get:

$$[\mathfrak{S}_{\alpha_t V} \circ \alpha_t](\hat{P}) = \mathfrak{S}_{\alpha_t V}\big(\alpha_t(\hat{P})\big) = \underline{S}_{(\alpha_t\hat{P})_{\alpha_t V}} = \big\{\lambda \in \underline{\Sigma}_{\alpha_t V} \,\big|\, \lambda\big(\alpha_t(\hat{P})\big) = 1\big\} \tag{19.17}$$

thus the diagram commutes. Using this, we can now prove conditions C1 and C2 in Proposition 19.1:

Proof of C1 Given a KMS state ρ, for any $V \in \mathcal{V}(\mathcal{N})$, we have:

$$\begin{aligned}
\big[\mu^\rho\big(\alpha(t)^*\underline{S}\big)\big](V) &= \mu^\rho_{|V}\big(\underline{S} \circ \alpha(t)|_V\big) := \rho_{|V} \circ \mathfrak{S}^{-1}_{|V}\big(\alpha(t)^*\underline{S}\big)(V) \\
&= \rho_{|V}\big(\hat{P}_{(\alpha(t)^*\underline{S})(V)}\big) = \rho_{|V} \circ \alpha(t)\hat{P}_{\underline{S}_V}
\end{aligned}$$

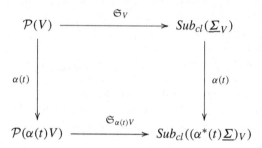

$$
\begin{aligned}
&= \alpha(t)^{-1}\rho_{|V} \circ \mathfrak{S}^{-1}_{|V}(\underline{S}_V) \\
&= \rho_{|V} \circ \mathfrak{S}^{-1}_{|V}(\underline{S}_V) = \left[\mu^\rho(\underline{S})\right](V).
\end{aligned} \tag{19.18}
$$

□

Proof of C2 Let us consider the expression $\mu^\rho(\alpha_t^*\underline{S} \wedge \underline{T})$ and evaluate in some context V_α. This gives

$$
\begin{aligned}
\left[\mu^\rho\left(\alpha_t^*\underline{S} \wedge \underline{T}\right)\right](V_\alpha) &:= \left(\rho_{|V_\alpha} \circ \mathfrak{S}^{-1}_{|V_\alpha}\right)\left(\alpha_t^*(\underline{S}) \wedge \underline{T}\right)(V_\alpha) \\
&= \rho_{|V_\alpha}\left(\mathfrak{S}^{-1}_{|V_\alpha}\alpha_t^*(\underline{S})(V_\alpha) \wedge \mathfrak{S}^{-1}_{|V}\underline{T}(V_\alpha)\right) \\
&= \rho_{|V_\alpha}(\alpha_t \hat{P}_{\underline{S}_{V_\alpha}} \wedge \hat{P}_{\underline{T}_{V_\alpha}}) \\
&= \rho_{|V_\alpha}(\hat{P}_{\underline{T}_{V_\alpha}} \wedge \alpha_{t+i\beta}\hat{P}_{\underline{S}_{V_\alpha}}) \\
&= \left[\mu^\rho\left(\underline{T} \wedge \alpha^*_{(t+i\beta)}\underline{S}\right)\right](V_\alpha)
\end{aligned} \tag{19.19}
$$

where the fourth equality follows from the second condition of KMS state on a suitable σ-weak sub-algebra $V_\alpha \subseteq V$.

We thus obtain, $\forall V_\alpha \in \mathcal{V}(\mathcal{N})$

$$
\left[\mu^\rho\left(\alpha_t^*\underline{S} \wedge \underline{T}\right)\right](V_\alpha) = \left[\mu^\rho\left(\underline{T} \wedge \alpha^*_{(t+i\beta)}\underline{S}\right)\right](V_\alpha). \tag{19.20}
$$

which is precisely the boundary condition on the measure.

The fact that $[\mu^\rho(\underline{S} \wedge \alpha(z)^*\underline{T})](V_\alpha)$ is analytic in the complex strip and continuous on the boundary follows from the definition of V_α.[4] □

We now want to relate condition C1 to the invariance of the measure with respect to the automorphisms group. To this end we will utilise Lemma 23 in [97] which shows that $\alpha(t) \cdot \mu^\rho = \mu^{\alpha(t)\rho} = \mu^{\rho\circ\alpha(t)^{-1}}$ is the measure associated to the state $\alpha(t)\rho$. With this in mind condition C1 becomes

$$
\begin{aligned}
\left[\mu^\rho\left(\alpha(t)^*\underline{S}\right)\right](V) = \mu^\rho_{|V}\left(\underline{S} \circ \alpha(t)_{|V}\right) &:= \rho_{|V} \circ \mathfrak{S}^{-1}_{|V}\left(\alpha(t)^*\underline{S}\right)(V) \\
&= \rho_{|V}(\hat{P}_{(\alpha(t)^*\underline{S})(V)}) = \rho_{|V} \circ \alpha(t)\hat{P}_{\underline{S}_V} \\
&= \alpha(t)^{-1}\rho_{|V} \circ \mathfrak{S}^{-1}_{|V}\underline{S}_V \\
&= \left[\alpha(t)^{-1}.\mu^\rho(\underline{S})\right](V) \\
&= \left[\mu^{\alpha(t)^{-1}\rho}(\underline{S})\right](V)
\end{aligned}
$$

[4] From now on, the analyticity and continuity properties of condition C2 will be immediate. Furthermore, the notation V_α will always refer to a suitable dense *-sub-algebra of any context V so that the automorphism $\alpha_{t+i\beta}$ finds a sense.

$$= \left[\mu^{\rho}(\underline{S})\right](V) \tag{19.21}$$

which is clearly satisfied for KMS states since $\alpha(t)^{-1}\rho = \rho$.[5]

If we now consider the topos formulation of quantum theory defined in [32] and take $\mathcal{N} = B(\mathcal{H})$, then the automorphisms group α becomes automatically internal and the KMS state has unique solution as Gibbs state. Thus we see that in the case of topos quantum theory defined for $\mathcal{N} = \mathcal{B}(\mathcal{H})$ our definition of general KMS state reduces to that of Gibbs state as it should do.

So far we have only defined the KMS condition for the topos measure associated to a canonical. However, since there is an injective correspondence between states and truth objects in the topos $\mathbf{Sets}^{\mathcal{V}(\mathcal{N})^{op}}$, we can translate the above conditions to conditions on the truth object, which should represent the topos analogue of the KMS state. In order to do this we need some extra definitions. Note that now, since we are dealing with truth objects, we need to work in the topos $\mathbf{Sets}^{(\mathcal{V}(\mathcal{N})\times(0,1)_L)^{op}}$ (see Sect. 15.7).

Definition 19.5 (Topos State μ-Equivalence) Two truth objects $\underline{\mathbb{T}}$ and $\underline{\mathbb{T}}'$ are said to be μ-equivalent (or equivalent under the measure μ) iff, at each context (V, r), for each $\underline{S} \in \underline{\mathbb{T}}_{(V,r)}$ there exists a $\underline{S}' \in \underline{\mathbb{T}}'_{(V,r)}$, such that $[\mu(\underline{S})](V') = [\mu(\underline{S}')](V')$ for all $V' \subseteq V$. Then, we write $\underline{\mathbb{T}} \simeq_{\mu} \underline{\mathbb{T}}'$.

This is trivially an equivalence relation. Given a truth object $\underline{\mathbb{T}}$, we will then denote its μ-equivalence class by $[\underline{\mathbb{T}}]_{\mu}$.

It is also possible to define a stronger notion of equivalence. To this end let us consider the notion of μ-invariant natural transformation.

---

[5]Note that in the case of von-Neumann algebras, as stated in Definition 19.1, a KMS state ρ is actually normal thus it can be defined in terms of a trace, i.e. $\rho := \text{tr}(\varrho -)$. Given such a state conditions C1 and C2 are easily derived. In fact the KMS measure condition C1 becomes

$$\begin{aligned}\left[\mu^{\rho}(\underline{S})\right](V) &:= \text{tr}(\varrho \hat{P}_{\underline{S}_V}) \\ &= \left[\mu^{\rho}\left(\alpha(t)^*\underline{S}\right)\right](V) := \text{tr}(\varrho \hat{P}_{(\alpha(t)^*\underline{S})_V}) \\ &= \text{tr}(\varrho \hat{P}_{\underline{S}_{\alpha(t)V}}).\end{aligned} \tag{19.22}$$

This is trivially satisfied for a KMS state ρ in fact, we obtain

$$\left[\mu^{\rho}(\underline{S})\right](V) := \text{tr}(\varrho \hat{P}_{\underline{S}_V}) = \text{tr}(\alpha_{-t}\varrho \hat{P}_{\underline{S}_V}) = \text{tr}(\varrho\alpha_t \hat{P}_{\underline{S}_V}) = \text{tr}(\varrho \hat{P}_{\underline{S}_{\alpha_t(V)}}) = \left[\mu^{\rho}\left(\alpha_t^*\underline{S}\right)\right](V). \tag{19.23}$$

On the other hand, given a KMS state ρ condition C2 is shown to hold:

$$\begin{aligned}\left[\mu^{\rho}\left(\alpha_t^*\underline{S} \wedge \underline{T}\right)\right](V_\alpha) &= \text{tr}(\varrho \hat{P}_{(\alpha_t^*\underline{S})_{V_\alpha}} \hat{P}_{\underline{T}_{V_\alpha}}) = \text{tr}(\varrho\alpha_t \hat{P}_{\underline{S}_{V_\alpha}} \hat{P}_{\underline{T}_{V_\alpha}}) \\ &= \text{tr}(\varrho \hat{P}_{\underline{T}_{V_\alpha}} \alpha_{(t+i\beta)} \hat{P}_{\underline{S}_{V_\alpha}}) = \text{tr}(\varrho \hat{P}_{\underline{T}_{V_\alpha}} \hat{P}_{(\alpha^*_{(t+i\beta)}\underline{S})_{V_\alpha}}) \\ &= \left[\mu^{\rho}\left(\underline{T} \wedge \alpha^*_{(t+i\beta)}\underline{S}\right)\right](V_\alpha).\end{aligned} \tag{19.24}$$

The third equality follows from the second KMS condition on ρ. The fact that $[\mu^{\rho}(\underline{S} \wedge \alpha(z)^*\underline{T})](V_\alpha)$ is analytic in the complex strip and continuous on the boundary follows from the trace properties in the case of normal states.

Definition 19.6 (μ-Invariant Natural Transformation) Given a μ-equivalence topos state class $[\underline{\mathbb{T}}]_\mu$, a μ-invariant natural transformation f_μ is a natural transformation between a pair of elements $\underline{\mathbb{T}}'$ and $\underline{\mathbb{T}}''$ (representatives of $[\underline{\mathbb{T}}]_\mu$) which is defined, for each context (V, r), as follows:

$$\begin{aligned} f_{\mu,V} : \underline{\mathbb{T}}'_{(V,r)} &\to \underline{\mathbb{T}}''_{(V,r)} \\ \underline{S} &\mapsto f_\mu \underline{S} \end{aligned} \tag{19.25}$$

where $f_\mu \underline{S}$ is the unique element such that $[\mu(f_\mu \underline{S})](V) = [\mu(\underline{S})](V)$.[6]

To show that, indeed, f_μ as, defined above, is a natural transformation, we must show that given any two contexts (V, r) and (V', r'), such that $i : (V, r) \geq (V', r')$ (see ordering defined in (15.73)), the following diagram commutes:

$$\begin{array}{ccc} \underline{\mathbb{T}}'_{(V,r)} & \xrightarrow{f_{\mu,V}} & \underline{\mathbb{T}}''_{(V,r)} \\ \Big\downarrow{\scriptstyle \underline{\mathbb{T}}'(i)} & & \Big\downarrow{\scriptstyle \underline{\mathbb{T}}''(i)} \\ \underline{\mathbb{T}}'_{(V',r')} & \xrightarrow{f_{\mu,V'}} & \underline{\mathbb{T}}''_{(V',r')} \end{array}$$

In one direction we have

$$\underline{\mathbb{T}}''(i) \circ f_{\mu,V}(\underline{S}) = \underline{\mathbb{T}}''(i)(f_\mu \underline{S}) = (f_\mu \underline{S})_{|\downarrow V'}. \tag{19.26}$$

In the other one we get

$$f_{\mu,V'} \circ \underline{\mathbb{T}}'(i)(\underline{S}) = f_{\mu,V'}(\underline{S})_{|\downarrow V'} = f_\mu\big((\underline{S})_{|\downarrow V'}\big). \tag{19.27}$$

Here, $\underline{S} \in Sub_{cl}(\underline{\Sigma}_{|\downarrow V})$, therefore $\underline{S} = \underline{S}_{\downarrow V}$ the latter implies that $(f_\mu \underline{S})_{|\downarrow V'} = (f_\mu \underline{S}_{|\downarrow V})_{|\downarrow V'} = f_\mu((\underline{S})_{|\downarrow V'})$.

We now join Definitions 19.5 and 19.6 in order to define the notion of strongly μ-equivalence as follows:

Definition 19.7 (Topos State Strong μ-Equivalence) Two objects $\underline{\mathbb{T}}'$ and $\underline{\mathbb{T}}$ are strongly μ-equivalent iff they are μ-equivalent and there exists a μ-invariant natural transformation between them. We will denote a strong μ-equivalence class as $|[\underline{\mathbb{T}}]|_\mu$.

[6] For notational simplicity we have written $f_{\mu,V}$ as f_μ. These two notations will be used interchangeably.

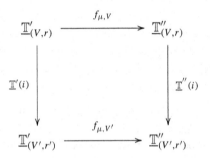

Using this notion, we can write the KMS condition as:

Definition 19.8 (Topos State KMS Conditions) Given a KMS state ρ, the truth object $\underline{\mathbb{T}}^{\rho}$ is the topos analogue of the KMS state if the following conditions are satisfied:

1. For all $\alpha_{\rho}(t) \in \mathrm{Aut}_{\rho}$,

$$\big|\big[\underline{\mathbb{T}}^{\rho}\big]\big|_{\mu^{\rho}} = \big|\big[\alpha_{\rho}(t)^{*}\underline{\mathbb{T}}^{\rho}\big]\big|_{\mu^{\rho}}, \tag{19.28}$$

with the μ-invariant natural transformation defined as

$$\begin{aligned} \underline{\mathbb{T}}^{\rho}_{(V,r)} &\to \alpha_{\rho}(t)^{*}\underline{\mathbb{T}}^{\rho}_{(V,r)} \\ \underline{S} &\mapsto \alpha_{\rho}(t)^{*}(\underline{S}). \end{aligned} \tag{19.29}$$

2. For each context $(V, r) \in (\mathcal{V}(\mathcal{N}) \times (0,1)_L)$, given any two elements $\underline{S}, \underline{T} \in \underline{\mathbb{T}}^{\rho}_{(V,r)}$, the function

$$F_{\underline{T},\underline{S}}(t) = \big[\mu^{\rho}\big(\underline{T} \wedge \alpha(t)^{*}\underline{S}\big)\big](V_{\alpha}) \tag{19.30}$$

for all $t \in \mathbb{R}$ admits an extension $F_{\underline{T},\underline{S}}(z)$ analytic in the complex strip $D_{\beta} = \{z = (t + iy) |\, t \in \mathbb{R},\ y \in]0, \beta[\}$ and continuous at the boundary of D_{β} such that

$$F_{\underline{T},\underline{S}}(t + i\beta) = \big[\mu^{\rho}\big(\alpha_t^{*}\underline{S} \wedge \underline{T}\big)\big](V) \geq r. \tag{19.31}$$

Alternatively condition (2) above can be stated as follows

(2′) For each context $(V_{\alpha}, r) \in (\mathcal{V}(\mathcal{N}) \times (0,1)_L)$, given any two elements $\underline{S}, \underline{T} \in \underline{\mathbb{T}}^{\rho}_{(V_{\alpha},r)}$ the function

$$F_{\underline{T},\underline{S}}(t + i\beta) := \big[\mu^{\rho}\big(\alpha(t)^{*}\underline{T} \wedge \underline{S}\big)\big]\big(V'_{\alpha}\big) = \big[\mu^{\rho}\big(\underline{S} \wedge \alpha(t + i\beta)^{*}\underline{T}\big)\big]\big(V'_{\alpha}\big) \geq r \tag{19.32}$$

where $V'_{\alpha} \subseteq V_{\alpha} \cdot F_{T,S}(t + i\beta)$ is analytic in the complex strip $D_{\beta} = \{z = (t + iy) |\, t \in \mathbb{R},\ y \in]0, \beta[\}$ and continuous at the boundary of D_{β}.

19.3 Deriving the Canonical KMS State from the Topos KMS State

We are now interested in proving that, given a topos KMS state it is possible to derive the 'canonical' KMS state. To this end, we will use the fact that there is a 1:1 correspondence between states ρ and truth objects $\underline{\mathbb{T}}^{\rho}$. Since $\underline{\mathbb{T}}^{\rho}$ is defined through the measure μ^{ρ} on $\underline{\Sigma}$, we will first concentrate on the link between the topos KMS measure and the canonical KMS state.

Theorem 19.1 *Consider a measure μ and a one parameter automorphism group* $\text{Aut}_\rho = \{\alpha(t) | t \in \mathbb{R}\}$ *on $\mathcal{N}$, whose action on the category $\mathcal{V}(\mathcal{N})$ is*

$$\begin{aligned} \text{Aut}_\rho \times \mathcal{V}(\mathcal{N}) &\to \mathcal{V}(\mathcal{N}) \\ \big(\alpha(t), V\big) &\mapsto \big[\alpha(t)\big](V) \end{aligned} \tag{19.33}$$

and such that Aut_ρ *can be extended to a complex strip* $\{\alpha(t+i\gamma) | t \in \mathbb{R}, \gamma \in [0, \beta]\}$. *If μ satisfies the following conditions*

$$\widetilde{C1}: \quad \mu(\underline{S}) = \mu\big(\alpha(t)^* \underline{S}\big) \tag{19.34}$$

and $\widetilde{C2}$ *: for all $\underline{S}$, $\underline{T}$ in Sub_{cl} and $V_\alpha \in \mathcal{V}(\mathcal{N})$, if $F_{\underline{T},\underline{S}}(z) := [\mu(\underline{T} \wedge \alpha(z)^* \underline{S})](V_\alpha)$ is analytic in the complex open strip D_β and continuous at the boundary of D_β such that*

$$F_{\underline{T},\underline{S}}(t+i\beta) := \big[\mu\big(\underline{T} \wedge \alpha(t+i\beta)^* \underline{S}\big)\big](V_\alpha)\big[\mu\big(\alpha(t)^* \underline{S} \wedge \underline{T}\big)\big](V_\alpha), \tag{19.35}$$

then the state ρ^μ associated to such a topos measure is the KMS state associated to the algebra $\mathcal{N}$.

Proof In order to define a state from a measure we apply Theorem 15.2 and (15.62), which can be written

$$m(\hat{P}) := \big[\mu(\underline{S}_{\hat{P}})\big](V). \tag{19.36}$$

Since our measure satisfies conditions (19.34) and (19.35), it can be shown that

$$m(\hat{P}) := \big[\mu(\underline{S}_{\hat{P}})\big](V) = \big[\mu\big(\alpha(t)^* \underline{S}_{\hat{P}}\big)\big](V) = \big[\mu(\underline{S}_{\hat{P}})\big]\big(\alpha(t)(V)\big) = m\big(\alpha(t)\hat{P}\big). \tag{19.37}$$

Given that μ is a finitely additive probability measure, by construction $m : \mathcal{P}(\mathcal{N}) \to [0, 1]$ will be finitely additive on the projections in $\mathcal{N}$, hence we can apply the generalised version of Gleason's theorem[7] [85] and obtain the unique state

---

[7]Gleason Theorem tells us that the only possible probability measures on Hilbert spaces of dimension at least 3 are measures of the form $\mu(P) = \text{Tr}(\rho P)$, where ρ is a positive semidefinite self-adjoint operator of unit trace. This theorem was extended to a von Neumann algebra $\mathcal{N}$ in [73] where the author shows that, provided $\mathcal{N}$ contains no direct summand of type I_2, then every finitely additive probability measure on $\mathcal{P}(\mathcal{N})$ can be uniquely extended to a normal state on $\mathcal{N}$. The general form of Gleason theorem is as follows

Theorem 19.2 Assume that $\dim((\mathcal{H}) \geq 3)$ and let μ be a σ-additive probability measure on $P(\mathcal{B}(\mathcal{H}))$ then the following three statement hold

1. μ is completely additive.
2. μ has support.
3. There exists a positive operator $x \in \mathcal{B}(\mathcal{H})$ of trace class such that $\text{tr}(x) = 1$ and $\mu(e) = \text{tr}(xe)$ for $e \in P(\mathcal{B}(\mathcal{H}))$.

$\rho_m : \mathcal{N} \to \mathbb{C}$ such that $\rho_m|_{\mathcal{P}(\mathcal{N})} = m$. Hence

$$m(\hat{P}) = \rho_m|_{\mathcal{P}(\mathcal{N})}(\hat{P}) = m(\alpha_t \hat{P}) = \rho_m|_{\mathcal{P}(\mathcal{N})}(\alpha_t \hat{P}). \tag{19.38}$$

Similarly, for the second condition, we have, for any $\hat{P}, \hat{R} \in \mathcal{P}(\mathcal{N}_\alpha)$

$$m(\alpha_t \hat{P}\hat{R}) = \big[\mu\big(\alpha_t^* \underline{S}_{\hat{P}} \wedge \underline{T}_{\hat{R}}\big)\big](V_\alpha) = \big[\mu\big(\underline{T}_{\hat{R}} \wedge \alpha^*_{t+i\beta}\underline{S}_{\hat{P}}\big)\big](V_\alpha) = m(\hat{R}\alpha_{t+i\beta}\hat{P}) \tag{19.39}$$

which translates to

$$\rho_m|_{\mathcal{P}(\mathcal{N}_\alpha)}(\alpha_t \hat{P}\hat{R}) = \rho_m|_{\mathcal{P}(\mathcal{N}_\alpha)}(\hat{R}\alpha_{t+i\beta}\hat{P}). \tag{19.40}$$

□

We are in position to derive the canonical KMS state from the topos KMS state.

Theorem 19.3 *Given a topos state $\underline{\mathbb{T}}^\rho$ defined via the topos measure μ^ρ satisfying $\widetilde{C}1$ and $\widetilde{C}2$, the unique state $\tilde{\rho}$ corresponding to μ^ρ is a KMS state in the canonical sense.*

Proof The first condition on $\underline{\mathbb{T}}^\rho$ is

$$[[\underline{\mathbb{T}}^\rho]]_{\mu^\rho} = [[\alpha_t^* \underline{\mathbb{T}}^\rho]]_{\mu^\rho}, \tag{19.41}$$

where the μ-invariant natural transformation is defined for each α_t as

$$\underline{\mathbb{T}}^\rho \to \alpha_t^* \underline{\mathbb{T}}^\rho \tag{19.42}$$

$$\underline{S} \mapsto \alpha_t^*(\underline{S}). \tag{19.43}$$

Given two elements $\underline{\mathbb{T}}^\rho$ and $\alpha_t^* \underline{\mathbb{T}}^\rho$ in $[[\underline{\mathbb{T}}^\rho]]_{\mu^\rho}$, for each context (V, r) the local components are, respectively,

$$\underline{\mathbb{T}}^\rho_{(V,r)} = \big\{\underline{S} \in Sub_{cl}(\underline{\Sigma}_{|\downarrow V}) \big| \forall V' \subseteq V,\ \big[\mu^\rho(\underline{S})\big](V) \geq r\big\} \tag{19.44}$$

and

$$\alpha_t^* \underline{\mathbb{T}}^\rho_{(V,r)} = \big\{\alpha_t^* \underline{S} \in Sub_{cl}\big(\alpha_t^* \underline{\Sigma}_{|\downarrow V}\big) \big| \forall V' \subseteq V, \big[\mu^\rho\big(\alpha_t^* \underline{S}\big)\big](V) \geq r\big\}. \tag{19.45}$$

From the definition of the μ-invariant natural transformation, we now know how to identify objects with the same measure:

$$\big[\mu^\rho\big(\alpha_t^* \underline{S}\big)\big](V) = \big[\mu^\rho(\underline{S})\big](V). \tag{19.46}$$

The second condition on the topos state tells us that, for all $V_\alpha \in \mathcal{V}(\mathcal{N})$,

$$\big[\mu^\rho\big(\alpha_t^* \underline{S} \wedge \underline{T}\big)\big](V_\alpha) = \big[\mu^\rho\big(\underline{T} \wedge \alpha^*_{t+i\beta}\underline{S}\big)\big](V_\alpha) \tag{19.47}$$

Applying Theorem 19.1 to conditions (19.47) and (19.46), one infers that ρ is indeed a KMS state.

□

19.4 The Automorphisms Group

Although the topos analogue of the KMS state reproduces the ordinary canonical KMS state, its definition still requires a collection of μ-equivalent twisted presheaves. Luckily, it is possible to avoid twisted presheaves by *internalizing* the group Aut_ρ (henceforth, for simplicity, denoted by H) and defining presheaves in terms of it. The tools used to achieve this were introduced in Chap. 16.

The first step is to define the topos analogue of the automorphisms group $H = \{\alpha(t) | t \in \mathbb{R}\}$ of the algebra $\mathcal{N}$. For each element, in this group, we obtain the induced geometric morphism

$$\alpha(t)^* : Sh(\mathcal{V}(\mathcal{N})) \to Sh(\mathcal{V}(\mathcal{N})) \tag{19.48}$$

$$\underline{S} \mapsto \alpha(t)^* \underline{S} \tag{19.49}$$

such that $\alpha(t)^* \underline{S}(V) := \underline{S}(\alpha(t)V)$. This action, however, gives rise to twisted presheaves [32]. In order to avoid this feature, the same recipe, as described in Chap. 16, can be used. We start with the base category $\mathcal{V}_f(\mathcal{N})$ which is fixed, i.e. the group H is not allowed to act on it. Then, we define the *internal* group $\underline{H}$ over this new base category as follows:

Definition 19.9 The internal group $\underline{H}$ is the presheaf defined on:

1. Objects: for each $V \in \mathcal{V}_f(\mathcal{N})$, $\underline{H}_V = H$;
2. Morphisms are simply identity maps.

It is straightforward to realize that the global sections of this presheaf reproduce the group, i.e. $\Gamma(\underline{H}) = H$.

Next we define the fixed point group presheaf $\underline{H_F}$ as follows:

Definition 19.10 The presheaf $\underline{H_F}$ is defined on:

1. Objects: for each $V \in \mathcal{V}_f(\mathcal{N})$, $\underline{H_F}_V = H_{FV}$ is the subgroup of H called the fixed point group of V, i.e. $H_{FV} = \{\alpha \in \underline{H} | \forall a \in V \alpha a = a\}$.
2. Morphisms: given a morphism $i_{V'V} : V' \subseteq V$ the corresponding morphism is

$$\begin{aligned} \underline{H_F}(i_{V'V}) : (\underline{H_F}_V = H_{FV}) &\to (\underline{H_F}_{V'} = H_{FV'}) \\ \alpha &\mapsto \underline{H_F}(i_{V'V})[\alpha] = \alpha|_{V'} \end{aligned} \tag{19.50}$$

i.e. $\underline{H_F}_V \to \underline{H_F}_{V'}$ is given in terms of subgroup inclusion, i.e. $\underline{H_F}_V \subseteq \underline{H_F}_{V'}$.

Similarly, as above, we have that $\Gamma(\underline{H}_F) = H_F$.

We will now define the main presheaf of our analysis:

Definition 19.11 The presheaf $\underline{H/H_F} \in \mathbf{Sets}^{\mathcal{V}_f(\mathcal{N})^{op}}$ is defined on

1. Objects: for each $V \in \mathcal{V}_f(\mathcal{N})$ there corresponds the set

$$\underline{H/H_F}_V := H_V/H_{FV} = H/H_{FV}$$
$$= \{[g]_V \mid g \sim g_1 \text{ iff } hg_1 = g \text{ for } h \in H_{FV}\}. \tag{19.51}$$

2. Morphisms: given a morphism $i_{V'V} : V' \subseteq V$, the corresponding presheaf morphism is the map $\underline{H/H_F}(i_{V'V}) : H/H_{FV} \to H/H_{FV'}$, defined as the bundle map of the bundle $H_{FV'}/H_{FV} \to H/H_{FV} \to H/H_{FV'}$.

We can also define the following presheaf:

Definition 19.12 The presheaf $\underline{H}/\underline{H_F}$ over $\mathcal{V}_f(\mathcal{N})$ is defined on

- Objects: for each $V \in \mathcal{V}_f(\mathcal{N})$ we obtain $(\underline{H}/\underline{H_F})_V := H/H_{FV}$, since the equivalence relation is computed context-wise.
- Morphisms: for each map $i : V' \subseteq V$ we will obtain the morphisms

$$(\underline{H}/\underline{H_F})_V \to (\underline{H}/\underline{H_F})_{V'} \tag{19.52}$$
$$H/H_{FV} \to H/H_{FV'}. \tag{19.53}$$

These are defined to be the projection maps $\pi_{V'V}$ of the fibre bundles

$$H_{FV'}/H_{FV} \to H/H_{FV} \to H/H_{FV'} \tag{19.54}$$

with fibre isomorphic to $H_{FV'}/H_{FV}$.

We then obtain the analogue of Theorem 16.1.

Theorem 19.4 [45]

$$\underline{H/H_F} \simeq \underline{H}/\underline{H_F}. \tag{19.55}$$

Since $\mathcal{V}_f(\mathcal{N})$ is a poset and, as such, it is equipped with the lower Alexandroff topology (which we denote as $\mathcal{V}_f(\mathcal{N})^-$), we then have the ordinary result

$$\mathbf{Sets}^{\mathcal{V}_f(\mathcal{N})^{op}} \simeq Sh(\mathcal{V}_f(\mathcal{N})^-), \tag{19.56}$$

where $Sh(\mathcal{V}_f(\mathcal{N})^-)$ is the topos of sheaves over $\mathcal{V}_f(\mathcal{N})^-$ (see Chap. 14). Thus, each presheaf in $\mathbf{Sets}^{\mathcal{V}_f(\mathcal{N})^{op}}$ is a sheaf in $Sh(\mathcal{V}_f(\mathcal{N})^-)$. From now on, we will simply denote $\mathcal{V}_f(\mathcal{N})^-$ by $\mathcal{V}_f(\mathcal{N})$. Moreover, using the 1:1 correspondence between sheaves and étale bundles, the presheaf $\underline{H/H_F}$ yields the associated étale bundle $p : \Lambda(\underline{H/H_F}) \to \mathcal{V}_f(\mathcal{N})$, whose bundle space $\Lambda(\underline{H/H_F})$ can be given a poset structure, as follows:

Definition 19.13 Given two elements $[g]_{V'} \in H/H_{FV'}$, $[g]_V \in H/H_{FV}$ we define the partial ordering by

$$[g]_{V'} \leq [g]_V \quad \text{iff} \quad p([g]_{V'}) \subseteq p([g]_V) \quad \text{and} \quad [g]_V \subseteq [g]_{V'}. \tag{19.57}$$

In addition, since each element $[g]$ corresponds uniquely to a faithful[8] automorphisms $\alpha^g_\rho(t) \in H$, it is also possible to define the ordering in terms of such automorphisms. We denote the set of faithful automorphisms from $V \to \mathcal{N}$ by $\mathrm{Aut}_{fth}(V, \mathcal{N})$.

Definition 19.14 Given two faithful automorphisms $\alpha_1 \in \mathrm{Aut}_{fth}(V, \mathcal{N})$ and $\alpha_2 \in \mathrm{Aut}_{fth}(V', \mathcal{N})$, we define the partial ordering by

$$\alpha_2 \le \alpha_1 \quad \text{iff} \quad p(\alpha_2) \subseteq p(\alpha_1) \quad \text{and} \quad \alpha_2 = \alpha_1|_{V'}. \tag{19.58}$$

In the following we will use, interchangeably, $[g]$ and its representative $\alpha^g_\rho(t)$, which we will denote as l_g.

Theorem 19.5 *The map $I : Sh(\mathcal{V}_f(\mathcal{N})) \to Sh(\Lambda(\underline{H/H_F}))$ is a functor defined as follows:*

(i) *Objects*: $(I(\underline{A}))_{[g]_V} := \underline{A}_{l_g(V)} = ((l_g)^*(\underline{A}))(V)$. *If* $[g]_{V'} \le [g]_V$, *then*

$$\big(I\underline{A}(i_{[g]_{V'},[g]_V})\big) := \underline{A}_{l_g(V),l_g(V')} : \underline{A}_{l_g(V)} \to \underline{A}_{l_g(V')} \tag{19.59}$$

where $V = p([g]_V)$ *and* $V' = p([g]_{V'})$.

(ii) *Morphisms: given a morphism* $f : \underline{A} \to \underline{B}$ *in* $Sh(\mathcal{V}_f(\mathcal{N}))$, *define the associated morphism in* $Sh(\Lambda(\underline{H/H_F}))$ *as*

$$I(f)_{[g]_V} : I(\underline{A})_{[g]_V} \to I(\underline{B})_{[g]_V} \tag{19.60}$$

$$f_{[g]_V} : \underline{A}_{l_g(p([g]_V))} \to \underline{B}_{l_g(p([g]_V))}. \tag{19.61}$$

The proof of this theorem is analogous to the proof of Theorem 16.2, hence, we will not report it.

Using the I functor, we are able to map all the sheaves in $Sh(\mathcal{V}_f(\mathcal{N}))$ to sheaves in $Sh(\Lambda(\underline{H/H_F}))$. Applying the functor $p! : Sh(\Lambda(\underline{H/H_F})) \to Sh(\mathcal{V}_f(\mathcal{N}))$, we finally obtain sheaves on our fixed category $\mathcal{V}_f(\mathcal{N})$ (for details the reader should refer back to Chap. 16).

In this context, the object $\underline{H}$ is indeed a group object and, as shown in [49], is the topos analogue of a one parameter group taking its values in $\underline{\mathbb{R}}$. Moreover, by defining the relevant objects in our formalism through the above method, i.e. in terms of the composite functor $p! \circ I$, the action of $\underline{H}$ does not induce twisted presheaves.

Having defined the automorphism group $\underline{H}$, we can define the KMS condition in terms of such a group. This is the purpose of the next section.

[8]Here, by faithful, we mean that, for each V, we will only consider automorphisms of the algebra $\mathcal{N} \supseteq V$, which do not leave any element $A \in V$ unchanged.

19.5 Internal KMS Condition

Now we work in the topos $Sh(\mathcal{V}_f(\mathcal{N}))$, for which the state space is defined as follows (see Sect. 16.5):

Definition 19.15 The spectral presheaf $\underline{\breve{\Sigma}} := p! \circ I(\underline{\Sigma})$ is defined on

(i) Objects: for each $V \in \mathcal{V}_f(\mathcal{N})$ we have

$$\underline{\breve{\Sigma}}_V := \coprod_{[g]_V \in H/H_{FV}} \underline{\Sigma}(l_g V) \simeq \coprod_{\alpha \in \mathrm{Aut}_{fth}(V,\mathcal{N})} \underline{\Sigma}(\alpha V) \tag{19.62}$$

which represents the disjoint union of the Gel'fand spectrum of all algebras related to V, via a faithful group transformation.

(ii) Morphisms: given a morphism $i_{V'V} : V' \to V$ $(V' \subseteq V)$, in $\mathcal{V}_f(\mathcal{N})$ the corresponding spectral presheaf morphism is

$$\underline{\breve{\Sigma}}(i_{V'V}) : \underline{\breve{\Sigma}}_V \to \underline{\breve{\Sigma}}_{V'} \tag{19.63}$$

$$\coprod_{\alpha_1 \in \mathrm{Aut}_{fth}(V,\mathcal{N})} \underline{\Sigma}_{\alpha_1(V)} \to \coprod_{\alpha_2 \in \mathrm{Aut}_{fth}(V',\mathcal{N})} \underline{\Sigma}_{\alpha_2(V')} \tag{19.64}$$

such that, given $\lambda \in \underline{\Sigma}_{\alpha_1(V)}$, we obtain $\underline{\breve{\Sigma}}(i_{V'V})(\lambda) := \underline{\Sigma}_{\alpha_1(V),\alpha_2(V')}\lambda = \lambda|_{\alpha_2(V')}$. Thus, $\underline{\breve{\Sigma}}(i_{V'V})$ is actually a coproduct of morphism $\underline{\Sigma}_{\alpha_1(V),\alpha_2(V')}$, one for each $\alpha_1 \in \mathrm{Aut}_{fth}(V,\mathcal{N})$.

Similarly, analogous to the previous $\underline{[0,1]}$, the presheaf of OR functions can be defined as follows:

Definition 19.16 The presheaf $\underline{[0,\breve{1}]} := p! \circ I(\underline{[0,1]})$ is defined on

(i) Objects: for each $V \in \mathcal{V}_f(\mathcal{N})$, we have

$$\underline{[0,\breve{1}]}_V := \coprod_{[g] \in H/H_{FV}} \underline{[0,1]}_{l_g V} \simeq \coprod_{\alpha \in \mathrm{Aut}_{fth}(V,\mathcal{N})} \underline{[0,1]}_{\alpha V}. \tag{19.65}$$

(ii) Morphisms: given $i_{V'V} : V' \subseteq V$, the corresponding presheaf morphism is

$$\underline{[0,\breve{1}]}(i_{V'V}) : \underline{[0,\breve{1}]}_V \to \underline{[0,\breve{1}]}_{V'} \tag{19.66}$$

$$\coprod_{\alpha_1 \in \mathrm{Aut}_{fth}(V,\mathcal{N})} \underline{[0,1]}_{\alpha_1(V)} \to \coprod_{\alpha_2 \in \mathrm{Aut}_{fth}(V',\mathcal{N})} \underline{[0,1]}_{\alpha_2(V')}. \tag{19.67}$$

Applying the composite functor $p! \circ I$ to the truth object $\underline{\mathbb{T}}^{\rho,r}$, we obtain the presheaf $\underline{\breve{\mathbb{T}}}^{\rho,r} := p! \circ I(\underline{\mathbb{T}}^{\rho,r})$.

Definition 19.17 The truth object presheaf $\underline{\breve{\mathbb{T}}}^{\rho,r}$ is defined on

(i) Objects: for each $V \in \mathcal{V}_f(\mathcal{N})$ we have

$$\breve{\underline{\mathbb{T}}}^{\rho,r}_V := \coprod_{[g]_V \in H/H_{FV}} \underline{\mathbb{T}}^{\rho,r}(l_g V) \simeq \coprod_{\alpha \in \mathrm{Aut}_{fth}(V,\mathcal{N})} \underline{\mathbb{T}}^{\rho,r}(\alpha V) \tag{19.68}$$

which represents the disjoint union of the truth object defined for all algebras related to V, via a faithful group transformation.

(ii) Morphisms: given a morphism $i : V' \to V$ $(V' \subseteq V)$, in $\mathcal{V}_f(\mathcal{N})$ the corresponding morphism is

$$\breve{\underline{\mathbb{T}}}^{\rho,r}(i_{V'V}) : \breve{\underline{\mathbb{T}}}^{\rho,r}_V \to \breve{\underline{\mathbb{T}}}^{\rho,r}_{V'} \tag{19.69}$$

$$\coprod_{\alpha_1 \in \mathrm{Aut}_{fth}(V,\mathcal{N})} \underline{\mathbb{T}}^{\rho,r}_{\alpha_1(V)} \to \coprod_{\alpha_2 \in \mathrm{Aut}_{fth}(V',\mathcal{N})} \underline{\mathbb{T}}^{\rho,r}_{\alpha_2(V')} \tag{19.70}$$

such that, given $\underline{S} \in \underline{\mathbb{T}}^{\rho,r}_{\alpha_1(V)}$, we have $\breve{\underline{\mathbb{T}}}^{\rho,r}(i_{V'V})(\underline{S}) := \underline{\mathbb{T}}^{\rho,r}_{\alpha_1(V),\alpha_2(V')}\underline{S} = \underline{S}_{|\downarrow\alpha_2(V')}$. $\breve{\underline{\mathbb{T}}}^{\rho,r}(i_{V'V})$ is actually a coproduct of morphism $\underline{\mathbb{T}}^{\rho,r}_{\alpha_1(V),\alpha_2(V')}$, one for each $\alpha_1 \in \mathrm{Aut}_{fth}(V,\mathcal{N})$.

In this setting, given a sate ρ, the associated measure is

$$\breve{\mu}^\rho : \breve{\underline{\Sigma}} \to \Gamma\underline{[0,1]}^{\breve{}} \tag{19.71}$$

$$\breve{\underline{S}} \mapsto \breve{\mu}^\rho(\breve{\underline{S}}) \tag{19.72}$$

such that, for each $V \in \mathcal{V}_f(\mathcal{N})$, we have

$$\breve{\mu}^\rho(\breve{\underline{S}})(V) := \coprod_{[g] \in H/H_{FV}} \mu^\rho \underline{S}(l_g V) = \coprod_{[g] \in H/H_{FV}} \mu^\rho(l_g^* \underline{S})(V). \tag{19.73}$$

If ρ is a KMS state, then $\mu^\rho(l_g^*\underline{S})(V) = \mu^\rho(\underline{S})(V)$. It follows that $\mu^\rho(\underline{S}) : \mathcal{V}_f(\mathcal{N}) \to [0,1]$ is constant on all H related V, i.e. such a measure is constant on the orbits H/H_{FV}, defined for each V. It is then reasonable, first of all, to define a measure on $Sub(I(\underline{\Sigma}))$, where $I : Sh(\mathcal{V}_f(\mathcal{N})) \to Sh(\Lambda(\underline{H/H_F}))$ was introduced above. Such a measure would be

$$\begin{aligned} \bar{\mu}^\rho : Sub\big(I(\underline{\Sigma})\big) &\to \Gamma\big(I\underline{[0,1]}\big) \\ \underline{S} &\mapsto \bar{\mu}^\rho \underline{S} \end{aligned} \tag{19.74}$$

where $\bar{\mu}^\rho\underline{S} : \Lambda(\underline{H/H_F}) \to [0,1]$ is such that, $\forall[g] \in \Lambda(\underline{H/H_F})$ we have $\bar{\mu}^\rho\underline{S}[g] := \mu^\rho\underline{S}(l_g(V)) = \rho(\hat{P}_{\underline{S}(l_g(V))})$.

Since for each $V \in \mathcal{V}_f(\mathcal{N})$ corresponds the orbit H/H_{FV}, each global element $\bar{\mu}^\rho(\underline{S})$, when restricted to such an orbit, gives the constant *local* section. In other words,

$$\bar{\mu}^\rho\underline{S}_{|H/H_{FV}} : \Lambda(\underline{H/H_F})(V) \to [0,1] \tag{19.75}$$

is constant on all $[g] \in \Lambda(\underline{H/H_F})(V)$. It is possible to consider $\bar{\mu}_{|H/H_{FV}}(I(\underline{S}))$ as a constant *global* section if we introduce the presheaf $\overline{[0,1]_V} \in \mathbf{Sets}^{\Lambda(\underline{H/H_{FV}})^{op}}$. Here $\underline{H/H_{FV}}$ is the category, whose elements are the equivalence classes $[g]$, as in $\Lambda(\underline{H/H_F})$, but whose morphisms are now given by group multiplication, i.e. $g:[h]\to[h']$ iff $h'=gh$ and $h\notin H_{FV}$.

Definition 19.18 The presheaf $\overline{[0,1]_V} \in \mathbf{Sets}^{\Lambda(\underline{H/H_{FV}})^{op}}$ has as

1. Objects: for each $[h]$, the respective presheaf set is

$$\overline{[0,1]_V}([h]) := I(\underline{[0,1]})[h] = \underline{[0,1]}l_h(V) = \{f\colon \downarrow l_h(V)\to[0,1] | f \text{ is } OR\}. \tag{19.76}$$

2. Morphisms: given a map $g:[h]\to[h']$ such that $h'=gh$, then we get a corresponding presheaf map

$$\begin{aligned} \overline{[0,1]_V}(g):\overline{[0,1]_V}([h'])&\to\overline{[0,1]_V}([h]) \\ f&\mapsto l_{g-1}(f) \end{aligned} \tag{19.77}$$

where $f\colon \downarrow l_h(V)\to[0,1]$ and $l_{g-1}f\colon \downarrow l_{g^{-1}h}V\to[0,1]$.

To show that this is indeed a presheaf, we need to show that, given two maps $g_j:[h'']\to[h']$ and $g_i:[h']\to[h]$, then the following equality holds:

$$\overline{[0,1]_V}(g_j\circ g_i)=\overline{[0,1]_V}(g_i)\circ\overline{[0,1]_V}(g_j). \tag{19.78}$$

Considering first the left hand side we obtain

$$\begin{aligned} \overline{[0,1]_V}(g_j\circ g_i):\overline{[0,1]_V}[h]&\to\overline{[0,1]_V}[h''] \\ f&\mapsto l_{(g_jg_i)^{-1}}(f)=l_{g_i^{-1}}l_{g_j^{-1}}f \end{aligned} \tag{19.79}$$

while the right hand side is

$$\begin{aligned} \overline{[0,1]_V}(g_i)\circ\overline{[0,1]_V}(g_j):\overline{[0,1]_V}[h]&\to\overline{[0,1]_V}[h']\to\overline{[0,1]_V}[h'] \\ f&\mapsto l_{g_j^{-1}}f\mapsto l_{g_i^{-1}}l_{g_j^{-1}}f. \end{aligned} \tag{19.80}$$

Thus, indeed, $\overline{[0,1]_V}$ is a well defined presheaf.

Obviously, for each $V_i\in\mathcal{V}_f(\mathcal{N})$, one gets a presheaf $\overline{[0,1]_{V_i}}\in\mathbf{Sets}^{\Lambda(\underline{H/H_{FV_i}})^{op}}$. Using these new presheaves we can, then, define the first condition on the measure associated to a KMS state to be the following:

$$\breve{\mu}^\rho(\breve{\underline{S}})(V)=\Gamma^c\big(\overline{[0,1]_V}\big), \tag{19.81}$$

where $\Gamma^c(\overline{[0,1]_V})$ denotes the global element with constant value c, which, in this case is $\mu^\rho(\underline{S})(l_gV)$. This condition on the measure translates to the following condition on the truth object representing a KMS:

$$\underline{\breve{\mathbb{T}}}^{\rho,r}_V = \{\underline{\breve{S}} \in Sub_{cl}(\underline{\breve{\Sigma}}_{|\downarrow V})| \quad \forall V' \subseteq V, \breve{\mu}(\underline{\breve{S}})(V') = \Gamma^c\big(\boxed{[0,1]_{V'}}\big)\text{for } c = \mu^\rho(\underline{S})(V) \geq r\big\}. \tag{19.82}$$

Thus, the truth object associated to the KMS state represents the collection of sub-objects for which the measure is constant on the orbits H/H_{FV}, for each $V \in \mathcal{V}_f(\mathcal{N})$. A moment of thought reveals that this is nothing but

$$\underline{\breve{\mathbb{T}}}^{\rho,r}_V = \big|\big[\alpha(t)^*\underline{\mathbb{T}}^{\rho,r}\big]\big|_{\mu^\rho,V}. \tag{19.83}$$

The action of the group $\underline{H}$ on $\underline{\breve{\mathbb{T}}}^{\rho,r}$, for each $V \in \mathcal{V}_f(\mathcal{N})$, is defined

$$\underline{H}_V \times \underline{\breve{\mathbb{T}}}^{\rho,r}_V \to \underline{\breve{\mathbb{T}}}^{\rho,r}_V \tag{19.84}$$

$$\big(\alpha(t), \underline{S}\big) \mapsto \alpha(t)^*(\underline{S}). \tag{19.85}$$

Restricting only to the first KMS condition, we can define a truth object satisfying such a condition as follows:

Definition 19.19 Given a truth object $\underline{\breve{\mathbb{T}}}^{\rho,r}$, it satisfies the first KMS condition iff the following diagram commutes:

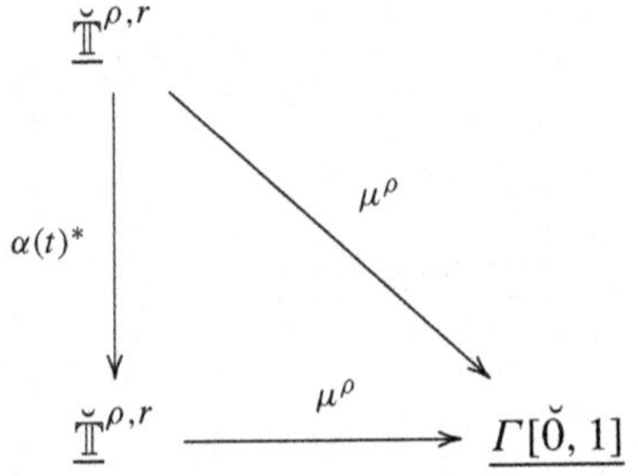

for all $\alpha(t) \in \Lambda(\underline{H})$.

We are still missing the second condition for a KMS state, namely for any two elements $\hat{A}, \hat{B} \in \mathcal{N}_\alpha$, then $\langle\rho, \hat{A}\alpha_{t+i\gamma}\hat{B}\rangle = \langle\rho, \alpha_t\hat{B}\hat{A}\rangle$. Here $\alpha_{t+i\gamma}$ belongs to the set $E = \{\alpha_{t+i\gamma} \mid t \in \mathbb{R}, \gamma \in [0,\beta]\}$ which represents an extension of the group H to a strip on the complex plane. Since we are also considering $\gamma = 0$, it follows that H is contained in the above set. Is it possible to internalise E as we did for H? The answer is yes, and it is done through the construction of the following trivial presheaf $\underline{E}$ for which, at each context, we simply assign the set E itself and the maps are identity maps. Note that such presheaf is not a group, nevertheless we get $\underline{H} \subset \underline{E}$. Recalling the definition of the presheaf $\underline{H}_F$, it is possible to define its action on $\underline{E}$ as follows:

$$\underline{H_F} \times \underline{E} \to \underline{E} \tag{19.86}$$

such that, for each $V \in \mathcal{V}_f(\mathcal{N})$, we obtain

$$\underline{H_{F}}_V \times \underline{E}_V \to \underline{E}_V \tag{19.87}$$

$$\big(\alpha(t_1), \alpha(t_2 + i\gamma)\big) \mapsto \alpha_\rho(t_1 + t_2, i\gamma). \tag{19.88}$$

Given this action, it is meaningful to define the quotient presheaf $\underline{E/H_F}$. This is a well defined presheaf since $H_{FV} \subseteq H_{FV'}$ for $V'' \subseteq V$. Therefore, for each $V \in \mathcal{V}_f(\mathcal{N})$ we obtains the set E/H_{FV}, where two elements $\alpha(t_1 + i\gamma) \simeq \alpha(t_2 + i\gamma)$ iff there exists an element $\alpha(t_3)$, such that $t_2 = t_1 + t_3$. Clearly, $\underline{H/H_F} \subset \underline{E/H_F}$. We can then extend what has been done with respect to the presheaf $\underline{H/H_F}$ to the presheaf $\underline{E/H_F}$.

First, we define the new truth object for each $V \in \mathcal{V}_f(\mathcal{N})$ as

$$\underline{\tilde{\mathbb{T}}}^{\rho,r}_V := \coprod_{[g] \in E/H_{FV}} \mathbb{T}^{\rho,r}(l_g V). \tag{19.89}$$

Here $\underline{\tilde{\mathbb{T}}}^{\rho,r}$ is defined as[9] $\underline{\tilde{\mathbb{T}}}^{\rho,r} := p_E! \circ \mathcal{I}(\underline{\mathbb{T}}^{\rho,r})$ where $\mathcal{I} : Sh(\mathcal{V}_f(\mathcal{N})) \to Sh(\Lambda(\underline{E/H_F}))$ and $p_E : \Lambda(\underline{E/H_F}) \to \mathcal{V}_f(\mathcal{N})$. Since $\underline{H} \subset \underline{E}$, the above truth object contains all the elements of $\underline{\breve{\mathbb{T}}}^{\rho}_{(V,r)}$. Hence, the first KMS condition keeps its form. On the other hand, the second KMS condition seems a little more involved since it concerns measures on the conjunction of two sub-objects of the state space. To implement the second KMS condition, the following natural transformation is needed:

$$F : \underline{E} \times \underline{\tilde{\mathbb{T}}}^{\rho} \to \underline{\tilde{\mathbb{T}}}^{\rho}. \tag{19.90}$$

However, we need to pay attention to domain issues. In particular we know that $F_{\underline{T},\underline{S}}(t + i\beta) = [\mu^\rho(\underline{T} \wedge \alpha(t + i\beta)^* \underline{S}]\mathcal{N}_\alpha$ only for $\mathcal{N}_\alpha$ being a σ-weakly α-invariant sub-algebra of $\mathcal{N}$. Therefore we can construct a category $\mathcal{V}(\mathcal{N}_\alpha)$ of abelian sub-algebras of $\mathcal{N}_\alpha$ ordered by inclusion. Clearly $\mathcal{V}(\mathcal{N}_\alpha)$ is a full subcategory of $\mathcal{V}(\mathcal{N})$, thus we can define the continuous identity map $i : \mathcal{V}(\mathcal{N}_\alpha) \to \mathcal{V}(\mathcal{N})$. This gives rise to the corresponding geometric morphisms between $i : Sh(\mathcal{V}(\mathcal{N}_\alpha)) \to Sh(\mathcal{V}(\mathcal{N}))$ whose associated inverse image is $i^* : Sh(\mathcal{V}(\mathcal{N})) \to Sh(\mathcal{V}(\mathcal{N}_\alpha))$ such that $(i^*(\underline{A}))_{V_\alpha} := \underline{A}_{i(V_\alpha)} = \underline{A}_{V_\alpha}$.

We then utilise these inverse image morphisms to correctly define the natural transformation in (19.90), which we still call F, as

$$F : i^*\big(\underline{E} \times \underline{\tilde{\mathbb{T}}}^{\rho}\big) \to i^*\big(\underline{\tilde{\mathbb{T}}}^{\rho}\big) \tag{19.91}$$

$$F : i^*(\underline{E}) \times i^*\big(\underline{\tilde{\mathbb{T}}}^{\rho}\big) \to i^*\big(\underline{\tilde{\mathbb{T}}}^{\rho}\big) \tag{19.92}$$

[9] Note that we are using the same method that was used to define $\underline{\breve{\mathbb{T}}}^{\rho,r}$, but now we are replacing I by $\mathcal{I}$ and $p!$ by $p_E!$.

where the last equation holds since the left adjoint f^* preserves finite limits. Given a context $V_\alpha \in \mathcal{V}(\mathcal{N}_\alpha)$ we then have

$$F_V : \big(i^*(\underline{E})\big)_{V_\alpha} \times \big(i^*(\underline{\tilde{\mathbb{T}}}^\rho)\big)_{V_\alpha} \to \big(i^*(\underline{\tilde{\mathbb{T}}}^\rho)\big)_{V_\alpha} \tag{19.93}$$

$$F_V : \underline{E}_{V_\alpha} \times \underline{\tilde{\mathbb{T}}}^\rho_{V_\alpha} \to \underline{\tilde{\mathbb{T}}}^\rho_{V_\alpha} \tag{19.94}$$

$$\big(\alpha(t+i\beta), \underline{S}\big) \mapsto F\big(\big(\alpha(t+i\beta), \underline{S}\big)\big) := \alpha^*(t+i\beta)\underline{S}. \tag{19.95}$$

To show that this is a well defined natural transformation, we prove that, given a pair of contexts V' and V, such that $i_{V'V} : V' \subseteq V$, the following diagram commutes:[10]

$$\begin{array}{ccc}
\underline{E}_{V_\alpha} \times \underline{\tilde{\mathbb{T}}}^\rho_{V_\alpha} & \xrightarrow{F_{V_\alpha}} & \underline{\tilde{\mathbb{T}}}^\rho_{V_\alpha} \\
\Big\downarrow {\scriptstyle \langle \underline{E}(i_{V'_\alpha V_\alpha}), \underline{\tilde{\mathbb{T}}}^\rho(i_{V'_\alpha V_\alpha})\rangle} & & \Big\downarrow {\scriptstyle \underline{\tilde{\mathbb{T}}}^\rho(i_{V'_\alpha V_\alpha})} \\
\underline{E}_{V'_\alpha} \times \underline{\tilde{\mathbb{T}}}^\rho_{V'_\alpha} & \xrightarrow{F_{V'_\alpha}} & \underline{\tilde{\mathbb{T}}}^\rho_{V'_\alpha}
\end{array}$$

Chasing the diagram round one way, we obtain:

$$\big[\underline{\tilde{\mathbb{T}}}^\rho(i_{V'_\alpha V_\alpha}) \circ F_V\big]\big(\alpha(t), \underline{S}\big) = \underline{\tilde{\mathbb{T}}}^\rho(i_{V'_\alpha V_\alpha})\big(\alpha(t)^*\underline{S}\big) = \big(\alpha(t)^*\underline{S}\big)_{|\downarrow V'_\alpha} = \underline{S}_{|\alpha(t)V'_\alpha} \tag{19.96}$$

while going round the other way it yields:

$$F_{V'} \circ \big\langle \underline{E}(i_{V'_\alpha V_\alpha}), \underline{\tilde{\mathbb{T}}}^\rho(i_{V'_\alpha V_\alpha})\big\rangle\big(\alpha(t), \underline{S}\big) = F_{V'_\alpha}\big(\alpha(t)_{|V'_\alpha}, \underline{S}_{|\downarrow V'_\alpha}\big) = \alpha(t)^*\underline{S}_{|\downarrow V'_\alpha} = \underline{S}_{|\alpha(t)V'_\alpha}. \tag{19.97}$$

Hence, F is a natural transformation.

We can now write the second condition for the KMS state as follows:

[10] Note that, since $(i^*(\underline{A}))_{V_\alpha} = \underline{A}_{i(V_\alpha)} = \underline{A}_{V_\alpha}$, form now on we will simply write $\underline{A}_{V_\alpha}$ and the action of the i^* functor will be assumed.

Definition 19.20 The truth object $\underline{\tilde{\mathbb{T}}}^{\rho}$ represents the topos analogue of the KMS state iff for all $V_\alpha \in \mathcal{V}(\mathcal{N})$, the following diagram commutes

$$\begin{array}{ccccc}
\underline{\tilde{\mathbb{T}}}^{\rho,r}_{V_\alpha} \times \underline{\tilde{\mathbb{T}}}^{\rho,r}_{V_\alpha} & \xrightarrow{\langle \alpha_t^*, id\rangle} & \underline{\tilde{\mathbb{T}}}^{\rho,r}_{V_\alpha} \times \underline{\tilde{\mathbb{T}}}^{\rho,r}_{V_\alpha} & \xrightarrow{f} & \underline{\tilde{\mathbb{T}}}^{\rho,r}_{V_\alpha} \times \underline{\tilde{\mathbb{T}}}^{\rho,r}_{V_\alpha} \\
\Big\downarrow{\scriptstyle \langle \alpha^*_{t+i\gamma}, id\rangle} & & & & \Big\downarrow{\scriptstyle \wedge} \\
\underline{\tilde{\mathbb{T}}}^{\rho,r}_{V_\alpha} \times \underline{\tilde{\mathbb{T}}}^{\rho,r}_{V_\alpha} & & \xrightarrow{\mu^\rho} & & \underline{\tilde{\mathbb{T}}}^{\rho,r}_{V_\alpha} \\
\Big\downarrow{\scriptstyle \wedge} & & & & \Big\downarrow{\scriptstyle \mu^\rho} \\
\underline{\tilde{\mathbb{T}}}^{\rho,r}_{V_\alpha} & & \xrightarrow{\mu^\rho} & & \underline{\Gamma[\tilde{0},1]}_{V_\alpha}
\end{array}$$

for all $\alpha_{t+i\gamma} \in \Lambda(\underline{E})$. Here f is a switching function, i.e. $f : \underline{\tilde{\mathbb{T}}}^{\rho,r}_{V_\alpha} \times \underline{\tilde{\mathbb{T}}}^{\rho,r}_{V_\alpha} \to \underline{\tilde{\mathbb{T}}}^{\rho,r}_{V_\alpha} \times \underline{\tilde{\mathbb{T}}}^{\rho,r}_{V_\alpha}$; $f(\underline{S}, \underline{T}) := (\underline{T}, \underline{S})$.

Thus given a pair $(\underline{S}, \underline{T}) \in \underline{\tilde{\mathbb{T}}}^{\rho,r} \times \underline{\tilde{\mathbb{T}}}^{\rho,r}$, going one way round the diagram, we obtain

$$\begin{aligned}
\mu^\rho \circ \wedge \circ f \circ \langle \alpha_t^*, id\rangle(\underline{S}, \underline{T})_{V_\alpha} &= \mu^\rho \circ \wedge \circ f\big((\alpha_t^* \underline{S})_{V_\alpha}, \underline{T}_{V_\alpha}\big) = \mu^\rho \circ \wedge\big(\underline{T}_{V_\alpha}, \alpha_t^* \underline{S}_{V_\alpha}\big) \\
&= \mu^\rho\big(\underline{T} \wedge \alpha_t^* \underline{S}\big)_{V_\alpha}. \qquad (19.98)
\end{aligned}$$

Going the other way instead, it can be proved that

$$\begin{aligned}
\mu^\rho \circ \wedge \circ \langle \alpha^*_{t+i\gamma}, id\rangle(\underline{S}, \underline{T})_{V_\alpha} &= \mu^\rho \circ \wedge\big((\alpha^*_{t+i\gamma} \underline{S})(V), \underline{T}_{V_\alpha}\big) \\
&= \mu^\rho \big(\alpha^*_{t+i\gamma} \underline{S} \wedge \underline{T}\big)_{V_\alpha}. \qquad (19.99)
\end{aligned}$$

This is precisely the second KMS condition.

It is straightforward to see that the above definition incorporates Definition 19.19 by setting $\gamma = 0$. So far, we have considered the one parameter family of truth objects $\underline{\tilde{\mathbb{T}}}^{\rho,r}_{V}$, one for each $r \in (0,1)_L$. As explained in Chap. 15, they can be combined to define an object in the topos $Sh(\mathcal{V}_f(\mathcal{N}) \times (0,1)_L)$, as follows:

Definition 19.21 The truth object $\underline{\tilde{\mathbb{T}}}^{\rho} \in Sh(\mathcal{V}_f(\mathcal{N}) \times (0,1)_L)$ is defined on:

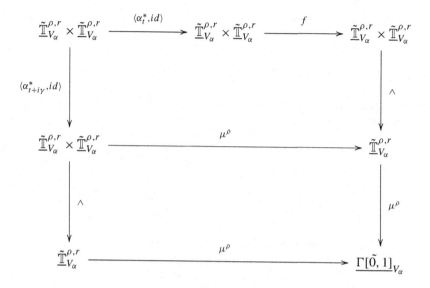

(i) Objects: for each pair $(V, r) \in \mathcal{V}_f(\mathcal{N}) \times (0,1)_L$, we have

$$\underline{\tilde{\mathbb{T}}}^{\rho}_{(V,r)} := \coprod_{[g]\in E/H_{FV}} \underline{\mathbb{T}}^{\rho,r}_{l_g(V)} \simeq \coprod_{\alpha_\rho(t)\in \mathrm{Aut}_{fth}(V,\mathcal{N})} \underline{\mathbb{T}}^{\rho,r}_{\alpha_\rho(t)(V)} \tag{19.100}$$

where each

$$\underline{\mathbb{T}}^{\rho,r}_{\alpha(t)(V)} = \left\{\underline{S} \in Sub_{cl}(\underline{\Sigma}_{\downarrow\alpha(t)(V)}) \,\middle|\, \forall V' \subseteq \alpha(t)(V), \rho(\hat{P}_{\underline{S}_{V'}}) \geq r\right\}. \tag{19.101}$$

(ii) Morphisms: given a map $i : (V', r') \leq (V, r)$ (iff $V' \subseteq V$ and $r' \leq r$), then the associated morphism is

$$\underline{\tilde{\mathbb{T}}}^{\rho}(i) : \coprod_{[g]\in H/H_{FV}} \underline{\mathbb{T}}^{\rho,r}_{l_g(V)} \to \coprod_{[h]\in H/H_{FV'}} \underline{\mathbb{T}}^{\rho,r}_{l_h(V')} \tag{19.102}$$

$$\coprod_{\alpha_1(t)\in \mathrm{Aut}_{fth}(V,\mathcal{N})} \underline{\mathbb{T}}^{\rho,r}_{\alpha_1(t)(V)} \to \coprod_{\alpha_2(t)\in \mathrm{Aut}_{fth}(V',\mathcal{N})} \underline{\mathbb{T}}^{\rho,r}_{\alpha_2(t)(V')} \tag{19.103}$$

such that, given a sub-object $\underline{S} \in \underline{\mathbb{T}}^{\rho,r}_{\alpha_1(t)(V)}$, we get

$$\underline{\breve{\mathbb{T}}}^{\rho}(i)\underline{S} := \underline{\mathbb{T}}^{\rho,r}(i_{\alpha_1(t)(V),\alpha_2(t)(V')})\underline{S} = \underline{S}_{|\alpha_2(t)(V')} \tag{19.104}$$

where $\alpha_2(t) \leq \alpha_1(t)$ thus $p(\alpha_2(t)) \subseteq p(\alpha_1(t))$ and $\alpha_2(t)(V') = (\alpha_1(t))|_{V'}(V')$. Obviously, now, the condition on the restricted sub-object is $\rho(\hat{P}_{S_{V''}}) \geq r'$ where $V'' \subseteq \alpha_2(t)(V')$. However, such a condition is trivially satisfied, since $r' \leq r$.

The derivation of the standard KMS state from the topos KMS state is done similarly as for the internal situation.

Chapter 20
One-Parameter Group of Transformations and Stone's Theorem

In this chapter we will define the topos notion of a one-parameter group taking values in the complex number object and in the real number objects. To this end we first of all need to upgrade the monoids $\underline{\mathbb{C}}^{\leftrightarrow}$ and $\underline{\mathbb{R}}^{\leftrightarrow}$ to groups. This can be done using a standard method called *Grothendieck k-Construction* already mentioned in [32]. Having done that the construction of a one-parametr group can be defined. This in turn allows us to define the topos analogue of the Stone's theorem which uniquely associates to each self adjoint operator $\breve{\delta}(\hat{A}) : \underline{\Sigma} \to \underline{\mathbb{R}}^{\leftrightarrow}$ a one parameter group $\underline{Q}^{\hat{A}}$. This is of particular importance in the view of defining a unique time evolution. In fact, given a Hamiltonian operator $\underline{H}$, the topos analogue would be $\breve{\delta}(\hat{H})$ with associated the unique one-parameter group of transformations $\underline{Q}^{\hat{H}}$. This group would represent the group of time evolutions in topos quantum theory. The detailed analysis of such a group and how it acts had not yet been carried out but it would be of particular interest to do so.

20.1 Topos Notion of a One Parameter Group

Since we now have the topos definitions of both the real and complex quantity value objects, we can define the topos notion of a one parameter group with the parameter taking values either in the real value object or in the complex value object. We will start with the former. Let us consider a one parameter group $\{\alpha(t)|t \in \mathbb{R}\}$ which defines an automorphisms of $\mathcal{H}$. We would now like to internalise such an object, i.e. to define the topos analogue of the automorphisms group $H = \{\alpha(t)|t \in \mathbb{R}\}$. We know that for each element in this group we obtain the induced geometric morphisms

$$\alpha(t)^* : Sh\big(\mathcal{V}(\mathcal{N})\big) \to Sh\big(\mathcal{V}(\mathcal{N})\big) \tag{20.1}$$

$$\underline{S} \mapsto \alpha(t)^*\underline{S} \tag{20.2}$$

such that $\alpha(t)^*\underline{S}(V) := \underline{S}(\alpha(t)V)$.

C. Flori, *A First Course in Topos Quantum Theory*, Lecture Notes in Physics 868,
DOI 10.1007/978-3-642-35713-8_20, © Springer-Verlag Berlin Heidelberg 2013

Such an action, however, gives rise to twisted presheaves. To solve this problem we need to apply the methods defined in [45] and use as the new base category the category $\mathcal{V}_f(\mathcal{H})$ which is fixed, i.e. we do not allow any group to act on it. In Sect. 20.2 we describe in more details how sheaves on the new category $\mathcal{V}_f(\mathcal{H})$ are defined. Clearly $\mathcal{V}_f(\mathcal{H}) \simeq \mathcal{V}(\mathcal{H})$. We now define the internal group $\underline{H}$ over the new base category $\mathcal{V}(\mathcal{H}) \simeq \mathcal{V}_f(\mathcal{H})$ as follows:

Definition 20.1 The internal group $\underline{H}$ is the presheaf defined on

1. Objects: for each $V \in \mathcal{V}(\mathcal{N})$ we obtain $\underline{H}_V = H$.
2. Morphisms: These are simply the identity maps.

It is straightforward to see that $\Gamma(\underline{H}) = H$.

We now would like to define the group $\underline{H}$ as a one parameter group of transformations, with parameter taking values in the quantity value object $\underline{\mathbb{R}}^{\leftrightarrow}$.

Generally, a one parameter group of transformations $\{\alpha(t)|\forall t \in \mathbb{R}\}$ is a representation of the additive abelian group $(\mathbb{R}, +)$. However, as shown in [32], $\underline{\mathbb{R}}^{\leftrightarrow}$ is only a commutative monoid, not an abelian group since, although addition $(+ : \underline{\mathbb{R}} \times \underline{\mathbb{R}} \to \underline{\mathbb{R}})$ is well defined, subtraction is not. Fortunately, this difficulty is not insurmountable.

In order to extent a semigroup with unit to a full group, one strategy to use is the well known Grothendieck k-Construction already mentioned in [32]. Such a construction is defined as follows:

Definition 20.2 A group completion of a monoid M is an abelian group $k(M)$ together with a monoid map $\theta : M \to k(M)$ which is universal. Therefore, given any monoid morphism $\phi : M \to G$, where G is an abelian group, there exists a unique group morphism $\phi' : k(M) \to G$ such that ϕ factors through ϕ', i.e., the following diagram commutes

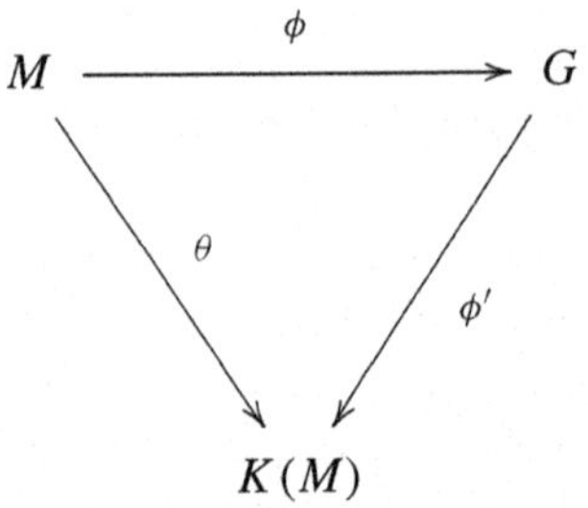

It is easy to show that any $k(M)$ is unique up to isomorphisms. As showed in [32] the construction of $k(M)$ is via an equivalence class. This is because what is missing is the inverse (subtraction) operation, however, given two elements $(a, b) \in M \times M$, if we think of them as meaning $a - b$, then we notice that $a - b = c - d$ iff $a + d = c + b$. Thus one defines an equivalence relation on $M \times M$ as follows:

$$(a, b) \simeq (c, d) \quad \text{iff} \quad \exists e \in M \quad \text{such that} \quad a + d + e = b + c + e. \tag{20.3}$$

Using this definition of equivalence we then equate $k(M)$ to precisely such a collection of equivalence classes where, again, each of them should be thought of as representing the subtraction of the two terms involved. This leads to the following definition:

Definition 20.3 The Grothendieck completion of an abelian monoid M is the pair $(k(M), \theta)$ defined as follows:

1. $k(M)$ is the set of equivalence classes $[a, b]$, where the equivalence relation is defined in (20.3). A group law on $k(M)$ is defined by
   (i) $[a, b] + [c, d] := [a + c, b + d]$
   (ii) $0_{k(M)} := [0_M, 0_M]$
   (iii) $-[a, b] := [b, a]$
   where 0_M is the unit in the abelian monoid M.
2. The map $\theta : M \rightarrow k(M)$ is defined by $\theta(a) := [a, 0]$ for all $a \in M$.

For the case at hand we then define the equivalence relation on $\underline{\mathbb{R}}^{\leftrightarrow} \times \underline{\mathbb{R}}^{\leftrightarrow}$ as follows: for each context V we have

$$((\mu_1, \nu_1), (\mu_2, \nu_2)) \equiv ((\mu_1', \nu_1'), (\mu_2', \nu_2')) \quad \text{iff} \quad \exists (\mu, \nu) \in \mathbb{R}_V \quad \text{such that}$$

$$(\mu_1, \nu_1) + (\mu_2', \nu_2') + (\mu, \nu) = (\mu_2, \nu_2) + (\mu_1', \nu_1') + (\mu, \nu) \tag{20.4}$$

where the summation in $\underline{\mathbb{R}}^{\leftrightarrow}$ was defined in (18.41). Given such an equivalence class we can now define the object $k(\underline{\mathbb{R}}^{\leftrightarrow})$ as follows:

Definition 20.4 The presheaf $k(\underline{\mathbb{R}}^{\leftrightarrow}) \in \mathbf{Sets}^{\mathcal{V}(\mathcal{H})}$ is defined on:

1. Objects: for each $V \in \mathcal{V}(\mathcal{H})$ we obtain

$$k(\underline{\mathbb{R}}^{\leftrightarrow})_V := \{[(\mu, \nu), (\mu', \nu')] \mid \mu, \mu' \in OP(\downarrow V, \mathbb{R}), \nu, \nu' \in OR(\downarrow V, \mathbb{R})\} \tag{20.5}$$

   where $[(\mu, \nu), (\mu', \nu')]$ denotes the k-equivalence class of $(\mu, \nu), (\mu', \nu')$.
2. Morphisms: for each $i_{V'V} : V' \subseteq V$ we obtain the arrow $k(\underline{\mathbb{R}}^{\leftrightarrow})(i_{V'V}) : k(\underline{\mathbb{R}}^{\leftrightarrow})_V \rightarrow k(\underline{\mathbb{R}}^{\leftrightarrow})_{V'}$ defined by

$$k(\underline{\mathbb{R}}^{\leftrightarrow})(i_{V'V})[(\mu, \nu), (\mu', \nu')]$$
$$:= [(\mu_{|V'}, \nu_{|V'}), (\mu_{|V'}', \nu_{|V'}')] \quad \text{for all} \quad [(\mu, \nu), (\mu', \nu')] \in k(\underline{\mathbb{R}}^{\leftrightarrow})_V. \tag{20.6}$$

In this way we have obtained an abelian group object $k(\underline{\mathbb{R}}^{\leftrightarrow})$. Is it now possible to define a one parameter group of automorphisms in terms of such an abelian group?
   Due to the cumbersome notation we will use $k(\underline{\mathbb{R}}^{\geq})$ instead of the full $k(\underline{\mathbb{R}}^{\leftrightarrow})$. Here $k(\underline{\mathbb{R}}^{\geq})$ is the k-extention of the presheaf $\underline{\mathbb{R}}^{\geq}$ which is defined as follows:

Definition 20.5 The presheaf $\underline{\mathbb{R}}^{\geq} \in \mathbf{Sets}^{\mathcal{V}(\mathcal{H})}$ is defined on:

1. Objects: for each $V \in \mathcal{V}(\mathcal{H})$ we obtain

$$\underline{\mathbb{R}}^{\geq}_{V} := \{\mu | \mu \in OR(\downarrow V, \mathbb{R})\}. \tag{20.7}$$

2. Morphisms: for each $i_{V'V} : V' \subseteq V$ we obtain the arrow $\underline{\mathbb{R}}^{\geq}(i_{V'V}) : \underline{\mathbb{R}}^{\geq}_{V} \to \underline{\mathbb{R}}^{\geq}_{V'}$ defined by

$$\underline{\mathbb{R}}^{\geq}(i_{V'V})(\mu) := \mu_{V'} \quad \text{for all } \mu \in \underline{\mathbb{R}}^{\geq}_{V}. \tag{20.8}$$

It follows that the k-extension of $\underline{\mathbb{R}}^{\geq}$ is:

Definition 20.6 The presheaf $k(\underline{\mathbb{R}}^{\geq}) \in \mathbf{Sets}^{\mathcal{V}(\mathcal{H})}$ is defined on:

1. Objects: for each $V \in \mathcal{V}(\mathcal{H})$ we obtain

$$k(\underline{\mathbb{R}}^{\geq})_V := \{[\mu, \nu] | \mu, \nu \in OR(\downarrow V, \mathbb{R})\} \tag{20.9}$$

where $[\mu, \nu]$ denotes the k-equivalence class of (μ, ν).
2. Morphisms: for each $i_{V'V} : V' \subseteq V$ we obtain the arrow $k(\underline{\mathbb{R}}^{\geq})(i_{V'V}) : k(\underline{\mathbb{R}}^{\geq})_V \to k(\underline{\mathbb{R}}^{\geq})_{V'}$ defined by

$$k(\underline{\mathbb{R}}^{\geq})(i_{V'V})[\mu, \nu] := [\mu_{|V'}, \nu_{|V'}] \quad \text{for all } [\mu, \nu] \in k(\underline{\mathbb{R}}^{\geq})_V. \tag{20.10}$$

This restriction to $k(\underline{\mathbb{R}}^{\geq})$ instead of $k(\underline{\mathbb{R}}^{\leftrightarrow})$ cases no trouble since $k(\underline{\mathbb{R}}^{\geq}) \subset k(\underline{\mathbb{R}}^{\leftrightarrow})$ and any result obtained for $k(\underline{\mathbb{R}}^{\geq})$ has an easy generalisation to $k(\underline{\mathbb{R}}^{\leftrightarrow})$. The advantage of using $k(\underline{\mathbb{R}}^{\geq})$ is that the notation is much more clear to understand.

20.1.1 One Parameter Group Taking Values in the Real Valued Object

With the above discussion in mind we attempt the following definition:

Definition 20.7 The presheaf $\underline{K} \in \mathbf{Sets}^{\mathcal{V}(\mathcal{H})}$ is defined on

1. Objects: for each context V we define $\underline{K}_V := \{\alpha_{[\mu,\nu]} | [\mu, \nu] \in k(\underline{\mathbb{R}}^{\geq})_V\}$.[1]
2. Morphisms: given the inclusion $i_{V'V} : V' \subseteq V$ we define $\underline{K}(i_{V'V}) : \underline{K}_V \to \underline{K}_{V'}$ as $\alpha_{[\mu,\nu]} \mapsto \alpha_{[\mu_{|V'}, \nu_{|V'}]}$.

This is clearly a presheaf since given $V'' \xrightarrow{i} V' \xrightarrow{j} V$,

$$\underline{K}(i_{V'V} \circ j_{V''V}) : \underline{K}_V \to \underline{K}_{V''} \tag{20.11}$$

$$\alpha_{[\mu,\nu]} \mapsto \alpha_{[\mu_{|V''}, \nu_{|V''}]} \tag{20.12}$$

[1] Here each $\alpha_{[\mu,\nu]}$ represents an abstract automorphisms on $\mathcal{H}$ parametrised by $[\mu, \nu]$.

while

$$\underline{K}(j_{V''V}) \circ \underline{K}(i_{V'V}) : \underline{K}_V \to \underline{K}_{V'} \to \underline{K}_{V''} \tag{20.13}$$

$$\alpha_{[\mu,\nu]} \mapsto \alpha_{[\mu_{|V'},\nu_{|V'}]}$$

$$\to \alpha_{[(\mu_{V'})|_{V''},(\nu_{V'})|_{V''}]} = \alpha_{[\mu_{|V''},\nu_{|V''}]}. \tag{20.14}$$

The presheaf $\underline{K}$ can be turned into a group by defining the additive operation, for all $V \in \mathcal{V}(\mathcal{H})$, as follows:

$$+_V : \underline{K}_V \times \underline{K}_V \to \underline{K}_V \tag{20.15}$$

$$(\alpha_{[\mu_1,\nu_1]}, \alpha_{[\mu_2,\nu_2]}) \mapsto +_V(\alpha_{[\mu_1,\nu_1]}, \alpha_{[\mu_2,\nu_2]}) := \alpha_{[\mu_1+\mu_2,\nu_1+\nu_2]}. \tag{20.16}$$

From now on we will denote $+_V(\alpha_{[\mu_1,\nu_1]}, \alpha_{[\mu_2,\nu_2]})$ as $\alpha_{[\mu_1,\nu_1]} \circ \alpha_{[\mu_2,\nu_2]}$. The presheaf $\underline{K}$ is clearly closed under such additive structure. The inverse is defined as follows:

Definition 20.8 For each $V \in \mathcal{V}(\mathcal{N})$ we have

$$-_V : \underline{K}_V \to \underline{K}_V \tag{20.17}$$

$$\alpha_{[\mu_1,\nu_1]} \mapsto -_V(\alpha_{[\mu_1,\nu_1]}) := \alpha_{-[\mu_1,\nu_1]} = \alpha_{[\nu_1,\mu_1]}. \tag{20.18}$$

The unit element at each V is defined as $\alpha_{[0,0]}$ where each (0) is the constant map with value 0, hence it is both order reversing and order preserving. We now want to show that the group axioms hold.

Associativity

$$(\alpha_{[\mu_1,\nu_1]} \circ \alpha_{[\mu_2,\nu_2]}) \circ \alpha_{[\mu_3,\nu_3]} = \alpha_{[\mu_1+\mu_2,\nu_1+\nu_2]} \circ \alpha_{[\mu_3,\nu_3]} = \alpha_{[\mu_1+\mu_2+\mu_3,\nu_1+\nu_2+\nu_3]}. \tag{20.19}$$

On the other hand

$$\alpha_{[\mu_1,\nu_1]} \circ (\alpha_{[\mu_2,\nu_2]} \circ \alpha_{[\mu_3,\nu_3]}) = \alpha_{[\mu_1,\nu_1]} \circ \alpha_{[\mu_2+\mu_3,\nu_2+\nu_3]} = \alpha_{[\mu_1+\mu_2+\mu_3,\nu_1+\nu_2+\nu_3]}. \tag{20.20}$$

Identity Axiom

$$\alpha_{[\mu_1,\nu_1]} \circ \alpha_{[0,0]} = \alpha_{[\mu_1+0,\nu_1+0]} = \alpha_{[0,0]} \circ \alpha_{[\mu_1,\nu_1]}. \tag{20.21}$$

Inverse Axiom

$$\alpha_{[\mu_1,\nu_1]} \circ \alpha_{-[\mu_1,\nu_1]} = \alpha_{[\mu_1-\mu_1,\nu_1-\nu_1]} = \alpha_{[0,0]}. \tag{20.22}$$

Moreover it is straight $\underline{K}$ is abelian.[2] In fact

$$\alpha_{[\mu_1,\nu_1]} \circ \alpha_{[\mu_2,\nu_2]} = \alpha_{[\mu_1+\mu_2,\nu_1+\nu_2]} = \alpha_{[\mu_2+\mu_1,\nu_2+\nu_1]} = \alpha_{[\mu_2,\nu_2]} \circ \alpha_{[\mu_1,\nu_1]}. \tag{20.23}$$

From the above it follows that:

[2]Essentially the abelian property is inherited by the group $K(\underline{\mathbb{R}^{\geq}})$.

Proposition 20.1 *The abelian group $\underline{K}$ is a one parameter group with the parameter taking values in $k(\underline{\mathbb{R}}^{\geq})$.*

Proof To prove the above theorem we need to define a continuous group homomorphism between $k(\underline{\mathbb{R}}^{\geq})$ and $\underline{K}$. This, for each $V \in \mathcal{V}(\mathcal{H})$, is defined as follows

$$p_V : k(\underline{\mathbb{R}}^{\geq})_V \to \underline{K}_V \tag{20.24}$$

$$[\mu, \nu] \mapsto \alpha_{[\mu,\nu]}. \tag{20.25}$$

Recalling that $k(\underline{\mathbb{R}}^{\geq})$ is equipped with the additive operation $+ : k(\underline{\mathbb{R}}^{\geq}) \times k(\underline{\mathbb{R}}^{\geq}) \to k(\underline{\mathbb{R}}^{\geq})$, for each context $V \in \mathcal{V}(\mathcal{H})$ we have

$$+_V : k(\underline{\mathbb{R}}^{\geq})_V \times k(\underline{\mathbb{R}}^{\geq})_V \to k(\underline{\mathbb{R}}^{\geq})_V \tag{20.26}$$

$$([\mu_1, \nu_1], [\mu_2, \nu_2]) \mapsto +_V([\mu_1, \nu_1], [\mu_2, \nu_2]) := [\mu_1 + \mu_2, \nu_1 + \nu_2]. \tag{20.27}$$

Thus p_V is clearly a homomorphisms.

In order to prove continuity we need to equip both sheaves with a topology. We simply choose the discrete topology. This makes the above map p_V continuous. □

We now consider the presheaf $\underline{\mathbb{R}}$ which is defined as follows:

Definition 20.9 The presheaf $\underline{\mathbb{R}} \in \mathbf{Sets}^{\mathcal{V}(\mathcal{H})}$ is defined on

1. Objects: for each $V \in \mathcal{V}(\mathcal{H})$ we obtain $\underline{\mathbb{R}}_V = \mathbb{R}$.
2. Morphisms: given the inclusion $i : V' \subseteq V$ the corresponding presheaf map is simply the identity map.

It is possible to embed $\underline{\mathbb{R}}$ in $\underline{\mathbb{R}}^{\geq}$ since each real number $r \in \underline{\mathbb{R}}_V$ can be identified with the constant function $c_{r,V} : \downarrow V \to \mathbb{R}$, which has constant value r for all $V \in \mathcal{V}(\mathcal{H})$. Such a function is trivially order-reversing, hence it is an element of $\underline{\mathbb{R}}^{\geq}$. Moreover, the global sections of $\underline{\mathbb{R}}$ are given by constant functions $r : \mathcal{V}(\mathcal{H}) \to \mathbb{R}$ which are also global sections of $\underline{\mathbb{R}}^{\geq}$, thus $\underline{\mathbb{R}} \subset \underline{\mathbb{R}}^{\geq}$. However $\underline{\mathbb{R}}^{\geq}$ can be seen as a sub-object of $k(\underline{\mathbb{R}}^{\geq})$ by sending each $\mu \in \underline{\mathbb{R}}^{\geq}_V$ to $[\mu, 0] \in k(\underline{\mathbb{R}}^{\geq}_V)$, thus $\underline{\mathbb{R}} \subseteq k(\underline{\mathbb{R}}^{\geq})$. Our claim is that $\underline{H}$ as defined in Definition 20.1 is isomorphic to the subgroup of $\underline{K}$ generated by $\underline{\mathbb{R}} \subset k(\underline{\mathbb{R}}^{\geq})$.

Theorem 20.1 *The group $\underline{H}$ is isomorphich to the one-parameter subgroup of $\underline{K}$ generated by $\underline{\mathbb{R}}$.*

Proof We want to show that there exists a functor

$$f : \underline{\mathbb{R}} \to \underline{K} \tag{20.28}$$

such that for each context V, f_V is a continuous injective group homomorphism and $Im(f) \simeq \underline{H} \subset \underline{K}$.

We thus define, for each V

$$f_V : \underline{\mathbb{R}}_V \to \underline{K}_V \tag{20.29}$$

$$r \mapsto \alpha_{[c_{r,V},0]}. \tag{20.30}$$

This is a well defined functor since for $V' \subseteq V$ the following diagram commutes

$$\begin{array}{ccc} \underline{\mathbb{R}}_V & \xrightarrow{f_V} & \underline{K}_V \\ \Big\downarrow{\scriptstyle \underline{\mathbb{R}}(i_{V'V})} & & \Big\downarrow{\scriptstyle \underline{K}(i_{V'V})} \\ \underline{\mathbb{R}}_{V'} & \xrightarrow[f_{V'}]{} & \underline{K}_{V'} \end{array}$$

In fact we have

$$\underline{K}(i_{V'V}) \circ f_V(r) = \underline{K}(i_{V'V})\alpha_{[c_{r,V},0]} = \alpha_{[c_{r,V'},0]} \tag{20.31}$$

while

$$f_{V'} \circ \underline{\mathbb{R}}(i_{V'V})(r) = f_{V'}(r) = \alpha_{[c_{r,V'},0]}. \tag{20.32}$$

Clearly f_V is injective and continuous on the image with respect to the discrete topologies.

We now need to check whether f is a group homomorphism, i.e. we need to show that $f_V(r_1 + r_2) = f_V(r_1) + f_V(r_2)$. We know that the left hand side is

$$f_V(r_1 + r_2) = \alpha_{[c_{(r_1+r_2),V},0]}. \tag{20.33}$$

However

$$c_{(r_1+r_2),V} : \downarrow V \to \mathbb{R} \tag{20.34}$$

$$V' \mapsto c_{(r_1+r_2),V}\big(V'\big) = r_1 + r_2 \tag{20.35}$$

while

$$+_V(c_{r_1,V}, c_{r_2,V}) : \downarrow V \to \mathbb{R} \tag{20.36}$$

$$V' \mapsto c_{r_1,V}\big(V'\big) + c_{r_2,V}\big(V'\big) = r_1 + r_2. \tag{20.37}$$

Hence

$$c_{(r_1+r_2),V} = c_{r_1,V} + c_{r_2,V} \tag{20.38}$$

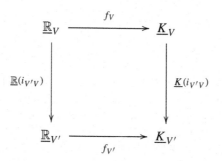

which implies that

$$\alpha_{[c_{(r_1+r_2)},V,0]}=\alpha_{[c_{r_1},V+c_{r_2},V,0]}=\alpha_{[c_{r_1},V,0]}\circ\alpha_{[c_{r_2},V,0]}=f_V(r_1)+f_V(r_2). \quad (20.39)$$

We now want to show that $im(f)\simeq \underline{H}$. We therefore construct the map $i: im(f)\to \underline{H}$, such that for each V we obtain

$$i_V: im(f)_V \to \underline{H}_V \quad (20.40)$$

$$\alpha_{[c_{r,V},0]}\mapsto i_V(\alpha_{[c_{r,V},0]}):=\alpha\big(c_{r,V}(V)\big). \quad (20.41)$$

This is clearly an isomorphism.[3] □

A real number $r\in\underline{\mathbb{R}}_V$ defines the pair $(c_{r,V},c_{r,V})$ given by two copies of the constant function $c_{r,V}:\downarrow V\to\mathbb{R}$. Clearly such a function is both order-preserving and order-reversing, hence $(c_{r,V},c_{r,V})\in\underline{\mathbb{R}}^{\leftrightarrow}$. However $\underline{\mathbb{R}}^{\leftrightarrow}\subset k(\underline{\mathbb{R}}^{\leftrightarrow})$, therefore $\underline{\mathbb{R}}\subset k(\underline{\mathbb{R}}^{\leftrightarrow})$. All the results proved for $\underline{\mathbb{R}}^{\succeq}$ hold for $\underline{\mathbb{R}}^{\leftrightarrow}$, however, since the constructions for $\underline{\mathbb{R}}^{\leftrightarrow}$ are more cumbersome, we will avoid reporting them here.

20.1.2 One Parameter Group Taking Values in Complex Number Object

We would now like to apply the same analysis but for the complex number object $\underline{\mathbb{C}}^{\leftrightarrow}$. As before, for simplicity of notation, we will consider the object $\underline{\mathbb{C}}^{\succeq}$ (defined below). All results will then translate in a simple way to $\underline{\mathbb{C}}^{\leftrightarrow}$.

Definition 20.10 The presheaf $\underline{\mathbb{C}}^{\succeq}\in\mathbf{Sets}^{\mathcal{V}(\mathcal{H})}$ is defined on

1. Objects: for each $V\in\mathcal{V}(\mathcal{H})$ we obtain the set $\underline{\mathbb{C}}^{\succeq}_V:=\{\mu|\text{ s.t. }\mu\in OR(\downarrow V,\mathbb{C})\}$.
2. Morphisms: given $i_{V'V}:V'\subseteq V$ the presheaf morphism is defined by the restriction: $\underline{\mathbb{C}}^{\succeq}_V\to\underline{\mathbb{C}}^{\succeq}_{V'}$; $\mu\mapsto\mu_{|V'}$.

We then define the k-extension $k(\underline{\mathbb{C}}^{\succeq})$ as follows:

Definition 20.11 The presheaf $k(\underline{\mathbb{C}}^{\succeq})\in\mathbf{Sets}^{\mathcal{V}(\mathcal{H})}$ is defined on:

1. Objects: for each $V\in\mathcal{V}(\mathcal{H})$ we obtain

$$k\big(\underline{\mathbb{C}}^{\succeq}\big)_V:=\big\{[\mu,\nu]|\mu,\nu\in OR(\downarrow V,\mathbb{C})\big\} \quad (20.44)$$

[3] We could have defined the map

$$h_V:\underline{K}_V\to\underline{H}_V \quad (20.42)$$

$$\alpha_{[\mu,\nu]}\mapsto\alpha_\rho\big(\mu(V)+\nu(V)\big) \quad (20.43)$$

but this would not have been 1:1.

where $[\mu, \nu]$ denotes the k-equivalence class of (μ, ν).

2. Morphisms: for each $i_{V'V} : V' \subseteq V$, we obtain the arrow $k(\underline{\mathbb{C}}^{\geq})(i_{V'V}) : k(\underline{\mathbb{C}}^{\geq})_V \to k(\underline{\mathbb{C}}^{\geq})_{V'}$ defined by

$$k\big(\underline{\mathbb{C}}^{\geq}\big)(i_{V'V})[\mu, \nu] := [\mu_{|V'}, \nu_{|V'}] \quad \text{for all } [\mu, \nu] \in k\big(\underline{\mathbb{C}}^{\geq}\big)_V. \tag{20.45}$$

It is interesting to note how in $k(\underline{\mathbb{C}}^{\geq})$ it is possible to define complex conjugation. In particular we define, for each context V

$$*_V : k\big(\underline{\mathbb{C}}^{\geq}\big)_V \to k\big(\underline{\mathbb{C}}^{\geq}\big)_V \tag{20.46}$$

$$[\mu, \nu] \mapsto \big[\mu^*, \nu^*\big] \tag{20.47}$$

where $\mu^*(V) := (\mu(V))^*$. Now if $(\mu, \nu) \simeq (\eta, \beta)$ then there exists an element $\gamma \in k(\underline{\mathbb{C}}^{\geq})_V$ such that $\mu + \beta + \gamma = \nu + \eta + \gamma$, which is defined for each $V' \in \downarrow V$ as $\mu(V') + \beta(V') + \gamma(V') = \nu(V') + \eta(V') + \gamma(V')$, thus obtaining the equality of complex numbers $(a + b + c) + i(d + e + f) = (a_1 + b_1 + c) + i(d_1 + e_1 + f)$. Applying the complex conjugation map we obtain $*_V[\mu, \nu] = [\mu^*, \nu^*]$ such that $(\mu^*, \nu^*) \simeq (\eta^*, \beta^*)$ iff $\mu^* + \beta^* + \gamma^* = \nu^* + \eta^* + \gamma^*$ which, by applying the same reasoning, translates to $(a+b+c) - i(d+e+f) = (a_1+b_1+c) - i(d_1+e_1+f)$. It follows that if $(\mu, \nu) \in [\mu, \nu]$ then $(\mu^*, \nu^*) \in [\mu^*, \nu^*]$. Thus $*_V$ is well defined.

We now want to show that $k(\mathbb{C}^{\geq})$ is a vector space over $\underline{R}$. To this end we need to define multiplication with respect to an element in $\underline{\mathbb{R}}$. We recall that each element $r \in \mathbb{R}$ is represented in $\underline{\mathbb{R}}$ as the global element $c_r \in \Gamma(\underline{\mathbb{R}})$ which, at each context $V \in \mathcal{V}(\mathcal{H})$, defines a constant function $c_{r,V} : \downarrow V \to \mathbb{R}$. We can then define multiplication with respect to such constant functions. Thus, given a context $V \in \mathcal{V}(\mathcal{H})$ we consider an element $[\mu, \nu] \in k(\underline{\mathbb{C}}^{\geq})_V$, and we define multiplication by $c_{r,V}$ as

$$\big(c_{r,V}[\mu, \nu]\big) := \begin{cases} [c_{r,V}\mu, c_{r,V}\nu] = [r\mu, r\nu] & \text{if } r \geq 0 \\ -[c_{-r,V}\mu, c_{-r,V}\nu] = [-r\nu, -r\mu] & \text{if } r < 0 \end{cases} \tag{20.48}$$

where for each $V' \in \downarrow V$ the above is defined as

$$\big(c_{r,V}[\mu, \nu]\big)\big(V'\big) = c_{r,V}\big(V'\big)\big[\mu\big(V'\big), \nu\big(V'\big)\big] = \big[r\mu\big(V'\big), r\nu\big(V'\big)\big] = [r\mu, r\nu]\big(V'\big) \tag{20.49}$$

for $r \geq 0$, while for $r < 0$ we have

$$\begin{aligned} \big(c_{r,V}[\mu, \nu]\big)\big(V'\big) &= c_{r,V}\big(V'\big)\big[\mu\big(V'\big), \nu\big(V'\big)\big] \\ &= -|r|\big[\mu\big(V'\big), \nu\big(V'\big)\big] = |r|\big[\nu\big(V'\big), \mu\big(V'\big)\big] \qquad (20.50) \\ &= -[c_{-r,V}\mu, c_{-r,V}\nu]\big(V'\big) = [-r\nu, -r\mu]\big(V'\big). \qquad (20.51) \end{aligned}$$

Similarly we can also define multiplication with respect to a constant complex number. In fact, given a complex number $z = x + iy \in \mathbb{C}$ this represents a global element in $\Gamma(\underline{\mathbb{C}})$ such that, for each context $V \in \mathcal{V}(\mathcal{H})$, we obtain the constant function $c_{z,V} : \downarrow V \to \mathbb{C}$. Thus, given an element $[\mu, \nu] \in k(\underline{\mathbb{C}}^{\geq})_V$, when $x + iy \geq 0$ with respect to the ordering defined in Definition 18.3, we define, for each $V' \in \downarrow V$

$$\big(c_{z,V}[\mu,\nu]\big)(V') = c_{z,V}(V')\big[\mu(V'), \nu(V')\big] = \big[z\mu(V'), z\nu(V')\big] = [z\mu, z\nu](V'). \tag{20.52}$$

On the other hand, when $x + iy < 0$ we have

$$\begin{aligned}\big(c_{z,V}[\mu,\nu]\big)(V') &= c_{z,V}(V')\big[\mu(V'), \nu(V')\big]\\ &= -\big[|z|\mu(V'), |z|\nu(V')\big] = [-z\nu, -z\mu](V').\end{aligned} \tag{20.53}$$

It is straight forward to see how this definition reduces to definition (20.48).

We now define the presheaf $\underline{Q}$ as follows:

Definition 20.12 The presheaf $\underline{Q} \in \mathbf{Sets}^{\mathcal{V}(\mathcal{H})}$ is defined on

1. Objects: for each $V \in \mathcal{V}(\mathcal{H})$ we obtain the set $\underline{Q}_V := \{\alpha_{[\mu,\nu]} | [\mu,\nu] \in k(\underline{\mathbb{C}}^{\geq})_V\}$.[4]
2. Morphisms: Given the inclusion map $V' \subseteq V$ the corresponding presheaf map is $\underline{Q}(i_{V'V}) : \underline{Q}_V \to \underline{Q}_{V'}; \alpha_{[\mu,\nu]} \mapsto \alpha_{[\mu_{|V'},\nu_{|V'}]}$.

$\underline{Q}$ can be given a group structure in exactly the same way as was done for $\underline{K}$. It then follows that $\underline{Q}$ is the one parameter group defined via the group homomorphisms $h : k(\underline{\mathbb{C}}^{\geq}) \to \underline{Q}$, which has components for each context

$$h_V : k(\underline{\mathbb{C}}^{\geq})_V \to \underline{Q}_V \tag{20.54}$$

$$[\mu,\nu] \mapsto \alpha_{[\mu,\nu]}. \tag{20.55}$$

Proposition 20.2 *The group $\underline{K}$ is a subgroup of $\underline{Q}$.*

Proof $\underline{K}$ is the one-parameter subgroup generated by $k(\underline{\mathbb{R}}^{\geq}) \subset k(\underline{\mathbb{C}}^{\geq})$. In fact we have the following continuous group homomorphisms for each V:

$$k(\underline{\mathbb{R}}^{\geq})_V \to k(\underline{\mathbb{C}}^{\geq})_V \to \underline{Q}_V \tag{20.56}$$

$$[\mu,\nu] \mapsto [\mu + i0, \nu + i0] \mapsto \alpha_{[\mu+i0,\nu+i0]} = \alpha_{[\mu,\nu]} \tag{20.57}$$

where again we are assuming the discrete topology. □

We now analyse the relation between $\underline{\mathbb{C}}$ and $k(\underline{\mathbb{C}}^{\geq})$. In particular, similarly to the real number object, also in this case we have that $\underline{\mathbb{C}} \subset k(\underline{\mathbb{C}}^{\geq})$. This inclusion is given by the following chain of inclusions for each V:

$$\gamma_V : \underline{\mathbb{C}}_V \to \underline{\mathbb{C}}^{\geq}_V \to k(\underline{\mathbb{C}}^{\geq})_V \tag{20.58}$$

[4] Similarly as for $\underline{K}$ even here the elements $\alpha_{[\mu,\nu]}$ should be understood as automorphisms on $\mathcal{H}$ parameterised by $[\mu,\nu]$.

$$t \mapsto c_{t,V} \to [c_{t,V}, 0]. \tag{20.59}$$

The proof is straightforward. As was done for the real valued number case we would like to define the topos analogue of the group $R := \{\alpha(a+ib)|a+ib \in \mathbb{C}\}$, which takes values in $\mathbb{C}$. We first construct the following presheaf:

Definition 20.13 The presheaf $\underline{R} \in \mathbf{Sets}^{\mathcal{V}(\mathcal{H})}$ is defined on:

1. Objects: for each $V \in \mathcal{V}(\mathcal{H})$ we obtain the set $\underline{R}_V := \{\alpha(a+ib)|a+ib \in \mathbb{C}\}$.
2. Morphisms: for any map $i_{V'V} : V' \subseteq V$, $\underline{R}(i_{V'V})$ is simply the identity.

The fact that this presheaf is a group comes from the fact that for each V, $\underline{R}_V$ is a group. We now would like to show that such a group object is a one parameter subgroup of $\underline{Q}$ taking its values in $\underline{\mathbb{C}}$.

To this end we construct the map $\phi : \underline{\mathbb{C}} \to \underline{Q}$, whose definition requires the factorisation via the maps γ and h above. Thus, for each V, we have

$$\phi_V : \underline{\mathbb{C}}_V \to \underline{Q}_V \tag{20.60}$$

$$t \mapsto \phi_V(t) := h_V \circ \gamma_V(t) = \alpha_{[c_{t,V},0]}. \tag{20.61}$$

Clearly such a map is injective. We need to show that it is also an homomorphism. In particular we need to show that

$$\phi_V(t_1 + t_2) = \phi_V(t_1) \circ \phi_V(t_2). \tag{20.62}$$

By applying the definition we have

$$\phi_V(t_1 + t_2) = \alpha_{[c_{t_1+t_2},V,0]} = \alpha_{[c_{t_1},V,0]} \circ \alpha_{[c_{t_2},V,0]} \tag{20.63}$$

where the last equation follows from the group laws in $\underline{Q}_V$.

On the other hand

$$\phi_V(t_1) \circ \phi_V(t_2) = \alpha_{[c_{t_1},V,0]} \circ \alpha_{[c_{t_2},V,0]}. \tag{20.64}$$

We thus obtain the one parameter subgroup of $\underline{Q}$ as the image of ϕ, i.e. $im(\phi) \subset \underline{Q}$.

We can then define the map $m : im(\phi) \to \underline{R}$ such that for each context V we have

$$m_V : im(\phi)_V \to \underline{R}_V \tag{20.65}$$

$$\alpha[c_{t_2,V}, 0] \mapsto m_V(\alpha_{[c_{t_2},V,0]}) := \alpha\big(c_{t_2,V}(V)\big). \tag{20.66}$$

This is clearly an isomorphism.

20.2 Stone's Theorem in the Language of Topos Theory

In the previous section we managed to define the topos analogue of the one parameter group of transformations. Since we are also able to define the topos analogue of self-adjoint operators, it is natural to ask whether it is possible to formulate Stone's theorem in the language of topos theory. The "standard" definition of Stone's theorem is the following:

Theorem 20.2 *Every strongly continuous*[5] *one-parameter group* $\{U_t\}$, $(-\infty < t < \infty)$ *of unitary transformations admits a spectral representation*

$$U_t = \int_{\infty}^{\infty} e^{i\lambda t} d\hat{E}_\lambda \tag{20.67}$$

where $\{\hat{E}_\lambda\}$ *is the spectral family such that*[6] $(\hat{E}_\lambda)\smile\smile\{U_t\}$.

Equivalently one can write $U_t = e^{i\lambda t\hat{A}}$ for the self adjoint operator

$$\hat{A} = \int_{\infty}^{\infty} \lambda \hat{E}_\lambda. \tag{20.68}$$

We are now interested in translating the above theorem into the topos language, that is, we are interested in finding a correspondence between self-adjoint operators $\breve{\delta}(\hat{A})$ and unitary one parameter groups.

First of all we need to specify what a unitary one parameter group is in a topos. We already have the definition of a one parameter subgroup, thus all we need to do is to add the property of unitarity. We thus consider the one parameter group $U := \{\alpha(t)|t \in \mathbb{R} \,\&\, \alpha(t)\alpha(-t) = 1\}$ of transformations on $\mathcal{H}$. These transformations can be extended to functors:

$$\alpha(t) : \mathcal{V}(\mathcal{H}) \to \mathcal{V}(\mathcal{H}) \tag{20.69}$$

$$V \mapsto l_{\alpha(t)}V := \big\{\alpha(t)\hat{A}\alpha(-t)|\hat{A} \in V\big\}. \tag{20.70}$$

We then define the associated presheaf $\underline{U}$ which has, as objects, for each $V \in \mathcal{V}(\mathcal{H})$ the entire group $\underline{U}_V = U$. The maps are simply the identity maps. The group $\underline{U}$ represents a one parameter sub-group of $\underline{K}$ of unitary transformations. The proof is similar as the proof given above for the subgroup $\underline{H}$ while the unitarity is derived directly from U. We should also add the property of strong continuity which, in terms of operators, can be stated as follows: for any $\hat{A}$ and $t \to t_0$ then $\alpha(t)\hat{A}\alpha(-t) \to \alpha(t_0)\hat{A}\alpha(-t_0)$.

Given such a strongly continuous one-parameter sub-group of transformations we want to somehow define a unique self-adjoint operator associated to it and, vice versa, given a self adjoint operator we want to associate to it a unique strongly continuous one-parameter sub-group of transformations. We will start from the latter. Since we will be employing group transformations we need to work with the sheaves

[5]Here strongly continuous means that, for any $\psi \in \mathcal{H}$ and $t \to t_0$, then $U_t(\psi) \to U(t_0)(\psi)$.

[6]We will now introduce the following notations (i) $\hat{A}\smile\hat{B}$ indicates that $\hat{A}$ and $\hat{B}$ commute; (ii) $\hat{A}\smile\smile\hat{B}$ means that $\hat{A}$ commutes with $\hat{B}$ and any other operator which commutes with $\hat{B}$.

$\underline{\breve{\Sigma}}$ and $\underline{\breve{\mathbb{R}}^{\leftrightarrow}}$ which are defined using the method introduced in [45] and described in details in Chap. 16. In this case, given the presheaf $\underline{K}$ we define the presheaf $\underline{K/K_F}$ as follows

Definition 20.14 The presheaf $\underline{K/K_F} \in \mathbf{Sets}^{\mathcal{V}_f(\mathcal{H})}$[7] is defined on

1. Objects: for each $V \in \mathcal{V}_f(\mathcal{H})$ we obtain the set $K/K_{FV} = \{[g]_V | g \sim g_1$ iff $hg_1 = g$ for $h \in K_{FV}\}$ where K_{FV} is the fixed point group of V.[8]
2. Morphisms: given a morphism $i_{V'V} : V' \subseteq V$ the corresponding presheaf morphism is the map $K/K_{FV} \to K/K_{FV'}$, defined as the bundle map of the bundle $K_{FV'}/K_{FV} \to K/K_{FV} \to K/K_{FV'}$.

From the above presheaf we obtain the associated étale bundle $p : \Lambda(\underline{K/K_F}) \to \mathcal{V}_f(\mathcal{N})$, whose bundle space $\Lambda(\underline{K/K_F})$ can be given a poset structure as follows:

Definition 20.15 Given two elements $[g]_{V'} \in K/K_{FV'}$, $[g]_V \in K/K_{FV}$ we define the partial ordering by

$$[g]_{V'} \leq [g]_V \quad \text{iff} \quad p\big([g]_{V'}\big) \subseteq p\big([g]_V\big) \quad \text{and} \quad [g]_V \subseteq [g]_{V'}. \tag{20.71}$$

Next one defines the functor $I : Sh(\mathcal{V}(\mathcal{H})) \to Sh(\Lambda(\underline{K/K_F}))$ as follows:

Theorem 20.3 *The map $I : Sh(\mathcal{V}(\mathcal{H})) \to Sh(\Lambda(\underline{K/K_F}))$ is a functor defined on*

(i) *Objects*: $(I(\underline{A}))_{[g]_V} := \underline{A}_{l_g(V)} = ((l_g)^*(\underline{A}))(V)$. *If* $[g]_{V'} \leq [g]_V$, *then*

$$\big(I\underline{A}(i_{[g]_{V'},[g]_V})\big) := \underline{A}_{l_g(V),l_g(V')} : \underline{A}_{l_g(V)} \to \underline{A}_{l_g(V')}$$

where $V = p([g]_V)$ *and* $V' = p([g]_{V'})$.

(ii) *Morphisms*: *given a morphism* $f : \underline{A} \to \underline{B}$ *in* $Sh(\mathcal{V}(\mathcal{H}))$ *we then define the corresponding morphism in* $Sh(\Lambda(\underline{K/K_F}))$ *as*

$$I(f)_{[g]_V} : I(\underline{A})_{[g]_V} \to I(\underline{B})_{[g]_V} \tag{20.72}$$

$$f_{[g]_V} : \underline{A}_{l_g(p([g]_V))} \to \underline{B}_{l_g(p([g]_V))}. \tag{20.73}$$

Such a functor was already defined in [45] where it was shown to be a functor. We are now able to map all the sheaves in $Sh(\mathcal{V}(\mathcal{H}))$ to sheaves in $Sh(\Lambda(\underline{K/K_F}))$. By then applying the functor $p! : Sh(\Lambda(\underline{K/K_F})) \to Sh(\mathcal{V}_f(\mathcal{H}))$ we finally obtain sheaves on our fixed category $\mathcal{V}_f(\mathcal{H})$. The advantage of this construction is that

[7] Recall that $\mathcal{V}_f(\mathcal{H})$ is the poset $\mathcal{V}(\mathcal{H})$ but were the group is not allowed to act.

[8] Recall that, given a context V, the fixed point group K_{FV} is defined as $K_{FV} := \{g \in K | \forall \hat{A} \in V g\hat{A}g^{-1} = \hat{A}\}$.

now the actions of both groups $\underline{K}$ and $\underline{H}$ do not induce twisted presheaves (see Chap. 16). In this context, self adjoint operators are defined as

$$\overline{\delta}(\hat{A})_V : \coprod_{g\in K/K_{FV}} \underline{\Sigma}_{l_g V} \to \coprod_{g\in K/K_{FV}} \underline{\mathbb{R}}_{l_g V} \tag{20.74}$$

such that for $\lambda \in \underline{\Sigma}_{l_g V}$ then $\overline{\delta}(\hat{A})_V(\lambda) := \breve{\delta}(\hat{A})_{l_g V}\lambda$, i.e. as co-products of the originally defined self-adjoint operators $\breve{\delta}(\hat{A})$. By denoting all the maps $\underline{\breve{\Sigma}} \to \underline{\breve{\mathbb{R}}}^{\leftrightarrow}$ by $(\underline{\breve{\mathbb{R}}}^{\leftrightarrow})^{\underline{\breve{\Sigma}}}$ we can then define the sub-object $\underline{Ob} \subseteq (\underline{\breve{\mathbb{R}}}^{\leftrightarrow})^{\underline{\breve{\Sigma}}}$ of observables, i.e. all maps $\overline{\delta}(\hat{A}) : \underline{\breve{\Sigma}} \to \underline{\breve{\mathbb{R}}}^{\leftrightarrow}$ associated to self adjoint operators $\hat{A}$. Next we define the collection of strongly continuous unitary subgroups of $\underline{K}$ which we denote $Sub_u(\underline{K})$. Given these ingredients we attempt the partial definition of Stone's theorem

Theorem 20.4 *The map $f : \underline{Ob} \to Sub_u(\underline{K})$ defined for each $V \in \mathcal{V}(\mathcal{H})$ as*

$$f_V : \underline{Ob}_V \to Sub_u(\underline{K})_V \tag{20.75}$$

$$\overline{\delta}(\hat{A})_{|\downarrow V} \mapsto \underline{U}_V^{\hat{A}} \tag{20.76}$$

where $\underline{U}_V^{\hat{A}} := \{e^{it\hat{A}} | t \in \mathbb{R}\}$ is injective.

The proof of injectivity is trivial, thus what remains to show is that indeed $\underline{U}_V^{\hat{A}}$, as defined above, is a strongly continuous unitary subgroup of $\underline{K}$. The proof is again similar to the one done for the sub-group $\underline{H}$, so we will not report it here. We now come to the more interesting part of Stone's theorem, namely showing that any strongly continuous unitary subgroups of $\underline{K}$ uniquely determines a self-adjoint operator. To this end we first construct the map $g : Sub_u(\underline{K}) \to \underline{Ob}$ such that, for each $V \in \mathcal{V}(\mathcal{H})$

$$g_V : Sub_u(\underline{K})_V \to \underline{Ob}_V \tag{20.77}$$

$$\underline{U}_V \mapsto g_V(\underline{U}_V) := \overline{\delta}(\hat{A}^U)_{|\downarrow V}. \tag{20.78}$$

The self-adjoint operator $\overline{\delta}(\hat{A}^Q)_{|\downarrow V}$ is defined by the following properties:

(a) For all $\alpha(t) \in \underline{U}_V$, the diagram

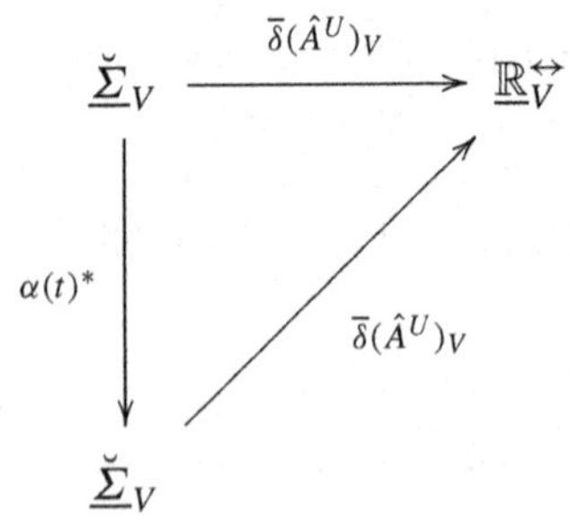

commutes. What this means is that, given an element $\lambda \in \underline{\Sigma}_V \in \coprod_{g \in K/K_{FV}} \underline{\Sigma}_{l_g V}$ we require

$$(\overline{\delta}(\hat{A}^U)_V \circ \alpha(t)^*)\lambda = (\overline{\delta}(\hat{A}^U)_V) l_{\alpha(t)}\lambda = \breve{\delta}(\hat{A}^U)_{l_{\alpha(t)}V}(l_{\alpha(t)}\lambda) \tag{20.79}$$

to be equal to

$$\overline{\delta}(\hat{A}^U)_V(\lambda) = \breve{\delta}(\hat{A}^U)(\lambda). \tag{20.80}$$

We can generalise such a condition for all elements of U at once by requiring that the following diagram commutes for all $V \in \mathcal{V}(\mathcal{H})$:

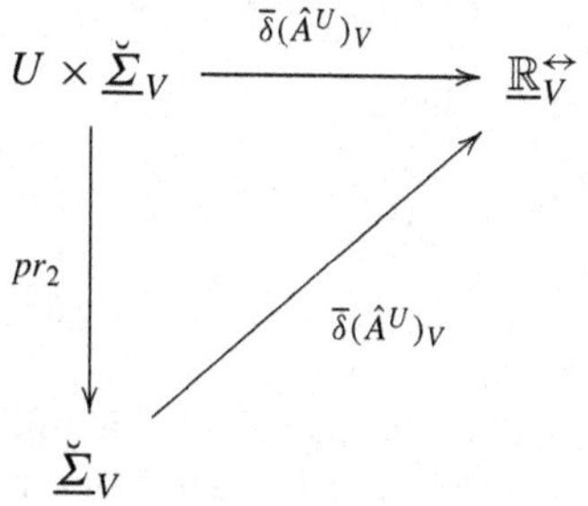

(b) For any $\hat{B}$ such that $\hat{B} \smile \hat{A}^U$ then

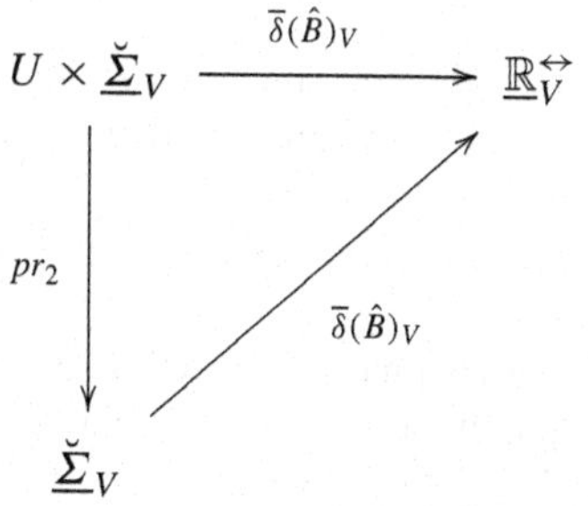

The correspondence between strongly continuous unitary groups and self adjoint operators is given by they following theorem:

Theorem 20.5 *The map g is injective.*

Before proving the theorem let us first analyse in more details what the two conditions (a) and (b) imply. To this end we introduce the following corollary

Corollary 20.1 *Given a self adjoint operator $\hat{A}$ with spectral projection $\{\hat{E}_\lambda\}$ and any $\alpha(t) \in U$, then*

$$\{\hat{E}_\lambda\} \smile \{\alpha(t)\} \quad \Leftrightarrow \quad \hat{A} \textit{ satisfies condition } (a). \tag{20.81}$$

On the other hand

$$\{\hat{E}_\lambda\}\smile\smile\{\alpha(t)\} \quad \Leftrightarrow \quad \hat{A} \textit{ satisfies condition } \text{(a) \& (b)}. \tag{20.82}$$

We now prove the above corollary.

Proof We assume that $\hat{A}$ satisfies condition (a), i.e. $(\overline{\delta}(\hat{A})_V \circ \alpha(t)^*)\lambda = \overline{\delta}(\hat{A})_V(\lambda)$. We recall that

$$\begin{aligned}(\overline{\delta}(\hat{A})_V \circ \alpha(t)^*)\lambda(\cdot) &= \breve{\delta}(\hat{A})_{l_{\alpha(t)}V}(l_{\alpha(t)}\lambda)(\cdot)\\ &= \big(\breve{\delta}^i(\hat{A})_{l_{\alpha(t)}V}(l_{\alpha(t)}\lambda), \breve{\delta}^o(\hat{A})_{l_{\alpha(t)}V}(l_{\alpha(t)}\lambda)\big)(\cdot)\end{aligned} \tag{20.83}$$

such that, for any $V' \in \downarrow V$ we have

$$\begin{aligned}\breve{\delta}^i(\hat{A})_{l_{\alpha(t)}V}(l_{\alpha(t)}\lambda)(V') &= \breve{\delta}^i(\hat{A})_{l_{\alpha(t)}V'}(l_{\alpha(t)}\lambda)\\ &= l_{\alpha(t)}\lambda\big(\delta^i(\hat{A})_{l_{\alpha(t)}V'}\big) = \lambda\delta\big(\alpha(-t)(\hat{A})\alpha(t)\big)_{V'}.\end{aligned} \tag{20.84}$$

Therefore for condition (a) to be satisfied it implies that $\{\hat{A}_\lambda\}\smile\{\alpha(t)\}$. Since $\{\hat{E}_\lambda\}\smile\smile\{\hat{A}\}$ it follows that $\{\hat{E}_\lambda\}\smile\{\alpha(t)\}$. The converse is trivial to prove. If we now assume that $\hat{A}$ also satisfies condition (b) we then have that, given any other operator $\hat{B}\smile\hat{A}$ then

$$\lambda\delta\big(\alpha(-t)(\hat{B})\alpha(t)\big)_{V'} = \lambda\delta(\hat{B})_{V'} \tag{20.85}$$

which implies that $\hat{B}\smile\{\alpha(t)\}$. But since $\hat{A}\smile\hat{B}$ and $\hat{A}\smile\{\alpha(t)\}$ it follows that $\hat{A}\smile\smile\{\alpha(t)\}$ hence $\{\hat{E}_\lambda\}\smile\smile\{\alpha(t)\}$. Again the converse is easy to prove. □

We are now ready to prove Theorem 20.5.

Proof We want to show that the map g is injective. In particular, given two strongly continuous unitary one parameter groups U and U', we want to show that if $g_V(U) = g_V(U')$ for all $V \in \mathcal{V}(\mathcal{H})$ then $U' = U$. Now if $g_V(U) = g_V(U')$ it follows that $\overline{\delta}(\hat{A}^U) = \overline{\delta}(\hat{A}^{U'}) = \overline{\delta}(\hat{A})$ are such that they satisfy conditions (a) and (b). Therefore, given the spectral family $\{\hat{E}_\lambda\}$ of $\hat{A}$, for each $\alpha(t) \in U$, $(\hat{E}_\lambda)\smile\smile\{\alpha(t)\}$ and similarly, for each $\beta(t) \in U'$, $(\hat{E}_\lambda)\smile\smile\{\beta(t)\}$. However this is precisely the condition for $\{\hat{E}_\lambda\}$ to be the spectral family of each $\alpha(t)$ and of each $\beta(t)$. It follows[9] that $\alpha(t) = \beta(t)$. □

Corollary 20.2 *$f \circ g = id_{Sub_u(\underline{K})}$ and $g \circ f = id_{\underline{Ob}}$.*

Proof We want to show that f and g are inverse of each other. First of all we recall that the composition of injective maps is itself an injective map, thus both $f \circ g$ and

[9] Recall that the spectral family is uniquely specified by the operator it decomposes.

$g \circ f$ are injective. We then consider the group $\underline{U}_V$ and apply the composite map $f_V \circ g_V$ for any $V \in \mathcal{V}(\mathcal{H})$, obtaining

$$f_V \circ g_V(\underline{U}_V) = f_V\big(\overline{\delta}\big(\hat{A}^U\big)\big) = \big\{e^{it\hat{A}^U} \big| t \in \mathbb{R}\big\}. \tag{20.86}$$

However, for the group $\{e^{it\hat{A}^U} | t \in \mathbb{R}\}$ we obtain

$$f_V \circ g_V\big(\big\{e^{it\hat{A}^U} \big| t \in \mathbb{R}\big\}\big) = f_V\big(\overline{\delta}\big(\hat{A}^U\big)\big) = \big\{e^{it\hat{A}^U} \big| t \in \mathbb{R}\big\} \tag{20.87}$$

and since $f_V \circ g_V$ is injective it follows that $\{e^{it\hat{A}^U} | t \in \mathbb{R}\} = \underline{U}_V$.

On the other hand

$$g_V \circ f_V\big(\overline{\delta}(\hat{A})\big) = g_V\big(\big\{e^{it\hat{A}} \big| t \in \mathbb{R}\big\}\big) = \overline{\delta}(\hat{A}). \tag{20.88}$$

□

Chapter 21
Future Research

In this chapter we would like to describe, briefly, the current topics of interest in the topos quantum theory field of research. Some of these topics have been partly investigated and are slowly developing, while others are still open issues which need to be addressed.

21.1 Quantisation

In Chap. 16 we have described a possible formalism for group actions in a topos. Recently, an alternative formulation was given in [97, 98]. In Chap. 16 we introduced all unitarily equivalent faithful 'representations' $\phi_g :\downarrow V \to \mathcal{V}(\mathcal{H})$ for all $g \in G$ and each algebra $V \in \mathcal{V}(\mathcal{H})$. These representations comprised the elements of our intermediate category.

Such a construct is reminiscent of the concept of unitary equivalent quantisations of a given classical system. It comes natural to ask wether it is possible to give a topos analogue of the concept of quantisation.

Bearing in mind that we are trying to avoid twisted presheaves, if we were to define the concept of quantisation, we should choose the base category to be a collection of labels of physical quantities with some structure and, such that, no group can act upon them. In this context quantisation would involve associating these labels with specific operators in a concrete Hilbert space.

The first issue would then be to define what sort of mathematical structure such a collection of labels should have. The choice will depend on whether or not we are considering an underlying classical system with a symplectic manifold S as state space or not. Assuming that, indeed, there is an underlying classical system, the base category can be chosen to be the poset of sub-algebras of $C^{\infty}(S, \mathbb{R})$[1] that are abelian, as computed with the Poisson bracket on S. The idea then is that each element of $C^{\infty}(S, \mathbb{R})$ is labelled by the physical quantity to which it corresponds, and

[1] When we consider $C^{\infty}(S, \mathbb{R})$ to be a Lie algebra we will use the notation $C^{\infty}_{Lie}(S, \mathbb{R})$.

C. Flori, *A First Course in Topos Quantum Theory*, Lecture Notes in Physics 868,
DOI 10.1007/978-3-642-35713-8_21,

that this labelling is fixed, i.e., one cannot start performing symplectic covariance transformations on the whole system.

Quantisation will then involve defining various 'topos representations' of this base poset. Of particular importance will be the construction of sheaves/presheaves over the poset, on which the Dirac 'covariance group', G, acts. This would be done in a similar way as done in Chap. 16, where the group acted only on the intermediate category $\Lambda(\underline{G/G_F})$. However, when following such a strategy for defining quantisation, certain issues need to be addressed.

1. The van Hove theorem shows that the naive interpretation of quantisation, namely associating commutators with Poisson brackets, is not possible. Thus it seems necessary to restrict $C^\infty_{Lie}(S,\mathbb{R})$ in some way. When S is finite-dimensional, and admits a finite-dimensional transitive group G of symplectic transformations,[2] then we will restrict our attention to the sub-algebra $L(G) \subseteq C^\infty_{Lie}(S,\mathbb{R})$ spanned by $L(G)$. The corresponding base category (poset) of abelian Lie sub-algebras of $Lie(G)$ will be denoted[3] PG. However, it will generally be the case that one needs to add 'by hand' certain physical quantities (e.g., the Hamiltonian), which are not already contained in $Lie(G)$. Of course, for some very limited situation this will not be the case, for example in the simple harmonic oscillator the Hamiltonian itself belongs to the Lie algebra.
   When there is no classical Lie algebra chosen, we will write the poset of labels as PS. In this situation, to get the poset structure it is necessary to use the Lie bracket on one particular quantisation. Thus $PS \simeq \mathcal{V}(\mathcal{H})$ although we should remember that the unitary group $U(\mathcal{H})$ does not act on PS but it does act on $\mathcal{V}(\mathcal{H})$.
2. It will be necessary to understand whether a topology should be assigned to $C^\infty_{Lie}(S,\mathbb{R})$. If this turns out to be the case, the question to ask will be, then, if the abelian sub-algebras should be closed subsets. Of course, for a finite-dimensional group G this problem does not arise.
3. Is $C^\infty_{Lie}(S,\mathbb{R})$ general enough? The reason C^∞-functions are chosen is because the Poisson bracket, between any two of them, is well-defined and belongs to the same space. However, this clearly excludes certain functions on S that, arguably, are of physical interest, but which are not of this type. For example, the characteristic function, χ_A, of a (measurable) subset $A \subseteq S$ corresponds to a proposition. This is not C^∞, however, it is represented in the quantum theory by a projection operator. So how is χ_A to be defined?
4. One might want to enlarge the classical space of observables to the space, $Meas(S,\mathbb{R})$, of measurable real-valued functions. The problem, of course, is that

[2]There is always an infinite-dimensional transitive group, namely, the group of symplectic transformations of S.

[3]Note that the use of PG raises the interesting question as to the extent to which the non-commutative structure of the Lie algebra $L(G)$ can be recovered from knowing the poset structure of its abelian Lie sub-algebras. Work in this direction has been done in [34], where it was shown that the Jordan structure of certain von Neumann algebras (without type I_2 summand) is determined by the poset of its abelian sub-algebras.

Poisson brackets are not defined on such functions unless, maybe, one generalises from functions to distributions. Could this be an acceptable strategy?

To our knowledge the only attempt in defining a possible quantisation in a topos was done in [37]. However, even in this attempt, many open questions remain unanswered, in particular:

(i) how to incorporate in the schema unitarily equivalent quantisations;
(ii) how to introduce irreducibility of the quantisation map;
(iii) a clear connection with the topos formalism defined in this book.

21.2 Internal Approach

A different approach for defining quantum theory in terms of topos theory was put forward in [47]. Here the authors start with a non-commutative C^*-algebra $\mathcal{A}$ and, then, construct the category (poset) of commutative sub-algebras of A, which we will denote by $\mathcal{V}(\mathcal{A})$. They, then, construct the topos $\mathbf{Sets}^{\mathcal{V}(\mathcal{A})}$ of covariant functors (i.e. co-presheaves) over $\mathcal{V}(\mathcal{A})$. In order to obtain a state space they first introduce the co-presheaf $\overline{\mathcal{A}}$ defined as follows:

Definition 21.1 The co-presheaf $\overline{\mathcal{A}}$ is defined on:

1. Objects: for each $V \in \mathcal{V}(\mathcal{A})$, then $\overline{\mathcal{A}}(V) = V$.
2. Morphisms: given a map $i_{V'V} : V' \subseteq V$ the corresponding co-presheaf map $\overline{\mathcal{A}}(i_{V'V}) : \overline{\mathcal{A}}(V') \to \overline{\mathcal{A}}(V)$ is simply the inclusion map.

$\overline{\mathcal{A}}$ is, then, shown to have the structure of a C^*-algebra inside the topos $\mathbf{Sets}^{\mathcal{V}(\mathcal{A})}$. The spectrum of this internal C^*-algebra, as computed internally, is then the state space object, i.e. the co-presheaf analogue of $\underline{\Sigma}$.

There are many similarities between this approach and the topos approach delineated in this book. An in-depth analysis of this is given in [46].

However, the notion of group and group transformation is missing in the internal approach, thus it would be very interesting to internalise the analysis carried out in Chap. 16 and the results obtained in [97, 98]. In particular, an insight into how to define unitary transformations and time evolution might be obtained in this way.

21.3 Configuration Space

One of the main problem to be resolved in topos quantum theory is to define a potential state object $\underline{Q}$. One way of approaching this issue is to define some general axioms at the level of the language $l(S)$ (see Appendix A) for some range of systems S, which would, somehow, identify the linguistic state term Q. The sought after configuration object $\underline{Q}$ would, then, be obtained by representing Q in the topos $Sh(\mathcal{V}(\mathcal{H}))$.

This is a rather abstract and difficult strategy to pursue. A more pragmatic approach would be to consider a classical system, whose state space is T^*M for some differentiable manifold M. The quantum theory we are interested in would, then, arise by quantising such a classical system. Given such a quantum system, we would like to define the topos configuration object $\underline{Q}_M$, such that the state space $\underline{\Sigma}_M$, defined in Chap. 9, can de derived as some sort of topos analogue of the cotangent bundle over $\underline{Q}_M$. Clearly, what this means, is yet to be defined.

In the naive quantisation of such a system the state vectors would be square-integrable functions on Q, with a measure that is determined by the volume form on the manifold Q. The group-theoretical quantisation suggests that this needs to be augmented, by looking at cross-sections of vector bundles over Q whose fibres carry an irreducible representation of a 'little group' H. There may also be 'singular' representations that come from orbits of G on $W \subset C^\infty(M,\mathbb{R})$ other than $M \simeq G/H$. However, whether or not, such representations should be regarded as a quantisation of the classical system with state space T^*Q, is still open to debate.

Irrespectively of the above issues, a generic quantum analogue of the configuration space would be any maximal commutative sub-algebra A of $\mathcal{B}(\mathcal{H})$. This is because, roughly, the state vectors can be thought of as functions on the 'simultaneous eigenvalues' of the self-adjoint operators in the 'complete commuting set' A. This would suggest that there should be some sort of 'configuration' object Q_A for each maximal sub-algebra A.

The set of all commutative von Neumann sub-algebras of A is $\downarrow A$. This partially-ordered set is a category in the usual way and, hence, we can consider the topos $\mathbf{Sets}^{\downarrow A^{op}}$. It is easy to define the spectral presheaf $\underline{\Sigma}_{\downarrow A}$ of this topos. Clearly it is determined by the spectrum of A alone, since everything else is obtained by restriction to a sub-algebra of A.

A possible way to define the configuration object would then be to 'extend' $\underline{\Sigma}_{\downarrow A}$, so that it becomes an object in $\mathbf{Sets}^{\mathcal{V}(\mathcal{H})^{op}}$. This would then be the $\underline{Q}$ we are looking for.

However the issue of defining a topos analogue of the state space is still an open problem, whose resolution is of paramount importance for the topos quantum theory program.

21.4 Composite Systems

The mathematical tools developed in [41], suggest new ways of analysing composite systems in the topos formulation of (single-time) quantum mechanics. Such an investigation would be a vital step in formulating a complete, self-contained analogue of quantum theory in the language of topos theory.

In classical physics, given the state spaces Σ_1 and Σ_2 of two distinct system, the sate space of the composite system is $\Sigma_1 \times \Sigma_2$.

In the case of quantum theory, as explained in Chap. 17, the state spaces of two distinct systems belong to two different topoi, namely $\underline{\Sigma}^{\mathcal{V}(\mathcal{H}_1)} \in \mathbf{Sets}^{\mathcal{V}(\mathcal{H}_1)}$ and

$\underline{\Sigma}^{\mathcal{V}(\mathcal{H}_2)} \in \mathbf{Sets}^{\mathcal{V}(\mathcal{H}_2)}$. In order to be able to compare and combine these two topoi, they have to be pushed back to the intermediate topos $\mathbf{Sets}^{\mathcal{V}(\mathcal{H}_1)\times\mathcal{V}(\mathcal{H}_2)}$. However, in this topos it is not possible to account for entangled contexts, thus a 'bigger' topos needs to be defined. Clearly the topos $\mathbf{Sets}^{\mathcal{V}(\mathcal{H}_1\otimes\mathcal{H}_2)^{op}}$ would be able to account for entanglement but, then, we would lose the temporal logic defined in terms of the tensor product of Heyting algebras (see Sect. 17.4). A possibility would be, somehow, to generalise the notion of tensor product of Heyting algebras so as to account for entanglement, obtaining a kind of 'quantum tensor product'.

Another way of tackling the issue of composite systems might be the following: let us assume that our base category is PG, as defined in Sect. 21.1. If we then were to consider two distinct systems S_1 and S_2, their state spaces would be two distinct sheaves defined over $PG_1 \subset C^{\infty}_{Lie}(S_1, \mathbb{R})$ and $PG_2 \subset C^{\infty}_{Lie}(S_1, \mathbb{R})$, respectively.

When considering the composite system one could, then, start by first considering the composite system at the classical level and, then, applying the (yet to be defined) quatisation sheaf. In this case, one would consider the classical state space $S_1 \times S_2$, whose physical quantities would be identified with $C^{\infty}_{Lie}(S_1 \times S_2, \mathbb{R})$. If this indeed represents the total Lie algebra, we would need to identify the Lie sub-algebras and define the poset of such sub-algebras. However, defining what the lie abelian sub-algebras of the composite system are and what the Lie algebras of a composite system are, is not an easy task.

This problem is reminiscent of the following situation: consider a 2 particle system, whose Hilbert space is $\mathcal{H}_1 \otimes \mathcal{H}_2$. Assuming that the particles are spin $s = 1$ particles, then each Hilbert space has dimension $(2s + 1)$. For each particle the Lie group of transformations is given by $SU(2)$, thus for the entire system we have $SU(2) \times SU(2)$. However, the full group of transformations for each particle is given by $SU(3) = SU(2s + 1)$, such that the lie algebras have an inclusion relation $su(2) \subseteq su(3)$ and the same holds for the respective groups. So the question is: what is the full symmetry of the composite system? In particular, we are looking for a group G, such that $SU(2) \times SU(2) \subset G$ and G represents the full algebra of our composite system, thus still acts on the state space.

21.5 Differentiable Structure

It would be very interesting to understand whether it is possible to define some differentiable structure on the state space $\underline{\Sigma}$. We know that this is a topological space where each $\underline{\Sigma}_V$ is a Hausdorff space, since it represents the Gel'fand spectrum of V, however there is no differentiable structure on it yet. Would it be conceivable to define such a structure?

A possibility would be to use the differentiable structure of the G-orbits on the base category $\mathcal{V}(\mathcal{H})$ and, somehow, 'pull up' this structure to the level of the bundle space $\Lambda(\underline{\Sigma})$. In particular we know that the unitary group $U(\mathcal{H})$ acts on the abelian sub-algebras via $l_{\hat{U}}(V) = \{\hat{U}\hat{A}\hat{U}^{-1} | \hat{A} \in V\}$. It is, then, possible to construct various fibre bundles associated with the orbit spaces of this action. At this point a question to be answered is if one should consider the whole group $U(\mathcal{H})$ or some subgroup G, which could either be (i) a canonical quantisation group; or (ii) a group of symmetries of the system. However, if we are interested in the covariance of the quantum formalism, then the entire $U(\mathcal{H})$ should be used.

Since the action of G preserves the ordering in $\mathcal{V}(\mathcal{H})$, it is possible to equip the space of orbits $\mathcal{V}(\mathcal{H})/G$ with a poset ordering. The Alexandroff topology on $V(\mathcal{H})$ will induce the usual identification topology on $\mathcal{V}(\mathcal{H})/G$. This will enable us to construct sheaves over $\mathcal{V}(\mathcal{H})/G$.

Moreover, we can construct the bundle $p : \mathcal{V}(\mathcal{H}) \rightarrow \mathcal{V}(\mathcal{H})/G$, where each fibre in the orbit space Λp carries the structure of a differentiable manifold with the appropriate topology. If one then defines, in an appropriate way, sheaves over such a bundle, the differentiable structure could then be 'lifted up' to the bundle space. In particular, if we had an étale bundle of the form $\Lambda \underline{\Sigma} \rightarrow \Lambda p$, then we could lift the differentiable structure to the bundle space $\Lambda \underline{\Sigma}$.

Appendix A
Topoi and Logic

In this section, we will explore the tight connection between topos theory and logic. In particular, to each topos there is associated a language for expressing the internal language of the topos. The converse is also true: given a language one can define a corresponding topos.

A.1 First Order Languages

A language, in its most raw definition, comprises a collection of atomic variables, and a collection of primitive operations called logical connectives, whose role is to combine together such primitive variables transforming them into formulas or sentences. Moreover, in order to reason with a given language, one also requires rules of inference, i.e. rules which allow you to generate other valid sentences from the given ones.

The semantics or meaning of the logical connectives, however, is not given by the formulae and sentences themselves, but it is defined through a so called evaluation map, which is a map from the set of atomic variables and sentences to a set of truth values. Such a map enables one to determine when a formula is true and, thus, defines its semantics/meaning.

In this perspective it turns out that the meaning of the logical connectives is given in terms of some set of objects which represent the truth values. The logic that a given language will exhibit will depend on what the set of truth values is considered to be. In fact, what was said above is a very abstract characterisation of what a language is. To actually use it as a deductive system of reasoning, one needs to define a mathematical context in which to represent this abstract language. In this way the elementary and compound propositions will be represented by certain mathematical objects and the set of truth values will itself be identified with an algebra.

For example, in standard classical logic, the mathematical context used is **Sets** and the algebra of truth values is the Boolean algebra of subsets of a given set.

C. Flori, *A First Course in Topos Quantum Theory*, Lecture Notes in Physics 868,
DOI 10.1007/978-3-642-35713-8,

However, in a general topos, the internal logic/algebra will not be Boolean but will be a generalisation of it, namely a Heyting algebra.

In order to get a better understanding of what a language is, we will start with a very simple language called propositional language $P(l)$.

A.2 Propositional Language

The propositional language $P(l)$ contains a set of symbols and a set of formation rules.

Symbols of $P(l)$

(i) An infinite list of symbols $\alpha_0, \alpha_1, \alpha_2, \dots$ called *primitive propositions*.
(ii) A set of symbols $\neg, \vee, \wedge, \Rightarrow$ which, for now, have no explicit meaning.
(iii) Brackets), (.

Formation Rules

(i) Each primitive proposition $\alpha_i \in P(l)$ is a sentence.
(ii) If α is a sentence, then so is $\neg\alpha$.
(iii) If α_1 and α_2 are sentences, then so are $\alpha_1 \wedge \alpha_2$, $\alpha_1 \vee \alpha_2$ and $\alpha_1 \Rightarrow \alpha_2$.

Note also that $P(l)$ does not contain the quantifiers $\forall$ and $\exists$. This is because it is only a propositional language. To account for quantifiers one has to go to more complicated languages called higher-order languages, which will be described later.

The inference rule present in $P(l)$ is the modus ponens (the 'rule of detachment') which states that, from α_i and $\alpha_i \Rightarrow \alpha_j$, the sentence α_j may be derived. Symbolically this is written as

$$\frac{\alpha_i, \alpha_i \Rightarrow \alpha_j}{\alpha_i}. \tag{A.1}$$

We will see, later on, what exactly the above expression means.

In order to use the language $P(l)$ one needs to represent it in a mathematical context. The choice of such a context will depend on what type of system we want to reason about. For now we will consider a classical system, thus the mathematical context in which to represent the language $P(l)$ will be **Sets**. In **Sets**, the truth object (object in which the truth values lie) will be the Boolean set $\{0, 1\}$, thus the truth values will undergo a Boolean algebra. This, in turn, implies that the logic of the language $P(l)$, as represented in **Sets**, will be Boolean.

The rigorous definition of a representation of the language $P(l)$ is as follows:

Definition A.1 Given a language $P(l)$ and a mathematical context τ, a representation of $P(l)$ in τ is a map π from the set of primitive propositions to elements in the algebra in question: $\alpha \to \pi(\alpha)$.

As we will see, the specification of the algebra will depend on what type of theory we are considering, i.e. classical or quantum.

In classical physics, propositions are represented by the Boolean algebra of all (Borel) subsets of the classical state space, thus, given a representation π, we can define the semantics of $P(l)$ as follows:

$$\begin{aligned}\pi(\alpha_i \vee \alpha_j) &:= \pi(\alpha_i) \vee \pi(\alpha_j)\\ \pi(\alpha_i \wedge \alpha_j) &:= \pi(\alpha_i) \wedge \pi(\alpha_j)\\ \pi(\alpha_i \Rightarrow \alpha_j) &:= \pi(\alpha_i) \Rightarrow \pi(\alpha_j)\\ \pi(\neg\alpha_i) &:= \neg\big(\pi(\alpha_i)\big)\end{aligned} \tag{A.2}$$

where, on the left hand side, the symbols $\{\neg, \wedge, \vee, \Rightarrow\}$ are elements of the language $P(l)$, while on the right hand side they are the logical connectives in algebra, in which the representation takes place. It is in such an algebra that the logical connectives acquire meaning.

For the classical case, since the algebra in which a representation lives, is the Boolean algebra of subsets, the logical connectives on the right hand side of (A.2) are defined in terms of set-theoretic operations. In particular, we have the following associations:

$$\pi(\alpha_i) \vee \pi(\alpha_j) := \pi(\alpha_i) \cup \pi(\alpha_j) \tag{A.3}$$

$$\pi(\alpha_i) \wedge \pi(\alpha_j) := \pi(\alpha_i) \cap \pi(\alpha_j)$$

$$\neg\big(\phi(\alpha_i)\big) := \pi(\alpha_i)^c$$

$$\pi(\alpha_i) \Rightarrow \pi(\alpha_j) := \pi(\alpha_i)^c \cup \pi(\alpha_j). \tag{A.4}$$

So far, we have seen how logical connectives are represented in the topos **Sets**. However, it is possible to give a general definition of logical connectives in terms of arrows. Such a definition would then be valid for any topos. To retrieve the logical connectives for the classical case, in which the topos is **Sets**, we then simply replace, in the definitions that will follow, the general truth object Ω with the set $\{0, 1\} = 2$.

Logical connectives in a general topos τ are defined as follows:

- **Negation**
  We will now describe how to represent negation as an arrow in a given topos τ. Let us assume that the τ-arrow representing the value true is $\top : 1 \to \Omega$, which is the arrow used in the definition of the sub-object classifier. Given such an arrow, negation is identified with the unique arrow $\neg : \Omega \to \Omega$, such that the following

diagram is a pullback:

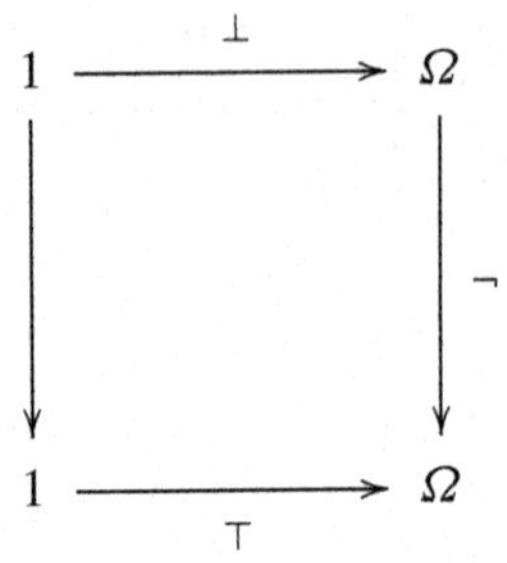

Where $\bot$ is the topos analogue of the arrow *false* in **Sets**, i.e. $\bot$ is the character of $!_1 : 0 \to 1$:

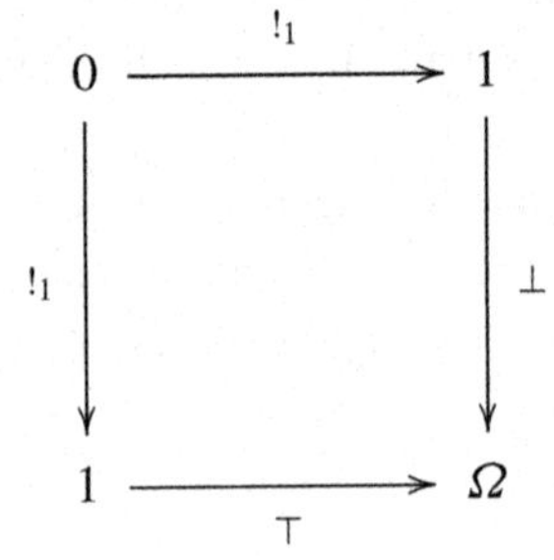

- **Conjunction**
  Conjunction is identified with the following arrow:

$$\cap : \Omega \times \Omega \to \Omega$$

which is the character of the product arrow $\langle \top, \top \rangle : 1 \to \Omega \times \Omega$, such that the following diagram is a pullback:

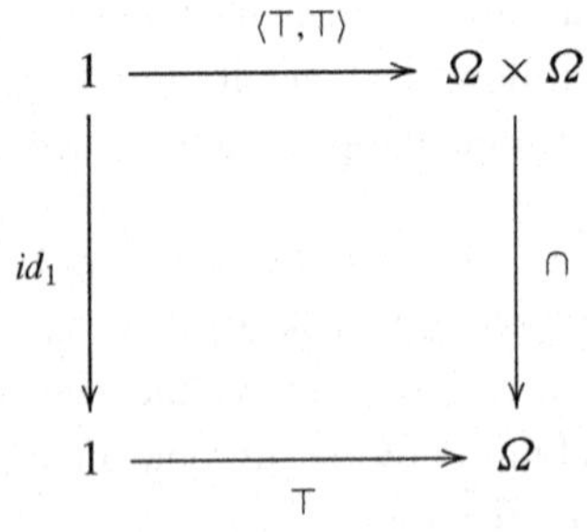

where $\langle \top, \top \rangle$ is defined as follows:

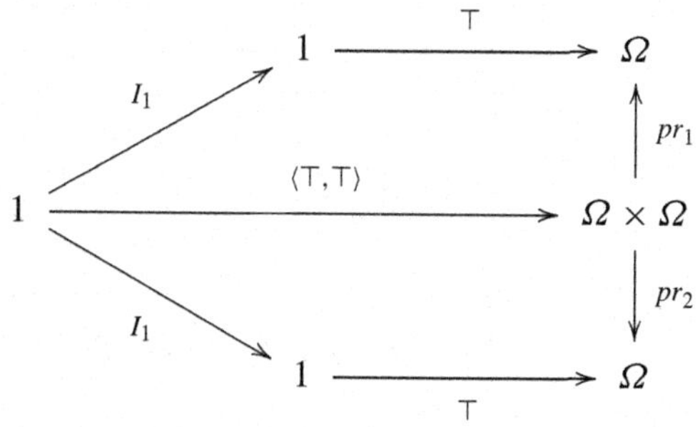

- **Disjunction**
  Disjunction is identified with the arrow

$$\cup : \Omega \times \Omega \rightarrow \Omega \tag{A.5}$$

which is the character of the image of the arrow

$$\big[\langle \top, id_{\Omega} \rangle, \langle id_{\Omega}, \top \rangle\big] : \Omega + \Omega \rightarrow \Omega \times \Omega \tag{A.6}$$

such that the following diagram commutes:

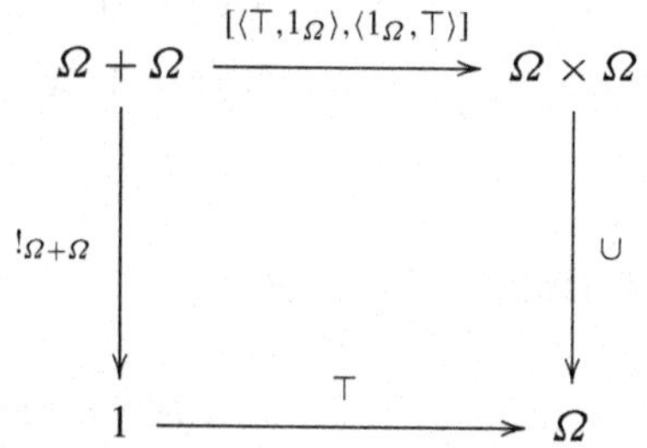

- **Implication**
  Given two arrows

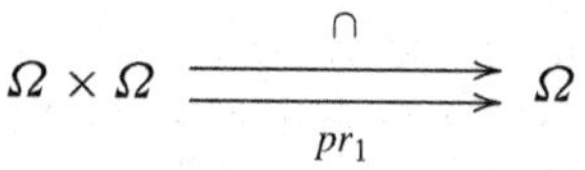

their equaliser is some map

$$e : \big(\leq := \big\{\langle x, y \rangle | x \leq y \text{ in } \Omega\big\}\big) \rightarrow \Omega \times \Omega \tag{A.7}$$

such that $\cap \circ e = pr_1 \circ e$.
Implication is then defined as the character of e, i.e. as the map

$$\Rightarrow : \Omega \times \Omega \rightarrow \Omega \tag{A.8}$$

such that the following diagram is a pullback:

$$\begin{array}{ccc} \leq & \xrightarrow{\ e\ } & \Omega \times \Omega \\ {\scriptstyle !_e}\downarrow & & \downarrow{\scriptstyle \Rightarrow} \\ 1 & \xrightarrow{\ \top\ } & \Omega \end{array}$$

In order to complete the definition of a propositional language in a general topos τ, we also need to define the valuation functions (which give us the semantics) in terms of arrows in that topos. We recall from the definition of the sub-object classifier that a truth value in a general topos τ is given by a map $1 \rightarrow \Omega$ (in **Sets** we have $1 \rightarrow \{0, 1\} = 2 = \Omega$). The collection of such arrows $\tau(1, \Omega)$ represents the collection of all truth values. Thus, a valuation map in a general topos is defined to be a map $V : \{\pi(\alpha_i)\} \rightarrow \tau(1, \Omega)$ such that the following equalities hold:

$$V\big(\neg\big(\pi(\alpha_i)\big)\big) = \neg \circ V\big(\pi(\alpha_i)\big) \tag{A.9}$$

$$V\big(\pi(\alpha_i) \vee \pi(\alpha_j)\big) = \vee \circ \big\langle V\big(\pi(\alpha_i)\big), V\big(\pi(\alpha_j)\big)\big\rangle \tag{A.10}$$

$$V\big(\pi(\alpha_i) \wedge \pi(\alpha_j)\big) = \wedge \circ \big\langle V\big(\pi(\alpha_i)\big), V\big(\pi(\alpha_j)\big)\big\rangle \tag{A.11}$$

$$V\big(\pi(\alpha_i) \Rightarrow \pi(\alpha_j)\big) = \Rightarrow \circ \big\langle V\big(\pi(\alpha_i)\big), V\big(\pi(\alpha_j)\big)\big\rangle. \tag{A.12}$$

A.2.1 Example in Classical Physics

We have stated above that classical physics uses the topos **Sets**. We now want to represent, in **Sets**, the propositional language $P(l)$, as defined for a classical system S. So now the elementary propositions will be propositions pertaining the physical system S. From now on, we will denote a language referred to a particular system S by $P(l)(S)$. Since S is a (classical) physical system, the elementary propositions which $P(l)(S)$ will contain will be of the form $A \in \Delta$ meaning " the quantity A which represents some physical observable, has value in a set Δ". We now define the representation map for this language as follows:

$$\begin{aligned} \pi_{cl} &: P(l)(S) \rightarrow \mathcal{O}(S) \\ A \in \Delta &\mapsto \big\{s \in S | \tilde{A}(s) \in \Delta\big\} = \tilde{A}^{-1}(\Delta) \end{aligned} \tag{A.13}$$

where S is the classical state space, $\mathcal{O}(S)$ is the Boolean algebra of subsets of S which lives in **Sets** and $\tilde{A} : S \rightarrow \mathbb{R}$ is the map from the state space to the reals, which identifies the physical quantity A. We now define the truth values for propositions.

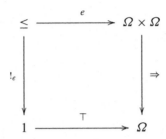

Normally, such truth values are state-dependent, i.e. they depend on the state with respect to which we are preforming the evaluation. In classical physics, states are simply identified with elements s of the state space S. Thus, for all $s \in S$, we define the truth value of the proposition $\tilde{A}^{-1}(\Delta)$ as follows:

$$v(A \in \Delta; s) = \begin{cases} 1 & \text{iff } s \in \tilde{A}^{-1}(\Delta) \\ 0 & \text{otherwise.} \end{cases} \tag{A.14}$$

Therefore the truth values lie in the Boolean algebra $\Omega = \{0, 1\}$.

It is interesting to note that the application of the propositional language $P(l)$ for quantum theory, fails. This is because in quantum theory propositions are identified with projection operators, thus the representation map will be

$$\begin{aligned} \pi_q : \{\alpha_i\} &\to P(\mathcal{H}) \\ A \in \Delta &\mapsto \pi_q(A \in \Delta) := \hat{E}[A \in \Delta] \end{aligned} \tag{A.15}$$

where $\hat{E}[A \in \Delta]$ is the projection operator which projects onto the subset Δ of the spectrum of $\hat{A}$. Now the problem with this construction is that the set of all projection operators undergoes a logic which is not distributive, but the logic of the propositional language is distributive with respect to the logical connectives $\vee$ and $\wedge$. Therefore, such a representation will not work. To solve this problem we need to introduce a higher order language which we will examine in the next section.

A.3 The Higher Order Type Language l

We will now define a more complex language called higher order type language and denoted by l. Such a language consists of a set of symbols and terms.

Symbols

1. A collection of "sorts" or "types". If $T_1, T_2, \ldots, T_n$, $n \geq 1$, are type symbols, then so is $T_1 \times T_2 \times \cdots \times T_n$. If $n = 0$ then $T_1 \times T_2 \times \cdots \times T_n = 1$.
2. If T is a type symbol, then[1] so is PT.
3. Given any type T there are countable many variables of type T.
4. There is a special symbol $*$.
5. A set of function symbols for each pair of type symbols, together with a map which assigns to each function its type. This assignment consists of a finite, non-empty list of types. For example, if we have the pair of type symbols (T_1, T_2), the associated function symbol will be $F_l(T_1, T_2)$. An element $f \in F_l(T_1, T_2)$ has type T_1, T_2. This is indicated by writing $f : T_1 \to T_2$.

[1] PT indicates the collection of all subobjects of T.

6. A set of relation symbols R_i together with a map which assigns the type of the arguments of the relation. This consists of a list of types. For example, a relation taking an argument $x_1 \in T_1$ of type T_1 and an argument $x_2 \in T_2$ of type T_2 is denoted as $R = R(x_1, x_2) \subseteq T_1 \times T_2$.

Terms

1. The variables of type T are terms of type $T, \forall T$.
2. The symbol $*$ is a term of type 1.
3. A term of type Ω is called a formula. If the formula has no free variables, then we call it a sentence.
4. Given a function symbol $f : T_1 \to T_2$ and a term t of type T_1, then $f(t)$ is a term of type T_2.
5. Given $t_1, t_2, \ldots, t_n$ which are terms of type $T_1, T_2, \ldots, T_n$, respectively, then $\langle t_1, t_2, \ldots, t_n \rangle$ is a term of type $T_1 \times T_2 \times \cdots \times T_n$.
6. If x is a term of type $T_1 \times T_2 \times \cdots \times T_n$, then for $1 \leq i \leq n$, x_i is a term of type T_i.
7. If ω is a term of type Ω and α is a variable of type T, then $\{\alpha|\omega\}$ is a term of type PT.
8. If x_1, x_2 are terms of the same type, then $x_1 = x_2$ is a term of type Ω.
9. If x_1, x_2 are terms of type T and PT respectively, then $x_1 \in x_2$ is a term of type Ω.
10. If x_1, x_2 are terms of type PT and PPT respectively, then $x_1 \in x_2$ is a term of type Ω.
11. If x_1, x_2 are both terms of type PT, then $x_1 \subseteq x_2$ is a term of type Ω.

The entire set of formulas in the language l are defined, recursively, through repeated applications of formation rules, which are the analogues of the standard logical connectives. In particular, we have *atomic formulas* and *composite formulas*. The former are:

1. The set of relation symbols.
2. Equality terms defined above.
3. Truth $\top$ is an atomic formula with no free variables.
4. False $\bot$ is an atomic formula with no of free variables.

We can now build more complicated formulas through the use of the logical connectives $\vee$, $\wedge$, $\Rightarrow$ and $\neg$. These are the *composite formulas*:

1. Given two formulas α and β, then $\alpha \vee \beta$ is a formula for which, the set of free variables is defined to be the union of the free variables in α and β.
2. Given two formulas α and β, then $\alpha \wedge \beta$ is a formula for which, the set of free variables is defined to be the union of the free variables in α and β.
3. Given a formula α its negation $\neg\alpha$ is still a formula with the same number of free variables.
4. Given two formulas α and β, then $\alpha \Rightarrow \beta$ is a formula with free variables given by the union of the free variables in α and β.

It is interesting to note that the logical operations just defined can actually be expressed in terms of the primitive symbols as follows:

1. $\top := (* = *)$.
2. $\alpha \wedge \beta := (\langle \alpha, \beta \rangle = \langle \top, \top \rangle) = \langle * = *, * = * \rangle$.
3. $\alpha \Leftrightarrow \beta := (\alpha = \beta)$.
4. $\alpha \Rightarrow \beta := ((\alpha \wedge \beta) \Leftrightarrow \alpha) := (\langle \alpha, \beta \rangle = \langle \top, \top \rangle = \alpha)$.
5. $\forall x \alpha := (\{x : \alpha\} = \{x : \top\})$.[2]
6. $\bot := \forall w w = (\{w : w\} = \{w : \top\})$.
7. $\neg \alpha := \alpha \Rightarrow \bot$.
8. $\alpha \vee \beta := \forall w[(\alpha \Rightarrow w \wedge \beta \Rightarrow w) \Rightarrow w]$.
9. $\exists x \alpha := \forall w[\forall x(\alpha \Rightarrow w) \Rightarrow w]$.[3]

A.4 Representation of l in a Topos

We now want to show how a representation of the first order language l takes place in a topos. The main idea is that of identifying each of the terms in l an arrow in a topos. In particular we have:

Definition A.2 Given a topos τ the interpretation/representation (M) of the language l in τ consists of the following associations:

1. To each type $T \in l$ an object $T^{\tau M} \in \tau$.
2. To each relation symbol $R \subseteq T_1 \times T_2 \times \cdots \times T_n$ a sub-object $R^{\tau M} \subseteq T_1^{\tau M} \times T_2^{\tau M} \times \cdots \times T_n^{\tau M}$.
3. To each function symbol $f : T_1 \times T_2 \times \cdots \times T_n \to X$ a τ-arrow $f^{\tau M} : T_1^{\tau M} \times T_2^{\tau M} \times \cdots \times T_n^{\tau M} \to X$.
4. To each constant c of type T a τ-arrow $c^{\tau M} : 1^{\tau M} \to T^{\tau M}$.
5. To each variabe x of type T a τ-arrow $x^{\tau M} : T^{\tau M} \to T^{\tau M}$.
6. The symbol Ω is represented by the sub-object classifier $\Omega^{\tau M}$.
7. The symbol 1 is represented by the terminal object $1^{\tau M}$.

Now that we understand how the basic symbols of the abstract language l are represented in a topos, we can proceed to understand also how the various terms and formulas are represented. Needless to say, these are all defined in a recursive manner.

Given a term $t(x_1, x_2, \ldots, x_n)$ of type Y with free variables x_i of type T_i, i.e. $t(x_1, x_2, \ldots, x_n) : T_1 \times \cdots \times T_n \to Y$, then the representative in a topos of this term

[2] $\forall x \alpha$ means:“ for all x with the property α”, while $\{x : y\}$ indicates the set of all x, such that y

[3] $\exists x \alpha$ means:“ there exists an x with the property α”.

would be a τ-map

$$t(x_1, x_2, \dots, x_n) : T_1^{\tau M} \times \dots \times T_n^{\tau M} \to Y^{\tau M}. \tag{A.16}$$

Formulas in the language are interpreted as terms of type Ω. In the topos τ this type Ω is identified with the sub-object classifier $\Omega^{\tau M}$. In particular, a term of type Ω of the form $\phi(t_1, t_2, \dots, t_n)$ with free variables t_i of type T_i is represented by an arrow

$$\phi(t_1, t_2, \dots, t_n)^{\tau M} : T_1^{\tau M} \times \dots \times T_n^{\tau M} \to \Omega^{\tau M}.$$

On the other hand, a term ϕ of type Ω with no free variables is represented by a global element $\phi : 1^{\tau M} \to \Omega^{\tau M}$. As we will see, these arrows will represent the truth values.

The reason why, in a topos, formulas are identified with arrows with codomain Ω rests on the fact that sub-objects, of a given object in a topos, are in bijective correspondence with maps from that object to the sub-object classifier. In fact, by construction, formulas single out sub-objects of a given object X in terms of a particular relation which they satisfy, i.e. they define elements of $Sub(X)$. Such sub-objects are in bijective correspondence with maps $X \to \Omega$.

In particular, given a formula $\phi(x_1, \dots, x_n)$ with free variables x_i of type T_i, which in the language l is associated with the subset $\{x_i|\phi\} \subseteq \prod_i T_i$, we obtain the topos representation

$$\big\{(x_1, \dots, x_n)|\phi\big\}^{\tau M} \subseteq T_1^{\tau M} \times \dots \times T_n^{\tau M} \tag{A.17}$$

which, through the *Omega Axiom* (see Axion 8.4), gets identified with the map

$$\big\{(x_1, \dots, x_n)|\phi\big\}^{\tau M} \to T_1^{\tau M} \times \dots \times T_n^{\tau M} \xrightarrow{\chi_{\{(x_1,\dots,x_n)|\phi\}^{\tau M}}} \Omega^{\tau M}. \tag{A.18}$$

To understand how formulas are represented in a topos τ, let us consider the formula stating that two terms are the same, i.e. $t(x_1, x_2, \dots, x_n) = t'(x_1, x_2, \dots, x_n)$. The representation of such a formula in a topos τ is identified with the equalizer of the two τ-arrows representing the terms $t(x_1, x_2, \dots, x_n)$ and $t'(x_1, x_2, \dots, x_n)$. In particular, we have

$$\big\{x_1, x_2, \dots, x_n|t = t'\big\}^{\tau M} \rightarrowtail T_1^{\tau M} \times \dots \times T_n^{\tau M} \underset{t'^{\tau M}}{\overset{t^{\tau M}}{\rightrightarrows}} Y^{\tau M}$$

Instead, if we consider a relation $R(t_1, \dots, t_n)$ of terms t_i of type Y_i with variables x_j of type T_j, then the formula pertaining this relation $\{(x_1 \dots x_n)|R(t_1, \dots, t_n)\}$ is represented, in τ, by pulling back the sub-object $R^{\tau M} \subseteq Y_1^{\tau M} \times \dots \times Y_n^{\tau M}$ (representing the relation $R(t_1, \dots, t_n)$) along the term arrow $\langle t_1^{\tau M}, \dots, t_n^{\tau M}\rangle : T_1^{\tau M} \times \dots \times T_n^{\tau M} \to$

$Y_1^{\tau M} \times \cdots \times Y_n^{\tau M}$:

$$\begin{array}{ccc}
\{(x_1, x_2, \ldots, x_n) | R(t_1, \ldots, t_n)\}^{\tau M} & \longrightarrow & R^{\tau M} \\
\Big\downarrow & & \Big\downarrow \\
T_1^{\tau M} \times \cdots \times T_n^{\tau M} & \xrightarrow{\langle t_1^{\tau M}, \ldots, t_n^{\tau M} \rangle} & Y_1^{\tau M} \times \cdots \times Y_n^{\tau M}
\end{array}$$

The atomic formulas meaning truth and false ($\top$ and $\bot$, respectively) will be represented in a topos τ by the greatest and lowest elements of the Heyting algebra of the sub-objects of any object in the topos. For example we have that

$$\{x_1 \ldots x_n | \top\}^{\tau M} = T_1^{\tau M} \times T_2^{\tau M} \times \cdots \times T_n^{\tau M} \tag{A.19}$$

$$\{x_1 \ldots x_n | \bot\}^{\tau M} = \emptyset^{\tau M}. \tag{A.20}$$

So far we have established how to represent formulas in a topos. Next, we will explain how to represent logical connectives between formulas in a topos. In particular, given a collection of formulas represented as sub-objects of the type object $\prod_i T_i$, the logical connectives between these are represented by the corresponding operations in the Heyting algebra of sub-objects of the object $\prod_i T_i^{\tau M}$ in τ. Since we are dealing with sub-objects, we can also represent the logical connectives in terms of τ-arrows with codomain Ω as follows: For example, consider two formulas ϕ, ρ, of type Ω with free variables x_1 and x_2 of type $T_1^{\tau M}$ and $T_2^{\tau M}$, respectively. The conjunction $\phi \wedge \rho$ is represented by the arrow

$$\phi \wedge \rho : T_1^{\tau M} \times T_2^{\tau M} \xrightarrow{\langle \phi, \rho \rangle} \Omega^{\tau M} \times \Omega^{\tau M} \xrightarrow{\wedge} \Omega^{\tau M}. \tag{A.21}$$

Similarly, we have

$$\phi \vee \rho : T_1^{\tau M} \times T_2^{\tau M} \xrightarrow{\langle \phi, \rho \rangle} \Omega^{\tau M} \times \Omega^{\tau M} \xrightarrow{\vee} \Omega^{\tau M} \tag{A.22}$$

$$\phi \Rightarrow \rho : T_1^{\tau M} \times T_2^{\tau M} \xrightarrow{\langle \phi, \rho \rangle} \Omega^{\tau M} \times \Omega^{\tau M} \xrightarrow{\Rightarrow} \Omega^{\tau M} \tag{A.23}$$

$$\neg \rho : T_2^{\tau M} \xrightarrow{\rho} \Omega^{\tau M} \xrightarrow{\neg} \Omega^{\tau M}. \tag{A.24}$$

We now give two very important notions: that of a *theory in l* and that of a *model of l*.

Definition A.3 Given a language l, a theory $\mathcal{T}$ in l is a set of formulas which are called the axioms of $\mathcal{T}$.

Definition A.4 A model of a theory is a representation M in which all the axioms of $\mathcal{T}$ are valid. Such axioms are, then, represented by the arrow true : $1 \to \Omega$.

An example of this is given by the theory of abelian groups which can be seen as model of a theory in a given language. The language required will only contain one type of elements G, no relations, two function symbols

$$+ : G \times G \to G \tag{A.25}$$

$$- : G \to G \tag{A.26}$$

and a constant 0. A representation of this language, which will lead us to the theory of groups, will be defined in the topos **Sets**. Such a representation of G will be identified as a set G^M, on which the function symbols act upon:

$$+^M : G^M \times G^M \to G^M \tag{A.27}$$

$$\langle g_1, g_2 \rangle \mapsto g_1 g_2 \tag{A.28}$$

and

$$-^M : G^M \to G^M \tag{A.29}$$

$$g \mapsto -g. \tag{A.30}$$

The constant 0 will be an element $0^M \in G^M$.

Such a representation will be a model for the theory of abelian groups if the function symbols satisfy the axioms of abelian groups, i.e. if the following should hold in the representation M:

$$(g_1 + g_2) + g_3 = g_1 + (g_2 + g_3) \tag{A.31}$$

$$g_1 + g_2 = g_2 + g_1 \tag{A.32}$$

$$g_1 + 0 = g_1 \tag{A.33}$$

$$g_1 + (-g_1) = 0. \tag{A.34}$$

Given two models M and M' of a theory $\mathcal{T}$ in a language l, we say that these two models are homomorphic if there is a homomorphism of the respective interpretations of the model, i.e. for each symbol type X in l, these maps are homomorphisms:

$$H_X : X^M \to X^{M'} \tag{A.35}$$

where X^M and $X^{M'}$ are the representations of the symbol type X of l in the representation M and M', respectively. Such a map is called a homomorphism if it respects all relation symbols, function symbols and constants.

In the example of abelian groups, model homomorphisms would simply be group homomorphisms.

The definition of homomorphic representations gives rise to a category $\mathcal{I}$, whose objects are all possible representations of a given language l in a topos τ, and whose morphisms are the above mentioned homomorphisms of representations. Given such a category, each theory $\mathcal{T}$ gives rise to a full subcategory of $\mathcal{I}$ called $Mod(\mathcal{T}, \tau)$, whose objects are models of the theory $\mathcal{T}$ in the topos τ, and whose morphisms are homomorphisms of models.

In this section, we have seen how, given a first order type language l, it is possible to represent such a language in a topos τ. However, interestingly enough, the converse is also true, namely: given a topos τ, it has associated to it an internal first order language l, which enables one to reason about τ in a set theoretic way, i.e. using the notion of elements.

Definition A.5 Given a topos τ, its internal language $l(\tau)$ has a type symbol $\ulcorner A \urcorner$ for each object $A \in \tau$, a function symbol $\ulcorner f \urcorner : \ulcorner A_1 \urcorner \times \ulcorner A_2 \urcorner \times \cdots \times \ulcorner A_n \urcorner \to \ulcorner B \urcorner$ for each map $f : A_1 \times A_2 \times \cdots \times A_n \to B$ in τ and a relation $\ulcorner R \urcorner \subseteq \ulcorner A_1 \urcorner \times \ulcorner A_2 \urcorner \times \cdots \times \ulcorner A_n \urcorner$ for each sub-object $R \subseteq A_1 \times A_2 \times \cdots \times A_n$ in τ.

A.5 A Language l for a Theory of Physics and Its Representation in a Topos τ

We will now try to construct a physics theory for a system S. The construction of such a theory is defined by an interplay between a language $l(S)$, associated to the system S, a topos and the representation of the theory in the topos. In particular, we can say that a theory of the system S is defined by choosing a representation/model, M, of the language $l(S)$ in a topos τ_M. The choice of both topos and representation depend on the kind of theory being used, i.e. if it is classical or quantum theory.

As we have seen above, since each topos τ has an internal language $l(\tau)$ associated to it, constructing a theory of physics consists in translating the language, $l(S)$, of the system to the local language $l(\tau)$ of the topos.

As a first step in constructing a theory of physics we need to specify, exactly, what $l(S)$ is. In particular we need to analyse which primitive type terms and formulas should be present in $l(S)$ for it to be a language that will enable us to talk about the physical system S.

A.5.1 The Language $l(S)$ of a System S

The minimum set of type symbols and formulas, which are needed for a language to be used as a language to talk about a physical system S, are the following:

1. *The state space object and the quantity value object.* These objects are represented in $l(S)$ by the ground type symbols Σ and $\mathcal{R}$.
2. *Physical quantities.* Given a physical quantity A, it is standard practice to represent such a quantity in terms of a function from the state space to the quantity value object. Thus, we require $l(S)$ to contain the set function symbols $F_{l(S)}(\Sigma, \mathcal{R})$ of signature $\Sigma \to \mathcal{R}$, such that the physical quantity is $A : \Sigma \to \mathcal{R}$.

3. *Values*. We would like to have values of physical quantities. These are defined in $l(S)$ as terms of type $\mathcal{R}$ with free variables s of type Σ, i.e. they are the terms $A(s) \in \mathcal{R}$, where $A : \Sigma \to \mathcal{R} \in F_{l(S)}(\Sigma, \mathcal{R})$.
4. *Propositions*. Imagine we would like to talk about collections of states of the system with a particular property. Such a collection is represented in terms of sub-objects of the state space, which comprises the states with that particular property in question. Thus we have terms $Q = \{s|A(s) \in \Delta\}$ which are of type $P\Sigma$ with a free variable Δ of type $P\mathcal{R}$.
5. *Truth values*. We generally would like to talk about values of physical quantities for a given state of the system, thus we require the presence of formulas of the type $A(s) \in \Delta$, where Δ is a variable of type $P\mathcal{R}$ and s is a variable of type Σ. Such a formula is a term of type Ω.

   A formula w with no free variables, called a sentence, is a special element of Ω which is represented, in a topos, by a global element of Ω, i.e.

$$[w] : 1 \to \Omega. \tag{A.36}$$

   These, as we will see later on, will represent truth values for propositions about the system.
6. *States*. There are three options for describing a state, which we will analyse separately.
   (i) *Microstate option*. The microstate option is the one used in classical physics where a state is identified with an element of the state space. Hence in the context of the language $l(S)$, a micro-state is a term t of type Σ, i.e. $t \in \Sigma$. To understand how the micro-state option is utilised to evaluate proposition consider the term $A(s) \in \Delta$, this is a term of type $\underline{\Omega}$ with free variables s and Δ of type $\underline{\Sigma}$ and $P(\mathcal{R})$, respectively. On the other hand $\{s|A(s) \in \Delta\}$ is a term of type $P(\underline{\Sigma})$ with free variable Δ of type $P(\mathcal{R})$. Given a state $t \in \underline{\Sigma}$ we can then form a term of type $\underline{\Omega}$ as follows: $t \in \{s|A(s) \in \Delta\}$. This term has free variables t and Δ of type Σ and $P\mathcal{R}$, respectively.

   Intuitively, $A(t) \in \Delta$ represents the proposition stating: "the value of A, given the state t, lies in the range Δ". However semantically[4] we have the following equivalence

$$t \in \big\{s|A(s) \in \Delta\big\} \quad \Leftrightarrow \quad A(t) \in \Delta. \tag{A.37}$$

   Therefore the proposition "the value of A, given the state t, lies in the range Δ" becomes the term $A(t) \in \Delta$ of type Ω with free variable $t \in \Sigma$ and $\Delta \subseteq \mathbb{R}$.
   (ii) *Pseudo-state option*. This method consists in defining a term $\mathfrak{w}$ of type $P(\Sigma)$. Then, given the term $\{s|A(s) \in \Delta\}$, which is of type $P\Sigma$ with a

[4]Note that there are two distinct notions of equivalence: (i) syntactical (ii) semantical. The first one is defined in terms of inference rules as discussed in Sect. A.6, while two propositions are semantically equivalence whenever, in each topos τ, they are represented by the same element in $\underline{\Omega}_\tau$.

free variable Δ of type $P\mathcal{R}$, we want to know whether the elements in $\mathfrak{w}$ have the property $A(s) \in \Delta$, i.e. we want to know whether the proposition $A(s) \in \Delta$ is true, given the pseudo-state $\mathfrak{w}$. To this end we need to check the assertion

$$\mathfrak{w} \subseteq \{s | A(s) \in \Delta\}. \tag{A.38}$$

This is a term of type Ω.

(iii) *Truth object option.* This method consists in defining a term $\mathbb{T}$ of type $P(P(\Sigma))$. The simplest choice is a variable of type $P(P(\Sigma))$ defined as

$$\mathbb{T} : P(P(\Sigma)) \to P(P(\Sigma)). \tag{A.39}$$

A term of type Ω is then obtained by

$$\{s | A(s) \in \Delta\} \in \mathbb{T} \tag{A.40}$$

which has as free variable $\Delta \in P(\mathcal{R})$ and whatever free variables are contained in $\mathbb{T}$.

Intuitively we can think of $\mathbb{T}$ as a collection of subsets of the state space that have a particular property which we know to be true. Then we consider another subset of the state space, namely $Q \subseteq S$ and we would like to know if the collection of states in Q have the property $A(s) \in \Delta$. Since we know that there is a $\mathbb{T}$ to which all collection of sets of states with the property $A(s) \in \Delta$ belong, we simply check if $\{s | A(s) \in \Delta\} \in \mathbb{T}$.

7. Any axioms added to the language have to be represented by the arrow *true* : $1 \to \Omega$.

A.5.2 Representation of $l(S)$ in a Topos

Given a topos τ with representation M, we now want to know how $l(S)$ is represented in τ.

1. *State space and quantity value object.* The objects Σ and $\mathcal{R}$ are represented by the objects $\Sigma^{\tau M}$ and $\mathcal{R}^{\tau M}$ in τ, which take the role of the state object and the quantity value object.
2. *Physical quantities.* Physical quantities are defined in terms of τ-arrows between the τ-objects $\Sigma^{\tau M}$ and $\mathcal{R}^{\tau M}$.

   We will generally require the representation to be faithful, i.e. the map $A \mapsto A^{\tau M}$ is one-to-one.
3. *Values.* Values are represented in τ by terms of type $\mathcal{R}^{\tau M}$, i.e. $A^{\tau M}(s) \in \mathcal{R}^{\tau M}$ where $A^{\tau M} : \Sigma^{\tau M} \to \mathcal{R}^{\tau M}$.
4. *Truth values.* A formula $A(s) \in \Delta$ is a term of type Ω, thus it is represented in a topos τ by an arrow

$$[A(s) \in \Delta]^{\tau M} : \Sigma^{\tau M} \times P\mathcal{R}^{\tau M} \to \Omega^{\tau M}.$$

Such an arrow gets factored as follows:

$$[A(s) \in \Delta]^{\tau_M} = e_{\mathcal{R}^{\tau_M}} \circ \langle [A(s)]^{\tau_M}, [\Delta]^{\tau_M} \rangle \tag{A.41}$$

where $e_{\mathcal{R}^{\tau_M}} : \mathcal{R}^{\tau_M} \times P\mathcal{R}^{\tau_M} \to \Omega^{\tau_M}$ is the evaluation map, $[A(s)]^{\tau_M} : \Sigma^{\tau_M} \to \mathcal{R}^{\tau_M}$ is the arrow representing the physical quantity A and $[\Delta]^{\tau_M} : P\mathcal{R}^{\tau_M} \to P\mathcal{R}^{\tau_M}$ is simply the identity arrow. Putting the two results together we have

$$\Sigma^{\tau_M} \times P\mathcal{R}^{\tau_M} \xrightarrow{[A(s)]^{\tau_M} \times [\Delta]^{\tau_M}} \mathcal{R}^{\tau_M} \times P\mathcal{R}^{\tau_M} \xrightarrow{e_{\mathcal{R}^{\tau_M}}} \Omega^{\tau_M}. \tag{A.42}$$

Truth values are terms of type Ω with no free variables. Hence in the topos τ, they will be represented by elements of the sub-object classifier Ω^{τ_M}, i.e. global elements $\gamma^{\tau_M} : 1^{\tau_M} \to \Omega^{\tau_M}$, $\gamma^{\tau_M} \in \Gamma\Omega^{\tau_M}$.

5. *Propositions*. A proposition is a term of type $P(\Sigma)$, hence in a topos it will be defined as an element in $P(\Sigma^{\tau_M})$. In particular, consider a term of type $P(\Sigma^{\tau_M})$ with free variable Δ of type $P(\mathcal{R}^{\tau_M})$. In a topos this is represented by an arrow

$$\big[\{s|A(s) \in \Delta\}\big]^{\tau_M} : P\mathcal{R}^{\tau_M} \to P\Sigma^{\tau_M}. \tag{A.43}$$

Using this term of type $P(\Sigma^{\tau_M})$, which is represented in τ by the arrow $[\Xi]^{\tau_M} : 1^{\tau_M} \to P(\mathcal{R}^{\tau_M})$, a proposition $A \in \Delta$ is represented as:

$$\big[\{s|A(s) \in \Delta\}\big]^{\tau_M} \circ [\Xi]^{\tau_M} : 1^{\tau_M} \to P\big(\Sigma^{\tau_M}\big). \tag{A.44}$$

6. *States*. We will now analyse how the different 'types' of states described above are represented in a topos.
   (a) *Micro-state option*. We have seen that a micro-state is essentially a term of type Σ, hence in a topos it is represented by a global element (if it exists) of Σ^{τ_M}, i.e.

$$s : 1^{\tau_M} \to \Sigma^{\tau_M}. \tag{A.45}$$

   Moreover, given a term of type $P(\mathcal{R})$, which is represented in τ by an arrow $[\Xi]_{\tau_M} : 1^{\tau_M} \to P(\mathcal{R}^{\tau_M})$ it is possible to define a map $\langle s, [\Xi]\rangle : 1^{\tau_M} \to \Sigma^{\tau_M} \times P(\mathcal{R}^{\tau_M})$, which, if combined with the arrow $[A(s) \in \Delta]^{\tau_M} : \Sigma^{\tau_M} \times P(\mathcal{R}^{\tau_M}) \to \Omega^{\tau_M}$ gives

$$1_{\tau_M} \xrightarrow{\langle s, [\Xi]_{\tau_M}\rangle} \Sigma^{\tau_M} \times P\big(\mathcal{R}^{\tau_M}\big) \xrightarrow{[A(s) \in \Delta]^{\tau_M}} \Omega^{\tau_M}. \tag{A.46}$$

   This is the global element of Ω^{τ_M} representing the truth value of the proposition $(A(s) \in \Delta)$.
   (b) *Psuedo-state object*. Pseudo-states are identified with terms of type $P(\Sigma)$, so in a topos they are represented by elements

$$\mathfrak{w}^{\tau_M} : 1^{\tau_M} \to P\big(\Sigma^{\tau_M}\big). \tag{A.47}$$

Given a proposition

$$[\{s|A(s)\in\Delta\}]^{\tau M}\circ[\Xi]^{\tau M}:1\rightarrow P(\Sigma^{\tau M}) \tag{A.48}$$

we combine the two maps to give

$$(\mathfrak{w}^{\tau M},[\{s|A(s)\in\Xi\}]^{\tau M}\circ[\Delta]^{\tau M}):1^{\tau M}\rightarrow P(\Sigma^{\tau M})\times P(\Sigma^{\tau M}). \tag{A.49}$$

Considering the arrow $[\mathfrak{w}\subseteqq[\{s|A(s)\in\Delta\}]^{\tau M}]:P(\Sigma^{\tau M})\times P(\Sigma^{\tau M})\rightarrow\Omega^{\tau M}$, which represents the term $(\mathfrak{w}\subseteqq\{s|A(s)\in\Delta\})$ of type Ω, we can define the truth value of the proposition (A.48) given the pseudo-state (A.47) as

$$1^{\tau M}\xrightarrow{(\mathfrak{w}^{\tau M},[\{s|A(s)\in\Delta\}]^{\tau M}\circ[\Xi]^{\tau M})}P(\Sigma^{\tau M})\times P(\Sigma^{\tau M})\xrightarrow{[\mathfrak{w}\subseteqq[\{s|A(s)\in\Delta\}]^{\tau M}]}\Omega^{\tau M}. \tag{A.50}$$

(c) *Truth object*. A truth object is a term $\mathbb{T}$ of type $P(P(\Sigma))$ such that, given the proposition $\{s|A(s)\in\Delta\}$, the term $(\{s|A(s)\in\Delta\}\in\mathbb{T})$ is of type Ω. Such a term has free variables Δ of type $P(\mathcal{R})$ and $\mathbb{T}$ of type $P(P(\Sigma))$. Therefore, its representation in a topos τ is

$$[\{s|A(s)\in\Delta\}\in\mathbb{T}]^{\tau M}:P(\mathcal{R}^{\tau M})\times P(P(\Sigma^{\tau M}))\rightarrow\Omega^{\tau M} \tag{A.51}$$

which can be factored as follows:

$$[\{s|A(s)\in\Delta\}\in\mathbb{T}]^{\tau M}=e^{P(\Sigma_{\tau M})}\circ([\{s|A(s)\in\Delta\}]^{\tau M}\times[\mathbb{T}]^{\tau M}). \tag{A.52}$$

Here $e_{P(\Sigma^{\tau M})}:P(\Sigma^{\tau M})\times P(P(\Sigma^{\tau M}))\rightarrow\Omega^{\tau M}$ is the evaluation map and

$$[\{s|A(s)\in\Delta\}]^{\tau M}:P(\mathcal{R}^{\tau M})\rightarrow P(\Sigma^{\tau M}) \tag{A.53}$$

$$[\mathbb{T}]^{\tau M}:P(P(\Sigma^{\tau M}))\stackrel{id}{\rightarrow}P(P(\Sigma^{\tau M})). \tag{A.54}$$

Given the above, the truth value of the proposition $A\in\Delta$ is represented as

$$v(A\in\Delta;\mathbb{T}):[\{s|A(s)\in\Delta\}\in\mathbb{T}]^{\tau M}\circ\langle[\Delta]^{\tau M},[\mathbb{T}]^{\tau M}\rangle \tag{A.55}$$

where $\langle[\Delta^{\tau M}],[\mathbb{T}^{\tau M}]\rangle:1^{\tau M}\rightarrow P(\mathcal{R}^{\tau M})\times P(P(\Sigma^{\tau M}))$.

For example in classical physics the topos in which we represent the language $l(S)$ is **Sets**, thus the truth object Ω is simply the set $\{0,1\}$. In this context we have

$$v(A\in\Delta;\mathbb{T}):=[\{s|A(s)\in\Delta\}\in\mathbb{T}^s]:P(\mathcal{R})\times P(P(\Sigma))\rightarrow\{0,1\} \tag{A.56}$$

such that

$$v(A\in\Delta;\mathbb{T})(\Delta,\mathbb{T})=\begin{cases}1 & \text{if } \{s\in\Sigma|A(s)\in\Delta\}\in\mathbb{T}\\ 0 & \text{otherwise}\end{cases} \tag{A.57}$$

$$=\begin{cases}1 & \text{if } A^{-1}(\Delta)\in\mathbb{T}\\ 0 & \text{otherwise.}\end{cases} \tag{A.58}$$

.

A.6 Deductive System of Reasoning for First Order Logic

So far, we have defined the symbols and formation rules for the first order language l. However, in order to actually use l as a language that enables us to talk about things, we also require rules of inference. Such rules will allow us to derive true statements from other true statements.

In order to describe this better we need to introduce the notion of a sequent.

Definition A.6 Given two formulae ψ and ϕ a sequent is an expression $\psi \vdash_{\vec{x}} \phi$ which indicates that ϕ is a logical consequence of ψ in the context $\vec{x}$.[5]

What this means is that any assignment of values of the variables in $\vec{x}$ which makes ψ true, will also make ϕ true.

The deduction system will then be defined as a sequent calculus, i.e. a set of inference rules which will allow us to infer a sequent from other sequents. Symbolically, a rule of inference is written as follows:

$$\frac{\Gamma}{\psi \vdash_{\vec{x}} \phi} \tag{A.59}$$

which means that the sequent $\psi \vdash_{\vec{x}} \phi$ can be inferred by the collection of sequents Γ. We can also have a double inference as follows:

$$\frac{\Gamma}{\overline{\psi \vdash_{\vec{x}} \phi}}.$$

This can be read in both directions, thus it means that $\psi \vdash_{\vec{x}} \phi$ can be inferred from the collection of sequents Γ, but also that the collection of sequents Γ can be inferred from $\psi \vdash_{\vec{x}} \phi$.

We will now define a list of inference rules. In the following, the symbol Γ will represent a collection of sequents, the letters γ, β, α will represent formulae, the letters $\sigma, \tau, \ldots$ will represent terms of some type and $\alpha \cup \Gamma$ represent the collections of formulas in both Γ and the formula α.

Variables in a term can be either *free* or *bounded*. We say that a variable α of a term σ is *bounded* if it appears within a context of the form $\{\alpha \vdash_{\vec{x}} \sigma\}$, otherwise it will be called *free*. The inference rules are:

- *Thinning*

$$\frac{\beta \cap \Gamma \vdash_{\vec{x}} \alpha}{\overline{\Gamma \vdash_{\vec{x}} \alpha}}.$$

- *Cut*

$$\frac{\Gamma \vdash_{\vec{x}} \alpha, \alpha \cup \Gamma \vdash_{\vec{x}} \beta}{\overline{\Gamma \vdash_{\vec{x}} \beta}}.$$

[5] A context $\vec{x}$ is a list of distinct variables. When applied to a formula α it indicates that α has variables only within that context.

For any free variable of α free in Γ or β.

- *Substitution*

$$\frac{\Gamma \vdash_{\vec{x}} \alpha}{\Gamma(x/\sigma) \vdash_{\vec{x}} \alpha(x/\sigma)}$$

when σ is free in Γ and α. The term $\Gamma(x/\sigma)$ indicates the term obtained from Γ by substituting σ (which is a term of some type) for each free occurrence of x.

- *Extensionality*

$$\frac{\Gamma \vdash_{\vec{x}} x \in \sigma \Leftrightarrow x \in \rho}{\Gamma \vdash_{\vec{x}} \sigma = \rho}$$

where x is not free in either Γ, σ or ρ.

- *Equivalence*

$$\frac{\alpha \cup \Gamma \vdash_{\vec{x}} \beta \quad \beta \cup \Gamma \vdash_{\vec{x}} \alpha}{\Gamma \vdash_{\vec{x}} \alpha \Leftrightarrow \beta}.$$

- *Finite Conjunction*

The rules for finite conjunction are the following:

$$\alpha \vdash_{\vec{x}} \alpha = \top, \qquad \alpha \wedge \beta \vdash_{\vec{x}} \beta, \qquad \alpha \wedge \beta \vdash_{\vec{x}} \alpha. \tag{A.60}$$

Note that we have used part of the definition of the logical connective '*if then*'.

A consequence of these rules is

$$\frac{\alpha \vdash_{\vec{x}} \beta \quad \alpha \vdash_{\vec{x}} \gamma}{\alpha \vdash_{\vec{x}} \gamma \wedge \beta}. \tag{A.61}$$

Proof

$$\dfrac{\alpha \vdash_{\vec{x}} \beta \qquad \dfrac{\alpha \vdash_{\vec{x}} \gamma \qquad \dfrac{\dfrac{}{\beta \vdash_{\vec{x}} \beta = \top}\,(3) \qquad \dfrac{\dfrac{}{\gamma \vdash_{\vec{x}} \gamma = \top}\,(1) \qquad \dfrac{}{\gamma = \top \cup \beta = \top \vdash_{\vec{x}} \gamma \wedge \beta}\,(2)}{\gamma \cup \beta = \top \vdash_{\vec{x}} \gamma \wedge \beta}\,(4)}{\gamma \cup \beta \vdash_{\vec{x}} \gamma \wedge \beta}\,(5)}{\alpha \cup \beta \vdash_{\vec{x}} \alpha \wedge \beta}\,(6)}{\alpha \vdash_{\vec{x}} \gamma \wedge \beta}\,(7)$$

□

This proof should be read from top to bottom and consists, as one can see, of a finite collection of sequents called a finite *tree*, in which the bottom vertex represents the *conclusion* of the proof. All the sequents of the proof are correlated to each other in the following way:

1. A sequent belonging to a node[6] which has nodes above it is derived by applying a rule of inference to the sequents belonging to the above nodes.
2. Every top most node is either a basic axiom or a premise of the proof.

[6] A node is an inference step: $\frac{\Gamma_1}{\Gamma_2}$.

In the proof above we have that

$$\overline{\gamma \vdash_{\vec{x}} \gamma = \top}$$

is derived by the thinning axiom, the equivalence axiom and the axioms $\gamma \vdash_{\vec{x}} \gamma$ and $\vdash_{\vec{x}}$ true as follows:

Proof

$$\cfrac{\cfrac{\cfrac{\vdash_{\vec{x}} \text{true}}{\gamma \vdash_{\vec{x}} \top}\,(1)}{\gamma \cup \gamma \vdash_{\vec{x}} \top}\,(3) \quad \cfrac{\overline{\gamma \vdash_{\vec{x}} \gamma}}{\top \cup \gamma \vdash_{\vec{x}} \gamma}\,(2)}{\gamma \vdash_{\vec{x}} \gamma = \top}\,(4)$$

□

Where the lines (1), (2) and (3) are an application of the thinning axiom, while line (4) is the application of the equivalence axiom where the equivalence $\alpha \Leftrightarrow \beta := \alpha = \beta$ was used.

Going back to the proof of the *conjunction axiom* the remaining lines are derived as follows:

(i) Line (2) is the definition of the logical connective $\wedge$.
(ii) All the other lines are derived from applications of the *cut axiom*.

It should be noted that it is also possible to form a more general version of the *conjunction axiom* by replacing the single sequent α by a collection of sequents Γ, thus the *conjunction Axiom* becomes:

$$\frac{\Gamma \vdash_{\vec{x}} \beta \quad \Gamma \vdash_{\vec{x}} \gamma}{\Gamma \vdash_{\vec{x}} \beta \wedge \gamma}.$$

- *Finite Disjunction*
  The rules for finite disjunction consist of the following axioms:

$$\bot \vdash_{\vec{x}} \alpha \ \alpha \vdash_{\vec{x}} \alpha \vee \beta \ \beta \vdash_{\vec{x}} \alpha \vee \beta \tag{A.62}$$

  and the following rule of inference:

$$\frac{\alpha \vdash_{\vec{x}} \gamma \quad \beta \vdash_{\vec{x}} \gamma}{\alpha \vee \beta \vdash_{\vec{x}} \gamma} \tag{A.63}$$

  whose generalisation is

$$\frac{\alpha \cup \Gamma \vdash_{\vec{x}} \gamma \quad \beta \cup \Gamma \vdash_{\vec{x}} \gamma}{\alpha \vee \beta \cup \Gamma \vdash_{\vec{x}} \gamma}.$$

- *Implication*
  For implication we have the double inference rule

$$\frac{\beta \wedge \alpha \vdash_{\vec{x}} \gamma}{\overline{\alpha \vdash_{\vec{x}} \beta \Rightarrow \gamma}}.$$

Again the general form of which the above is a specification is

$$\frac{\beta \cup \Gamma \vdash_{\vec{x}} \gamma}{\Gamma \vdash_{\vec{x}} \beta \Rightarrow \gamma}.$$

To see why that is the case we will prove the above generalisation, but only one way:

Proof

$$\cfrac{\cfrac{\cfrac{\beta \vdash_{\vec{x}} \beta}{\beta \cup \Gamma \vdash_{\vec{x}} \beta}(1)}{\beta \wedge \gamma \cup \Gamma \vdash_{\vec{x}} \beta}(2) \quad \cfrac{\beta \cup \Gamma \vdash_{\vec{x}} \beta \quad \beta \cup \Gamma \vdash_{\vec{x}} \gamma}{\beta \cup \Gamma \vdash_{\vec{x}} \beta \wedge \gamma}(3)}{\Gamma \vdash_{\vec{x}} \beta \Rightarrow \gamma}(4).$$

where in (4) we used the definition of implication given in Sect. A.3: $\alpha \Rightarrow \beta := (\alpha \wedge \beta) \Leftrightarrow \alpha$. □

- *Negation*
  For negation we only have one axiom

$$\bot \vdash_{\vec{x}} \alpha \tag{A.64}$$

  while the inference rules are

$$\frac{(\alpha \cup \Gamma) \vdash_{\vec{x}} \bot}{\Gamma \vdash_{\vec{x}} \neg\alpha}$$

  and

$$\frac{\Gamma \vdash_{\vec{x}} \alpha}{(\neg\alpha \cup \Gamma) \vdash_{\vec{x}} \bot}.$$

- *Universal Quantification*
  We have the following double inference rule

$$\frac{\alpha \vdash_{\vec{x}y} \beta}{\alpha \vdash_{\vec{x}} \forall y \beta}$$

  where y is not free in either β or α.
  Again the generalisation is

$$\frac{\Gamma \vdash_{\vec{x}y} \beta}{\Gamma \vdash_{\vec{x}} \forall y \beta}.$$

- *Existential Quantifier*
  We have the double inference rule

$$\frac{\alpha \vdash_{\vec{x}y} \beta}{(\exists y)\alpha \vdash_{\vec{x}} \beta}$$

where y is a free variable in β.

Again the generalization would be

$$\frac{\alpha \cup \Gamma \vdash_{\vec{x}y} \beta}{(\exists y)\alpha \cup \Gamma \vdash_{\vec{x}} \beta}.$$

- *Distributive Axiom*

$$\big(\alpha \wedge (\beta \vee \gamma)\big) \vdash_{\vec{x}} \big((\alpha \wedge \beta) \vee (\alpha \wedge \gamma)\big). \tag{A.65}$$

- *Frobenius Axiom*

$$\big(\alpha \wedge (\exists y)\beta\big) \vdash_{\vec{x}} (\exists y)(\alpha \wedge \beta) \tag{A.66}$$

where $y \notin \vec{x}$.

- *Law of Excluded Middle*

$$\top \vdash_{\vec{x}} \alpha \vee \neg\alpha. \tag{A.67}$$

It should be noted that, for intuitionistic type of higer order languages the law of excluded middle does not hold. All the rest does.

With this we end our definition of higher order languages $\mathcal{L}$, which are comprised of a set of term types, a set of logical connectives and a set of rules of inference which determine the logic.

Appendix B
Worked out Examples

B.1 Category Theory

Example B.1 An iso arrow is always epic. In fact, consider an iso f such that $g \circ f = h \circ f$ ($f : a \to b$ and $g, h : b \to c$)

$$g = g \circ id_b = g \circ \left(f \circ f^{-1}\right) = (g \circ f) \circ f^{-1} = (h \circ f) \circ f^{-1}$$
$$= h \circ \left(f \circ f^{-1}\right) = h$$

therefore f is right cancellable.

Example B.1 Given a category $\mathcal{C}$, isomorphism is an equivalence relation on $\mathcal{C}$. This is defined as follows:

Given any two objects $A, B \in \mathcal{C}$ we say that they are equivalent iff there exists an iso $i : A \to B$. This is a well defined equivalence relation since it satisfies the following properties:

1. Reflexivity. $\forall A \in \mathcal{C}$, $id_A : A \to A$ is an iso since $id_A \circ id_A = id_A$.
2. Transitive. Assume that we have two isos $i_1 : A \to B$ and $i_2 : B \to C$. We define the composite $h := i_2 \circ i_1$. For this to be an iso it has to have an inverse.

$$(i_2 \circ i_1) \circ (i_2 \circ i_1)^{-1} = i_2 \circ i_1 \circ i_1^{-1} \circ i_2^{-1} \overset{\text{associativity}}{=} i_2 \circ id_B \circ i_2^{-1}$$
$$\overset{\text{identity}}{=} i_2 \circ i_2^{-1} = id_C. \quad \text{(B.1)}$$

   Similarly we can show that $(i_1 \circ i_2)^{-1} \circ (i_1 \circ i_2) = id_B$
3. Symmetry. If we have an iso $i : A \to B$ then there exists a unique inverse $i^{-1} : B \to A$. Since $i^{-1} : B \to A$ is unique with inverse i then i^{-1} is the desired iso.

Example B.2 In any category the following are true:

1. $g \circ f$ is monic if both g and f are monic.
2. If $g \circ f$ is monic then so is f.

C. Flori, *A First Course in Topos Quantum Theory*, Lecture Notes in Physics 868,
DOI 10.1007/978-3-642-35713-8, © Springer-Verlag Berlin Heidelberg 2013

In fact we have that

(1) If $f : A \to B$ and $g : B \to C$ are monic then, given two arrows $h, j : D \to A$ such that $(g \circ f) \circ h = (g \circ f) \circ j$ we have:

$$\begin{aligned} &(g \circ f) \circ h = (g \circ f) \circ j \\ &\xrightarrow{\text{associativity}} g \circ (f \circ h) = g \circ (f \circ j) \\ &\xrightarrow{\text{monic}} f \circ h = f \circ j \\ &\xrightarrow{\text{monic}} h = j. \end{aligned} \tag{B.2}$$

It follows that $(g \circ f)$ is monic.

(2) Given $f : A \to B$, and $h, k : D \to A$, assume that $f \circ h = f \circ k$. Then consider $g : B \to C$ such that $(g \circ f)$ is monic. It follows that

$$(g \circ f) \circ h = g \circ (f \circ h) = g \circ (f \circ k) = (g \circ f) \circ k \tag{B.3}$$

implies that $h = k$.

Example B.3 Given any category $\mathcal{C}^{op}$, a map $f \in \mathcal{C}^{op}$ is monic in $\mathcal{C}^{op}$ if and only if it is epic in $\mathcal{C}$. In fact $f : B \to A$ is monic in $\mathcal{C}^{op}$ iff for all $g, h : C \to B$ in $\mathcal{C}^{op}$ then

$$f \circ g = f \circ h \quad \text{implies that} \quad g = h. \tag{B.4}$$

However from the definition of dual category the above holds iff for all $g, h : B \to C$ in $\mathcal{C}$ we have

$$g \circ f = h \circ f \quad \text{implies} \quad g = h \tag{B.5}$$

where now $f : A \to B$. However, this is true iff f is epic in $\mathcal{C}$. We then say that monic and epic are dual notions.

Example B.4 The category $(\mathbb{R}, \leq)^{op}$ has as:

1. Objects: $r \in \mathbb{R}$.
2. Morphisms: given any two elements $x, y \in (\mathbb{R}, \leq)$ such that $x \leq y$ then there exists a unique arrow $f : y \to x$ in $(\mathbb{R}, \leq)^{op}$.

   These morphisms undergo the following properties:

   (i) Composition: if $x \leq y$ and $y \leq z$ then in $(\mathbb{R}, \leq)$, $x \leq z$. This implies that there exist three maps in $(\mathbb{R}, \leq)^{op}$, namely $f : y \to x$, $g : z \to y$ and $h : z \to x$ such that $h := f \circ g$.
   (ii) Associativity: given that[1] $f' : x \leq y$, $g' : y \leq z$ and $k' : z \leq w$ in $(\mathbb{R}, \leq)$ we then have that $k' \circ (g' \circ f') = (k' \circ g') \circ f'$, therefore in $(\mathbb{R}, \leq)^{op}$ we have $f' \circ (g' \circ k') = (f' \circ g') \circ k'$.

[1] Here $x \leq y$ means that there is a map $f' : x \to y$ in $(\mathbb{R}, \leq)^{op}$ or equivalently a map $f : y \to x$ in $(\mathbb{R}, \leq)$.

(iii) Identity element: in $\mathbb{R}$ we have that $x \leq x$, by duality $i_x : x \to x$ is the identity morphism on x in $(\mathbb{R}, \leq)^{op}$.

Example B.5 Given a category $\mathcal{C}$ with initial object 0, the following hold:

(i) If $A \simeq 0$ (i.e. there is an iso map between them) then A is an initial object.
(ii) If there exists a monic arrow $f : A \to 0$, then f is an iso.

Proof (i) An initial object 0 is such that, given any other object $A \in \mathcal{C}$ there exists one and only one map $i : 0 \to A$. If such an arrow i is iso, then $i^{-1} : A \to 0$ exists and is the unique inverse.

Now consider any other object $C \in \mathcal{C}$, we know that there exists a unique arrow $f : 0 \to C$. We then assume that we have two maps $h, g : A \to C$, we then obtain the following diagram

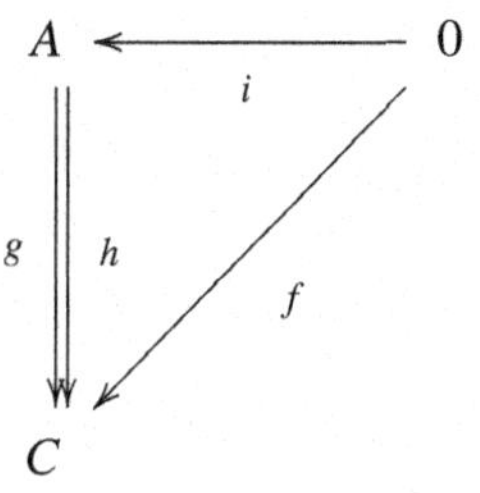

From the property of the initial object we have that

$$h \circ i = g \circ i = f. \tag{B.6}$$

However since i is iso, it is right cancellable, therefore $h = g$. This implies that given any object $C \in \mathcal{C}$ there exists one and only one map $f : A \to C$. Thus A is an initial object.

(ii) $f : A \to 0$ is monic. Since 0 is the initial object we have the unique arrow $i : 0 \to A$. From the property of 0 being an initial object it follows that $f \circ i = id_0$ (since there is one and only one arrow from any object to 0, including from 0 to itself). On the other hand

$$f \circ (i \circ f) = f \circ id_C \quad \text{implies} \quad i \circ f = id_C \quad \text{(monic property).} \tag{B.7}$$

□

Example B.6 Given a category $\mathcal{C}$ with terminal object 1, if there exists an arrow $g : 1 \to A$ with domain the terminal object, then g must be monic. In fact, if we assume that there exists an arrow $g : 1 \to A$, since 1 is the terminal object then, given any other object $A \in \mathcal{C}$, $h : A \to 1$ is unique, including $1 \to 1$. It follows that $h \circ g = id_1$. Now consider two maps $f, k : B \to 1$ such that $g \circ k = g \circ f$ we then have

$$f = id_1 \circ f = (h \circ g) \circ f = h \circ (g \circ f) = h \circ (g \circ k) = (h \circ g) \circ k = id_1 \circ k = k \tag{B.8}$$

which implies that g is monic.

Example B.7 Given two objects A, B in some category $\mathcal{C}$, consider the category $\mathbf{Pair}(A, B)$, whose objects are triplets (P, p_1, p_2) where $p_1 : P \to A$ and $p_2 : P \to B$. The morphisms in $\mathbf{Pair}(A, B)$ are maps

$$f : (P, p_1, p_2) \to (Q, q_1, q_2) \tag{B.9}$$

such that $f : P \to Q$ is a morphism in $\mathcal{C}$ and $q_1 \circ f = p_1$; $q_2 \circ f = p_2$, i.e. the following diagram commutes

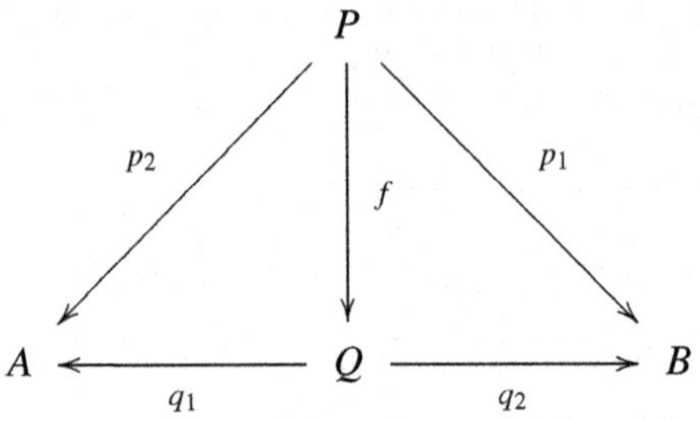

We then have that

1. $\mathbf{Pair}(A, B)$ is a category.
2. $(A \times B, \pi_1, \pi_2)$ is a product if it is a terminal object in $\mathbf{Pair}(A, B)$.

Proof 1. To show that $\mathbf{Pair}(A, B)$ is a category, we need to show that the following properties hold:

(a) *Composition.* Given $f : (P, p_1, p_2) \to (Q, q_1, q_2)$ and $g : (Q, q_1, q_2) \to (S, s_1, s_2)$ we need to define composition. Thus we say that $g \circ f$ is the map which makes the following diagram commute:

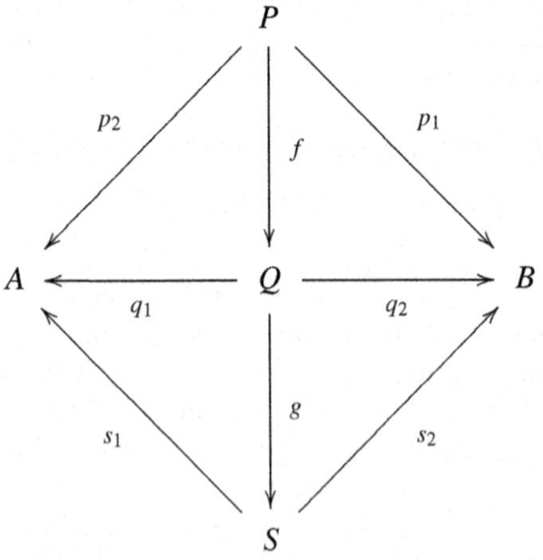

(b) *Identity morphism.* $Id_{(P,p_1,p_2)} : (P, p_1, p_2) \to (P, p_1, p_2)$

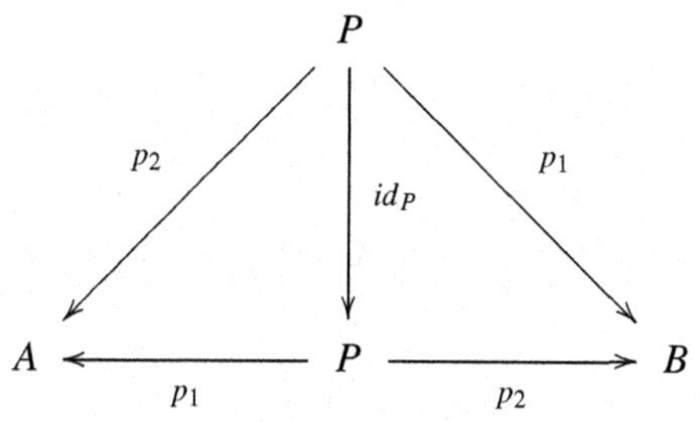

(c) *Associativity*

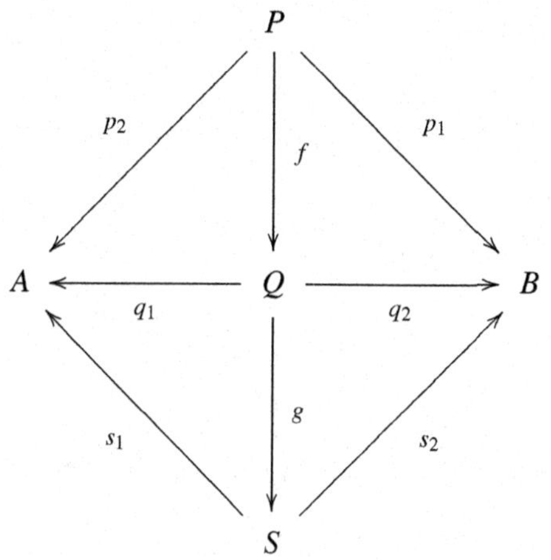

composed with

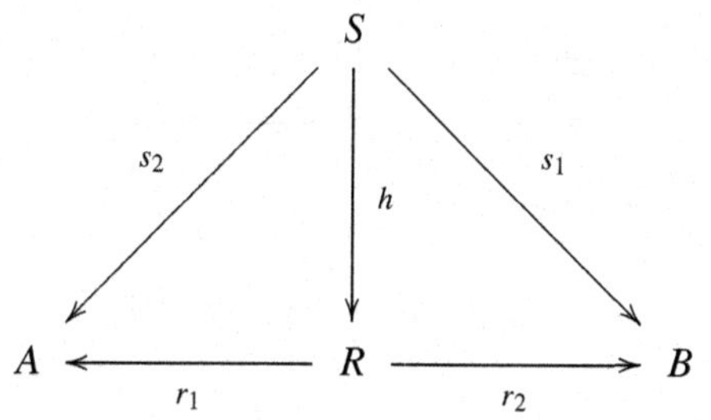

gives

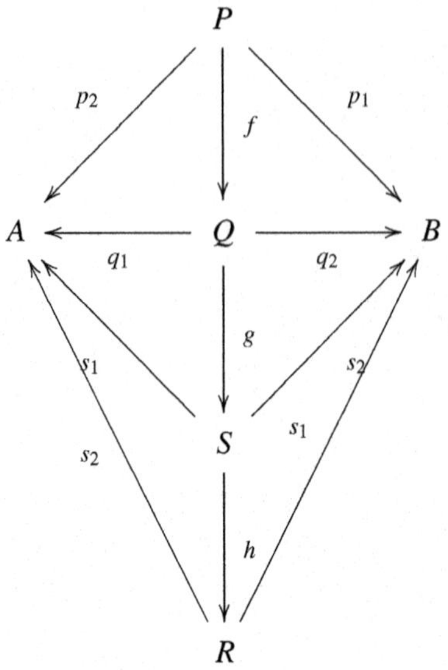

which is the same as

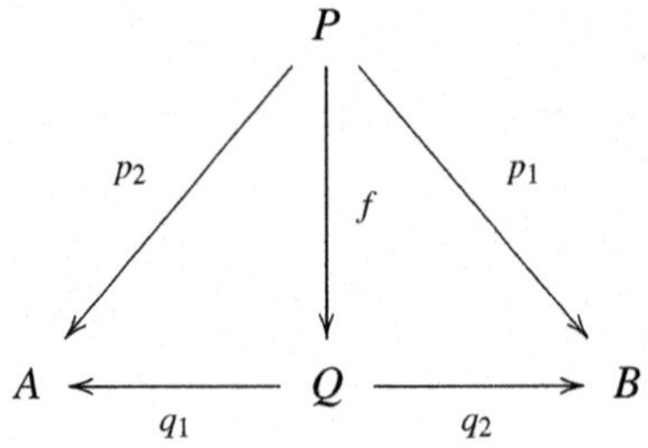

composed with

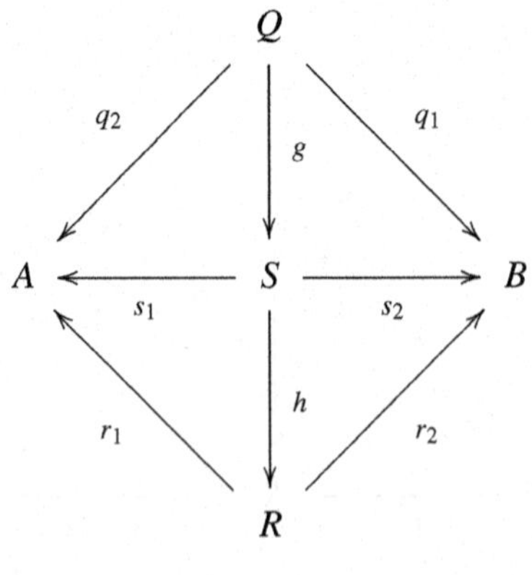

which gives

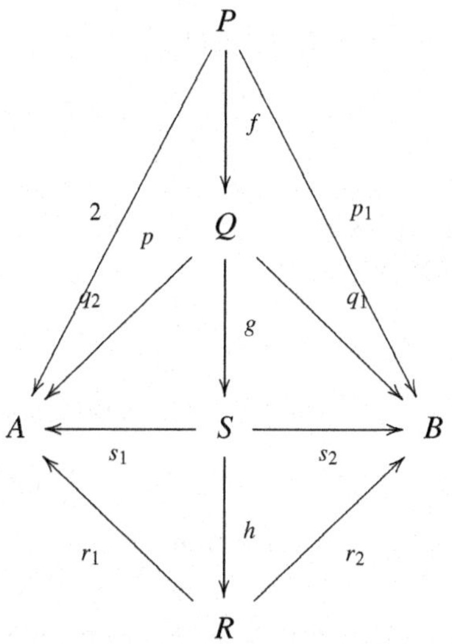

2. If $(A \times B, \pi_1, \pi_2)$ is a terminal object then, for all elements $(P, p_1, p_2) \in$ $\mathbf{Pair}(A, B)$ there exists one and only one arrow

$$(P, p_1, p_2) \to (A \times B, \pi_1, \pi_2) \tag{B.10}$$

which, by definition implies that the following diagram commutes:

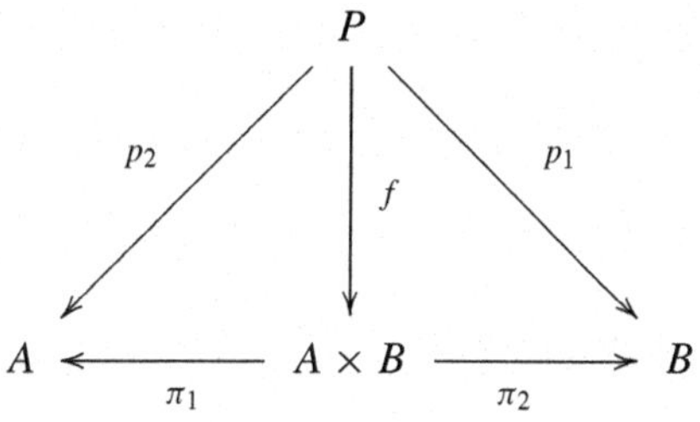

But this is precisely the condition satisfied by the product. □

Example B.8 For any triple $A \overset{\pi_1}{\longleftarrow} A \times B \overset{\pi_2}{\longrightarrow} B$ the following propositions are equivalent.

1. Given any triple $A \overset{f}{\leftarrow} C \overset{g}{\rightarrow} B$, there exists a unique morphism $\langle f, g\rangle : C \to A \times B$, such that

$$\pi_1 \circ \langle f, g\rangle = f \quad \text{and} \quad \pi_2 \circ \langle f, g\rangle = g. \tag{B.11}$$

2. For any triple $A \overset{f}{\leftarrow} C \overset{g}{\rightarrow} B$ there exists a morphism $\langle f, g\rangle : C \to A \times B$, such that

$$\pi_1 \circ \langle f, g\rangle = f \quad \text{and} \quad \pi_2 \circ \langle f, g\rangle = g \tag{B.12}$$

and, moreover, for any $h : C \to A \times B$,

$$h = \langle \pi_1 \circ h, \pi_2 \circ h\rangle. \tag{B.13}$$

Proof (a) (1) $\Rightarrow$ (2). Given any $h : C \to A \times B$, we need to show that $h = \langle \pi_1 \circ h, \pi_2 \circ h\rangle$. From (1) we have that

$$A \xleftarrow{\pi_1 \circ h} C \xrightarrow{\pi_2 \circ h} B. \tag{B.14}$$

Thus there exists a unique arrow $k : C \to A \times B$ such that

$$\pi_1 \circ k = \pi_1 \circ h \quad \text{and} \quad \pi_1 \circ k = \pi_1 \circ h. \tag{B.15}$$

Since this equation holds for $k := h$ and from (1) $k := \langle \pi_1 \circ h, \pi_2 \circ h\rangle$, it follows that $h = \langle \pi_1 \circ h, \pi_2 \circ h\rangle$.

(b) (2) $\Rightarrow$ (1). For any triple $A \overset{f}{\leftarrow} C \overset{g}{\rightarrow} B$, (2) implies that there exists a map $\langle f, g\rangle : C \to A \times B$ such that

$$\pi_1 \circ \langle f, g\rangle = f \quad \text{and} \quad \pi_2 \circ \langle f, g\rangle = g. \tag{B.16}$$

What remains to show is that $\langle f, g\rangle$ is unique. To this end let $k : C \to A \times B$ be such that

$$\pi_1 \circ k = f \quad \text{and} \quad \pi_2 \circ k = g. \tag{B.17}$$

Then, from (2) it follows that

$$k = \langle \pi_1 \circ k, \pi_2 \circ k\rangle = \langle f, g\rangle. \tag{B.18}$$

□

Example B.9 Given the category $\mathcal{C}$:

$$0 \to 1, \qquad 1' \to 2 \tag{B.19}$$

and the category $\mathcal{D}$:

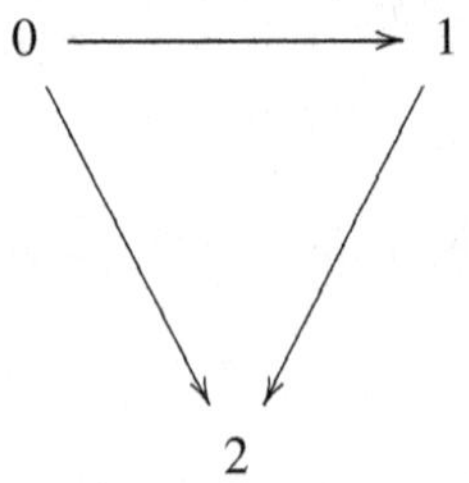

we define the functor $F : \mathcal{C} \to \mathcal{D}$ as

$$\begin{aligned} F : 2+2 &\to 3 \\ 0 &\mapsto 0 \\ 1 &\mapsto 1 \\ 1' &\mapsto 1 \\ 2 &\mapsto 2. \end{aligned} \tag{B.20}$$

It follows that

$$F(f : 0 \to 1) := 0 \to 1; \qquad F\big(g : 1' \to 2\big) := 1 \to 2. \tag{B.21}$$

The only arrow which does not lie in the image of the F functor is $0 \to 2$ which is $0 \to 2$ which is the composite $F(g) \circ F(f)$. Thus the image of F is not a subcategory.

Example B.10 We will now define the *bi-variant Hom functor* $\mathcal{C}(-,-) : \mathcal{C}^{op} \times \mathcal{C} \to$ *Set* and show that it is a functor. A possible definition would be:

$$\begin{aligned} \mathcal{C}(-,-) : \mathcal{C}^{op} \times \mathcal{C} &\to \mathit{Sets} \\ (A, B) &\mapsto \mathcal{C}(A, B) \\ (f, g) &\mapsto \mathcal{C}(f, g) \end{aligned} \tag{B.22}$$

where $f : C \to D$ and $g : E \to F$ are such that

$$(f, g) : (D, E) \to (C, F) \tag{B.23}$$

and

$$\begin{aligned} \mathcal{C}(f, g) : \mathcal{C}(D, E) &\to \mathcal{C}(C, F) \\ h &\to g \circ h \circ f. \end{aligned} \tag{B.24}$$

The requirements for $\mathcal{C}(-,-)$ to be a well defined functor are

$$\mathcal{C}(id_A, id_A) = id_{\mathcal{C}(A,A)} \tag{B.25}$$

which is trivially satisfied and

$$\mathcal{C}(g \circ f, g \circ f) = \mathcal{C}(f, g) \circ \mathcal{C}(f, g) \tag{B.26}$$

where

$$\begin{aligned} \mathcal{C}(g \circ f, g \circ f) : \mathcal{C}(C, A) &\to \mathcal{C}(A, C) \\ h &\mapsto g \circ f \circ h \circ g \circ f \end{aligned} \tag{B.27}$$

while

$$\mathcal{C}(f,g)\circ\mathcal{C}(f,g):\mathcal{C}(C,A)\to\mathcal{C}(B,B)\to\mathcal{C}(A,C)$$
$$h\mapsto f\circ h\circ g\mapsto g\circ f\circ h\circ g\circ f. \tag{B.28}$$

Thus (B.22) is a well defined functor.

Example B.11 The diagram

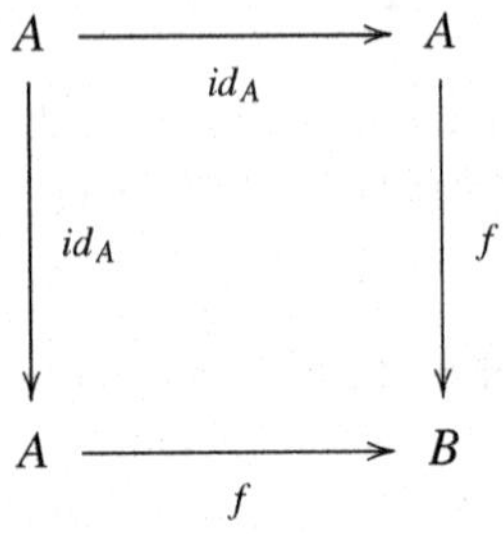

is a pullback iff f is monic. We start by assuming that f is monic. Given any pair of maps $h,g:C\to A$, such that $f\circ id_A\circ g=f\circ id_A\circ h$ then $g=h$, thus g will be the only map making the following diagram commute:

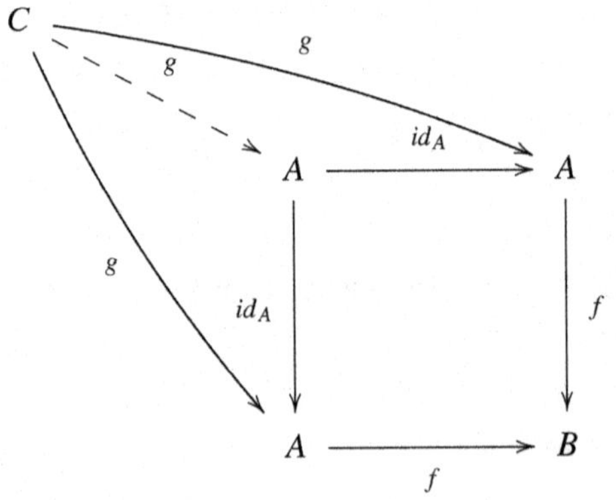

Hence the square is a pullback.

On the other hand, if the diagram is a pullback then for any other $h:C\to A$, such that $f\circ id_A\circ g=f\circ id_A\circ h$, by uniqueness of g it follows that $g=h$. Thus f is monic.

Example B.12 Given any category, if

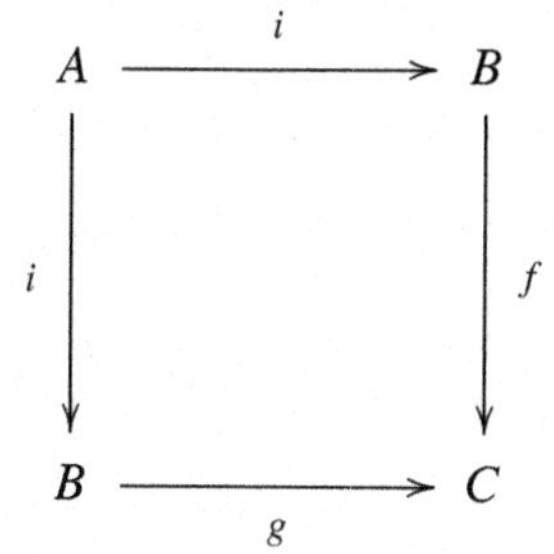

is a pullback, then i is an equaliser of f and g. In fact, let us assume that the diagram is a pullback,

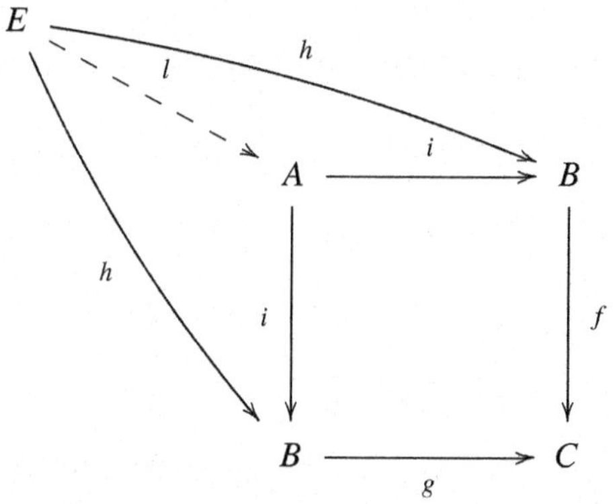

then $f \circ i = g \circ i$. Moreover for any arrow $h : E \to B$, such that $f \circ h = g \circ h$, there exists a unique l such that $h = i \circ l$, thus i is an equaliser.

On the other hand, if i is an equaliser, then $f \circ i = g \circ i$. Moreover from the universal property of an equaliser, given any arrow $h : E \to B$ there exists a unique arrow $l : E \to A$ which makes the outer square of the above diagram commute. It follows that the inner square is a pullback.

Example B.13 Given a category $\mathcal{C}$ with products and terminal object 1. For any two objects $A, B \in \mathcal{C}$ the pullback of $A \to 1 \leftarrow B$ is the product of A and B. To see this

we start by assuming that

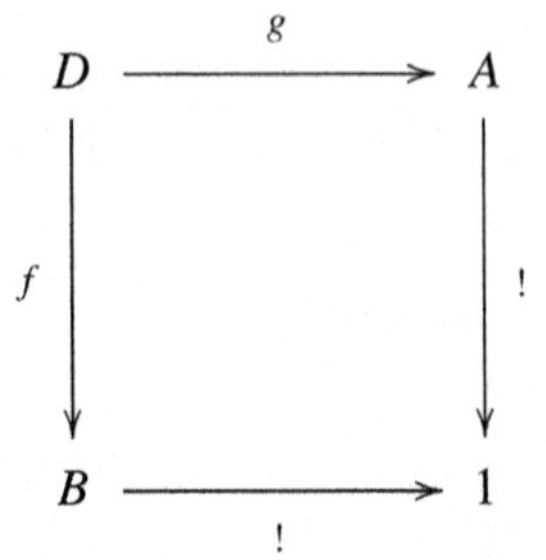

is indeed a pullback. Therefore, for each pair of map $h : E \to A$ and $k : E \to S$ such that $! \circ h =! \circ k$, there exists a unique $l : E \to D$ such that the following diagram commutes:

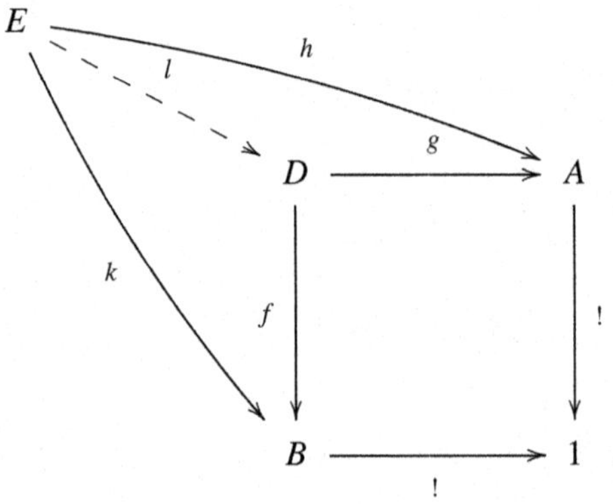

Thus $h = g \circ l, k = f \circ l$. Moreover since ! is the unique arrow to the terminal object, the condition $! \circ h =! \circ k$ is trivially satisfied (always satisfied), thus we end up with the following diagram:

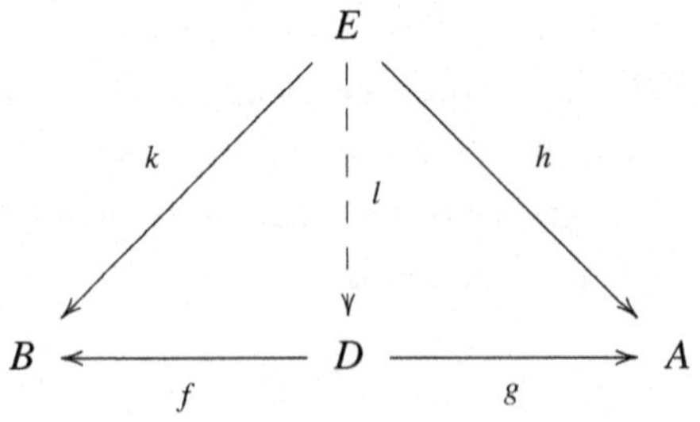

But this is precisely the definition of a product.

Example B.14 If A is an object in a category with a terminal object 1, then

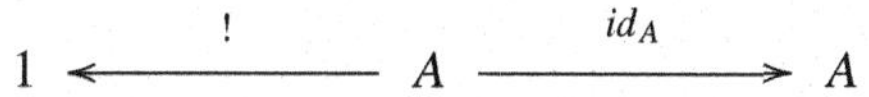

is a product diagram. In fact, given the maps $! : B \to 1$ and $f : B \to A$ we construct

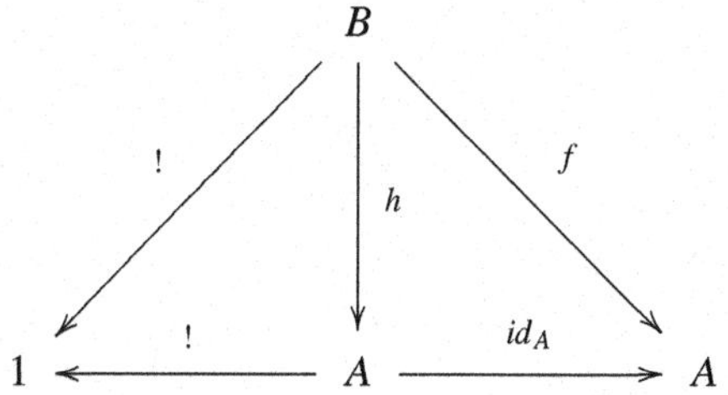

Obviously, the only arrow making the above diagram commute would be $h := f$.

Example B.15 Consider a pair of morphisms $A \xrightarrow{f} C \xleftarrow{g} B$. We then define a category $Con(f, g)$ whose objects are (f, g)-cone defined as a triple (D, p, q), such that the following diagram commutes:

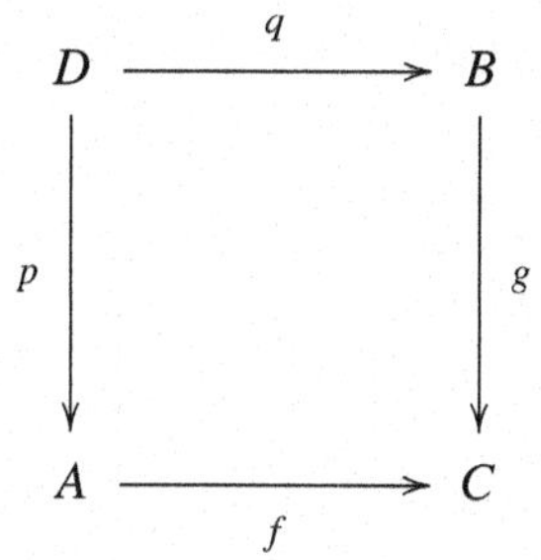

Given two (f, g)-cones (D, p, q) (D_1, p_1, q_1) a morphism $h : (D, p, q) \to (D_1, p_1, q_1)$ between them is a map $h : D \to D_1$, such that the following diagram commutes:

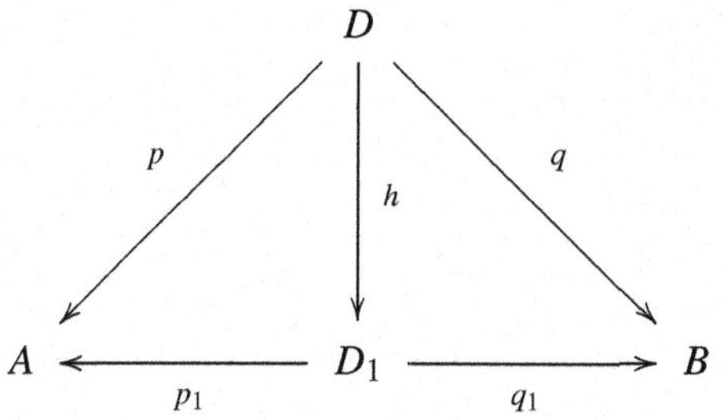

A pullback of f along g is defined via the diagram

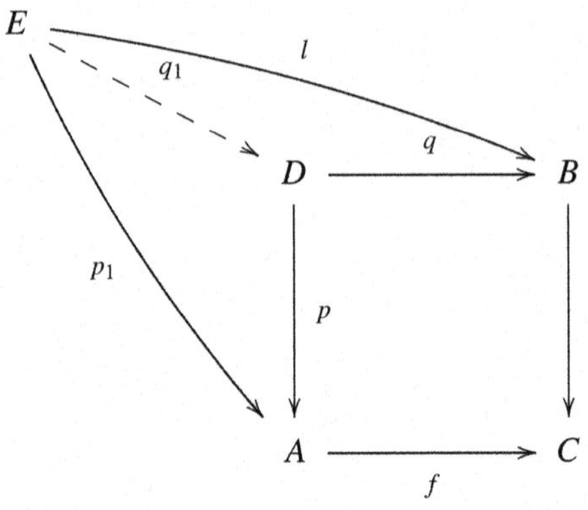

where the map l is unique for a given E. However the top part of the diagram is simply an object (E, p_1, q_1) in $Con(f, g)$. Thus the pullback property tells us that for each object $(E_i, p_i, q_i) \in Con(f, g)$ there exists a unique map $l : (E_i, p_i, q_i) \to (D, p, q)$. This means precisely that (D, p, q) is a terminal object

Example B.16 Given any map $f : A \to B$ then the characteristic functions of the identities id_A and id_B are such that $\chi_{id_B} \circ f = \chi_{id_A}$. To see this, consider the diagram

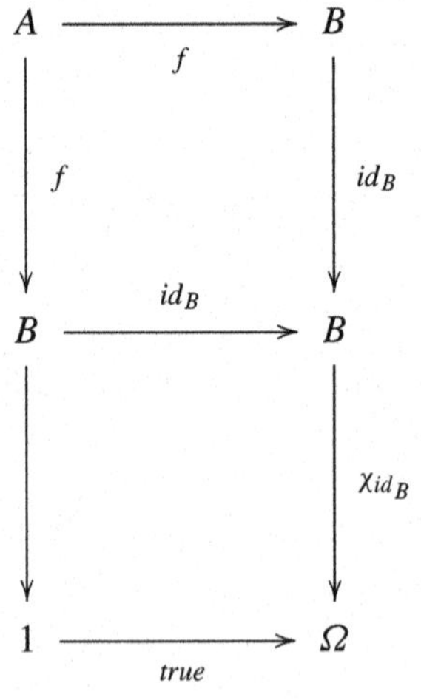

where $!_B : B \to 1$ and $!A : A \to 1$.

Since pullbackness implies commutativity, it follows that $\chi_{id_B} \circ f = true \circ !_B \circ f = true \circ !_A = \chi_{id_A}$.

B.2 Topos Quantum Theory

Example B.17 Let us consider a 4 dimensional Hilbert space $\mathbb{C}^4$, with basis $\psi_1 = (1,0,0,0)$, $\psi_2 = (0,1,0,0)$, $\psi_3 = (0,0,1,0)$, $\psi_4 = (0,0,0,1)$. We would like to define the proposition $S_z \in [-3,1] \wedge S_z \in [1,3]$, where S_z represents the value of the spin in the z direction of a two particle system. Total spin in the z direction can only have values -2, 0, 2 since the self-adjoint operator representing S_z is

$$\hat{S}_z = \begin{pmatrix} 2 & 0 & 0 & 0 \\ 0 & 0 & 0 & 0 \\ 0 & 0 & 0 & 0 \\ 0 & 0 & 0 & -2 \end{pmatrix}.$$

Thus the only value that S_z can take in the interval $[-3,1]$ is -2 while the only value it can take in the interval $S_z \in [1,3]$ is 2.

In this setting the proposition $\hat{S}_z \in [1,3]$ is represented by the projection operator $\hat{P}_1 = \text{diag}(1,0,0,0)$ (see Sect. 10.3). On the other hand the proposition $\hat{S}_z \in [-3,-1]$ is represented by the projection operator $\hat{P}_4 = \text{diag}(0,0,0,1)$ (see Sec. 12.3). Therefore all that remains to compute is $\underline{\delta(\hat{P}_1)} \wedge \underline{\delta(\hat{P}_4)}$. Let us compute this for each context $V \in \mathcal{V}(\mathcal{H})$. For the maximal algebra V and for $V_{\hat{P}_1,\hat{P}_4}$ we have

$$\underline{\delta(\hat{P}_1)}_{V_{\hat{P}_1,\hat{P}_4}} \wedge \underline{\delta(\hat{P}_4)}_{V_{\hat{P}_1,\hat{P}_4}} = \underline{\delta(\hat{P}_1)}_V \wedge \underline{\delta(\hat{P}_4)}_V = \{\lambda_1\} \cap \{\lambda_4\} = \emptyset. \tag{B.29}$$

For $V_{\hat{P}_1}$ we have

$$\underline{\delta(\hat{P}_1)}_{V_{\hat{P}_1}} \wedge \underline{\delta(\hat{P}_4)}_{V_{\hat{P}_1}} = \{\lambda_1\} \cap \{\lambda_{123}\} = \emptyset. \tag{B.30}$$

For $V_{\hat{P}_4}$ we have

$$\underline{\delta(\hat{P}_1)}_{V_{\hat{P}_4}} \wedge \underline{\delta(\hat{P}_4)}_{V_{\hat{P}_4}} = \{\lambda_{123}\} \cap \{\lambda_4\} = \emptyset. \tag{B.31}$$

For $V_{\hat{P}_2,\hat{P}_3}$ we have

$$\underline{\delta(\hat{P}_1)}_{V_{\hat{P}_2,\hat{P}_3}} \wedge \underline{\delta(\hat{P}_4)}_{V_{\hat{P}_2,\hat{P}_3}} = \{\lambda_{14}\} \cap \{\lambda_{14}\} = \{\lambda_{14}\}. \tag{B.32}$$

For $V_{\hat{P}_1,\hat{P}_j}$, $j \in \{2,3\}$ we have

$$\underline{\delta(\hat{P}_1)}_{V_{\hat{P}_1,\hat{P}_j}} \wedge \underline{\delta(\hat{P}_4)}_{V_{\hat{P}_1,\hat{P}_j}} = \{\lambda_1\} \cap \{\lambda_{4i}\} = \emptyset. \tag{B.33}$$

For $V_{\hat{P}_i,\hat{P}_4}$, $i \in \{2,3\}$ we have

$$\underline{\delta(\hat{P}_1)}_{V_{\hat{P}_i,\hat{P}_4}} \wedge \underline{\delta(\hat{P}_4)}_{V_{\hat{P}_i,\hat{P}_4}} = \{\lambda_{1j}\} \cap \{\lambda_4\} = \emptyset. \tag{B.34}$$

It is interesting to compare such a proposition with the proposition $S_z \in ([-3.1] \cap [1,3])$ which is represented by $\delta^o(\hat{P}_1 \wedge \hat{P}_4)$. This is clearly equal to $\hat{0}$ for all contexts, hence $\delta^o(\hat{P}_1 \wedge \hat{P}_4) \leq \delta^o(\hat{P}_1) \wedge \delta^o \hat{P}_4)$.

Example B.18 Given the same setting as above we would like to give the topos analogue of the proposition $S_z \in [-3, 3]$. This is represented by the projection operators $\hat{1}$. Therefore, for each context $V \in \mathcal{V}(\mathcal{H})$, we have that

$$\delta^o(\hat{1})_V = \hat{1} \tag{B.35}$$

which implies that

$$S_{\delta^o(\hat{1})_V} = \underline{\Sigma}_V. \tag{B.36}$$

Example B.19 Consider the algebra of bounded operators $\mathcal{B}(\mathcal{H})$ on a Hilbert space $\mathcal{H}$. We then define the category $\mathcal{V}(\mathcal{B})$ of abelian sub-algebras of $\mathcal{B}(\mathcal{H})$. This can be easily seen to be a category under sub-algebra inclusion. We now would like to define a covariant functor $F : \mathcal{V}(\mathcal{B}) \to \mathcal{V}(\mathcal{H})$. A first guess would be

$$\begin{aligned} F : \mathcal{V}(\mathcal{B}) &\to \mathcal{V}(\mathcal{H}) \\ B &\mapsto F(B) := B'' \end{aligned} \tag{B.37}$$

where B'' represents the double commutant of B. Is this a functor? First we need to show that, given two sub-algebras $i : B_i \subseteq B$, the following diagram commutes:

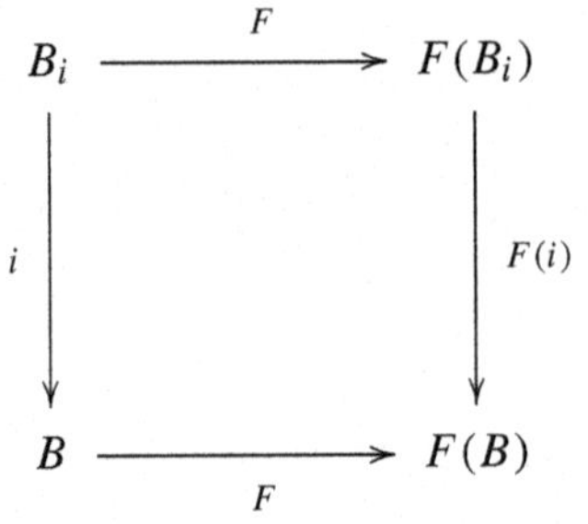

The fact that the above diagram commutes follows trivially from the fact that if $B_i \subseteq B$ then $B_i'' \subseteq B''$. Given the commutativity of the above diagram it follows at once that $F(i \circ j) = F(i) \circ F(j)$ for $j : B_j \subseteq B_i$.

Moreover

$$F(id_B) := \big(F(B) \to F(B)\big) = id_{F(B)}. \tag{B.38}$$

Example B.20 Consider a set of classical observables $\mathcal{O}$ which you want to quantized. Such a set forms a Lie algebra with respect to an appropriately defined commutator. For example the Poisson algebra of a set of functions on phase space such that two elements $A, B \in \mathcal{O}$ are considered to be non-commuting when $\{A, B\} = 0$.

We now want to define a possible quantisation funtor. Usually quantization of $\mathcal{O}$ is defined via an irreducible map $v : A \to \hat{A}$ for all $A \in \mathcal{O}$. The assignment of this map is such that the Lie-non-commutativity in $\mathcal{O}$ is reflected by the non-commutativity in the operator algebra on $\mathcal{H}$. Moreover v has to be faithful, i.e. $A \neq B \Rightarrow \hat{A} \neq \hat{B}$.

To ensure that the operator associated to each observable is bounded one refines the quantization map as follows

$$A \to \hat{A} \to e^{i\hat{A}}. \tag{B.39}$$

Next we define the category $\mathbf{C}_0$ of Lie-abelian sub-algebras of the classical observables in $\mathcal{O}$ This category forma a posetunder sub-algebra inclusion. Moreover, we will assume that $\mathbf{C}_0$ is invariant under any symplectic covariance transformation. Given such a category it is possible to define a possible quantisation functor. In particular, given any classical observable (a) we can define a possible quantisation of (a) through the faithful map [37]

$$\tilde{v} : a \mapsto e^{i\hat{a}}. \tag{B.40}$$

If we then consider the collection of all Lie-abelian sub-algebras of classical observables,the above map translates to:

$$\begin{aligned} \phi : \mathbf{C}_0 &\to \mathcal{V}(\mathcal{H}) \\ C &\mapsto \phi(C) := \Upsilon(C)'' \end{aligned} \tag{B.41}$$

where $\Upsilon(C)'' = (\tilde{v}(C) \cup \tilde{v}(C)^*)''$ (here $''$ represents the double commutant operator) and $\mathcal{V}(\mathcal{H})$ is the category of abelian von Neumann sub-algebras in $\mathcal{H}$. Thus $\Upsilon(C)''$ is the smallest abelian von Neumann algebra containing $\Upsilon(C) := \tilde{v}(C) \cup \tilde{v}(C)^*$. However we are not only interested in a single quantisation, but in all possible unitary equivalent quantisations, thus we need to define the action of $G \subseteq U(\mathcal{H})$ on ϕ. In this way we define the notion of unitary equivalent quantisations implementing Dirac covariance[2] of quantum theory. So, for each $g \in G$ and $C \in \mathbf{C}_0$ we define

$$l_g\phi(C) := l_g\big(\phi(C)\big) = \hat{U}_g\phi(C)\hat{U}_g^{-1}. \tag{B.42}$$

Having defined the action of G on the quantisation functor we can define the *quantisation presheaf* over $\mathbf{C}_0$ as follows:

Definition B.7 The quantisation presheaf $\underline{Q} : \mathbf{C}_0 \to \mathit{Sets}$ is defined on

1. Objects: for each $C \in \mathbf{C}_0$ we assign the collection of unitary equivalent quantisation maps, i.e. $\underline{Q}(C) := \{l_g\phi : \downarrow C \to \mathcal{V}(\mathcal{H}) | g \in G\}$ where $l_g\phi(C) := l_g(\phi(C))$. We assume that there is no group action on $\mathbf{C}_0$.

[2] By Dirac covariance we mean the fact that given a unitary group G then considering an operator $\hat{A}$ and a state $|\psi\rangle$ is equivalent to considering $\hat{U}_g\hat{A}\hat{U}_g^{-1}$ and $\hat{U}_g|\psi\rangle$.

2. Morphisms: given a map $i_{C_1,C_2} : C_1 \subseteq C_2$ the corresponding presheaf map is

$$\begin{aligned} \underline{Q}(i_{C_1,C_2}) : \underline{Q}(C_2) &\to \underline{Q}(C_1) \\ \phi &\mapsto \phi_{|C_1}. \end{aligned} \tag{B.43}$$

Example B.21 Given a sheaf $\overline{\underline{A}}$ over a topological space with Alexandroff topology, we want to define the bundle of germs of $\overline{\underline{A}}$. This concept simplifies for our Alexandroff base spaces as, given any point $V \in \mathcal{V}(\mathcal{H})$, there is a unique smallest open set, namely $\downarrow V$, to which V belongs.

Let $\mathcal{O}_1$ and $\mathcal{O}_2$ be open neighbourhoods of $V \in \mathcal{V}(\mathcal{H})$ with $s_1 \in \bar{\underline{A}}(\mathcal{O}_1)$ and $s_2 \in \bar{\underline{A}}(\mathcal{O}_2)$. Then s_1 and s_2 have the same germ at V, if there is some open $\mathcal{O} \subseteq \mathcal{O}_1 \cap \mathcal{O}_2$ such that $s_1|_{\mathcal{O}} = s_2|_{\mathcal{O}}$. Since $\mathcal{V}(\mathcal{H})$ has the Alexandroff topology, we can see at once that s_1 and s_2 have the same germ at V iff

$$s_1|_{\downarrow V} = s_2|_{\downarrow V}. \tag{B.44}$$

It follows that if $V \in \mathcal{O}$, $s \in \bar{\underline{A}}(\mathcal{O})$, then $germ_V s = s_{|\downarrow V}$. Hence

$$(\Lambda\bar{\underline{A}})_V = \bar{\underline{A}}(\downarrow V). \tag{B.45}$$

Example B.22 Given the presheaf $\underline{\Sigma}$ we define the set $\Sigma := \coprod_{V\in\mathcal{V}(\mathcal{H})} \underline{\Sigma}_V = \bigcup_{V\in\mathcal{V}(\mathcal{H})} \{V\} \times \underline{\Sigma}_V$. Associated with this is the map $p_\Sigma : \Sigma \to \mathcal{V}(\mathcal{H})$ defined by $p_\Sigma(\lambda) = V$, where V is the context such that $\lambda \in \underline{\Sigma}_V$, therefore $p_\Sigma^{-1}(V) = \underline{\Sigma}_V$.

We now prove the following theorem:

Theorem B.1 *Given $\Sigma_{|\downarrow V} := \coprod_{V'\in\downarrow V} \underline{\Sigma}_{V'}$, a local section $\sigma : \downarrow V \to \Sigma_{|\downarrow V}$ of the bundle $p_\Sigma : \Sigma \to \mathcal{V}(\mathcal{H})$ is continuous with respect to the spectral topology on Σ, if and only if it is a local section of the presheaf $\underline{\Sigma}_{|\downarrow V}$. Therefore a continuous local section of the bundle $\Lambda\breve{\underline{\Sigma}}$ equipped with the étale topology.*

Proof Given a section σ of the presheaf $\underline{\Sigma}_{\downarrow V}$, let us consider $\sigma^{-1}(S \cap \Sigma_{\downarrow V})$ for any basis set, S, for the spectral topology of Σ. From the definition of S it is clear that if $\sigma(V_1) \in S_{V_1}$ for some $V_1 \in \mathcal{V}(\mathcal{H})$ then, $\sigma(V_2) \in S_{V_2}$ for all $V_2 \subseteq V_1$, i.e., $\sigma^{-1}(S_{|\downarrow V_1}) = \downarrow V_1$. It follows that:

$$\sigma^{-1}(S \cap \sigma_{\downarrow V}) = \bigcup \{\downarrow V_1 | V_1 \subseteq V, \sigma(V_1) \in S_{V_1}\}. \tag{B.46}$$

This is a union of lower sets and is, hence, open in the Alexandroff topology on $\mathcal{V}(\mathcal{H})$. It follows that σ is a continuous section of the bundle $\Sigma_{|\downarrow V}$.

Conversely, if σ is a continuous section of the bundle $\Sigma_{|\downarrow V}$, then, for any basis element $S := \coprod_{V\in\mathcal{V}(\mathcal{H})} \underline{S}_V$, $\sigma^{-1}(S \cap \Sigma_{\downarrow V})$ is a lower set. In particular, consider any topological base set $\underline{S}$ such that $\sigma(V) \in \underline{S}_V$. Then in order for $\sigma^{-1}(\underline{S} \cap \underline{\Sigma}_{\downarrow V})$ to be open we must have $\sigma(V_1) \in \underline{S}_{V_1}$ for all $V_1 \subseteq V$. Thus σ is 'collared' for any $\underline{S}$, such that $\sigma(V) \in \underline{S}_V$. Now the induced topology on each $\underline{\Sigma}_V$ is just the spectral topology on $\underline{\Sigma}_V$, and this is extremely disconnected. It follows that for any $\lambda \in \underline{\Sigma}_V$, $\lambda = \bigcap\{\underline{C} | \underline{C}$ is clopen and $\lambda \in \underline{C}\}$. Thus taking the intersection of the basis set, $\underline{S}$, that contain $\lambda \in \underline{\Sigma}_V$ implies that σ is a section of the presheaf $\underline{\Sigma}_{|\downarrow V}$. □

Example B.23 The space $\mathcal{V}(\mathcal{H})/G$ can be given the structure of a poset by defining, for each pair of orbits w_1, w_2 in $\mathcal{V}(\mathcal{H})/G$ the relation

$$w_1 \leq w_2 \quad \text{iff there exists} \quad V_1 \in O_{w_1} \text{ and } V_2 \in O_{w_2} \quad \text{such that} \quad V_1 \subseteq V_2 \tag{B.47}$$

where $O_{w_1} \subset \mathcal{V}(\mathcal{H})$ is the orbit associated to w_1.

Proof 1. Reflexivity: it is trivial that $w \leq w$ for all $w \in \mathcal{V}(\mathcal{H})/G$.

2. Transitivity: if $w_1 \leq w_2$ and $w_2 \leq w_3$ then there exists (i) $V_1 \in O_{w_1}$ and $V_2 \in O_{w_2}$ such that $V_1 \subseteq V_2$ and (ii) $V_3 \in O_{w_2}$ and $V_4 \in O_{w_3}$ such that $V_3 \subseteq V_4$. Now, since G acts transitively on the orbit w_2 there exists $g \in G$ such that $V_3 = l_g(V_2)$, and hence $l_g(V_2) \subseteq V_4$. Therefore, $V_2 \subseteq l_{g^{-1}}(V_4)$ and thus $V_1 \subseteq l_{g^{-1}}(V_4)$, and so $w_1 \leq w_3$, as required.

3. Antisymmetry: suppose $w_1 \leq w_2$ and $w_2 \leq w_1$. Then there exists $V_1 \in O_{w_1}$, $V_2 \in O_{w_2}$, $V_3 \in O_{w_2}$, $V_4 \in O_{w_1}$ such that $V_1 \subseteq V_2$ and $V_3 \subseteq V_4$. Since G acts transitively on w_2 there exists $g_1 \in G$ such that $V_3 = l_{g_1}(V_2)$ and hence $V_2 = l_{g_1^{-1}}(V_3) \subseteq l_{g_1^{-1}}(V_4)$. Hence we have $V_1, V_4 \in O_{w_1}$ such that $V_1 \subseteq l_{g_1^{-1}}(V_4)$. Now, because G acts transitively on the orbit w_1 there exists $g_2 \in G$ such that $V_4 = l_{g_2}(V_1)$ and hence $V_1 \subseteq l_g(V_1)$ where $g := g_1^{-1} g_2$. Now, the orbits of G form an anti-chain, since the algebra-map $V \to l_g(V) := \hat{U}(g)V\hat{U}(g)^{-1}$ is an isomorphism, then V cannot be a proper subset of $l_g(V)$. Therefore, $V_1 \subseteq l_g(V_1)$ implies that $V_1 = l_g(V_1) = l_{g_1^{-1}}(l_{g_2}(V_1)) = l_{g_1^{-1}}(V_4)$, and hence $V_2 \subseteq V_1$. Thus $V_1 = V_2$ and $w_1 = w_2$. □

Since $\mathcal{V}(\mathcal{H})/G$ is a poset, the Alexandroff topology is simply the topology whose basis are the lower sets $\downarrow w$ with the ordering defined above.

We now want to define the map

$$\begin{aligned} \pi : \mathcal{V}(\mathcal{H}) &\to \mathcal{V}(\mathcal{H})/G \\ V &\mapsto w_V \end{aligned} \tag{B.48}$$

which maps each algebra to its equivalence class. In order for this map to give rise to a geometric morphism we need to show that it is continuous. By placing the Alexandroff topology on $\mathcal{V}(\mathcal{H})/G$, then by Lemma 14.2, π is indeed continuous. However, the map $\pi : \mathcal{V}(\mathcal{H}) \to \mathcal{V}(\mathcal{H})/G$ although continuous is not étalé, since this would require π to be a local homeomorphism, and this cannot be true at those points in the base space where there is a 'sudden' change of stability group for the corresponding G-orbit. In fact even if $V' \subseteq V$ there is no obvious relation between the respective stability groups.

The fact that π is continuous implies that we can define a geometric morphism (see Theorem 14.4) whose direct and inverse image are, respectively:

$$\pi_* : Sh\mathcal{V}(\mathcal{H}) \to Sh\big(\mathcal{V}(\mathcal{H})/G\big) \tag{B.49}$$

$$\pi^* : Sh\big(\mathcal{V}(\mathcal{H})/G\big) \to Sh\mathcal{V}(\mathcal{H}) \tag{B.50}$$

such that $\pi^*(\underline{A})(\downarrow V) := \underline{A}(\pi(\downarrow V)) = \underline{A}\downarrow w$ where $V \in O_w$, while $\pi_*((\underline{S}))(\downarrow w) := \underline{S}(\pi^{-1}(\downarrow W))$. We now would like to define the sheaf of (continuous) section of π, which we denote as $\gamma(\pi) \in Sh(\mathcal{V}(\mathcal{H})/G)$. In order to do so we need the following lemma.

Lemma B.1 *Let $\alpha : P_1 \to P_2$ be a map between posets P_1 and P_2. Then α is order preserving, if and only if, for each lower set $L \in P_2$, we have that $\alpha^{-1}(L)$ is a lower subset of P_1.*

Proof Suppose α is order preserving and let $L \in P_2$ be lower. Now let $z \in \alpha^{-1}(L) \subseteq P_1$, i.e., $\alpha(z) = l$ for some $l \in L$, and suppose $y \in P_1$ is such that $y \leq z$. Since α is order preserving we have $\alpha(y) \leq \alpha(z) = l \in L$, which, since L is lower, means that $\alpha(y) \in L$ i.e., $y \in \alpha^{-1}(L)$. Hence $\alpha^{-1}(L)$.

Conversely, suppose that for any lower set $L \subseteq P_2$ we have that $\alpha^{-1}(L) \subseteq P_1$ is lower and consider a pair $x, y \in P_1$ such that $x \leq y$. Now $\downarrow\alpha(y)$ is lower in P_2 and hence $\alpha^{-1}(\downarrow\alpha(y))$ is a lower subset of P_1. However $\alpha(y) \in \downarrow\alpha(y)$ and hence $y \in \alpha^{-1}(\downarrow\alpha(y))$. Therefore the fact that $x \leq y$ implies that $x \in \alpha^{-1}(\downarrow\alpha(y))$, i.e., $\alpha(x) \in \downarrow\alpha(y)$, which means that $\alpha(x) \leq \alpha(y)$. Thus α is order preserving. □

We now would like to define the sheaf of (continuous) section of π, which we denote as $\gamma(\pi) \in Sh(\mathcal{V}(\mathcal{H})/G)$. In this context we recall that a section of a bundle $p_Y : Y \to X$ over an open subset $U \subseteq X$ is a map $s : U \to Y$ such that $p_Y \circ s = id_U$. In our case, using the Alexandroff topology on both $\mathcal{V}(\mathcal{H})$ and $\mathcal{V}(\mathcal{H})/G$ we know from Lemma 14.2 that $s : U \subseteq \mathcal{V}(\mathcal{H})/G \to \mathcal{V}(\mathcal{H})$ is continuous, if and only if, it is order preserving. Thus $\gamma(\pi)$ is the sheaf of order preserving local sections of the bundle $\pi : \mathcal{V}(\mathcal{H}) \to \mathcal{V}(\mathcal{H})/G$. In particular, for each open set $U \subseteq \mathcal{V}(\mathcal{H})/G$ we obtain $\gamma(\pi)(U) := \{s : U \to \mathcal{V}(\mathcal{H}) | s \text{ is order preserving}\}$, while the morphisms are given by restriction.

We can now define the bundle

$$\begin{aligned} p_{\gamma(\pi)} : \Lambda\big(\gamma(\pi)\big) &\to \mathcal{V}(\mathcal{H})/G \\ s &\mapsto w^s \end{aligned} \tag{B.51}$$

where $s : \downarrow w \to \mathcal{V}(\mathcal{H})$. It follows that the stalk is

$$\Lambda\big(\gamma(\pi)\big)_w := \gamma(\pi)(\downarrow w) \tag{B.52}$$

which is the set of all local sections $s : \downarrow w \subseteq \mathcal{V}(\mathcal{H})/G \to \mathcal{V}(\mathcal{H})$.

We now want to show that $p_{\gamma(\pi)}$ is étale also with respect to the Alexandroff topology (yet to be defined), i.e. we want to show that $p_{\gamma(\pi)}$ is a local homeomorphism with respect to the Alexandroff topology. To this end we first of all define the Alexandroff topology on $\Lambda(\gamma(\pi))$. This is easy since $\Lambda(\gamma(\pi))$ is actually a poset, thus it comes with an Alexandroff topology. The poset ordering is given as follows:

$$s_1 \leq s_2 \quad \text{iff} \quad p_{\gamma(\pi)}(s_1) \leq p_{\gamma(\pi)}(s_2) \quad \text{and} \quad s_1 = (s_2)_{|p_{\gamma(\pi)}(s_1)}. \tag{B.53}$$

That this is indeed a partial ordering is easy to prove so we will not report the proof here.

Given such an ordering, then a basis open in $\Lambda(\gamma(\pi))$ will be $\downarrow s$. We want to show that $p_{\gamma(\pi)}$, restricted to such an open, is a homeomorphism. In particular we obtain

$$(p_{\gamma(\pi)})_{|\downarrow s} : \downarrow s \rightarrow \mathcal{V}(\mathcal{H})/G \tag{B.54}$$

such that $(p_{\gamma(\pi)})_{|\downarrow s}(\downarrow s) = \downarrow(p_{\gamma(\pi)})_{|\downarrow s}(s) = \downarrow w^s$ is an open set in the Alexandroff topology. That this is a local homeomorphism is then easy to see. The only difficult part is showing injectivity. In particular given two section s_1 and s_2 in $\downarrow s$ assume that $w^{s_1} = w^{s_2}$. However, $s_1 \in \downarrow s$ then $s_1 \leq s$ therefore $s_1 = s_{|w^{s_1}} = s_{|w^{s_2}} = s_2$ since $s_2 \leq s$. It follows that the bundle $\gamma(\pi)$ in (B.51) is étale with corresponding sheaf $\gamma(\pi)$.

We know from Sect. 14.2 that $\Lambda : Sh(\mathcal{V}(\mathcal{H})/G) \rightarrow Bund(\mathcal{V}(\mathcal{H})/G)$ and $\gamma : Bund(\mathcal{V}(\mathcal{H})/G) \rightarrow Sh(\mathcal{V}(\mathcal{H})/G)$ are adjoints. We thus want to compute the co-unit ϵ as applied to the bundle $\pi : \mathcal{V}(\mathcal{H}) \rightarrow \mathcal{V}(\mathcal{H})/G$. This is the map

$$\epsilon_Y : \Lambda\gamma(p_Y) \rightarrow Y \tag{B.55}$$

where $p_Y : Y \rightarrow \mathcal{V}(\mathcal{H})/G$ is any bundle over $\mathcal{V}(\mathcal{H})/G$. Thus, considering the bundle $\pi : \mathcal{V}(\mathcal{H}) \rightarrow \mathcal{V}(\mathcal{H})/G$ we obtain the commutative triangle

$$\begin{array}{ccc} \Lambda\gamma(\pi) & \xrightarrow{\epsilon_{\mathcal{V}(\mathcal{H})}} & \mathcal{V}(\mathcal{H}) \\ & {\scriptstyle p_{\gamma(\pi)}} \searrow \quad \swarrow {\scriptstyle \pi} & \\ & \mathcal{V}(\mathcal{H})/G & \end{array}$$

such that

$$\begin{aligned} \epsilon_{\mathcal{V}(\mathcal{H})}(w) : \Lambda\big(\gamma(\pi)\big) &\rightarrow \mathcal{V}(\mathcal{H}) \\ s &\mapsto s(w) \end{aligned} \tag{B.56}$$

for $s : \downarrow w \rightarrow \mathcal{V}(\mathcal{H})$, i.e. $s \in \Lambda\gamma(\pi)_w$.

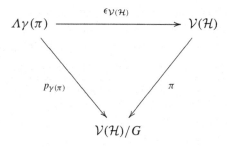

We can now pullback the bundle $\Lambda\underline{\Sigma} \to \mathcal{V}(\mathcal{H})$ via the co-unit to obtain

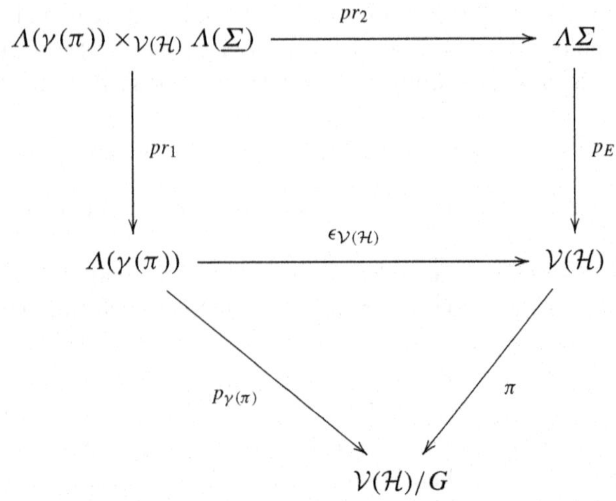

where

$$\Lambda\big(\gamma(\pi)\big) \times_{\mathcal{V}(\mathcal{H})} \Lambda(\underline{\Sigma}) := \big\{(s,\lambda) \subseteq \Lambda\big(\gamma(\pi)\big) \times \Lambda(\underline{\Sigma}) \big| \epsilon_{\mathcal{V}(\mathcal{H})}(s) = p_E(\lambda)\big\}. \tag{B.57}$$

Now, since p_E is étale and, since a pullback of an étale bundle by a continuous map (in this case $\epsilon_{\mathcal{V}(\mathcal{H})}$) is étale, it follows that $\Lambda(\gamma(\pi)) \times_{\mathcal{V}(\mathcal{H})} \Lambda(\underline{\Sigma}) \to \Lambda(\gamma(\pi))$ is étale. Moreover, since $p_{\gamma(\pi)}$ is étale, the combination $\Lambda(\gamma(\pi)) \times_{\mathcal{V}(\mathcal{H})} \Lambda(\underline{\Sigma}) \xrightarrow{pr_1} \Lambda(\gamma(\pi)) \xrightarrow{p_{\gamma(\pi)}} \mathcal{V}(\mathcal{H})/G$ is étale.

From the above reasoning it follows that we can define a functor $F : Sh(\mathcal{V}(\mathcal{H})) \to Sh(\mathcal{V}(\mathcal{H})/G)$ as follows

$$\begin{aligned} F : Sh\big(\mathcal{V}(\mathcal{H})\big) &\to Sh\big(\mathcal{V}(\mathcal{H})/G\big) \\ \underline{\Sigma} &\mapsto \gamma\big(\Lambda\underline{\Sigma} \xrightarrow{p_E} \mathcal{V}(\mathcal{H}) \xrightarrow{\pi} \mathcal{V}(\mathcal{H})/G\big) \end{aligned} \tag{B.58}$$

where $\gamma(\Lambda\underline{\Sigma} \to \mathcal{V}(\mathcal{H}) \to \mathcal{V}(\mathcal{H})/G)$ is the sheaf of sections of the composite bundle $\Lambda\underline{\Sigma} \to \mathcal{V}(\mathcal{H}) \to \mathcal{V}(\mathcal{H})/G$. We then obtain the following proposition:

Proposition B.1

$$\Lambda\big(\gamma(p_E \circ \pi)\big) \simeq \Lambda\big(\gamma(\pi)\big) \times_{\mathcal{V}(\mathcal{H})} \Lambda(\underline{\Sigma}) \tag{B.59}$$

as bundles over $\mathcal{V}(\mathcal{H})/G$.

Proof Given any $w \in \mathcal{V}(\mathcal{H})/G$ we have

$$\Lambda\big(\gamma(p_E \circ \pi)\big)_w = \{\sigma : \downarrow w \to \underline{\Sigma}\} \tag{B.60}$$

where σ is a local section of $\pi \circ p_E : \underline{\Sigma} \to \mathcal{V}(\mathcal{H})/G$ defined on $\downarrow w$, while

$$\Lambda\big(\gamma(\pi)\big) \times_{\mathcal{V}(\mathcal{H})} \Lambda(\underline{\Sigma})_w = \big\{(s, \lambda) | s(w) = p_E(\lambda)\big\}. \tag{B.61}$$

Now since $p_E \circ \sigma$ is a local section of $\pi : \mathcal{V}(\mathcal{H}) \to \mathcal{V}(\mathcal{H})/G$ on $\downarrow w$, we can define the map

$$\begin{aligned} i_w : \Lambda\big(\gamma(p_E \circ \pi)\big)_w &\to \Lambda\big(\gamma(\pi)\big) \times_{\mathcal{V}(\mathcal{H})} \Lambda(\underline{\Sigma})_w \\ \sigma &\mapsto \big(p_E \circ \sigma, \sigma(w)\big). \end{aligned} \tag{B.62}$$

To show that the pair $(p_E \circ \sigma, \sigma(w))$ satisfies the condition for a pullback over $\mathcal{V}(\mathcal{H})$ we simply note that

$$\epsilon_{\mathcal{V}(\mathcal{H})}(p_E \circ \sigma) = p_E\big(\sigma(w)\big) \tag{B.63}$$

as required. Moreover, for the étalce bundle $\Lambda \underline{\Sigma}$ there is an inverse in which $(s, \lambda) \in \Lambda(\gamma(\pi)) \times_{\mathcal{V}(\mathcal{H})} \Lambda(\underline{\Sigma})_w$ is taken to the unique lift of the local section $s : \downarrow w \to \mathcal{V}(\mathcal{H})$ that passes through λ, where $p_E(\lambda) = s(w)$. □

Now that we have defined the functor F we would like to analyse its action on certain constructs. In particular we will analyse its effect on the terminal object $\underline{1}_{Sh(\mathcal{V}(\mathcal{H}))}$. Our claim is that

$$F(\underline{1}_{Sh(\mathcal{V}(\mathcal{H}))}) = \gamma(\pi). \tag{B.64}$$

Applying the definition to the left hand side we obtain

$$F(\underline{1}_{Sh(\mathcal{V}(\mathcal{H}))}) = \gamma(\Lambda \underline{1} \xrightarrow{p_1} \mathcal{V}(\mathcal{H}) \xrightarrow{\pi} \mathcal{V}(\mathcal{H})/G). \tag{B.65}$$

We then recall that $\Lambda \underline{1}$ is simply a collection of singletons, one for each $V \in \mathcal{V}(\mathcal{H})$, thus $\Lambda \underline{1} \to \mathcal{V}(\mathcal{H})$ is simply $\mathcal{V}(\mathcal{H})$. Hence the result follows.

As a last step in our example we will construct a right adjoint to F, i.e. we will define a functor

$$G : Sh\big(\mathcal{V}(\mathcal{H})/G\big) \to Sh\big(\mathcal{V}(\mathcal{H})\big) \tag{B.66}$$

such that

$$F \dashv G. \tag{B.67}$$

There are two ways of approaching this. One is by explicitly constructing the G functor and then showing that it is indeed a right adjoint of F, while the second is to show that F is defined as a combination of left adjoint, such that G will be defined as the combination of the respective right adjoints. We will use the second method. To this end we note that

$$F = p_{\gamma(\pi)}! \circ \epsilon^* \tag{B.68}$$

where $\epsilon^*_{\mathcal{V}(\mathcal{H})} : Sh(\mathcal{V}(\mathcal{H})) \to Sh(\Lambda(\gamma(\pi)))$ is the inverse image part (hence left adjoint of ϵ_*) of the geometric morphism induced by the continuous map $\epsilon_{\mathcal{V}(\mathcal{H})}$:

$\Lambda(\gamma(\pi)) \to \mathcal{V}(\mathcal{H})$. On the other hand $p_{\gamma(\pi)}! : Sh(\Lambda(\gamma(\pi))) \to Sh(\mathcal{V}(\mathcal{H})/G)$ is the left adjoint of $p^*_{\gamma(\pi)}$. It follows that:

$$G := \epsilon_* \circ p^*_{\gamma(\pi)} \tag{B.69}$$

and $F \dashv G$.

References

1. J.S. Bell, *Against 'measurement'* (Physics World, Institute of Physics Publishing, Bristol and Philadelphia, 1990)
2. H. Everett, Relative state formulation of quantum mechanics. Rev. Mod. Phys. **29**, 141–149 (1957)
3. B. DeWitt, N. Graham, *The Many-Worlds Interpretation of Quantum Mechanics* (Princeton University Press, Princeton, 1973)
4. A. Einstein, B. Podolsky, N. Rosen, Can quantum-mechanical description of physical reality be considered complete? Phys. Rev. **47**(10), 777–780 (1935)
5. N. Bohr, Can quantum-mechanical description of physical reality be considered complete? Phys. Rev. **48**, 700 (1935)
6. D. Bohm, A suggested interpretation of the quantum theory in terms of "hidden" variables. I. Phys. Rev. **85**, 166–179 (1952)
7. D. Bohm, A suggested interpretation of the quantum theory in terms of "hidden" variables. II. Phys. Rev. **85**, 180–193 (1952)
8. M. Lockwood, *Mind, Brain and the Quantum: The Compound I* (Blackwell, Oxford, 1992)
9. N.G. van Kampen, The scandal of quantum mechanics. Am. J. Phys. **76**, 989 (2008)
10. R. Peierls, Selected Scientific Papers of Sir Rudolf Peierls. World Scientific Series in 20th Century Physics, vol. 19 (1997)
11. K. Gottfried, *Quantum Mechanics* (Benjamin, Elmsford, 1966)
12. V.F. Lenzen, in *Philosophical Problems of the Statistical Interpretation of Quantum Mechanics Proceedings*. Second Berkeley Symp. on Math. Statist. and Prob. (University of California Press, Berkeley, 1951), pp. 567–579
13. J.S. Bell, *Speakable and Unspeakable in Quantum Mechanics (Collected Papers on Quantum Philosophy)* (Cambridge University Press, Cambridge, 2004)
14. C.J. Isham, J. Butterfield, A topos perspective on the Kochen-Specker theorem: I. Quantum states as generalized valuations (1998). arXiv:quant-ph/9803055
15. J. Butterfield, C.J. Isham, A topos perspective on the Kochen-Specker theorem: II. Conceptual aspects, and classical analogues (1998). arXiv:quant-ph/9808067
16. J. Butterfield J. Hamilton, C.J. Isham, A topos perspective on the Kochen-Specker theorem: III. Von Neumann algebras as the base category (1999). arXiv:quant-ph/9911020
17. R. Goldblatt, *Topoi the Categorial Analysis of Logic* (North-Holland, London, 1984)
18. S. MacLane, I. Moerdijk, *Sheaves in Geometry and Logic: A First Introduction to Topos Theory* (Springer, London, 1968)
19. J.L. Bell, *Toposes and Local Set Theories* (Clarendon Press, Oxford, 1988)
20. S. MacLane, *Categories for the Working Mathematician* (Springer, London, 1997)
21. C.J. Isham, *Lectures on Quantum Theory, Mathematical and Structural Foundations* (Imperial College Press, London, 1995)

C. Flori, *A First Course in Topos Quantum Theory*, Lecture Notes in Physics 868,
DOI 10.1007/978-3-642-35713-8, © Springer-Verlag Berlin Heidelberg 2013

22. J. Bub, *Interpreting the Quantum World* (Cambridge University Press, Cambridge, 1997)
23. M. Kernaghan, Bell-Kochen-Specker theorem for 20 vectors. J. Phys. A **27** (1994)
24. A. Doering, C.J. Isham, *Classical and Quantum Probabilities as Truth Values* (2011). arXiv: 1102.2213v1
25. A. Doering, Quantum states and measures on the spectral presheaf (2008). arXiv:0809.4847v1 [quant-ph]
26. C.J. Isham, J. Butterfield, Some possible roles for topos theory in quantum theory and quantum gravity. Found. Phys. **30**, 1707 (2000). arXiv:gr-qc/9910005
27. A. Doring, C.J. Isham, A topos foundation for theories of physics. I. Formal languages for physics. J. Math. Phys. **49**, 053515 (2008). arXiv:quant-ph/0703060 [quant-ph]
28. J.B. Geloun, C. Flori, Topos analogues of the KMS state. arXiv:1207.0227v1 [math-ph]
29. A. Doring, C.J. Isham, A topos foundation for theories of physics. II. Daseinisation and the liberation of quantum theory. J. Math. Phys. **49**, 053516 (2008). arXiv:quant-ph/0703062 [quant-ph]
30. A. Doring, C.J. Isham, A topos foundation for theories of physics. III. The representation of physical quantities with arrows. J. Math. Phys. **49**, 053517 (2008). arXiv:quant-ph/0703064 [quant-ph]
31. A. Doring, C.J. Isham, A topos foundation for theories of physics. IV. Categories of systems. J. Math. Phys. **49**, 053518 (2008). arXiv:quant-ph/0703066 [quant-ph]
32. A. Doring, C. Isham, 'What is a thing?': topos theory in the foundations of physics. arXiv: 0803.0417 [quant-ph]
33. A. Doring, Topos theory and 'neo-realist' quantum theory. arXiv:0712.4003 [quant-ph]
34. J. Harding, A. Doering, Abelian subalgebras and the Jordan structure of a von Neumann algebra. arXiv:1009.4945 [math-ph]
35. C.J. Isham, Is it true or is it false; or somewhere in between? The logic of quantum theory. Contemp. Phys. **46**, 207 (2005)
36. C.J. Isham, A topos perspective on state-vector reduction. arXiv:quant-ph/0508225
37. K. Nakayama, Sheaves in quantum topos induced by quantization. arXiv:1109.1192 [math-ph]
38. C.J. Isham, Topos theory and consistent histories: the internal logic of the set of all consistent sets. Int. J. Theor. Phys. **36**, 785 (1997). arXiv:gr-qc/9607069
39. A. Doering, Topos quantum logic and mixed states. arXiv:1004.3561 [quant-ph]
40. A. Doering, The physical interpretation of daseinisation. arXiv:1004.3573 [quant-ph]
41. C. Flori, Concept of quantization in a topos. In preparation
42. C. Flori, Review of the topos approach to quantum theory. arXiv:1106.5660 [math-ph]
43. C. Flori, A topos formulation of history quantum theory. J. Math. Phys. **51**, 053527 (2010). arXiv:0812.1290 [quant-ph]
44. C.J. Isham, Topos methods in the foundations of physics. arXiv:1004.3564 [quant-ph]
45. C. Flori, Group action in topos quantum physics. arXiv:1110.1650 [quant-ph]
46. S. Wolters, A comparison of two topos-theoretic approaches to quantum theory. arXiv:1010.2031 [math-ph]
47. C. Heunen, N.P. Landsman, B. Spitters, A topos for algebraic quantum theory. Commun. Math. Phys. **291**, 63 (2009). arXiv:0709.4364 [quant-ph]
48. C. Heunen, N.P. Landsman, B. Spitters, S. Wolters, The Gelfand spectrum of a non-commutative C*-algebra: a topos-theoretic approach. J. Aust. Math. Soc. **90**, 39 (2011). arXiv:1010.2050 [math-ph]
49. W. Brenna, C. Flori, Complex numbers and normal operators in topos quantum theory. arXiv:1206.0809 [quant-ph]
50. C. Flori, Approaches to quantum gravity. arXiv:0911.2135 [gr-qc]
51. F. Dowker, A. Kent, On the consistent histories approach to quantum mechanics. J. Stat. Phys. **82**, 1575 (1996). arXiv:gr-qc/9412067
52. C.J. Isham, Topos theory and consistent histories: the internal logic of the set of all consistent sets. Int. J. Theor. Phys. **36**, 785 (1997). arXiv:gr-qc/9607069

53. C.J. Isham, Quantum logic and the histories approach to quantum theory. J. Math. Phys. **35**, 2157 (1994). arXiv:gr-qc/9308006
54. M. Gell-Mann, J.B. Hartle, in *Complexity, Entropy and the Physics of Information*, ed. by W. Zurek. SFI Studies in the Sciences of Complexity, vol. VIII (Addison Wesley, Reading, 1990)
55. M. Gell-Mann, J.B. Hartle, in *Proceedings of the 3rd International Symposium on the Foundations of Quantum Mechanics in the Light of New Technologies*, ed. by S. Kobayashi, H. Ezawa, Y. Murayama, S. Nomura (Physical Society of Japan, Tokyo, 1990)
56. M. Gell-Mann, J.B. Hartle, in Proceedings of the 25th International Conference on High Energy Physics, ed. by K.K. Phua, Y. Yamaguchi, Singapore, August 2–8 1990 (South East Asia Theoretical Physics Association and Physical Society of Japan) (Worlds Scientific, Singapore, 1990)
57. R.B. Griffiths, Consistent histories and the interpretation of quantum mechanics. J. Stat. Phys. **36**, 219 (1984)
58. R.B. Griffiths, Logical reformulation of quantum mechanics. I. Foundations. J. Stat. Phys. **53**, 893 (1988)
59. R.B. Griffiths, Logical reformulation of quantum mechanics. II. Interferences and the Einstein-Podolsky-Rosen experiment. J. Stat. Phys. **53**, 933 (1988)
60. R.B. Griffiths, Logical reformulation of quantum mechanics. III. Classical limit and irreversibility. J. Stat. Phys. **53**, 957 (1988)
61. R.B. Griffiths, Logical reformulation of quantum mechanics. IV. Projectors in semiclassical physics. J. Stat. Phys. **57**, 357 (1989)
62. R.B. Griffiths, The consistency of consistent histories: a reply to d'Espagnat. Found. Phys. **23**, 1601 (1993)
63. R. Omnes, Consistent interpretations of quantum mechanics. Rev. Mod. Phys. **64**, 339–382 (1992)
64. J.J. Halliwell, A review of the decoherent histories approach to quantum mechanics. Ann. N.Y. Acad. Sci. **755**, 726 (1995). arXiv:gr-qc/9407040
65. C.J. Isham, N. Linden, Quantum temporal logic and decoherence functionals in the histories approach to generalized quantum theory. J. Math. Phys. **35**, 5452 (1994). arXiv:gr-qc/9405029
66. P. Mittelstaedt, Quantum logic and decoherence. Int. J. Theor. Phys. **43**(6) (2004)
67. P. Mittelstaedt, Time dependent propositions and quantum logic. J. Philos. Log. **6**, 463–472 (1977)
68. E.W. Stachow, Logical foundations of quantum mechanics. Int. J. Theor. Phys. **19**(4) (1980)
69. E.W. Stachow, A Model Theoretic Semantics for Quantum Logic, in *Proceedings of the Biennial Meeting of the Philosophy of Science Association*, vol. 1 (1980), pp. 272–280
70. K. Savvidou, C. Anastopoulos, Histories quantisation of parameterised systems: I. Development of a general algorithm. Class. Quantum Gravity **17**, 2463 (2000). arXiv:gr-qc/9912077
71. N.K. Savvidou, Continuous time in consistent histories (1999). arXiv:gr-qc/9912076v1
72. S. Vickers, *Topology Via Logic* (Cambridge University Press, Cambridge, 1989)
73. A.M. Gleason, Measures on the closed subspaces of a Hilbert space. J. Math. Phys. **6**, 885 (1957)
74. G. Helmberg, *Introduction to Spectral Theory in Hilbert Space* (Dover Publications, New York, 2008)
75. R. Whitley, The spectral theorem for a normal operator. Am. Math. Mon. **75**, 856–861 (1968)
76. H.F. de Groote, On a canonical lattice structure on the effect algebra of a von Neumann algebra (2007). arXiv:math-ph/0410018
77. R. Kubo, Statistical mechanical theory of irreversible processes. 1. General theory and simple applications in magnetic and conduction problems. J. Phys. Soc. Jpn. **12**, 570 (1957)
78. P.C. Martin, J.S. Schwinger, Theory of many particle systems. 1. Phys. Rev. **115**, 1342 (1959)
79. H. Araki, P.D.F. Ion, On the equivalence of KMS and Gibbs conditions for states of quantum lattice systems. Commun. Math. Phys. **35**, 1 (1974)

80. T. Matsui, Purification and uniqueness of quantum Gibbs states. Commun. Math. Phys. **162**, 321 (1994)
81. G.L. Sewell, *Quantum Mechanics and Its Emergent Macrophysics* (Princeton University Press, Princeton, 2002)
82. S. Stratila, *Modular Theory in Operator Algebras* (Abacus Press, Tunbridge Wells, 1981)
83. M. Takesaki, *Tomita's Theory of Modular Hilbert Algebras and Its Applications* (Springer, New York, 1970)
84. M. Takesaki, *Theory of Operator Algebras I*. Encyclopaedia of Mathematical Sciences, vol. 124 (Springer, Berlin, 2002)
85. S. Maeda, Probability measures on projections in von Neumann algebras. Rev. Math. Phys. **1**, 235 (1989)
86. H.J. Borchers, On revolutionizing quantum field theory with Tomita's modular theory. J. Math. Phys. **41**, 3604 (2000)
87. S.J. Summers, Tomita-Takesaki modular theory. arXiv:math-ph/0511034
88. A. Connes, *Noncommutative Geometry* (Academic Press, San Diego, 1994)
89. B. Banaschewski, C.J. Mulvey, The spectral theory of commutative C^*-algebras: the constructive spectrum. Quaest. Math. **23**, 4 (2009)
90. B. Banaschewski, C.J. Mulvey, A globalisation of the Gelfand duality theorem. Ann. Pure Appl. Log., vol. 137 (2006)
91. C. Rousseau, Topos theory and complex analysis, in Applications of Sheaves. Lecture Notes in Mathematics, vol. 753 (1979)
92. A. Doering, B. Dewitt, Self-adjoint operators as functions I (2012). arXiv:1208.4724v1 [math-ph]
93. A. Doering, R.S. Barbosa, Unsharp values, domains and topoi (2011). arXiv:1107.1083v1 [quant-ph]
94. P.T. Johnstone, *Sketches of an Elephant A Topos Theory Compendium*, vols. 1, 2 (Oxford University Press, London, 2002)
95. J. Dugundji, *Topology* (Allyn and Bacon, Needham Heights, 1976)
96. C.J. Isham, *Modern Differential Geometry for Physicists*. Lecture Notes in Physics (World Scientific, Singapore, 1999)
97. A. Doering, Generalised gelfand spectra of Nonabelian unital C^*-algebras II: flows and time evolution of quantum systems (2012). arXiv:1212.4882 [math.OA]
98. A. Doering, Generalised gelfand spectra of Nonabelian unital C^*-algebras I: categorical aspects, automorphisms and Jordan structure (2012). arXiv:1212.2613 [math.OA]
99. A. Doering, Topos-based logic for quantum systems and bi-Heyting algebras (2012). arXiv:1202.2750 [quant-ph]

Index

C. Flori, *A First Course in Topos Quantum Theory*, Lecture Notes in Physics 868,
DOI 10.1007/978-3-642-35713-8, © Springer-Verlag Berlin Heidelberg 2013